Polarity in Plants

Annual Plant Reviews

A series for researchers and postgraduates in the plant sciences. Each volume in this series will focus on a theme of topical importance and emphasis will be placed on rapid publication.

Titles in the series:

1. Arabidopsis
Edited by M. Anderson and J. Roberts

2. Biochemistry of Plant Secondary Metabolism
Edited by M. Wink

3. Functions of Plant Secondary Metabolites and their Exploitation in Biotechnology
Edited by M. Wink

4. Molecular Plant Pathology
Edited by M. Dickinson and J. Beynon

5. Vacuolar Compartments
Edited by D.G. Robinson and J.C. Rogers

6. Plant Reproduction
Edited by S.D. O'Neill and J.A. Roberts

7. Protein-Protein Interactions in Plant Biology
Edited by M.T. McManus, W.A. Laing and A.C. Allan

8. The Plant Cell Wall
Edited by J. Rose

9. The Golgi Apparatus and the Plant Secretory Pathway
Edited by D.G. Robinson

10. The Plant Cytoskeleton in Cell Differentiation and Development
Edited by P.J. Hussey

11. Plant-Pathogen Interactions
Edited by N.J. Talbot

12. Polarity in Plants
Edited by K. Lindsey

Polarity in Plants

Edited by

KEITH LINDSEY
The Integrative Cell Biology Laboratory
University of Durham, UK

Blackwell
Publishing

CRC Press

Editorial Offices:
Blackwell Publishing Ltd, 9600 Garsington Road, Oxford OX4 2DQ, UK
Tel: +44 (0)1865 776868
Blackwell Publishing Asia Pty Ltd, 550 Swanston Street, Carlton, Victoria 3053, Australia
Tel: +61 (0)3 8359 1011

ISBN 1-4051-1432-0

Published in the USA and Canada (only) by CRC Press LLC, 2000 Corporate Blvd., N.W., Boca Raton, FL 33431, USA
Orders from the USA and Canada (only) to CRC Press LLC

USA and Canada only:
ISBN: 0-8493-2344-4

First published 2004

Library of Congress Cataloging-in-Publication Data:
A catalog record for this title is available from the Library of Congress

British Library Cataloguing-in-Publication Data:
A catalogue record for this title is available from the British Library

Set in 10/12 Times
by MHL Production Services Ltd, Coventry, UK
Printed and bound in Great Britain by MPG Books, Bodmin, Cornwall

The publisher's policy is to use permanent paper from mills that operate a sustainable forestry policy, and which has been manufactured from pulp processed using acid-free and elementary chlorine-free practices. Furthermore, the publisher ensures that the text paper and cover board used have met acceptable environmental accreditation standards.

For further information on Blackwell Publishing, visit our website:
www.blackwellpublishing.com

Contents

List of Contributors

Dr František Baluška, Department of Plant Cell Biology, Institute of Botany, Rheinische-Friedrich-Wilhelms-University of Bonn, Kirschallee 1, D-53115 Bonn, Germany

Dr Peter W. Barlow, School of Biological Sciences, University of Bristol, Woodland Road, Clifton, Bristol BS8 1UG, UK

Dr Thomas Berleth, Department of Botany, University of Toronto, 25 Willcocks Street, Toronto, Ontario M5S 3B2, Canada

Dr John Bowman, Section of Plant Biology, University of California at Davis, One Shields Avenue, Davis, CA 95616, USA

Professor Colin Brownlee, Marine Biological Association, The Laboratory, Citadel Hill, Plymouth, PL1 2PB, UK

Dr Brendan Davies, Centre for Plant Sciences, University of Leeds, Leeds LS2 9JT, UK

Dr Michael J. Deeks, The Integrative Cell Biology Laboratory, School of Biological and Biomedical Sciences, University of Durham, South Road, Durham DH1 3LE, UK

Dr Stefanie Falk, Botanical Institute III, Universität Köln, Gyrhofstrasse 15, D-50931 Köln, Germany

Dr Timothy J. Hawkins, The Integrative Cell Biology Laboratory, School of Biological and Biomedical Sciences, University of Durham, South Road, Durham DH1 3LE, UK

Dr Martin Hülskamp, Botanical Institute III, Universität Köln, Gyrhofstrasse 15, D-50931 Köln, Germany

Professor Patrick J. Hussey, The Integrative Cell Biology Laboratory, School of Biological and Biomedical Sciences, University of Durham, South Road, Durham DH1 3LE, UK

Dr Martin Kieffer, Centre for Plant Sciences, University of Leeds, Leeds LS2 9JT, UK

Dr Tijs Ketelaar, The Integrative Cell Biology Laboratory, School of Biological and Biomedical Sciences, University of Durham, South Road, Durham DH1 3LE, UK

Professor Keith Lindsey, The Integrative Cell Biology Laboratory, School of Biological and Biomedical Sciences, University of Durham, South Road, Durham DH1 3LE, UK

Dr Jaideep Mathur, Botanical Institute III, Universität Köln, Gyrhofstrasse 15, D-50931 Köln, Germany

Dr Arthur J. Molendijk, Institut für Biologie II, Zellbiologie, Universität Freiburg, Schänzlestrasse 1, D-79104 Freiburg, Germany

Professor Klaus Palme, Institut für Biologie II, Zellbiologie, Universität Freiburg, Schänzlestrasse 1, D-79104 Freiburg, Germany

Dr Ivan A. Paponov, Institut für Biologie II, Zellbiologie, Universität Freiburg, Schänzlestrasse 1, D-79104 Freiburg, Germany

Dr Benedetto Ruperti, Institut für Biologie II, Zellbiologie, Universität Freiburg, Schänzlestrasse 1, D-79104 Freiburg, Germany

Dr Enrico Scarpella, Department of Botany, University of Toronto, 25 Willcocks Street, Toronto, Ontario M5S 3B2, Canada

Dr Rüdiger Simon, Institut für Entwicklungsbiologie, Universität zu Köln, Gyrhofstrasse 17, D-50931 Köln, Germany

Dr Olaf Tietz, Institut für Biologie II, Zellbiologie, Universität Freiburg, Schänzlestrasse 1, D-79104 Freiburg, Germany

Dr Ramón A. Torres Ruiz, Institute for Genetics, University of Düsseldorf, Universitätstrasse 1, D-40225 Düsseldorf, Germany

Professor Dieter Volkmann, Botanisches Institut, Universität Bonn, Venusbergweg 23, D-53115 Bonn, Germany

Professor Przemysław Wojtaszek, Institute of Molecular Biology and Biotechnology, Adam Mickiewicz University, Fredry 10, 61-701 Poznañ, Poland

Preface

A fundamental facet of developmental biology is that of the establishment of polarity. It can be described at different levels: polarity of the organism, polarity in tissue patterning and organ development, and polarity of the cell. Plants begin life with a shoot and a primary root (or thallus and rhizoid), each the product of a localised region of cell division established at opposite poles of the embryo (the apical meristems of higher plants). Roots themselves have a clear apical-basal polarity, with linearly arranged zones of cell division, expansion and differentiation. Root hair-producing cells also exhibit different types of polarity, with the position of hair emergence from the trichoblast being one expression of polarity, normally occuring at one end of the cell; while the emerging hair itself exhibits an extreme polarity of growth, from the tip. Radial patterning across the root represents yet another form of polarity. This book describes these and other developmental situations in which polarity is evident, including leaf and flower asymmetries, trichome and pollen tube development, and vascular patterning.

Two key questions of course are: how is polarity established, and what are the consequences of that polarity? One important theme that emerges from the chapters in this volume is that the polarity of organs and organisms is a function of the polarity of individual cells. This is most clearly seen in the example of polar auxin transport. The direction of auxin flow is determined in certain cells by the polar distribution of transport proteins, and secretion systems (Chapter 2). This leads to the generation in tissues of signalling gradients, which in turn define cell identity and pattern in the embryo (Chapter 6) and in individual organs (Chapters 7 and 9). The cytoskeleton plays a critical role in regulating polarity: in the sorting of proteins and membranes, essential for this process, and in cell morphogenesis and division symmetry (Chapters 1 and 4). Therefore, we are beginning to unravel the molecular mechanisms that determine the polarity of cells and organs. Common features are evident between higher plants and lower, such as the fucoid algae (Chapter 5), for example in the role of targeted secretion of wall components (Chapter 3). The shoot apical meristem, like the root meristem, represents a source of stem cells (Chapter 8), the identity of which is regulated by a now well characterised set of interacting transcription factors. This meristem exhibits both radial and apical-basal polarity, and the lateral organs that are formed from it, be they leaves or floral organs, show dorsoventrality. Again, we are making rapid progress in understanding how this patterning is regulated (Chapters 10 and 11).

So we are beginning to find answers to the first question, of how polarity is established. The consequences of polarity are also becoming clear. On one level, the effects are obvious: the determination of leaf shape and the architecture of flowers; the asymmetric distribution of membranes, proteins and signals. The consequences of polarity are also becoming clearer in terms of the roles of asymmetries in cell division. Cell fate is, in several instances, known to be determined by division asymmetries, whether in the development of pollen or stomatal complexes, or (and this is perhaps the best characterised system) in the determination of radial patterning in the root via the SCARECROW and SHORTROOT transcription factors (Chapter 7). But we still have much to learn.

These are exciting times for plant biology. Plant genomes have been, and are being, sequenced, and the associated genomic and cell biology tools (mutants, transgenics, DNA microarray, proteomics, cell imaging) together have turned plant experimental systems, and notably *Arabidopsis*, of course, into key models for developmental studies. It is now possible to address fundamentally important biological questions, such as what controls polarity, as easily in plants as in yeast or animal cells (more easily, in some cases). And there are fascinating plant-specific questions to be asked. The chapters in this book illustrate very clearly that answers to these questions are coming forth in abundance.

I would like to take this opportunity to thank all the authors for their enormous efforts in producing their chapters in a timely fashion, and to such a high standard. I would also like to thank Graeme MacKintosh and David McDade at Blackwell Publishing for their support throughout the development of the book.

Keith Lindsey

1 Polar cell growth and the cytoskeleton biology

Patrick J. Hussey, Michael J. Deeks, Timothy J. Hawkins and Tijs Ketelaar

1.1 Introduction

Cell growth is defined as an increase in cell volume. Plant cells can only increase their volume by increasing the total surface area of their plasma membrane and cell wall. The ability of the plasma membrane to stretch is limited (2% of its surface area; Wolfe & Steponkus, 1981, 1983). Moreover, the cell wall in growing plant cells does not become thinner as the cell surface increases (Roberts, 1994; root hairs: Miller *et al.*, 2000). This implies that cell growth is dependent on the continuous delivery of plasma membrane and cell wall materials to the cell surface (Miller *et al.*, 1997). The building blocks are vesicles derived from Golgi bodies that consist of membrane and embedded cellulose-synthesising complexes encasing cell wall matrix materials and cell wall-synthesising enzymes. Fusion of the membranes of these vesicles with the plasma membrane results in the cell wall matrix materials being deposited into the cell wall. This exocytosis takes place at the points of cell growth.

Three different types of cell expansion are recognised in plant cells: (i) isodiametrical growth, where growth takes place over the whole cell surface; (ii) intercalary, diffuse or elongation growth, where more expansion takes place over selected wall facets than over others, resulting in anisotropically expanding cells; (iii) local or tip growth, where all cell expansion takes place in one area of the cell surface. In all cases, cell expansion can take place only if the cell wall is flexible.

There is a plethora of evidence demonstrating that the cytoskeleton coordinates, mediates, and controls cell expansion. We will discuss the evidence for the role of cytoskeletal networks in regulating cell expansion in three cell types that expand differently: root hairs, intercalary growing cells and trichomes. We will then go on to discuss the components of the cytoskeleton that mediate cytoskeletal structure and function: actin- and microtubule-binding proteins.

1.2 Role of the cytoskeleton in cell expansion

1.2.1 Tip-growing cells

Tip growing cells are often used as a model system for studying cell growth, since all growth is tightly localised to one region of the cell, the tip. The exocytosis, required for cell expansion, takes place in abundance at the tip (root hairs: Derksen & Emons, 1990; Shaw *et al*., 2000). This localised exocytosis is very reproducible, predictable and easy to manipulate. Root hairs and pollen tubes are the two most well-characterised tip growing plant cells. Each cell type has its advantages experimentally: pollen tubes are easier to handle and quick to grow, whilst root hairs are the preferred system for genetic analysis, as mutants in pollen tube tip growth are likely to be infertile. The two systems are similar, but differences are apparent (for a review see Geitmann & Emons, 2000). As an example of tip growth, we will focus on root hair morphogenesis in this section and the role of the cytoskeleton in this process.

1.2.1.1 Root hair initiation

Root hairs are uni-axial, tubular extensions produced by special root epidermal cells. They are the major point of contact between roots and the environment and they increase the surface area of roots dramatically (~75% in some major crop plants) so that water and nutrients can be taken up effectively (Bates & Lynch, 2000a,b). Root hair development initiates by the formation of a swelling on an epidermal cell that is committed to forming a root hair. This swelling itself is not considered to be tip-growth (for a review see Grierson & Ketelaar, 2003). In *Arabidopsis*, this swelling is formed at a precise location towards the bottom of the cell and swelling formation takes about 30 minutes under optimal conditions (Bibikova *et al*., 1998). Before and during the formation of the swelling or bulge, numerous proteins with a role in determining which cells will form root hairs and at what point they form root hairs have been identified (see Grierson & Ketelaar, 2003).

The formation of a swelling during root hair initiation takes place normally after depolymerisation or stabilisation of microtubules (Ketelaar *et al*., 2002). A similar observation was made in the temperature sensitive *microtubule organisation 1* (*mor1*) mutant. At the restrictive temperature microtubules in cells including root hair cells are disrupted and disorganised (Whittington *et al*., 2001). This indicates that microtubules do not play a role in root hair initiation. Cortical microtubules at the swelling site appear normal: for example in *Medicago truncatula* microtubules loop through the site where the bulge is formed just as they do throughout the rest of the cell (Sieberer *et al*., 2002).

Actin depolymerisation does not seem to have an effect on bulge formation, either on the size or the location of the bulge, as this also takes place normally in the presence of actin depolymerising drugs (Miller *et al*., 1999; Ovecka *et al.*, 2000; Molendijk *et al.*, 2001). However, mutations in *ACT2*, one of the actin genes that is expressed during root hair formation, causes an increased bulge diameter (Ringli *et al.*, 2002; Gilliland *et al.*, 2002). Gilliland *et al.* (2002) demonstrated that the increased bulge size in one of the mutants (*act2–1*) could be complemented by a different actin gene. This indicates that the increased bulge diameter is caused by a reduction of the total amount of G-actin in the cell and not the specific requirement of one actin isoform (Gilliland *et al.*, 2002). These observations do not fit with the observations made after drug-induced actin depolymerisation, as both drug treatment and mutations in *ACT2* lead to a decreased amount of F-actin. Therefore, the increased swelling size may be caused by a defect in focusing tip growth after the initial swelling formation, and not during the initial swelling itself. Although no effect on the positioning of bulges has been reported in plants treated with actin depolymerising drugs, Ringli *et al.* (2002) reported that the position of root hairs on root hair-forming cells was occasionally altered in *Arabidopsis der* (*ACTIN2*) mutants. This indicates that actin may also play a role in selecting the root hair initiation site.

ROP2 localises to the growth site before root hair growth initiates (Jones *et al.*, 2002; Molendijk *et al.*, 2001). ROPs (*R*ho small GTPases *o*f *p*lants) are known to affect actin localisation in many cell types (Jones *et al.*, 2002; Molendijk *et al.*, 2001; Fu *et al.*, 2001, 2002). Plants overexpressing ROP2 or a constituously active mutant form of ROP2, which is permanently in the GTP-bound form (CA-rop2), produce extra root hair initiation sites in root hair forming cells, indicating that ROPs may play a role in organising the actin cytoskeleton during bulge formation. The *crooked* mutant, that has a T-DNA insert in the ArpC5 subunit of the Arp2/3 complex resulting in a knock-out, also occasionally forms multiple root hair initiation sites in root hair forming cells (Mathur *et al.*, 2003b). The Arp2/3 complex may be involved in ROP2–regulated actin organisation during root hair initiation, as it gives a similar phenotype.

1.2.1.2 Root hair tip growth

Once a swelling has formed, tip growth initiates. During tip growth, the cytoplasm inside the root hair tube is highly polarised. Golgi-derived vesicles are delivered to the apical dome, which is packed with these vesicles (defined as the clear zone, because no organelles are present) and where highly localised exocytosis takes place. The (sub-)apical area of growing root hairs is cytoplasmically dense. Further towards the root hair base, the cytoplasm is increasingly vacuolated, and still further basewards, the root hair tube is completely vacuolated with a thin layer of cortical cytoplasm. *Arabidopsis* root hairs typically grow at ~1 μm.min^{-1} (Wymer *et al.*, 1997; Bibikova *et al.*, 1999; Ketelaar *et al.*, 2002). Cytoplasmic contents are moved rapidly through the hair by cytoplasmic streaming. In growing root hairs, cytoplasmic streaming takes place in a reversed fountain-like pattern (Sieberer & Emons, 2000). The nucleus enters the shaft of the growing root hair and follows the tip at the speed of growth at a fixed distance (~75 μm in *Arabidopsis*). Proper positioning of the nucleus is essential for root hair tip growth (Ketelaar *et al.*, 2002). At a typical growth rate (1 μm.min^{-1}), *Arabidopsis* root hairs increase their volume by approximately 50fL min^{-1} (Lew, 2000). During root hair tip growth, the cytoplasmic calcium concentration at the extreme tip increases to at least 1 μM, whereas the calcium concentration in the root hair tube remains unchanged (approximantely 200 nM), creating a tip-focused calcium gradient (Wymer *et al.*, 1997; Felle & Hepler, 1997; Bibikova *et al.*, 1999; De Ruijter *et al.*, 1998; Cárdenas *et al.*, 1999). Localised calcium influx at the apex (Herrman & Felle, 1995; Felle & Hepler, 1997) from the extracellular environment is a prerequisite for growth of root hairs (Schiefelbein *et al.*, 1992). Calcium influx presumably takes place by hyper-polarisation activated calcium channels that are activated by the negative potential across the membrane at the tip of the hair (−160 to −200 mV) (Lew, 1996; Very and Davies, 2000).

As in all plant cells (Green, 1962; Ledbetter & Porter, 1963), root hairs possess cortical microtubules (CMTs). CMTs in the root hair tube are oriented in a net-axial or helical orientation, depending on the species; and in the tips, either in a random orientation (for a review: see Ketelaar & Emons, 2000) or absent (Sieberer *et al.*, 2002). In legumes, abundant endoplasmic microtubules (i.e. non-cortical microtubules), surrounding the nucleus and radiating towards the root hair tip, have been reported (*Vicia hirsute*: Lloyd *et al.*, 1987; *Medicago truncatula*: Sieberer *et al.*, 2002). However, in *Arabidopsis* no abundant endoplasmic microtubules have been found (Ketelaar *et al.*, 2002).

Treatment of growing root hairs with the microtubule depolymerising drug oryzalin does not lead to changes in growth speed (Bibikova *et al.*, 1999; Ketelaar *et al.*, 2003) or to the diameter of root hairs (Ketelaar *et al.*, 2003). These drug treatments appear to phenocopy the *mor1* mutant at the restrictive temperature (Whittington *et al.*, 2001). This implies that the amount of exocytosis and the size of the area where exocytosis takes place do not change in the absence of microtubules. However, root hair growth becomes wavy, and branching occasionally occurs (Bibikova *et al.*, 1999; Ketelaar *et al.*, 2003), indicating that microtubules play a role in focusing the area where exocytosis takes place and thus the direction of root hair growth. When microtubules are stabilised with taxol, the wavy root hair phenotype is more severe, and branching occurs frequently (Bibikova *et al.*, 1999). Therefore, the ability of microtubules to focus the area where exocytosis takes place is more severely affected when the microtubules are stabilised then when they are absent. Bibikova *et al.* (1999) showed that taxol-stabilised microtubules indeed lose the ability to focus the area where exocytosis takes place by localised uncaging of calcium in the apical area of growing root hairs. In root hairs with a functioning microtubule cytoskeleton, local uncaging of calcium led to a transient reorientation or root hair growth towards the side where uncaging had been performed, whereas root hairs with stabilised microtubules reoriented their growth direction permanently.

Plants with reduced α-tubulin expression form branches immediately after bulge formation and during root hair tip growth (Bao *et al.*, 2001). This indicates that, not only is the focusing of the exocytic area dependent on normal microtubule dynamics, but also on a finely tuned level of tubulin subunits. Also, Tominaga *et al.* (1997) and Ketelaar *et al.* (2003) have shown, by depolymerisation of microtubules followed by depolymerisation and recovery of F-actin, that microtubules may orient the area where exocytosis takes place by directing actin filaments.

Although in *Arabidopsis*, microtubules do not seem to be involved in maintaining the distance between the growing tip and the nucleus, in legumes they do seem to play a role because the distance between nucleus and tip increases when microtubules are depolymerised (Lloyd *et al.*, 1987; Sieberer *et al.*, 2002). Sieberer *et al.* (2002) showed that, as well as the increase in distance between nucleus and tip, the growth speed of *Medicago truncatula* root hairs decreased in the absence of microtubules. The role that the microtubules play in the positioning of the nucleus and in cell growth may be peculiar to legume root hairs.

Plant cell walls, including those of root hairs, are reinforced with cellulose microfibrils (CMFs), which are synthesised by cellulose synthase complexes in the plasma membrane and deposited in the extracellular space between the plasma membrane and the existing cell wall. In many cases, the orientation of CMTs and of CMFs in plant cell walls is similar. Based on these observations, Heath (1974) hypothesised that CMTs orient CMFs and thus the direction of cell expansion. This could be true in the expanding part of the root hair cell (in the apical dome), although it is difficult to be certain because both microtubules and cellulose microfibrils are randomly arranged in this part of the hair (Emons, 1989; Ketelaar & Emons, 2000). Moreover, in the root hairs of some species CMTs are absent from the root hair tip, raising the possibility that CMF orientation is not CMT dependent. This is also true in several other systems where random CMF deposition does occur without CMT (Baskin, 2001). Recently, Sugimoto *et al.* (2003) showed that the organisation of CMF deposition after drug-induced microtubule depolymerisation is not altered in intercalary expanding cells in the root elongation zone. This is also the case in the same cells in the temperature sensitive *mor1* mutant. Heath's hypothesis is therefore currently under debate. In the root hair shaft, which is no longer expanding but is being reinforced with CMFs, CMTs are seen to co-align with newly deposited CMFs in some species but not in all (Ketelaar & Emons, 2000). Although this does not exclude a guidance mechanism by microtubules (Baskin, 2001), an alternative hypothesis has been proposed. Emons & Mulder (2000) explained helicoidal depositioning of

CMFs in root hair tubes by the geometrical properties of the root hair tube, which is theoretically sufficient for helicoidal CMF deposition.

Actin filaments are thought to mediate cytoplasmic streaming in all plant cells (Esseling *et al*., 2000). The actin organisation in any cell type reflects the cytoplasmic organisation, since the cytoplasm is organised around actin filaments (Valster *et al*., 1997). Root hairs are no exception to this rule. In the extreme tip there is a small zone which is devoid of organised F-actin, and this correlates with a cytoplasmic streaming speed of 0 μm.sec^{-1} in *Equisetum* root hairs (Emons, 1987). The subapical area is cytoplasmic dense, filled with net-axially oriented fine F-actin. Subapical fine F-actin has been hypothesised to deliver Golgi-derived vesicles to the tip, retain the vesicles in the tip and sieve out other organelles to inhibit their penetration into the tip (De Ruijter & Emons, 1999). In root hair tubes, further down towards the base, thick bundles of F-actin in the cortical cytoplasm surround a large vacuole (Miller *et al*., 1999; Ketelaar *et al*., 2002, 2003; Jones *et al*., 2002). This is where villin-mediated bundling takes place (Tominaga *et al*., 2000). In *Arabidopsis*, the actin cytoskeleton positions the nucleus at a fixed distance behind the tip in growing root hairs (Ketelaar *et al*., 2002).

Using actin antagonists and actin mutants, an understanding of how fine F-actin organisation dictates the location of cell growth has been gained. By continued application of 1 μM cytochalasin D to *Vicia sativa* root hairs, Miller *et al*. (1999) broke down the subapical fine F-actin and inhibited growth, whereas cytoplasmic streaming continued. This indicates that subapical fine F-actin is more sensitive to cytochalasin D than more basal, thicker bundles of F-actin and that the subapical fine F-actin is required for continued fusion of Golgi-derived vesicles with the apical plasma membrane. However, single applications of lower concentrations of cytochalasin D (minimum 0.1 μM) or latrunculin A to growing *Arabidopsis* root hairs do not inhibit exocytosis (i.e. the increase in surface area over time remained constant; Ketelaar *et al*., 2003). However, the exocytosis takes place over a larger surface area, which leads to an increase in root hair diameter. This correlates with a decrease in fluorescence of immunolabelled F-actin in the subapical area, indicating that fine F-actin delimits the area where exocytosis takes place, and is not necessarily involved in the vesicle fusion itself (Ketelaar *et al*., 2003). Root hairs of *Arabidopsis* plants, mutated in the *ACTIN2* gene (the *der* and *act2–1* mutants), produce root hairs with widely varying diameters (Ringli *et al*., 2002; Gilliland *et al*., 2002), suggesting that, similar to the results of Ketelaar *et al*. (2003), the amount of polymerised actin and its location is disturbed. This demonstrates that proper organisation of the subapical fine F-actin is extremely important for determining the area where exocytosis takes place. Mutants in the *ACTIN2* gene also produce multiple tip-growing structures from each swelling (Ringli *et al*., 2002; Gilliland *et al*., 2002), indicating that the amount of F-actin or the actin dynamics also play a role in focusing the area where exocytosis can take place during the start of tip-growth.

Actin organisation, essential for normal root hair growth, is controlled by actin binding proteins (ABPs). These ABPs can affect the amount of F-actin by regulating actin dynamics. The modes of action of different actin binding proteins are discussed below. Here we will discuss the reported effects of some of these proteins on root hair growth. Profilin is a G-actin binding protein. When G-actin is bound to profilin, nucleation and incorporation at the pointed end of F-actin is inhibited, whereas incorporation at the barbed ends continues at the normal rate (Pollard *et al.*, 2000). At least one of four *Arabidopsis* profilins, PFN-1, is expressed in root hairs (Ramachandran *et al.*, 2000). Although some papers report an accumulation of profilin in root hair apices (Braun *et al.*, 1999; Baluška *et al.*, 2000), this localisation may be caused by fixation artifacts. Profilin is likely to be evenly distributed through the cytoplasm in root hairs, as it is in pollen tubes (Vidali & Hepler, 1997; Hepler *et al.*, 2001). Transgenic plants overexpressing PFN-1 have root hairs that are twice as long as wild-type hairs (Ramachandran *et al.*, 2000), indicating that profilin does not play a role in positioning the fine F-actin, since the diameter of the root hairs overexpressing profilin does not change. However, it does contribute to the total amount of exocytosis, since the volume of the root hairs is larger when profilin is overexpressed. The growth speed was not measured, therefore it is not clear whether the amount of exocytosis per time unit or the duration of root hair growth increases.

ADF/cofilin binds both to G- and F-actin and enhances actin dynamics by severing actin filaments and increasing the depolymerisation from the pointed end (Carlier *et al.*, 1997; Gungabissoon *et al.*, 1998). An ADF from maize, ZmADF3, has been localised in maize root hairs (Jiang *et al.*, 1997). Recombinant ZmADF3 has the ability to sever F-actin at high pH (8.0), whereas it binds F-actin at a lower pH (6.0). Moreover, it enhances the actin bundling activity of maize elongation factor 1α (ZmEF1α) *in vitro* (Gungabissoon *et al.*, 2001). ZmADF3 activity is reduced by phosphorylation of Ser-6 (Smertenko *et al.*, 1998), and by binding to phosphatidylinositol 4,5-bisphosphate (PIP_2) or phosphatidylinositol 4-monophosphate (PIP) (Gungabissoon *et al.*, 1998). Phosphorylation of ZmADF3 is regulated by a calmodulin-like domain protein kinase (CDPK) (Smertenko *et al.*, 1998; Allwood *et al.*, 2001). The ADF activity that is demonstrated *in vitro* does reflect the actin organisation in root hairs. In the apical area, where the concentration of free cytoplasmic calcium is relatively high, CDPK may be active and ADF phosphorylated and therefore, inactive. The subapical zone of pollen tubes is alkaline and because similar proton fluxes are observed in root hairs, they are also likely to possess a subapical alkaline zone (for a review, see Hepler *et al.*, 2001). At high pH, ADF severs F-actin, thus creating many free barbed ends, where polymerisation can take place. This area is where the fine F-actin is present. Further, towards the root hair base, the pH is lower and ADF may stimulate villin and/or perhaps ZmEF-1α mediated bundling of actin filaments, which creates the thick bundles of F-actin in the root hair tube. Dong *et al.* (2001) over and underexpressed AtADF1 in *Arabidopsis.* In root hairs, overexpression led to a highly irregular F-actin organisation and the disappearance of thick bundles of actin. This resulted in an increase in the radial diameter. Underexpression led to an increased number of longitudinally-oriented actin cables and an increased root hair length. It is not clear whether the increased root hair length is caused by increased root hair growth speeds or a prolonged growth period. The increased root hair width in ADF overexpressing plants can be explained by increased actin depolymerisation by increased ADF activity. This phenotype could be compared to the increase in root hair width observed during actin depolymerisation (see above). Actin Interacting Protein 1 (AIP1) is a protein that enhances ADF activity *in vitro*, possibly by capping barbed ends that are generated by ADF mediated actin filament severing (Ono, 2003). *Arabidopsis* plants, harbouring an inducible AIP1 RNAi construct have been generated (Ketelaar *et al.*, 2004). Induction of AIP1 RNAi caused filamentous actin to ramify into the tips of growing root hairs and a decrease in root hair growth speed. This resulted in a decrease in root hair number and length (Ketelaar *et al.*, 2004). Interestingly, ADF under-expression affects root hair length in the opposite way. It appears that even though AIP1 and ADF act synergistically to increase the turnover of F-actin, in terms of root hair development, their activity is antagonistic.

The Arp2/3 complex consists of 7 subunits (Arp2, Arp3, ArpC1/p41, ArpC2/p31, ArpC3/p21, ArpC4/p20 and ArpC5/p16; Higgs & Pollard, 2001). Activated Arp2/3 complex nucleates actin by promoting barbed-end actin assembly and caps the pointed end. It also binds to existing actin filaments while nucleating a new filament, which results in the formation of a branch. Coding sequences for orthologues of the 7 components of the Arp2/3 complex are present in the *Arabidopsis* genome database (McKinney *et al.*, 2002). A complicating factor is that WASp/Scar families of proteins that activate the Arp2/3 complex in mammalian cells do not seem to be present in the *Arabidopsis* genome database (Deeks *et al.*, 2002; see below). T-DNA insertional mutants for Arp2, Arp3 and ArpC5/p16 cause a phenotype in a variety of cell types in *Arabidopsis* (Mathur *et al.*, 2003a,b; Le *et al.*, 2003; Li *et al.*, 2003). Root hair growth is disturbed in plants with mutations in any of the 3 subunits (Mathur *et al.*, 2003a,b). The phenotype varies from wavy growth in a widened diameter to wide root hairs with multiple tips. These observations indicate that the Arp2/3 complex is involved in the organisation of the subapical fine F-actin in root hair growth.

In both root hairs and pollen tubes ROPs have a strong effect on the direction and amount of root hair tip growth. Overexpression of ROP2 leads to the formation of very long root hairs that repeatedly branch (Jones *et al.*, 2002). Plants expressing CA-rop2 form partly or completely spherical root hairs, suggesting that inhibition of ROP deactivation (i.e. cycling between the

GTP-bound form to the GDP-bound form) leads to an unlimited plasma membrane area where exocytosis takes place, i.e. over the whole root hair surface (Jones *et al.*, 2002; Molendijk *et al.*, 2001). Therefore, localised activation of ROP seems to control the cell surface area where exocytosis occurs. Indeed, plants expressing a ROP that is permanently in the GDP-bound state (DN-rop2) have fewer, shorter root hairs. Sometimes, these root hairs are wavy (Jones *et al.*, 2002). When Green Fluorescent Protein (GFP), fused to the actin binding site of mouse talin (GFP-mTalin; Mathur *et al.*, 1999) was used to visualise F-actin in living root hairs, hairs expressing DN-rop2 had a reduced amount of fine F-actin in comparison to wild type hairs, whereas in hairs expressing CA-rop2, a dense network of fine F-actin was present (Jones *et al.*, 2002). This is consistent with the requirement of fine F-actin at the location where cell growth takes place. In pollen tubes, there is an indication that the tobacco pollen specific NtADF1 acts in the same pathway as ROPs, as NtADF1 overexpression reverts the isotropic expansion phenotype caused by NtROP1 overexpression (Chen *et al.*, 2003). There may be a similar mechanism by which ROPs control actin dynamics in root hairs.

1.2.1.3 Root hair growth termination

Root hair tip growth stops when hairs reach a predictable length, depending on the environment (Grierson & Ketelaar, 2003). During growth termination, root hairs lose their polarity: the calcium gradient dissipates (Wymer *et al.*, 1997), the (sub-)apical cytoplasmic dense area disappears (Sieberer & Emons, 2000) and the nucleus retreats from its fixed distance from the tip to a random position within the root hair (Ketelaar *et al.*, 2002). Growth termination effectively produces a rounded dome, with the same cytoplasmic organisation as root hair tubes: vacuolated with a thin layer of cortical cytoplasm.

Microtubules are not essential for root hair growth termination, demonstrated by the fact that termination takes place normally in oryzalin-treated root hairs (Ketelaar *et al.*, 2002) and also in the *mor1* mutant at the restrictive temperature (Whittington *et al.*, 2002). In *Medicago* root hairs, Sieberer *et al.* (2002) showed that CMTs protrude into the root hair tip and EMTs disappear during root hair growth termination. After growth termination, the density of CMTs gradually decreases. During growth termination, the fine F-actin that correlates with growth disappears as the cytoarchitecture at the tip changes. In fully-grown root hairs, thick bundles of actin filaments loop through the tip (Miller *et al.*, 1999; Ketelaar *et al.*, 2002). The cytoplasmic streaming shifts to rotational streaming through the tip of the root hair (Sieberer & Emons, 2000) which reflects the observed actin organisation. The movement of the nucleus during growth termination is F-actin dependent (Ketelaar *et al.*, 2002).

1.2.2 Intercalary growing cells

In developing intercalary growing cells, two different types of cell growth have been identified: (i) the early growth phase and (ii) the late growth phase (Fu *et al.*, 2002). These growth phases correlate with the transition zone and the elongation zone in roots, respectively (*Arabidopsis* root: e.g. Dolan *et al.*, 1993; Beemster & Baskin, 1998; Schindelman *et al.*, 2001; maize root: for a review see Baluška *et al.*, 2001c). In the early phase, more or less isodiametrical cell expansion occurs in root cells but this is not always the case, such as in mesophyl cells (Fu *et al.*, 2002). When cells enter the elongation zone, rapid anisotropic cell elongation occurs. Since cell growth occurs over a larger cell surface than in root hairs, the expansion rate per square micrometre surface area is lower in the expanding part of the cell. This makes it more difficult to observe changes in the direction and amount of cell expansion and explains why relatively little is known about cell expansion in intercalary growing cells.

During the early expansion phase, CMTs are organised in a random orientation, and in root cells, microtubules radiate through the cytoplasm from the nuclear membrane. In the elongation zone, or the late phase, CMTs are oriented more or less perpendicular to the growth axis in expanding areas of most cell types (for a review see Baskin, 2001; Ketelaar & Emons, 2001). During growth termination, the orientation of CMTs changes to a longitudinal or random orientation. Depolymerisation and stabilisation of microtubules in elongating cells leads to isodiametrical expansion (Green, 1962; Baskin *et al.*, 1994), as do cells in mutants with disorganised CMTs (*ton*: Traas *et al.*, 1995; *fass*: McClinton & Sung, 1997; *mor1*: Whittington *et al.*, 2001). These observations point to the importance of the presence of a functional CMT array for elongation growth. Heath (1974) hypothesised that cortical microtubules directed the deposition of the cellulose microfibrils in plant cell walls, and thus the direction of cell expansion. After reviewing available litterature, Baskin (2001) proposed a *templated incorporation model*, where nascent CMFs may be oriented by scaffolds from both within the cell (CMTs) and from CMFs in the existing cell walls. This explains why functional CMTs are essential, but not sufficient for controlling elongation growth (Baskin, 2001; Wasteneys, 2000). This fits with the results of Sugimoto *et al.* (2003). In this report it is shown that after CMT depolymerisation, and also in the *mor1* mutant, CMF deposition still takes place orderly. Moreover, in intercalary growing cells of the *radially swollen 4* and *7* (*rsw4* and *rsw7*) mutants, both CMTs and CMFs are transversely oriented. However, *rsw4* and *rsw7* cells still expand more isotropically than in wildtype plants (Wiedemeier *et al.*, 2002). These data indicate that in addition to transverse CMTs and CMFs, RSW4 and RSW7 are essential for growth anisotropy. Generally, it can be concluded that in elongating cells, as in root hairs, CMTs play a role in determining the direction of cell expansion. However, it is hard to tell if the size of the area where cell expansion takes place remains identical in intercalary growing cells as it does in root hair mainly because expansion takes place over most of the surface in these cells.

In intercalary growing cells, the actin cytoskeleton ramifies the cytoplasm. Fu *et al.* (2002) reported a diffuse cortical fine F-actin network at the points of cell growth in the early expansion phase cells of GFP-mTalin-expressing plants. During the early expansion phase of intercalary growing cells, as in root hairs, the location of fine F-actin is determined by ROP proteins. Expression of CA-rop2 leads to expansion over the whole cell surface, which correlates with cortical fine F-actin over the whole cell surface. In contrast, expression of DN-rop2 leads to inhibition of growth and a decrease in the amount of cortical fine F-actin (Fu *et al.*, 2002). These observations seem to be very similar to the observations made in root hairs: at the membrane surface where expansion takes place, ROPs are activated and fine F-actin is formed. Similarly, downregulation of AIP1 and mutations in the Arp2/3 complex subunits cause defects in actin organisation in intercalary growing cell types. AIP1 suppression leads to dramatic changes in the organisation of the actin cytoskeleton; actin is organised into thicker and more compact bundles, which results in a decrease in cell expansion and is eventually lethal (Ketelaar *et al.*, 2004). Mutations in Arp2/3 complex subunits result in defects in both cell expansion rate and direction in leaf pavement and cotyledon cells (Mathur *et al.*, 2003a,b; Le *et al.*, 2003; Li *et al.*, 2003). The actin filaments become organised into thick, transversely linked actin bundles which indicates that the Arp2/3 complex is required to generate fine F-actin. During the late stage of intercalary cell growth, Baluška *et al.* (2001a) demonstrated that actin is essential, as cells do not elongate properly when treated with high concentrations of actin depolymerising drugs. However, Fu *et al.* (2002) did not report the presence of cortical fine F-actin during this stage of unidirectional cell expansion. This could mean that fine F-actin is not very abundant over the whole surface of the cell, making detection technically challenging. Alternatively, fine F-actin may only play a role in localised expansion, such as constraining the Golgi-derived vesicles in a small area of the cell surface, and this has been suggested by Miller *et al.* (1999) for root hairs.

1.2.3 Trichomes

Trichomes are unicellular hair-like structures that extend from the surface of many aerial organs of the plant. *Arabidopsis* trichomes are highly polarised and consist of a stalk and multiple branches. They reach a final length of about 500μm. *Arabidopsis* trichome development is used as a model process for cell morphogenesis (see Chapter 4). Trichome initiation correlates with a switch from the mitotic to endomitotic cell cycle (DNA replication in the absence of cell division) and takes place in regions of the leaf that contain fields of dividing cells (Lloyd *et al.*, 1994; Larkin *et al.*, 1996). A number of genes have been identified that play a role in cell fate and patterning to form trichome precursors (see Beilstein & Szymanski, 2003). After a trichome precursor has been defined, trichome development takes place. Six developmental stages with characteristic morphological landmarks have been identified (Szymanski *et al.*, 1998, 1999):

1. Isodiametric expansion within the plane of the epidermis. During or after enlargement, the nucleus within the spherical precursor cell increases in size, suggesting that it undergoes additional rounds of DNA endoreduplication (Hülskamp *et al.*, 1994). Although endoreduplication occurs systemically in the leaf, it occurs to a greater extent (Melaragno *et al.*, 1993; Szymanski & Marks, 1998) and perhaps earlier (Hülskamp *et al.*, 1994) in trichomes.
2. Polarised expansion occurs perpendicular to the leaf surface, leading to the production of a stalk.
3. After reaching a length of approximately 30 μm, branches initiate sequentially on the developing stalk.
4. The branches elongate and have blunt buds.
5. After the branches reach a length of approximately 15 μm, the tip morphology becomes more pointed. Most of the cell growth takes place during this stage.
6. In this final stage papillae form on the surface of a fully grown trichome.

The microtubule cytoskeleton in developing trichomes has only been studied by expression of GFP fused to the microtubule binding domain of MAP4 (Marc *et al.*, 1998) and GFP, fused to α-tubulin 6 (Ueda *et al.*, 1999). Images of microtubules in fully grown trichomes reveal a dense population of microtubules, oriented net-axially relative to the long axis of the cell (Mathur & Chua, 2000). In addition, in abortive branch points of mature mutant trichomes, discordant or unaligned microtubules have been found. High resolution imaging of GFP-fusion constructs in developing branches suggests that CMTs do not reorient from transverse to longitudinal until well after branch formation (Beilstein & Szymanski, 2003; Folkers *et al.*, 2002). This is comparable to the transverse microtubules that have been reported to be present during bulge formation during root hair development (Sieberer *et al.*, 2002). In the apical region of unbranched stalks and in early phase branches, a wide band of transverse CMTs have been reported, although the extreme apex appears to be depleted of microtubules (Beilstein & Szymanski, 2003). Pharmacological data with microtubule stabilising and depolymerising drugs have given some insight into the function of microtubules during the different stages of trichome development. During stage 1, treatment with high concentrations of oryzalin inhibits establishment of polar expansion (Mathur & Chua, 2000). Trichomes at this stage respond to microtubule depolymerising drugs as do intercalary growing cells, i.e. they expand isodiametrically. In contrast, tip-growing root hairs do not lose polarity in response to treatment with microtubule depolymerising drugs. Application of lower concentrations of oryzalin does not completely inhibit polar expansion and the formation of branch initiation sites in stages 1 and 2 (Szymanski *et al.*, 1999). Application of oryzalin during later stages leads to reduced branch elongation and formation of swollen and twisted branches (Szymanski *et al.*, 1999). These observations suggest that intercalary growth occurs during the development of trichome branches because microtubule depolymerisation in tip-growing cells would not change their diameter. These observations fit with the observations made by Schwab *et al.* (2003), who applied latex beads to developing trichomes and showed that the expansion was indeed intercalary. Interestingly, root hair initiation cannot be inhibited by microtubule depolymerisation. Therefore, branch initiation in trichomes may depend on cellular processes different to those involved in root hair initiation.

The actin cytoskeleton in trichomes has been studied by both immunofluorescence (Szymanski *et al.*, 1999) and expression of GFP-mTalin (Mathur *et al.*, 1999). Both groups report dense actin networks during development. Immunofluorescence reveals filamentous actin during the first stages of development and a punctate and diffuse actin signal at sites where growth patterns are established. GFP-mTalin actin staining has been described as diffuse in stage (1) and stage (2) trichomes. In the later stages, the results obtained with both actin visualisation methods are similar. In stage (3), foci of diffuse F-actin are formed that mark the areas where growth patterns are being established, which correlates with fine F-actin at locations where cell expansion takes place in other cell types. In the the rapid expanding phase (stages 4 and 5), thick bundles of F-actin run through the cytoplasm, mostly aligned with the local axis of elongation, and cortical fine F-actin throughout the stalk and branches. Actin depolymerisation by cytochalasin D or latrunculin B causes trichome phenotypes that appear to phenocopy the trichome mutants with *distorted* phenotypes, indicating that these mutant phenotypes are caused by defects in actin organisation (see below; Mathur *et al.*, 1999; Szymanski *et al.*, 1999). The actin cytoskeleton does not seem to be essential during the first stages of trichome development, as stalk elongation (stage 2) and branch initiation take place normally (Mathur *et al.*, 1999; Szymanski *et al.*, 1999). These observations fit with the observation that trichomes expand by an intercalary process (Schwab *et al.*, 2003), since actin depolymerisation does not lead to changes in cell shape, although the growth speed and final size decrease dramatically. If the stalk were elongating by tip growth, an increase in tip width or growth inhibition would be expected. During later developmental stages (stage 4 and 5), actin depolymerisation does lead to morphological changes. Stalks appear swollen and branch morphology varies from aborted tubes to swollen and twisted branches that reach sizes comparable to untreated branches. Here, F-actin may function in determining the exact locations where exocytosis takes place. Fine F-actin localises throughout the cell cortex during these stages. Actin depolymerisation may change this location. This does not mean that the trichome expands by tip growth. On the contrary, the fine F-actin localises throughout the trichome during the rapid expansion stage, indicating that growth does not take place locally. Again, this fits with the observations of Schwab *et al.* (2003), who applied latex beads to the surface of expanding trichomes and show that the stalk expansion takes place over its whole surface.

Many genes have been identified that play a role in trichome morphogenesis. Some mutants in which cell expansion and/or division is defective, also have trichome phenotypes. Some of these mutations have been linked directly to cytoskeletal function. A number of mutants affected in trichome morphogenesis can be classified into two groups: the ones that resemble trichomes grown in the presence of microtubule depolymerising drugs (*stichel*, *zwichel*, mutations in the p60 subunit of katanin and mutations in tubulin folding cofactors), and actin depolymerising drugs (*gnarled*, *crooked* and *distorted*). Three genes from the distorted class have been cloned and have been shown to encode subunits of the Arp2/3 complex; *WURM* encodes the Arp2 subunit, *DISTORTED1* encodes the Arp3 subunit and *CROOKED* encodes the ArpC5/p16 subunit (Mathur *et al.*, 2003a,b; Le *et al.*, 2003; Li *et al.*, 2003). Other genes from this group still await characterisation. For a detailed review about the functions of the genes from the two classes as described above, see Beilstein & Szymanski (2003).

1.3 Components of the cytoskeleton

1.3.1 Actin binding proteins

We have described how the actin cytoskeleton is required to direct tip growth and facilitate intercalary growth during plant development. These processes require the spatial and temporal control of actin filament formation. A large variety of actin binding proteins (ABPs) identified in plants, fungi, and animals are known to regulate the assembly of G-actin into F-actin. The response of ABPs to cell signalling cascades allows the actin cytoskeleton to respond to developmental and environmental stimuli. A large number of identified plant ABPs have high sequence homology with animal and fungal ABPs. However, the interactions between these conserved proteins seem to draw a plant-exclusive control network where conserved actin binding proteins are linked to novel signal transduction cascades. In the post-genomic era it is possible to identify familiar proteins within the plant genome and move directly to the exploration of their role within the growing regulatory network. This section of the review will describe several major proteins believed to be involved in the developmental regulation of the plant actin cytoskeleton, and will also summarise the probable importance of homologues of several animal and fungal ABPs identified within plant genomes that are yet to be fully characterised. For detailed reviews on all known actin binding proteins in plants, see Staiger *et al.* (2001), Hussey *et al.* (2002), Staiger & Hussey (2004).

1.3.1.1 Profilin

Profilins are G-actin binding proteins of approximately 160 kDa. Profilin can act as a G-actin sequestering protein if actin filaments are capped (Gutsche-Perelroizen *et al.*, 1999), and consequently the microinjection of large amounts of profilin into some cells results in depolymerisation of actin fibres (Valster *et al.*, 1997). G-actin sequestering proteins allow a concentration of 50–200 μM to be maintained within the animal cell cytosol (Sohn & Goldschmidt-Clermont, 1994). Actin sequestering factors must play an analogous role in the plant cell cytosol, as the *in vitro* critical concentration of maize actin is approximately 0.6 μM (Gibbon *et al.*, 1998) but maize pollen can maintain a dynamic population of actin at 127 μM (Gibbon *et al.*, 1999). Thymosin β4 acts as a powerful G-actin sequestering protein in animal cells but it is absent from plant genomes (Hussey *et al.*, 2002). The absence of thymosin β4 could leave profilin to be the major plant G-actin sequestering protein. This certainly seems to be the case in pollen tubes, where the ratio of profilin to G-actin is believed to be 1:1 (Gibbon *et al.*, 1999). In animal cells, profilin catalyses the polymerisation of actin (Kang *et al.*, 1999) possibly by capturing monomers from thymosin β4. As long as the fast growing 'barbed' ends of actin filaments are uncapped, the activity of profilin will actually boost filament assembly. The balance between filament capping and profilin activity has been shown to be essential during the morphogenesis of *Drosophila* bristle cells (Hopmann & Miller, 2003). A reduction in capping protein activity increases profilin-mediated actin filament extension, while capped filaments are not so amenable to extension. The change in morphology of root hairs observed in plants over-expressing PFN1 (Ramachandran *et al.*, 2000) might be caused by changes in actin biochemistry analogous to that caused by profilin overexpression in *Drosophila* bristle cells.

Actin subunits are linked to either ADP or ATP, and the ATP form is more amenable to polymerisation. Animal and fungal profilin can accelerate the rate of G-actin exchange of ADP for ATP (Lu & Pollard, 2001). Surprisingly, plant profilins do not have this activity (Perelroizen *et al.*, 1996; Kovar *et al.*, 2000). Interestingly, the nucleotide exchange activity of plant actin is 10 to 20 times higher than animal actin (Kovar *et al.*, 2001), possibly substituting the requirement of similar activity from profilin. *Arabidopsis* has five profilin genes (Huang *et al.*, 1996; Christensen *et al.*, 1996); three of these are expressed in a wide range of tissue types (*PRO1, PRO2, PRO5*) while two are restricted to reproductive tissues (*PRO3, PRO4*). Individual plant profilin isoforms differ in their biochemical properties (Kovar *et al.*, 2000), meaning that the tissue-specific expression of a particular isoform will endow that tissue with a particular mode of actin polymerisation. Different isoforms may also be involved in specific signal transduction cascades.

The manipulation of profilin is likely to affect the growth and polarisation of the plant cell, as described earlier. In animal and fungal cells profilin can be localised to sub-cellular domains (Petersen *et al.*, 1998a; Finkel *et al.*, 1994). Sub-cellular local concentrations of profilin have been observed in plant cells – in microspores (von Witsch *et al.*, 1998), and root hairs (Baluška *et al.*, 2000) – although it has already been stated that such observations could be artefacts caused by differential penetration of the tissue. The activity of plant profilin is dictated by the local concentration of calcium ions (Kovar *et al.*, 2000). This provides a means by which profilin activity can be modified locally in environments where profilin concentration does not vary, such as the pollen tube (Mittermann *et al.*, 1995; Vidali & Hepler, 1997). Calcium ion influx at the tip of the pollen tubes and root hairs produces a gradient of calcium ion concentration along the distal portion of the cytoplasm. The unique actin arrays seen at the apex of tip growing cells might be formed in part by local calcium-stimulated profilin activity.

1.3.1.2 Actin depolymerising factor (ADF)

ADFs are actin depolymerisation factors that can bind to ADP associated F-actin subunits within actin fibres and encourage their release from the 'pointed' end of the fibre. The conformational changes imposed by ADF binding along the length of an actin fibre can induce shearing. Although ADF induces the depolymerisation of filaments from the pointed end, the shearing effect can produce multiple barbed ends that can enhance net polymerisation. As with most ABPs, plants have multiple isoforms of ADF, and individual ADF isoforms have been shown to have unique biochemical properties (Smertenko *et al.*, 2001; Allwood *et al.*, 2002). The divergence of the biochemical properties of three isoforms of ADF has also been observed in mice (Vartiainen *et al.*, 2002), supporting the notion that eukaryotes with multiple ADF isoforms can alter actin biochemistry by differentially expressing divergent forms of the same ABP. The regulation of ADF activity through phosphorylation and pH in relation to tip growth has already been described above. ADF is also regulated by phospholipids, as both phosphatidylinositol 4-phosphate (PIP) and phosphatidylinositol 4,5-bisphosphate (PIP_2) will bind to plant ADF and suppress its activity (Gungabissoon *et al.*, 1998). In animal cells, the phosphorylation of ADF is mediated by LIM kinases (Arber *et al.*, 1998) and TESK kinases (Toshima *et al.*, 2001). Despite the demonstration that maize ADF3 can be phosphorylated at ser6 by plant extract in a manner that regulates its biochemical activity (Smertentko *et al.*, 1998), no homologues of the animal kinases that regulate ADFs have been identified within plant genomes. Instead, CDPKs appear to be responsible for ADF phosphorylation (Smertenko *et al.*,

1998; Allwood *et al.*, 2001). The divergence of this signalling pathway between animals and plants to ADF is a trend repeated by other plant ABPs; the factors that directly affect the biochemistry of actin polymerisation are conserved while plants have employed novel signalling inputs.

The utilisation of ADF regulation by phosphorylation has been demonstrated through the manipulation of phosphorylation of animal cofilin at ser3. *In vivo,* phosphorylation at ser3 affects the growth of neural growth cones (Endo *et al.*, 2003). Overexpression of LIMK1 represses growth cone motility, while overexpression of the Slingshot phosphatase that removes the phosphate group from ser3 increases motility (Endo *et al.*, 2003), presumably by releasing the repression of ADF and increasing actin dynamics within the growth cone of the dorsal root ganglion. Manipulation of ADF activity by calcium-responsive kinases and antagonising phosphatases could regulate the activity of some plant ADF isoforms locally in tip-growing plant cells such as root hairs. Regulation by phosphorylation is further complicated by interactions of phosphorylated residues with 14-3-3 proteins (Birkenfeld *et al.*, 2003). Such an interaction has been observed to protect cofilin ser3 phosphorylation from phosphatases (Gohla & Bokoch, 2002), resulting in a net increase in the levels of phosphorylated cofilin. 14-3-3 proteins have been isolated from plants (reviewed by Roberts, 2003), thus providing another means of attenuating plant ADF activity.

An increase of ADF activity stimulated by dephosphorylation or changes in pH could increase actin dynamics in zones of polar growth. As detailed earlier, a proton gradient that could be exploited by ADF for such purposes has been observed emanating from the tip region of pollen tubes and root hairs. ADF preferentially binds F-actin at pH 6.0, but binds G-actin above pH 7.4 (Hawkins *et al.*, 1993; Gungabissoon *et al.*, 1998). Modulation of CDPK activity by the calcium ion gradient has the potential to play a major regulatory role, but experiments with lily pollen-specific ADF have shown that, *in vitro,* this particular isoform is not amenable to regulation by phosphorylation (Allwood *et al.*, 2002). Finally, the physical localisation of ADF protein is also under tight control, as pollen tubes in contact with a solid substrate specifically accumulate ADF within the tip region (Smertenko *et al.*, 2001). Like profilin, the manipulation of chemical environment and localisation, as well as the interaction of ADF with other protein factors, provide means to shape the actin cytoskeleton of developing cells.

Figure 1.1 highlights the relationship between ADF and profilin as being synergistic in nature. *In vitro* experiments show that profilin in isolation cannot increase the rate of actin filament treadmilling (Didry *et al.*, 1998). ADF in isolation can increase the rate by 25-fold. However, combining *Arabidopsis* ADF with bovine profilin raises the rate of rabbit muscle actin treadmilling by a factor of 125. G-actin released by ADF-induced depolymerisation at the filament pointed end drives profilin mediated F-actin growth at the barbed end, resulting in a dramatic increase in actin dynamics. Despite its lack of nucleotide

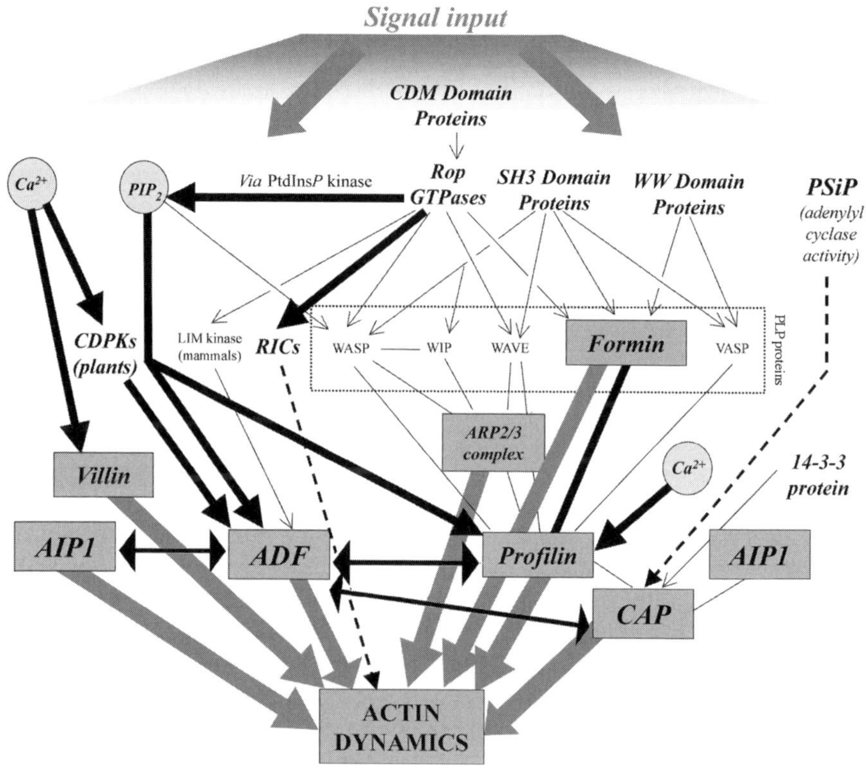

Figure 1.1 Diagram showing a selection of signalling pathways that affect actin dynamics. Factors in italics are conserved in plants. Circled factors are non-protein signalling molecules. Solid lines show characterised interactions, dashed lines represent hypothesised pathways. Fortified lines (with solid arrow heads) describe interactions applicable to plants while thin lines (with open arrow heads) represent pathways currently characterised only in animals or fungi. Synergistic interactions are shown using double arrow heads. Proteins in grey boxes interact directly with G-actin and F-actin to modulate polymerisation dynamics or filament bundling.

exchange activity, *Arabidopsis* profilin also enhances the treadmilling-action of ADF (Didry *et al.*, 1998). The coordination of regulation of these two ABPs is likely to be essential in plant cells to optimise local actin dynamics.

1.3.1.3 Actin interacting protein 1 (AIP1)

Actin interacting protein 1 (AIP1) was first identified as a binding partner of yeast actin (Amberg *et al.*, 1995). This WD repeat protein has two *Arabidopsis* homologues that are expressed in a tissue-specific manner (Allwood *et al.*, 2002). AIP1 has also been found to interact with CAP, and has been shown in animals to form a protein complex with CAP, ADF, and actin (Moriyama & Yahara, 2002). The involvement of AIP1 in actin-mediated processes has been confirmed *in vivo* (Konzok *et al.*, 1999). Biochemical studies suggest that AIP1 might increase ADF severing activity by preventing the ligation of actin filaments severed by ADF (Okada *et al.*, 2002), and can enhance lily pollen ADF activity by approximately 60% (Allwood *et al.*, 2002). Both lily ADF and *Arabidopsis* AIP1-1 show depolymerising activity when applied to F-actin individually, but they act synergistically to produce more dramatic increases in activity. The co-localisation of ADF and AIP1 to pollen tubes (Allwood *et al.*, 2002) supports a model where AIP1 is a major regulator of plant ADF activity.

1.3.1.4 Cyclase associated protein (CAP)

The gene encoding yeast adenylyl cyclase-associated protein (CAP) was identified using a mutant with reduced adenylyl cyclase activity. It has since been shown that this multifunctional protein binds actin monomers via its C-terminal region (Freeman *et al.*, 1995). This interaction with actin monomers may, in some organisms, partially overlap profilin function, as profilin mutants of *Dictyostelium* compensate by increasing CAP expression. Mammalian CAP appears to have a complex effect on actin dynamics. The N-terminus of CAP mediates an interaction between an ADF and F-actin complex that includes AIP1 (Moriyama & Yahara, 2002). *In vitro* the N-terminal domain of mammalian CAP1 accelerates depolymerisation at the pointed end. The addition of ADF *in vitro* leads to a synergistic increase in depolymerisation activity. In addition to this, the C-terminus of CAP was observed to accelerate actin monomer nucleotide exchange, re-cycling actin monomers for addition to the barbed end (Moriyama & Yahara, 2002).

CAP has many control inputs and interacting partners. The sequestering activity can be inhibited by PIP_2, (Gottswald *et al.*, 1996; Noegel *et al.*, 1999) but not calcium (Gottswald *et al.*, 1996). Yeast 2-hybrid and *in vitro* co-precipitation experiments demonstrated that a 14-3-3 protein interacts with the N-terminus but not the full length CAP from the fungi *L. edodes* and *S. pombe* (Zhou *et al.*, 2000). The interplay between the CAP N-terminus, C-terminus, and 14-3-3 proteins could constitute a regulatory mechanism. The local activity of CAP has the potential to be regulated through enrichment at sub-cellular polar sites. In *Dictyostelium*, CAP is localised in a sub-cellular fashion to the anterior and posterior poles (Gottswald *et al.*, 1996). The N-terminus is sufficient to achieve this polar localisation when fused to GFP (Noegel *et al.*, 1999). The N-terminus domain of mammalian CAP is also the domain required to localise CAP specifically to dynamic actin structures in fibroblasts (Moriyama & Yahara, 2002). Animal and fungal CAP contains two short stretches of poly-proline sequence towards the middle of the protein. The proline-rich regions of mammalian CAP are necessary for interactions *in vitro* with profilin (Lambrechts *et al.*, 1997), and it has recently been reported that *Drosophila* CAP interacts with Abl tyrosine kinase (Wills *et al.*, 2002), a SH3 domain-containing protein that has the potential to interact with proline rich motifs.

Arabidopsis contains a single CAP homologue (Barrero *et al*., 2002). The biochemistry of plant CAP has not been investigated in depth, but the overexpression of CAP under an inducible promoter has shown that excessive CAP can disrupt cytokinesis and cell expansion (Barrero *et al*., 2002) in a manner that phenocopies the treatment of *Arabidopsis* plants with the actin depolymerising drug latrunculin B (Baluška *et al*., 2001a). Overexpression of similar constructs in tobacco reduces cell expansion (Barrero *et al*., 2003). The C-terminus of plant CAP is more highly conserved between plants and animals than the N-terminus (Barrero *et al*., 2002), and the proline-rich region is much reduced in plant CAP isoforms. The divergence of these localisation and interaction domains might indicate that plant CAP has novel interaction partners. *Arabidopsis* CAP can complement CAP-deficient yeast (Barrero *et al*., 2002), but cotton CAP was unable to replace yeast CAP function (Barrero *et al*., 2003) despite sharing high homology with *Arabidopsis* CAP. It seems that some functions of plant CAP have diverged from those of its animal and yeast counterparts. A current enigma is the characterisation of a plant adenylyl cyclase-associated protein in the absence of a plant adenylyl cyclase. There is no evidence to suggest such an interaction would be conserved in plants, but Fig. 1.1 shows a hypothesised interaction with PsiP, a maize protein that could have *in planta* adenylyl cyclase activity (Moutinho *et al*., 2001).

1.3.1.5 Villin

Villin proteins crosslink and bundle actin filaments. They also have a calcium dependent F-actin capping and severing activity (Glenney *et al*., 1980; Janmey & Matsudaira, 1988). Villin contains six gelsolin domain repeats, but a single gelsolin domain has been shown to be sufficient to bind G-actin. However, the bundling activity of villin comes from its headpiece domain (Pope *et al*., 1994). Biochemical studies isolated a 135 kDa homologue of villin from lily pollen (Vidali *et al*., 1999) that has the ability to bundle F-actin *in vitro* (Yokota *et al*., 1998). Five homologues of plant villin exist within the *Arabidopsis* genome, and each isoform is expressed in a wide range of tissues (Klahre *et al*., 2000; Staiger & Hussey, 2003). This broad set of expression patterns contrasts that of mammalian villin, which is restricted to the microvilli of brush border cells. It has been hypothesised that plant villin functions to maintain actin bundles within transvacuolar strands. Disruption of these bundles in root hairs with an anti-villin antibody resulted in the disturbance of reverse-fountain cytoplasmic streaming (Tominaga *et al*., 2000).

Figure 1.1 illustrates that calcium can influence the activity of plant villin. The bundling activity of the lily pollen villin is calcium-dependent (Yokota *et al*., 2000). The modulation of the activities of profilin and ADF might be responsible for the dynamic actin at the distal end of tip growing cells, but it appears that the local activity of villin proximal to the growth zone has a role in bundling actin filaments. The unbundling of root hair filaments by the microinjection of anti-villin antibody results in an unusual movement of the root hair nucleus into the apex (Ketlaar *et al.*, 2002). Normally the nucleus is located a fixed distance from the growing tip, mechanically connected to the proximal section of the cytoplasm (Ketlaar *et al.*, 2002). Villin could function prominently in the segregation of tip growing plant cells into zones of contrasting actin dynamics.

1.3.1.6 Actin nucleators

The ABPs described so far modulate actin filament turnover or arrange filaments into higher order structures. Another class of ABPs is responsible for stimulating the initiation of filaments *de novo*. Two major classes of actin nucleators have been identified: the Arp2/3 complex, and formin proteins. Both groups have plant homologues.

1.3.1.6.1 Arp2/3 complex

The Arp2/3 complex consists of seven proteins (Mullins *et al.*, 1997). Two of these proteins (Arp2 and Arp3) are related in sequence to actin. It is thought that the spatial arrangement of Arp2 and Arp3 within the complex mimics an actin dimer seed from which new filaments can be nucleated (Robinson *et al.*, 2001). As the complex attaches itself to the flank of an existing filament, a new filament is initiated from this point at a precise angle of 70 degrees relative to the parental filament (Mullins *et al.*, 1998; Blanchoin *et al.*, 2000). In yeast, the complex is essential for the formation of dynamic actin patches that are associated with sites of polar growth (Evangelista *et al.*, 2002). It is hypothesised that these actin structures facilitate endocytosis (Pruyne & Bretscher, 2000), and the characterisation of a temperature-sensitive *S. cerevisiae arp2* mutant has demonstrated that a functioning ARO2 protein is required for endocytosis but not for exocytosis (Moreau *et al.*, 1997). In multicellular animals Arp2/3 is required for developmental processes such as the formation of *Drosophila* embryo pseudocleavage furrows (Stevenson *et al.*, 2002). Unicellular pathogens and cell organelles can exploit the explosive nucleating power of the Arp2/3 complex to propel themselves through the cell cytoplasm (Welch *et al.*, 1997; Taunton *et al.*, 2000), demonstrating the kinetic abilities of the complex.

Four publications have recently identified and characterised *Arabidopsis* homologues of the Arp2/3 complex. Essentially, mutant alleles of Arp2/3 complex components cause the loss of sub-cellular cortical domains of fine F-actin and the accumulation of denser actin bundles (Mathur *et al.*, 2003a,b; Li *et al.*, 2003; Le *et al.*, 2003). The absence of these domains disrupts the expansion of epidermal pavement cell lobes, the expansion of leaf trichomes, the coherence of the hypocotyl epidermis during cell expansion, and to some extent, the elongation of root hairs (Mathur *et al.*, 2003a,b). Organelle transport is also affected (Mathur *et al.*, 2003b). Only subtle pollen tube phenotypes have been identified (Li *et al.*, 2003), which is surprising considering that this growth process relies upon the cycling of fine actin filaments. In the absence of a functioning Arp2/3 complex, the actin of trichomes, hypocotyl cells, and epidermal pavement cells accumulates into thick bundles (Mathur *et al.*, 2003a,b). Mathur *et al.* (2003b) also establish that the vacuole of trichome cells becomes fragmented in a manner similar to the vacuoles of yeast Arp2/3 mutants suggesting a role for Arp2/3 in membrane fusion, and consequently in the maintenance of turgor pressure applied to plant cell walls from central vacuoles.

It has been established that Arp2/3 is required for actin organisation in some plant cell types, but the biochemical characteristics of the plant complex have not been investigated and the method of activity regulation has not been identified. The biochemistry of plant ABPs such as profilin is significantly different to their animal and fungal cousins, so it is quite feasible that the plant Arp2/3 complex will hold several surprises. With regards to cell biology beyond actin polymerisation; could the Arp2/3 complex play a role in plant cell endocytosis analogous to its role in fungal cells? Eighty-one per cent of the membrane delivered to the apex of tip growing *Arabidopsis* cells must be recovered by endocytosis, and the estimated number of endocytotic events at the plasma membrane of root hair and pollen tube tips vastly outnumbers exocytotic events (Ketelaar, 2002). The demand for membrane recovery has to be just as fierce in isotropically growing plant cells, and this transport must be somehow coordinated. The discovery of a family of plant intersectin1-like SH3 domain proteins that are involved in the assembly of clathrin-coated vesicles and bind to actin filaments *in vitro* associates plant endocytosis with the actin cytoskeleton (Lam *et al.*, 2001), providing a niche for Arp2/3 function in plant endocytosis. Mathur *et al.* (2003b) observed accumulations of unusually large Golgi stacks engaged in atypical motion in regions of *crooked* (*arpc5*) trichomes where actin had become bundled. It was also suggested that a reduction in exocytosis at the apical and basal ends of hypocotyl cells undergoing rapid cell expansion could explain the loss of adherence associated with the excessive local actin bundling that has been observed in *arp2/3* mutants (Mathur *et al.*, 2003b). Evidence is growing that plant Arp2/3 plays a major role in organising exocytosis and other membrane based events in addition to, or instead of, endocytosis.

In animals and fungi, the Arp2/3 complex interacts with a plethora of regulatory proteins that guide its local activity to form structures such as membrane ruffles. Very few of these regulatory proteins have been conserved in plant genomes (see section 1.3.2.1). Among the absentee regulators are: WASP, WIP, cortactin and caldesmon (Hussey *et al.*, 2002). Plant analogues of these proteins may exist, but they are not easy to identify by screening plant genome databases with primary protein sequence. One of two mammalian isoforms of the Arp2/3 complex component ArpC5 can be phosphorylated by

MAPKAPK2 (Singh *et al.*, 2003). The *Arabidopsis* homologue of this sub-unit is encoded by the *CROOKED* gene (Mathur *et al.*, 2003b). There is as yet no evidence that phosphorylation results in an alteration of Arp2/3 mediated actin polymerisation, but phosphorylation of Arp2/3 complex proteins provides another potential means for the attenuation of plant Arp2/3 activity. Such phosphorylation could be committed by plant-specific kinases as is the case for CPDKs and ADF phosphorylation.

The BRICK1 protein is a rare example of a conserved plant Arp2/3 complex interactor (Smith, 2003; Frank *et al.*, 2003; Frank & Smith, 2002). Its mammalian homologue, HSPC300, associates in a complex that includes the Arp2/3 activator WAVE (Eden *et al.*, 2002). The binding of activated Rac1 to the complex releases three proteins (Nap125, PIR121, and Abi2) that otherwise repress the Arp2/3 activating ability of WAVE1. HSPC300 remains associated with WAVE1, and the HSPC300–WAVE1 complex is as effective at activating Arp2/3 *in vitro* as the recombinant WAVE1 without HSPC300 (Eden *et al.*, 2002). There is currently no published evidence that HSPC300 interacts directly with Arp2/3, but Eden *et al.* (2002) report that preliminary evidence exists for HSPC300 to have a role in regulating actin dynamics. Figure 1.1 shows that WAVE is the only plant component missing from the interaction web to complete a pathway from signalling protein to Arp2/3 mediated actin nucleation. Manipulation of SPIKE1 (DOCK180 homologue), ROP2 (RAC-like GTPase), BRICK1 (homologous to a WAVE binding partner), or the components of the Arp2/3 complex itself, all result in epidermal cell lobe phenotypes connected with disorganisation of the actin cytoskeleton. The absence of plant WAVE1 necessitates one of the following explanations: (i) WAVE is too poorly conserved at the primary amino acid level to be detected easily within plant genomes but a plant orthologue does exist; (ii) BRICK1 does not require WAVE (but instead possibly an analogue of WAVE or no binding partner at all) to activate Arp2/3; (iii) BRICK1 does not activate Arp2/3 but instead organises actin by an independent pathway. The third option cannot be totally discounted until double mutants between BRICK1 and components of the Arp2/3 complex are found to have no greater severity than the single mutants, or until it is demonstrated biochemically that BRICK1/HSPC300 activates Arp2/3.

1.3.1.6.2 Formins

Formins represent a second major group of actin nucleating proteins. In yeast, formins operate in a pathway genetically and biochemically distinct from Arp2/3 (Evangelista *et al.*, 2002; Pring *et al.*, 2003). *S. cerevisiae* cells that are double mutants of formins *Bni1* and *Bnr1* retain actin patches but lose actin cables (Evangelista *et al.*, 2002). It appears that at least in yeast the two forms of actin filament nucleation are not redundant but are required for specific arrays of actin filaments. Recent publications have demonstrated that the FH2 (*F*ormin *H*omologous 2) domains of three individual formins, Bni1p

(*S. cerevisiae*), Cdc12p (*S. pombe*), and mDia1 (mammals), have the ability to nucleate actin filaments (Pruyne *et al*., 2002; Sagot *et al*., 2002b; Pring *et al*., 2003; Kovar *et al*., 2003; Li & Higgs, 2003) and associate with the barbed end of growing actin filaments (Pruyne *et al*., 2002; Sagot *et al*., 2002b; Kovar *et al*., 2003; Li & Higgs, 2003). The association of the FH2 domain with the barbed end of actin filaments leads to either a partial capping of the filament (Pruyne *et al*., 2002; Li & Higgs, 2003) or a strong capping interaction (Kovar *et al*., 2003) depending possibly on the identity of the formin protein or the exact borders of the protein truncation. The formin FH2 domain is located within the C-terminus of the majority of characterised formins (Waller & Alberts, 2003), covering a region of approximately 400 amino acids. In addition to its actin-organising activity, it has been proposed that the FH2 domain might also influence the patterning of the microtubule cytoskeleton (Ishizaki *et al*., 2001). The FH2 domain is highly conserved between the plant, fungal, and animal kingdoms, and a phylogenetic comparison of the FH2 domain from various formins places the formins into clades with boundaries largely defined by the eukaryote kingdom taxonomy (Deeks *et al*., 2002). Mutant alleles of the *Drosophila* formin *Cappuccino* (Emmons *et al*., 1995), mutant alleles of vertebrate *Formin1* (Woychik *et al*., 1990), truncations of mDIA1 (Watanabe *et al*., 1999) and of Cdc12p (Chang *et al*., 1997) all show that the integrity of the FH2 domain is a necessary requirement for formin biological function.

Most FH2 domains are preceded by a low-complexity FH1 domain that contains variable numbers of poly-L-proline residues. FH1 and FH2 domains in isolation can alone rarely achieve cytoskeletal rearrangements, and *in vivo* usually require the presence of the other domains (examples are mDIA1, Watanabe *et al*., 1999; and Cdc12p, Chang *et al*., 1997). The Bni1p FH1 domain must be attached for the FH2 domain to show activity *in vivo* (Evangelista *et al*., 2002; Sagot *et al*., 2002a). *In vitro* the Bni1p FH1–FH2 fragment is twice as effective at nucleating actin filaments (in the absence of profilin) as the FH2–only domain fragment (Pruyne *et al*., 2002). However, this is not the case with mDia1 (Li & Higgs, 2003), where it has been shown *in vitro* that the FH2 domain is just as effective a nucleator as the FH2 domain coupled to the FH1 domain. The FH1 domain in isolation does not show any nucleating activity (Pruyne *et al*., 2003). The FH1 domain of Bni1p has the ability to interact with actin in a yeast two-hybrid assay (Evangelista *et al*., 1997), providing an explanation for the enhancing effect of the Bni1p FH1 domain during actin polymerisation. However it is unknown whether the FH1 domains of other formins share a similar property.

The key function of the FH1 domain in regards to actin nucleation is its affinity for profilin-actin. Profilin binds to PLP with sufficient strength to be purified from eukaryotic cells by passing cell extracts through a PLP affinity column containing peptides with a minimum of eight chained prolines (Sohn & Goldschmidt-Clermont, 1994). NMR studies of the profilin molecule suggests that its PLP binding site actually has room for a six proline residue ligand (Metzler *et al.*, 1994), but some profilin interacting domains consist of proline interrupted by other residues (e.g. *Listeria* actA). The *profilin* section of this review described how the majority of G-actin within a living cell must be complexed with sequestering proteins such as profilin in order to maintain concentrations of G-actin above the critical concentration for polymerisation. Biochemical studies show that profilin-actin either enhances nucleation assays (Sagot *et al.*, 2002b), or slightly reduces the efficiency of nucleation (Pruyne *et al.*, 2002; Pring *et al.*, 2003; Li & Higgs, 2003), in comparison with a free G-actin substrate. Despite the drop in efficiency, the FH2 domain cannot utilise profilin-actin *in vitro* without the FH1 domain (Pring *et al.*, 2003; Li & Higgs, 2003). Unlike Bni1p and mDia1, it has been found that Cdc12p FH1–FH2 fragment nucleated filaments will only grow from their barbed ends in the presence of profilin, as profilin can relieve the strong barbed end capping activity of this formin fragment (Kovar *et al.*, 2003). The ability of the FH1 domain to interact with profilin-actin is necessary within the *in vivo* environment of sequestered G-actin to shuttle G-actin monomer from profilin to F-actin polymer.

The details of the mechanism of formin nucleation are currently unclear. Pring *et al.* (2003) calculated that the FH2 domain is likely to stabilise an actin dimer seed. Filaments are generated *de novo*, and not by the branching of existing filaments as is the case for the Arp2/3 complex, as incubation of FH2 fragments with pre-formed filaments does not enhance the rate of nucleation (Pruyne *et al.*, 2002). It is unknown whether a formin protein has to act as a monomer or as a multimer to nucleate an actin filament. Evidence is building for a formin multimer: both Bni1p and mDia1 show a lower efficiency of nucleation than the activated Arp2/3 complex. The number of F-actin filaments produced *in vitro* in the presence of Bni1p is approximately 3% of formin protein present (Pring *et al.*, 2003). This compares to almost 100% for the activated Arp2/3 complex (Higgs *et al.*, 1999). This figure for Cdc12p is slightly higher at 17% (at maximal conditions: Kovar *et al.*, 2003), and slightly higher still for mDia1 at 20% (Li & Higgs, 2003). Sagot *et al.* (2002b) suggested this could be caused by a requirement for formins to form multimers during nucleation. This hypothesis has since been supported by the report that fragments of mDia1 elute from a size exclusion column at a weight approximately five times than that expected (Li & Higgs, 2003).

The N-terminus of a formin generally contains the domains required to localise the protein, and in combination with the FH1 domain is important in linking the activity of the formin to cell signalling molecules. A poorly conserved FH3 domain within the N terminus has been identified within fungal and some animal formins that contributes to sub-cellular localisation (Petersen *et al.*, 1998b). The FH3 domain is frequently accompanied by a GTPase Binding Domain (GBD). The GBD in the N-terminus of Diaphanous-related formins binds to a Diaphanous Autoregulatory Domain (DAD) that lies at the very C-terminus of the protein (Watanabe *et al.*, 1999; Alberts, 2001). This inhibits the actin nucleation function of the protein (Li & Higgs, 2003). The binding of a Rho protein to the GBD releases the C-terminus and activates actin nucleation (Li & Higgs, 2003), but it is currently unclear whether the Rho protein is required to be in its active GTP state to achieve activation. The RhoA interaction with the mammalian Diaphanous GBD has been observed using either GTP-RhoA or GDP-RhoA (Li & Higgs, 2003; Watanabe *et al.*, 1999). Repression of actin nucleation is not lifted completely *in vitro* by RhoA (Li & Higgs, 2003), suggesting that the activation process *in vivo* may involve more factors. In addition to the Rho interaction, numerous regulatory proteins can bind to FH1 domains via SH3 and WW domains, and the FH2 domain also can court a selection of proteins such as the kinesin-like Smy or eF1α (for a comprehensive review of binding partners, see Wallar & Alberts, 2003).

Bioinformatics and cDNA library screens have identified a large family of plant formins numbering at least 21 (Deeks *et al.*, 2002; Cvrckova, 2000; Banno & Chua, 2000). Strikingly, the family is divided into two distinct clades that are as distantly related from each other as they are from animal and fungal formins (Deeks *et al.*, 2002). The division into two clades based on the sequence comparison of the FH2 domain might reflect a division in biochemical behaviour. Comparisons between Cdc12p, mDia1 and Bni1p demonstrate that formins can differ in their ability to nucleate and cap filaments. The concomitant divergence of N-termini between the two groups could relate to a bifurcation of signalling pathways to the two clades. Neither sub-clade show significant levels of homology to their animal and fungal cousins within their N-termini. It is impossible to predict whether these formins rely upon auto-inhibition, or whether a *trans*-acting repressor is required, as has been observed for WAVE. No interaction between a Rop small GTPase (or RIC) and a plant formin has yet been identified. Some plant formin interactors have been isolated. A yeast two-hybrid cDNA library screen identified a putative membrane protein of unknown function that could bind to the group-I formin AtFHP1 *in vivo* and *in vitro* (Banno & Chua, 2000). An intimate association with plasma membrane for group-1 formins is supported by predicted transmembrane regions at the N-terminus of most group-1 formins. Plant profilin does appear to interact with the FH1 domain (Banno & Chua, 2000; Deeks, Davies & Hussey, unpublished) as would be predicted from the nucleation models described above. Identifying the signal transduction connections to the plant formins may radically redraw the plant interaction web leading to the actin cytoskeleton.

1.3.1.6.3 The relationship between the actin nucleators

The presence of two nucleating proteins within yeast and other eukaryotes, including plants, asks the question: Why two nucleators? The mechanisms of Arp2/3 and formin nucleation are very different. Primarily, Arp2/3 nucleates from the pointed end of a filament while formin nucleates from the barbed end of the filament, partially capping the growing barbed end as a result (Pring *et al.*, 2003). Arp2/3 nucleated filaments are generated from the flanks of existing filaments by mimicking a G-actin dimer, while formins somehow have to assemble their own G-actin dimer seed to generate an independent filament. The corollary of these differences is the specialisation of function between the nucleators in yeast to create actin patches (Arp2/3) and cables (formin). It has been suggested that Arp2/3 actin structures exert an expansive force while formin nucleated cables resist tension (Pring *et al.*, 2003). Modulation of the sub-cellular localisation and activation of the opposing nucleator proteins, together with other factors including the ABPs described earlier in this chapter, will create a diversity of functionally unique F-actin structures within a cell.

The observation that there exist different modes of action of the nucleators needs to be applied to plant biology. The relatively light trauma experienced by plants with Arp2/3 mutant alleles is unexpected considering that analogous alleles in *Drosophila* are lethal (Hudson & Cooley, 2002). An argument that the epidermal phenotype alleles of Arp2/3 complex components do not represent null alleles is not supported by the publication of alleles showing complete absence of full length transcript or severely truncated reading frames that have epidermal specific phenotypes (Le *et al.*, 2003; Li *et al.*, 2003). Le *et al.* (2003) suggest two alternative explanations. First, Arp2/3 complexes in yeast can partially assemble in the absence of some components (Winter *et al.*, 1999) so that the plant Arp2/3 complex could retain some function even in the presence of a null allele for a single component. Second, the diversification of the plant formin family has led to a high level of functional redundancy between plant actin nucleating proteins. This concept contrasts with the situation in yeast where there is conspicuously little redundancy between the nucleators. Perhaps what needs to be considered is the applicability of Arp2/3 'dendritic' actin filament growth to plant development. The 'expansive' role of Arp2/3 nucleated filaments in an Arp2/3-formin tensegrity model could be (in most cases) superseded in plants by osmotic pressure applied by the plant cell vacuole. Also to be considered is the possibility that spontaneous nucleation, or generation of new barbed ends by filament severing, or even a plant specific nucleator, might provide sufficient cytoskeletal function for most *Arabidopsis* cell types under standard conditions. Biochemical data is urgently required to characterise the plant Arp2/3 complex and the formin nucleators that could be the key-stone to plant actin polymer organisation.

1.3.2 Signalling to the actin cytoskeleton

1.3.2.1 PLP proteins

PLP proteins encompass a broad class of proteins that link actin polymerisation to a wide variety of signalling inputs. Their common domain of stretches of poly-L-proline residues can interact with profilin and/or protein interaction domains such as SH3 and EVH1 domains. The presence of PLP domains and PLP-binding domains in the same protein permits cross-talk between members of the class such as WASP and VASP (Castellano *et al.*, 2001). Holt and Koffer (2001) have attempted to classify profilin-binding proline-rich domains as having the consensus sequence XPPPPP, where X can be G, S, A, I, or L, but this consensus does not apply to all profilin binding domains, only the majority.

Four major classes of eukaryotic protein have so far been implicated in exploiting this affinity of profilin for PLP to exert an influence over the eukaryote cytoskeleton. The WASP, WAVE, VASP and formin families all have PLP domains flanked by an N-terminal regulatory domain and a C-terminal cytoskeletal interaction domain. Within animals and fungi these four classes of protein provide multiple bridges between signalling networks and the actin cytoskeleton. The interactions include numerous small GTPases, kinases, SH3 adaptor proteins and WW domain proteins. These interactions have been summarised in Fig. 1.1. Only members of the formin family of PLP profilin regulators have been isolated from plants, and searches through the *Arabidopsis* genome sequencing data have not identified any conserved WASP, VASP or WAVE related genes. Formins may represent the only PLP profilin interactor conserved between animals and plants.

A large proportion of the PLP families interact with the Arp2/3 actin nucleation complex. WASP and WAVE activate Arp2/3 complex directly through their VCA (verprolin-cofilin-acidic) domains (Panchal *et al.*, 2003; Machesky & Insall, 1998; Machesky *et al.*, 1999; Rohatgi *et al.*, 1999), and WIP can enhance Arp2/3 nucleation via Cortactin (Kinley *et al.*, 2003). VASP can also modulate actin polymerisation but its relationship to Arp2/3 has not been clearly established (Boujeema-Paterski *et al.*, 2001; Castellano *et al.*, 2001). The binding of profilin by the PLP Arp2/3 activators could enhance the ability of the complex to shuttle actin monomers onto the Arp2/3 polymer seed. The absence of a PLP Arp2/3 activator might suggest that plant Arp2/3 does not require a WASP-style interactor to function as a plant nucleator. The absence of a large cohort of plant cytoskeletal PLP proteins leaves a gap in the understanding of signal transduction to the plant actin cytoskeleton. The reduction in variety of non-formin PLP proteins may be balanced by the expansion of the plant formin family into two distinct clades with totally novel interaction domains.

1.3.2.2 ROPs

ROPs (Rho of Plants) are members of the RAS GTPase super-family. All members of the Ras super-family exchange GDP for GTP to enter an active state in which they interact with effector proteins to perpetuate signalling cascades. The members of the small GTPase sub-family (often referred to as Rho-related GTPases) include the Rho, Rac, Cdc42, and ROP GTPases, but

ROPs represent the only small GTPases in plants. The Rho-related small-GTPases are commonly post-translationally modified to allow association with plasma membrane where they are involved in regulating multiple processes. The small GTPases have proven to be potent organisers of the actin cytoskeleton in animals and fungi, and it now seems that the ROP sub-family could exert a similar influence on plant actin. ROPs have been implicated in many plant processes including shoot apical meristem organisation (Trotochaud *et al.*, 1999), NADPH oxidase activation (Baxter-Burrell *et al.*, 2002), auxin-mediated gene activation (Tao *et al.*, 2002), guard cell closure (Lemichez *et al.*, 2001), tip growth (Lin & Yang, 1997), and epidermal cell morphogenesis (Fu *et al.*, 2002). The involvement of *Arabidopsis* ROP2 in the development of root hairs and epidermal cells has already been discussed. The *Arabidopsis* ROP1 isoform, along with ROPs 3 and 5 (Gu *et al.*, 2003; Fu *et al.*, 2001) play an actin organising role in pollen tubes.

The activity of ROP proteins in the apices of tip-growing cells has led to the hypothesis that ROP activity could direct calcium ion flux (Fu *et al.*, 2001). As ADF is phosphorylated by CDPKs, it is plausible that calcium could mediate a ROP-induced ADF response (Chen *et al.*, 2003). The regulation of animal ADF sets a precedent for small GTPase regulation, as LIM kinases are stimulated by active Rho, Rac, and Cdc42 GTPases (Sumi *et al.*, 1999; Edwards *et al.*, 1999). This demonstrates a change in the effectors of small GTPases for plants. The absence of many classes of PLP proteins has removed another class of classical small GTPase effector proteins, spurring a requirement for novel plant-specific GTPase effectors to achieve ROP mediated actin organisation. Rop Interacting CRIB motif proteins (RICs) are strong candidates for being the major class of ROP effector proteins. Wu and colleagues acquired RIC1 from a yeast two-hybrid library screen using a constitutively active form of ROP1 as a bait (Wu *et al.*, 2001). The domain responsible for the active ROP2-RIC1 interaction has homology to the CRIB domain utilised by small GTPase GAPs (Wu *et al.*, 2001). Searches of the *Arabidopsis* sequence database identified a further 10 RIC proteins. It was then shown that RICs 3 and 4 are localised by ROP activity to the pollen tube tip, and that they are true effectors of Rop1 (Wu *et al.*, 2001). Data is currently not available that could reveal any ABP targets of the RICs responsible for governing actin filaments in the pollen tube tip.

1.3.2.3 CDM domain proteins

CDM domain proteins are essential cytoskeletal organisers in some animal cell types (Fukui *et al.*, 2001), and a recently identified plant CDM homologue, SPIKE1, has a role in organising *Arabidopsis* cell polarity (Qiu *et al.*, 2002). The seedling-lethal *spike1* mutant shows reduced trichome branching and defects in epidermal cell morphology. Members of the CDM domain protein family have been found to interact with and activate small

GTPases (Kiyokawa *et al.*, 1998; Nolan *et al.*, 1998). The method of CDM activation of small GTPases could be exemplified by the ability of DOCK180 to load Rac with GTP (Brugnera *et al.*, 2002). *In vivo* this requires a third factor, ELMO, to be present within the same complex. BLAST searches of the complete *Arabidopsis* genome show that SPIKE1 is the only CDM protein within the genome, but there are at least 6 predicted proteins with ELMO-like domains. Curiously, ELMO has been found to share an SH3 domain containing interaction partner with WASP (Scott *et al.*, 2002). The GEF-like activity of the ELMO-CDM complex can be activated in mammals by another small GTPase, RhoG (Katoh & Negishi, 2003), providing a route by which small GTPases can activate other small GTPases. The Rac-like ROP proteins, and the Arp2/3 complex have been implicated in plant epidermal cell morphogenesis, so it seems reasonable to hypothesise that SPIKE1 acts through ROP proteins to stimulate cytoskeletal rearrangement possibly via the modulation of Arp2/3 activity. Figure 1.1 shows that in non-plant systems this pathway requires WASP, WIP, or WAVE proteins that are absent from plant genomes. SPIKE1 is another protein 'familiar' to animal signalling networks that, like ROPs and formins, is likely to be acting in an eccentric fashion in plant cells to achieve cytoskeletal organisation.

1.3.2.4 PIP$_2$

PIP$_2$ is a phosphoinositide signalling molecule implicated in cytoskeletal organisation. Both plant ADF and profilin are regulated by PIP$_2$ signalling (Gungabissoon *et al.*, 1998; Drøbak *et al.*, 1994). PIP$_2$ is thought to bind to these proteins and localise the cytoskeletal proteins to the plasma membrane in an inactive state. Signalling to phospholipase C stimulates the catalysis of PIP$_2$ to IP3 and DAG. IP3 feeds into the calcium ion pathways by liberating calcium from intracellular stores into the cytoplasm (Malhó, 1998). The cleavage of PIP$_2$ can release ADF repression and lead to the disassociation of profilin from the plasma membrane. Plant profilin has been reported to be localised to the plasma membrane in pollen grains (von Witsch *et al.*, 1998) and at the growing tips of root hairs (Baluška *et al.*, 2000). The wasp venom peptide mastoparan has been observed to stimulate phospholipase C activity in plants (Cho *et al.*, 1995), and the application of mastoparan to maize root cells results in a change of profilin localisation correlated with actin filament rearrangement (Baluška *et al.*, 2001b). This supports the hypothesis that the *in vitro* association of plant profilin with PIP$_2$ might be exploited *in vivo* to regulate cytoskeletal organisation. However, mastoparan binds to GTPases (Higashijima & Ross, 1991) and can stimulate plant GTPases (Irving, 1998; Iwasaki *et al.*, 1997). The range of mastoparan targets includes the ROP-related RAC GTPases (Koch *et al.*, 1991). It is quite possible that plant cytoskeletal changes stimulated by mastoparan occur directly through GTPase-mediated signalling and not via phospholipase C. PIP$_2$ is catalysed from PIP by PtdInsPkinase (phosphatidylinositol-4-phosphate 5-kinase). A plant PtdInsPkinase has been found to interact with the pollen tube tip localised AtRAC2 (Kost *et al.*, 1999). This provides a mechanism by which PIP_2 levels can be regulated within the pollen tube, having consequences on local profilin and ADF activity. In animal and fungal cells PIP_2 has a range of cytoskeletal targets beyond profilin and ADF that includes WASP (Miki *et al.*, 1996), an Arp2/3 activator. It is possible that in plants, PIP_2 interacts with and regulates as-yet unidentified plant specific ABPs.

1.3.3 Microtubule associated proteins

As we have described, cell growth is dependent on both a functional actin and microtubule cytoskeleton and the role of actin associated proteins has been discussed. Here we describe some of the microtubule associated proteins in plants that either crosslink or regulate the MT network in plant cells and have an impact on cell growth. Comparison of the *Arabidopsis* and human genome sequences has revealed that plants do not contain the equivalents of the well-characterised mammalian MAPs, MAPs 1, 2, 4 and tau. Of all the plant microtubule associated proteins and plant microtubule interacting proteins known there are functional equivalents in animal cells (with the exception so far of MAP190). However, the sequence identity can be relatively low (e.g. between plant MAP65, mammalian PRC1 and fungal Ase1p equivalents) which presumably indicates some tailoring to suit plant-specific requirements (Hussey *et al.*, 2002; Lloyd *et al.*, 2004). The microtubule motor proteins, the kinesins, are conserved in plants: the conserved motor domains can be N-terminal, central or C-terminal as in other organisms but tail domains are often divergent implicating plant specialised activities (Reddy & Day, 2001).

1.3.3.1 Tubulin and tubulin cofactors

The primary structure of the tubulins is highly conserved in plants but aspects of their biochemical properties are quite different (Lloyd *et al.*, 2004). Aberrant expression of tubulin in plants has a dramatic effect on plant growth morphology, with reduced tubulin expression causing randomised polar growth in root hairs resulting in extensive root hair branching (Bao *et al.*, 2001). Correct folding of α/β-tubulin requires chaperones plus additional proteins called cofactors to produce assembly competent α/β tubulin heterodimers (Cleveland, 1978; Gao *et al.*, 1992, 1993). There are several cofactors present in vertebrates, A, B, C, D and E (Tian *et al.*, 1996, 1997). In the correct folding pathway α-tubulin binds cofactor B whilst β-tubulin binds cofactor A. These are then replaced by cofactors E and D respectively. Two produce the heterodimer, the two pathways converge and a α/β-tubulin-cofactorE/B complex is formed. This complex is then bound by cofactor C and following the hydrolysis of GTP the α/β-tubulin heterodimer is released

(Lewis *et al.*, 1997). Yeast possess homologues of all of these cofactors with the exception of C, but the sequence of their involvement in folding, as identified by genetic analysis, appears to be inconsistent with other systems. Interestingly Alp1(D) and Alp21(E) have been shown to co-localise with microtubules. In addition it was found that Alp31(A) plays an important role in microtubule integrity and cell polarity. In deletion or overexpression lines of Alp31(A), cells showed bent or branched morphologies. In these cells, no intact microtubules were evident, but instead short, dotted or misorientated microtubules were observed.

TFCs have been shown to be essential for normal morphogenesis in plants. Embryo-lethal screens identified *Arabidopsis* homologues of all TFCs except TFC-B (Tzafrir *et al.*, 2002; Steinborn *et al.*, 2002). These *titan* and *piltz* mutants resulted in abortive embryos with severely reduced cell number, defective cell plates and enlarged nuclei. Mutant screens for reduced trichome branching identified the importance of TFCs in non-dividing cells (Kirik *et al.*, 2002). These *kisA* and *porC* plants display trichome branching, cell swelling, dwarfism, sporadic sterility and hypocotyl cell shape defects.

1.3.3.2 Katanin

Katanin is the only ATP-dependent microtubule severing protein. Originally isolated from sea urchin eggs (McNally & Vale, 1993), katanin is a heterodimer of 60kDa and 80kDa subunits. The 60kDa subunit alone is sufficient for ATP-dependent microtubule severing and exhibits ATPase activity (Hartman *et al.*, 1998). This ATPase activity can be inhibited at high ratios of microtubules to katanin, e.g. 500:1. Katanin localisation is concentrated at centrosomes throughout the cell cycle (McNally *et al.*, 1996) and a human homologue of the p80 subunit has also been found to concentrate at interphase centrosomes (Hartman *et al.*, 1998). The p80 subunit contains a WD40 repeat, which has been shown to act as a centrosome-targeting domain (Hartman *et al.*, 1998). Therefore it is believed that the function of the p80 subunit is to target the catalytically active p60 domain to centrosomes. In immunodepletion studies using antibodies raised to a *Xenopus* p60 homologue, the microtubule severing activity of *Xenopus* extracts was inhibited by 90% (McNally & Thomas 1998). In addition, this severing activity had previously been shown to be activated by cyclin B/cdc2 (Vale 1991). The presence of a consensus cdc2 phosphorylation site in the human p60 would therefore suggest that the activity of animal katanin p60 could be activated by cdc2. Such regulation could suggest that microtubule severing might play a role in the reorganisation of the mitotic spindle at the G2/M transition or spindle assembly. Furthermore the activity of katanin appears to be dependent on its localisation at the spindle pole. Here it co-localises with polo kinase 1 and Pix1 *in vivo* – purified Pix1 is capable of increasing microtubule severing *in vitro*. Therefore Pix1 is an activator of katanin at the spindle pole. In contrast, the microtubule severing activity of purified *Xenopus* katanin is inhibited by the *Xenopus* homologue of MAP4, XMAP230. Although both p80 and p60 homologues exist in the *Arabidopsis* genome only p60 has been analysed biochemically and genetically. The *Arabidopsis* katanin p60 subunit, AtKSS (Katanin Small Subunit), which shares 43% identity with sea urchin katanin, has been shown to be a functional ATPase *in vitro*. It associates with microtubules in the absence of ATP and severs MTs in an ATP-dependent manner (Stoppin-Mellet *et al*., 2002). Furthermore, the subcellular location of AtKSS is similar to other katanin p60 subunits, with concentration of the protein at the perinuclear regions of interphase cells and at the spindle poles during mitosis. The gene that encodes AtKSS has several mutant alleles: *fragile fibre 2 (fra2*; Bichet *et al*., 2001), *botero1* (*bot1*; Burk *et al*., 2001), *ectopic root hair 3* (*erh3*; Webb, 2002). Phenotypes consist of problems with cortical microtubule orientation, anisotropic cell expansion, cell wall biosynthesis, cell elongation and cell specification in the *Arabidopsis* root epidermis. The reduced anisotropic growth correlated with random microtubule and cellulose microfilbril organisation. Therefore the observed phenotypes are likely to be as a result of secondary effects on the organisation of the cell wall. Although here, cell shape and organ architecture are affected throughout the root and shoot, such genetic analysis has demonstrated a fundamental role for AtKSS, which appears to be limited to non-dividing cells. However examination of microtubule organisation in dividing cells of *fra2* root tips showed a delay in the redistribution of microtubules associated with the nucleus to the cortex, following exit from the cell cycle (Burk *et al*., 2001). In addition, as AtKSS has been shown to localise to the nuclear membrane, these data lead to the suggestion that here, at the nuclear membrane, AtKSS microtubule severing activity may cut back microtubules, allowing them to recycle and redistribute into the cortical array.

1.3.3.3 MAP65

MAP65 was first isolated from tobacco BY2 suspension cells, and MAP65 was found to decorate all four plant microtubule arrays and to bundle microtubules *in vitro* (Jiang & Sonobe, 1993). Subsequently, 65kDa MAPs, antigenically related to the tobacco proteins, were isolated from carrot cytoskeletons (Chan *et al*., 1996). Biochemically purified 65kDa MAPs were found to induce bundling of neuronal microtubules (Chan *et al*., 1999) and form cross-bridges *in vitro* which resemble the 25–30 nm spacing between cortical microtubules in plant cells (Lancelle, 1986). Anti-tobacco MAP65 antibodies were used to isolate MAP65 cDNAs from a tobacco expression library. Antibodies raised to the bacterially expressed tobacco MAP65, NtMAP65–1a, when used in immunohistochemical experiments, stain subsets of the cortical microtubule array and the preprophase band in interphase, the midzone of the spindle at late anaphase and the midzone of the phragmoplast (Smertenko, 2000). The localisation to these midzone regions where microtubules overlap suggests that MAP65 may play a role in the behaviour of anti-parallel microtubules at these sites. Further more, in this study recombinant NtMAP65–1a was shown to bind microtubules and stimulate microtubule polymerisation *in vitro*. The completion of the *Arabidopsis* genome allowed the identification of the complete *MAP65* gene family from *Arabidopsis*. This family encodes nine member proteins with molecular weights ranging from 62–80 kDa. Members of this *Arabidopsis* family have been expressed in bacteria and shown to bind and bundle microtubules like other plant members of the family (Hussey, unpublished data). Analysis of these sequences showed the presence of a 'destruction box' (R-XX-L-XXXX-N), a sequence present in all cell cycle-regulated proteins (Glotzer *et al.*, 1991; Hussey *et al.*, 2002). This sequence targets the protein to the ubiquitin degradation pathway, resulting in rapid degradation during cell cycle progression. There appear to be relatives of MAP65 in humans (PRC1) and yeast (Ase1p), although the primary sequence identity is very low. They do, however, share the domain architecture of MAP65, consisting of two to three coiled coil regions and the destruction box. Interestingly all of these MAPs have been located to the spindle midzone and therefore may share a common role. The yeast example, Ase1p, has been shown to have similar properties to the MAP65 family, including the binding and bundling of microtubules (Schuyler *et al.*, 2003). Furthermore, the loss of Ase1p results in premature spindle disassembly; and conversely, the overexpression of Ase1p is sufficient to trigger spindle elongation (Schuyler *et al.*, 2003). It remains to be seen whether perturbation of MAP65 levels have similar results but it is likely that the members of this extended family, MAP65/Ase1p/PRC1, all stabilise the spindle midzone through the production of microtubule cross-bridges, and their levels are controlled in a cell cycle-specific manner.

1.3.3.4 MAP190

In plant cells actin filaments and microtubules often co-distribute in all of the four plant cell microtubule arrays: the preprophase band, the spindle, the phragmoplast and cortical microtubules (Lloyd, 1987; Schmidt 2002). In addition, actin filaments and microtubules co-localise around the nuclei in tobacco BY-2 cells and play an important role in the migration and positioning of the nucleus (Katsuta & Shibaoka, 1998; Katsuta *et al.*, 1990). Therefore it has been suggested that factors may mediate interactions between these two cytoskeletal systems at these locations. In animal cells various proteins which interact with actin and microtubules have been reported (Gavin, 1997; Rodriguez *et al.*, 2003) and the 190 kDa microtubule-associated protein, MAP190, is a candidate for such a role in plant cells. Based on its ability to bind actin, MAP190 was isolated from a microtubule binding protein fraction of tobacco BY2 miniprotoplasts (Igarashi *et al.*, 2000). A mouse monoclonal antibody was then raised to the partially purified 190 kDa protein and used in immunohistochemical experiments with either purified microtubules or actin filaments. In both cases the crude MBP fraction was mixed with actin or microtubules and double stained with anti-MAP190 and the appropriate cytoskeletal antibody. In the presence of the MBP fraction, actin filaments formed bundles, which stained with the anti-MAP190 antibody. As with actin filaments in the presence of the MBP fraction, microtubules formed bundles, which stained with the MAP190 antibody. Further localisation experiments in BY2 cells showed that MAP190 localised to the nucleus but not the nucleolus during interphase and preprophase. Following the breakdown of the nuclear envelope, MAP190 localised to the spindle and in telophase co-localised to microtubules in the phragmoplast. Furthermore, during telophase MAP190 was shown to co-localise with actin filaments in the phragmoplast. The cDNA for MAP190 was cloned and sequence analysis revealed the presence of several motifs. These included three nuclear localisation signals (NLS), a ER membrane retention signal (ERMRS), calcium binding EF hand domains, a RED repeat and an unidentified motif 'GHKAEQQY' which repeats three times (Igarashi *et al.*, 2000; Hussey *et al.*, 2002). The presence of NLS is consistent with the protein's nuclear localisation at interphase, and one can speculate that MAP190 might cross-link ER, microtubules and/or actin in response to calcium ions for the organisation of the phragmoplast (Hussey *et al.*, 2002). In addition, RED repeats (arginine/glutamine or aspartate) have been reported to be essential for the production of aggregates in the nucleus (Assier *et al.*, 1999). Interestingly, MAP190 is observed to localise as dots within the nucleus.

1.3.3.5 MOR1/GEM1

MOR1/GEM1 is a a member of a group of conserved MAPs from taxonomically distinct organisms including *Xenopus laevis* XMAP215 (Tournebieze *et al.*, 2000), human Ch-TOGp (Charrasse *et al.*, 1998), *Dictyostelium discoideum* DdCP224 (Graef *et al.*, 2000), *Drosophila melanogaster* Msps (Cullen, 1999), *Caenorhabditis elegans* Zyg-9 (Hirsh & Vanderslice, 1976), *Schizosaccharomyces pombe* Dis1 (Ohkura *et al.*, 1988) and Alp14 (Garcia *et al.*, 2001), and *Saccharomyces cerevisiae* Stu2 (Wang & Huffaker, 1997). Various biochemical, genetic and cell biological experiments have shown that members of this class of MAP are involved in regulating microtubule dynamics and are essential for aspects of cell division and cell morphogenesis. The plant member of this group was identified by homology (Tournebieze *et al.*, 2000) and during a genetic screen for temperature-sensitive effects on microtubule organisation in *Arabidopsis* seedlings (Whittington *et al.*, 2001). Two mutant alleles of *MICROTUBULE ORGANISATION 1 (MOR1)* were identified, both of which result in a single amino acid substitution in the amino terminal HEAT repeat, a motif which is believed to be responsible for protein-protein interactions. At the permissive temperature of 21°C, CMT were similar to wild-type, but on shifting to a restrictive temperature of 29ºC they became progressively disorganised. Examination of *mor1* plants grown at the restrictive temperature revealed severe morphological defects including a left-handed twist of organs, isotropic cell expansion and impaired root hair polarity. Furthermore, seeds germinated at the restrictive temperature produce severely stunted plants that do not develop flowers. These observations suggested that MOR1 was essential in the maintenance of the interphase cortical array and for correct morphogeneis. Interestingly, an analysis of *mor1-1* and *rsw1-1* double mutants has led to the possibility that growth anisotropy is not wholly dependent on transversely oriented microtubules relative to cell expansion (Sugimoto *et al.*, 2003).

An essential role for this MAP in cytokinesis was discovered through genetic complementation of the pollen mitosis1 (PM1) cytokinesis-defective *gemini1-1 (gem1-1)* mutant with *MOR1* (Twell *et al.*, 2002). Phenotypic analysis of *gem1-1* and a T-DNA insertion allele *gem1-2* confirmed that MOR1/GEM1 was essential for the regular patterns of cytokinesis in PM1. Both mutations, *gem1-1* and *gem1-2*, cause the truncation of MOR1/GEM1. Moreover, the C-terminal domain absent in both mutants binds microtubules *in vitro*, indicating that the defect in *gem1* mutants was due to defective MT binding (Twell *et al.*, 2002).

1.3.3.6 Kinesins

The extensive kinesin family may be subdivided into a number of groups or subfamilies and all possess a well-conserved motor domain located at the N-terminus, C-terminus or in the centre of the protein. The *Arabidopsis* genome contains 61 kinesins, which can be divided into seven of the nine recognised subfamilies found in other organisms (Reddy & Day, 2001). Many kinesins use mechanochemical energy derived from the hydrolysis of ATP to translocate along microtubular tracks. However, other groups also directly effect the kinetics and stability of microtubules such as the MCAK/KIF2 or Kin I family (Desai *et al.*, 1999).

When plant kinesins have been located, although some do associate with the cortical and mitotic arrays, most have been found to associate with the cytokinetic phragmoplast. Kinesins are commonly implicated in the function of the plant-specific phragmoplast through their transport of vesicles required to form the new cell plate, or in the control of its constituent microtubules (see Lloyd *et al.*, 2004).

Arabidopsis hinkel mutants were identified as having abnormal embryos, consisting of enlarged cells with incomplete walls and multiple nuclei as a result of severely perturbed division patterns (Strompen *et al.*, 2002). In view of the cell division defects, the phragmoplasts in these *hik* mutants were studied. Here, the phragmoplasts were seen to nearly span the diameter of the cell during cytokinesis, suggesting that lateral expansion of the phragmoplast was unaffected. However, the *hik* phragmoplasts did exhibit microtubules across the entire division plane (Strompen *et al.*, 2002). Wild-type cells lack microtubules in the phragmoplast division plane, suggesting that HIK function is required for reorganisation of phragmoplast microtubules during cell plate formation. The *HIK* cDNA was sequenced and found to encode a 942 amino acid kinesin-related protein with an N-terminal motor domain. Analysis showed that the three *hik* alleles have premature stop codons, whilst a further two show conserved amino acid substitutions in the motor domain. Identical to HINKEL is AtNACK1, the *Arabidopsis* homologue of tobacco kinesin, NPK activating kinase-like protein 1 (NACK1), which binds NPK1 and stimulates its kinase activity *in vitro*. Although in *hik* mutants the lateral expansion of the phragmoplast was unaffected, overexpression of a NACK1 truncation, lacking the motor domain, produced cells in which initial cell plates failed to expand (Nishihama *et al.*, 2002). Further examples of phragmoplast-associated kinesins include TKRP125/DcKRP120 and AtPAKRP1/2. TKRP125 is a novel kinesin isolated from tobacco BY2 cells, and although it has been localised to all the plant microtubule arrays, an essential role for the kinesin during cytokinesis has been demonstrated (Asada *et al.*, 1997). Here antibodies to the motor domain of TKRP125 were found to inhibit the translocation of microtubules within the phragmoplast. The 145 kDa kinesin-related protein, AtPAKRP1 (*Arabidopsis* phragmoplast associated kinesin related protein), does not belong to any previously defined kinesin group but does resemble XKlp2. It associates with the positive ends of phragmoplast microtubule bundles and is believed to play a role in establishing and/or maintaining the phragmoplast microtubule array (Lee & Liu, 2000).

atk1-1 is an *Arabidopsis* mutant of the ATK or KATA kinesin, which locates to the spindle mid-zone during metaphase and anaphase (Liu *et al.*, 1996). *atk1-1* mutants have defective spindle organisation and chromosome segregation, with spindles exhibiting unfocused multi-axial poles (Chen *et al.*, 2002). As ATK is a minus end directed motor, it is believed that its role at the spindle is to draw together the minus ends of spindle microtubules.

Kinesins have also been shown to have important roles at the interphase cortical array and two good illustrations of this are the *Arabidopsis* mutants *zwichel and fragile fibre1*. The *zwichel* (*zwi*) mutant has trichomes with reduced branching and swollen stalks, suggesting that the *ZWI* gene is required for normal trichome branch formation and stalk elongation (Oppenheimer *et al.*, 1997). Cloning of the *ZWI* gene revealed that it encoded a minus-end directed kinesin-like motor protein named kinesin-like calmodulin binding protein (KCBP), initially identified in a screen of calmodulinbinding proteins (Reddy *et al.*, 1996a,b). The microtubule binding/bundling properties of this kinesin are inhibited in the presence of Ca^{2+} and calmodulin (Song *et al.*, 1997; Deavours *et al.*, 1998). The ZWI/KCBP protein has been located to the PPB, spindle and phragmoplast (Smirnova *et al.*, 1998) and plays a role in microtubule organisation during the metaphase-anaphase transition (Vos *et al.*, 2000). The *fragile fibre1* mutant has shorter stems than wild-type due to the disorganisation of cellulose microfibrils. However CMT organisation is maintained. The FRA1 protein is a kinesin-related protein, closely related to KIF4 and is located to the cortical cytoplasm. As KIF4 is implicated in vesicle transport it has been speculated that FRA1 could move membrane bound cellulose synthase along microtubules.

1.4 Conclusion

In this chapter we have described how the actin and microtubule networks mediate cell growth in a number of different cell types. The cytoskeletal networks within these cell types are stimulus responsive and have to be able to reorganise in response to developmental cues and environmental conditions. So arises complex pathways that signal the cytoskeleton and involve numerous intermediate factors, the relationships between which are beginning to be elucidated by biochemical, cell biological and genetic studies in plants. What lies ahead is the unravelling of the complex circuitry that relates signals to cell growth and plant morphology.

Note added in proof: recent bioinformatic analysis suggests that the *Arabidopsis* genome does contain proteins with a SCAR homology domain and a potential VCA region, suggesting WAVE homologues exist in plants (Deeks & Hussey, unpublished).

References

Alberts, A.S. (2001) Identification of a carboxyl-terminal diaphanous-related formin homology protein autoregulatory domain. *J. Biol Chem.*, **276**, 2824–2830.

Allwood E.G., Anthony, R.G, Smertenko, A.P., Reichelt, S., Drobak, B.K., Doonan, J.H., Weeds, A.G. & Hussey, P.J. (2002) Regulation of the pollen-specific actin-depolymerizing factor LIADF1 *Plant Cell*, **14**, 2915–2927.

Allwood, E.G., Smertenko, A.P. & Hussey, P.J. (2001) Phosphorylation of plant actin-depolymerising factor by calmodulin-like domain protein kinase. *FEBS Letters*, **499**, 97–100.

Amberg, D. C., Basart, E. & Botstein, D. (1995) Defining protein interactions with yeast actin in vivo, *Struct. Biol.*, **2**, 28–34.

Arber, S., Barbayannis, F. A., Hanser, H., Schneider, C., Stanyon, C. A., Bernard, O. & Caroni, P. (1998) Regulation of actin dynamics through phosphorylation of cofilin by LIM-kinase, *Nature*, **393**, 805–809.

Asada, T., Kuriyama, R. & Shibaoka, H. (1997) TKRP125, A kinesin-related protein involved in the centrosome-independent organisation of the cytokinetic apparatus in tobacco BY-2 cells, *J. Cell Sci.*, **110**, 179–189.

Assier Bouzinba-Segard, H., Stolzenberg, M.C., Stephens, R., Bardos, J., Freemont, P., Charron, D., Trowsdale, J. & Rich, T. (1999) Isolation, sequencing and expression of RED, a novel human gene encoding an acidic-basic dipeptide repeat. *Gene*, **230**, 145–154.

Baluška, F., Jasik, J., Edelmann, H.G., Šalajova, T. & Volkmann, D. (2001a). Latrunculin B-induced plant dwarfism: Plant cell elongation is F-actin-dependent. *Devel. Biol.*, **231**, 113–124.

Baluška, F., Salaj, J., Mathur, J., Braun, M., Jasper, F., Šamaj, J., Chua, N.H., Barlow, P.W. &

Volkmann, D. (2000) Root hair formation: F-actin-dependent tip growth is initiated by local assembly of profilin-supported F-actin meshworks accumulated within expansin-enriched bulges. *Dev. Biol.* **227**, 618–632.

Baluška, F., Volkmann, D. & Barlow, P.W. (2001c). A polarity crossroad in the transition growth zone of maize root apices: cytoskeletal and developmental implications. *J. Plant Growth Reg.*, **20**, 170–181.

Baluška, F., von Witsch, M., Peters, M., Hlavačka, A. & Volkmann, D. (2001b) Mastoparan alters subcellular distribution of profilin and remodels F-actin cytoskeleton in cells of maize root apices. *Plant Cell Physiol.*, **42**, 912–922.

Banno, H. & Chua, N.H. (2000) Characterization of the *Arabidopsis*, formin-like protein AFH1 and its interacting protein. *Plant Cell Physiol.*, **41**, 617–626.

Bao, Y.Q., Kost, B. & Chua, N.H. (2001) Reduced expression of a-tubulin genes in *Arabidopsis*, thaliana specifically affects root growth and morphology, root hair development and root gravitropism, *Plant J.*, **28**, 145–157.

Barrero, R.A., Umeda, M., Yamamura, S. & Uchimiya, H. (2002) *Arabidopsis*, CAP regulates the actin cytoskeleton necessary for plant cell elongation and division. *Plant Cell*, **14**, 149–163.

Barrero, R.A., Umeda, M, Yamamura, S. & Uchimiya, H. (2003) Over-expression of *Arabidopsis*, CAP causes decreased cell expansion leading to organ size reduction in transgenic tobacco plants. *Ann. Bot.*, **91**, 599–603.

Baskin, T.I. (2001) On the aligment of cellulose microfibrils by cortical microtubules: a review and a model. *Protoplasma*, **215**, 150–171.

Baskin, T.I., Wilson, J.E., Cork, A. & Williamson, R.E. (1994) Morphology and microtubule organization in *Arabidopsis*, roots exposed to oryzalin or taxol. *Plant Cell Physiol.*, **36**, 935–942.

Bates, T.R. & Lynch, J.P. (2000a). Plant growth and phosphorus accumulation of wild type and two root hair mutants of *Arabidopsis thaliana*, (Brassicaceae). *Am. J. Bot.* **87**, 958–963.

Bates, T.R. & Lynch, J.P. (2000b). The efficiency of *Arabidopsis thaliana*, (Brassicaceae) root hairs in phosphorus acquisition. *Am. J. Bot.*, **87**, 964–970.

Baxter-Burrell, A., Yang, Z.B., Springer, P.S. & Bailey-Serres, J. (2002) RopGAP4–dependent Rop GTPase rheostat control of *Arabidopsis*, oxygen deprivation tolerance. *Science*, **296**, 2026–2028.

Beemster, G.T. & Baskin, T.I. (1998) Analysis of cell division and elongation underlying the developmental acceleration of root growth in *Arabidopsis*, thaliana. *Plant Physiol.*, **116**, 1515–1526.

Beilstein, M. & Szymanski, D. (2003) Cytoskeletal requirements during *Arabidopsis*, trichome development, in: *The Plant Cytoskeleton in Cell Differentiation and Development* (Hussey P.J., ed.). Annual Plant Reviews Series, Blackwell Publishers, UK.

Bibikova, T.N., Blancaflor, E.B. & Gilroy, S. (1999) Microtubules regulate tip growth and orientation in root hairs of *Arabidopsis thaliana*. *Plant J.*, **17**, 657–665.

Bibikova, T.N., Jacob, T., Dahse, I. & Gilroy, S. (1998) Localized changes in apoplastic and cytoplasmic pH are associated with root hair development in *Arabidopsis thaliana*. *Development*, **125**, 2925–2934.

Bichet, A., Desnos, T., Turner, S., Grandjean, O. & Hofte, H. (2001) *BOTERO1* is required for normal orientation of cortical microtubules and anisotropic cell expansion in *Arabidopsis, Plant J.*, **25**, 137–148.

Birkenfeld, J., Betz, H. & Roth, D. (2003) Identification of cofilin and LIM-domain-containing protein kinase 1 as novel interaction partners of 14–3–3 zeta. *Biochem J.*, **369**, 45–54.

Blanchoin, L., Amann, K.J., Higgs, H.N., Marchand, J.B., Kaiser, D.A. & Pollard, T.D. (2000) Direct observation of dendritic actin filament networks nucleated by Arp2/3 complex and WASP/Scar proteins. *Nature.*, **404**, 1007–1011.

Braun, M., Baluška, F., Von Witsch, M. & Menzel, D. (1999) Redistribution of actin, profilin and phosphatidylinositol-4,5–bisphosphate in growing and maturing root hairs. *Planta*, **209**, 435–443.

Brugnera, E., Haney, L., Grimsley, C., Lu, M.J., Walk, S.F., Tosello-Trampont, A.C., Macara, I.G., Madhani, H., Fink, G.R. & Ravichandran, K.S. (2002) Unconventional Rac-GEF activity is mediated through the Dock180–ELMO complex. *Nature Cell Biol.*, **4**, 574–582.

Burk, D.H., Liu, B., Zhong, R.Q., Morrison, W.H. & Ye Z., (2001) A katanin-like protein regulates normal cell wall biosynthesis and cell elongation, *Plant Cell*, **13**, 807–827.

Cárdenas, L., Feijo, J.A., Kunkel, J.G., Sánchez, F., Holdaway-Clarke, T., Hepler, P.K. & Quinto, C. (1999) Rhizobium nod factors induce increases in intracellular free calcium and extracellular calcium influxes in bean root hairs. *Plant J.*, **9**, 347–352.

Cárdenas, L., Vidali, L., Domínguez, J., Pérez, H., Sánchez, F., Hepler, P. & Quinto, C. (1998) Rearrangement of actin microfilaments in plant root hairs responding to Rhizobium etli nodulation signals. *Plant Physiol.*, **116**, 871–877.

Carlier, M.F., Laurent, V., Santolini, J., Melki, R., Didry, D., Xia, G.X., Hong, Y., Chua ,N.H. & Pantaloni, D. (1997) Actin depolymerizing factor (ADF/cofilin) enhances the rate of filament turnover: Implication in actin-based motility. *J. Cell Biol.*, **136**, 1307–1322.

Castellano, F., Le Clainche, C., Patin, D., Carlier, M.F. & Chavrier, P. (2001) A WASp-VASP complex regulates actin polymerization at the plasma membrane. *EMBO J.*, **20**, 5603–5614.

Chan, J., Jensen, C.G., Jensen, L.C.W., Bush, M. & Lloyd, C.W. (1999) The 65-kDa carrot mictrotubule-associated protein regularly arranged filamentus cross-bridges between microtubules, *Proc. Natl Acad. Sci. USA*, **96**, 14931–14936.

Chan, J., Rutten, T. & Lloyd, C.W. (1996) Isolation of microtubule associated-proteins from carrot cytoskeletons: a 120 kDa MAP decorates all microtubule arrays and the nucleus, *Plant J.*, **10**, 251–259.

Chang, F., Drubin, D. & Nurse. P. (1997) cdc12p, a protein required for cytokinesis in fission yeast, is a component of the cell division ring and interacts with profilin. *J. Cell Biol.*, **137**, 169–82.

Charrasse, S., Schroeder, M., Gauthier-Rouviere, C., Ango, F., Cassimeris, L., Gard, D.L. & Larroque, C. (1998) The TOGp protein is a new human microtubule-associated protein homologous to the Xenopus XMAP215. *J. Cell Sci.*, **111**, 1371–1383.

Chen, C.Y., Cheung, A.Y. & Wu, H.M. (2003) Actin-depolymerizing factor mediates Rac/Rop GTPase-regulated pollen tube growth. *Plant Cell*, **15**, 237–249.

Chen, C., Marcus, A., Li, W., Hu, Y., Calzada, J.P., Grossniklaus, U., Cyr, R.J. & Ma, H. (2002) The *Arabidopsis*, *ATK1*, gene is required for spindle morphogenesis in male meiosis. *Development*, **129**, 2401–2409.

Cho, M.H., Tan, Z., Erneux, C., Shears, S.B. & Boss W.F. (1995) The effects of mastoparan on the carrot cell plasma-membrane polyphosphoinositide phospholipase-C. *Plant Physiol.*, **107**, 845–856.

Christensen, H.E.M., Ramachandran, S., Tan, C.T., Surana, U., Dong, C.H. & Chua, N.H. (1996) *Arabidopsis*, profilins are functionally similar to yeast profilins: Identification of a vascular bundle-specific profilin and a pollen-specific profilin. *Plant J.*, **10**, 269–279.

Cleveland, D. W., Kirschner, M. W. & Cowan N. J. (1978) Isolation of separate mRNA for – and ß-tubulin and characterization of the corresponding in vitro translation products. *Cell*, **15**, 1021–1031.

Cullen, C.F., Deah, P., Glover, D.M. & Ohkura, H. (1999) mini spindles: A gene encoding a conserved microtubule-associated protein required for the integrity of the mitotic spindle in Drosophila, *J. Cell Biol.*, **146**, 1005–1018.

Cvrckova, F. (2000) Are plant formins integral membrane proteins? *Genome Biol.* **1**, RESEARCH 001.

De Ruijter, N.C.A. & Emons, A.M.C. (1999) Actin-binding proteins in plant cells. *Plant Biol.*, **1**, 26–35.

De Ruijter, N.C.A., Rook, M.B., Bisseling, T. & Emons, A.M.C. (1998) Lipochito-oligosaccharides reinitiate root hair tip growth in *Vicia sativa*, with high calcium and spectrin-like antigen at the tip. *Plant J.*, **13**, 341–350.

Deavours, B.E., Reddy, A.S.N. & Walker, R.A. (1998) Ca2+/calmodulin regulation of the *Arabidopsis*, kinesin-like calmodulin-binding protein *Cell Motil. Cytoskel.*, **40**, 408–416.

Deeks, M.J., Hussey, P.J. & Davies B (2002) Formins: intermediates in signal-transduction cascades that affect cytoskeletal reorganization. *Trends Plant Sci.*, **7**, 492–498.

Derksen, J. & Emons, A.M.C. (1990) Microtubules in tip growth systems, in *Tip growth in plant and fungal cells* (Heath I.B., ed.), Academic Press, San Diego, 147–181.

Desai, A., Verma, S., Mitchison, T.J. & Walczak, C.E. (1999) Kin I kinesins are microtubule-destabilizing enzymes. *Cell*, **96**, 69–78.

Didry, D., Carlier, M.F. & Pantaloni, D. (1998) Synergy between actin depolymerizing factor/cofilin and profilin in increasing actin filament turnover. *J. Biol. Chem.*, **273**, 25602–25611.

Dolan, L., Janmaat, K., Willemsen, V., Linstead, P., Poethig, S., Roberts, K. & Scheres, B. (1993)

Cellular organisation of the *Arabidopsis*, *thaliana*, root. *Development*, **119**, 71–84.
Dong, C.H., Xia, G.X., Hong, Y., Ramachandran, S., Kost, B. & Chua, N.H. (2001) ADF proteins are involved in the control of flowering and regulate F-actin organization, cell expansion, and organ growth in *Arabidopsis*. *Plant Cell*, **13**, 1333–1346.
Drøbak, B.K., Watkins, P.A.C., Valenta, R., Dove, S.K., Lloyd, C.W. & Staiger, C.J. (1994) Inhibition of plant plasma-membrane phosphoinositide phospholipase-C by the actin-binding protein, profilin. *Plant J.*, **6**, 389–400.
Eden, S., Rohatgi, R., Podtelejnikov, A.V., Mann, M. & Kirschner, M.W. (2002) Mechanism of regulation of WAVE1–induced actin nucleation by Rac1 and Nck. *Nature*, **418**, 790–793.
Edwards, D.C., Sanders, L.C., Bokoch, G.M. & Gill, G.N. (1999) Activation of LIM-kinase by Pak1 couples Rac/Cdc42 GTPase signalling to actin cytoskeletal dynamics. *Nature Cell Biol.*, **1**, 253–259.
Emmons, S., Phan, H., Calley, J., Chen, W., James, B. & Manseau, L. (1995) Cappuccino, a Drosophila maternal effect gene required for polarity of the egg and embryo, is related to the vertebrate limb deformity locus. *Genes Dev.*, **9**, 2482–2494.
Emons, A.M.C. (1987) The cytoskeleton and secretory vesicles in root hairs of Equisetum and Limnobium and cytoplasmic streaming in root hairs of Equisetum. *Ann. Bot.*, **60**, 625–632.
Emons, A.M.C. (1989) Helicoidal microfibril deposition in a tip-growing cell and microtubule alignment during tip morphogenesis: a dry-cleaving and freeze-substitution study. *Canadian J. Botany*, **67**, 2401–2408.
Emons, A.M.C. & Mulder, B.M. (1998) The making of the architecture of the plant cell wall: how cells exploit geometry. *Proc. Natl. Acad. Sci. USA*, **95**, 7215–7219.
Endo, M., Ohashi, K., Sasaki, Y., Goshima, Y., Niwa, R., Uemura, T. & Mizuno, K. (2003) Control of growth cone motility and morphology by lim kinase and slingshot via phosphorylation and dephosphorylation of Cofilin. *J. Neurosci.*, **23**, 2527–2537.
Esseling, J., de Ruijter, N. & Emons, A.M.C. (2000) The root hair cytoskeleton as backbone, highway, morphogenetic instrument and target for signalling, in *Root Hairs. Cell and Molecular Biology*. (eds. Ridge R.W. & Emons A.M.C.) Tokyo, Berlin, Heidelberg, New York: Springer-Verlag, pp. 29–52.
Evangelista, M., Blundell, K., Longtine, M.S., Chow, C.J., Adames, N., Pringle, J.R., Peter, M. & Boone, C.B (1997) ni1p, a yeast formin linking cdc42p and the actin cytoskeleton during polarized morphogenesis. *Science*, **276**, 118–122.
Evangelista, M., Pruyne, D., Amberg, D.C., Boone, C. & Bretscher, A. (2002) Formins direct Arp2/3-independent actin filament assembly to polarize cell growth in yeast. *Nature Cell Biol.*, **4**, 32–41.
Felle, H.H. & Hepler, P.K. (1997) The cytosolic Ca^{2+} concentration gradient of *Sinapis alba*, root hairs as revealed by Ca^{2+}-selective microelectrode tests and fura dextran ration imaging. *Plant Physiol.*, **114**, 39–45.
Finkel, T., Theriot, J.A., Dise, K.R., Tomaselli, G.F. & Goldschmidt-Clermont, P.J. (1994) Dynamic actin structures stabilized by profilin. *Proc. Natl. Acad. Sci.USA*, **91**, 1510–1514.
Folkers, U., Kirik, V., Schöbinger, U., Falk, S., Krishnakumar, S., Pollock, M.A., Oppenheimer, D.G., Day, I., Reddy, A.R., Jürgens, G. & Hülskamp, M (2002) The cell morphogenesis gene *ANGUSTIFOLIA* encodes a CtBP/BARS-like protein and is involved in the control of the microtubule cytoskeleton. *EMBO J.*, **21**, 1280–1288.
Frank, M.J. & Smith, L.G. (2002) A small, novel protein highly conserved in plants and animals promotes the polarized growth and division of maize leaf epidermal cells. *Curr. Biol.*, **12**, 849–853.
Frank, M.J., Cartwright, H.N. & Smith, L.G. (2003) Three Brick genes have distinct functions in a common pathway promoting polarized cell division and cell morphogenesis in the maize leaf epidermis. *Development*, **130**, 753–762.
Freeman, N.L., Chen, Z., Horenstein, J., Weber, A. & Field, J. (1995) An actin monomer binding activity localizes to the carboxyl-terminal half of the *Saccharomyces cerevisiae* cyclase-associated protein. *J. Biol. Chem.*, **270**, 5680–5685.
Fu, Y., Li, H. & Yang, Z. (2002) The ROP2 GTPase controls the formation of cortical fine F-actin and the early phase of directional cell expansion during *Arabidopsis* organogenesis. *Plant Cell,* **14**, 777–794.
Fu, Y., Wu, G. & Yang, Z. (2001) Rop GTPase-dependent dynamics of tip-localized F-actin

controls tip growth in pollen tubes. *J. Cell Biol.*, **152**, 1019–1032.

Fukui, Y., Hashimoto, O., Sanui T., Oono, T., Koga, H., Abe, M., Inayoshi, A., Noda, M., Oike, M., Shirai, T. & Sasazuki, T. (2001) Haematopoietic cell-specific CDM family protein DOCK2 is essential for lymphocyte migration. *Nature*, **412**, 826–831.

Gao, Y., Thomas J.O., Chow, R.L, Lee, G.H. & Cowan, N.J. (1992) A cytoplasmic chaperonin that catalyzes ß-actin folding. *Cell*, **69**, 1043–1050.

Gao, Y., Vainberg, I.E., Chow, R.L. & Cowan, N.J. (1993) Two cofactors and cytoplasmic chaperonin are required for the folding of α- and β-tubulin. *Mol. Cell. Biol.*, **13**, 2478–2485.

Garcia, M.A., Vardy, L., Koonrugsa, N. & Toda, T. (2001) Fission yeast ch-TOG/XMAP215 homologue Alp14 connects mitotic spindles with the kinetochore and is a component of the Mad2-dependent spindle checkpoin. *EMBO J.*, **20**, 3389–3401.

Gavin, R.H. (1997) Microtubule microfilament synergy in the cytoskeleton. *Int. Rev. Cytol.*, **173**, 207–242.

Geitmann, A. & Emons, A.M.C. (2000) The cytoskeleton in plant and fungal cell tip growth. *J. Microscopy*, **198**, 218–245.

Gibbon, B.C., Zonia, L.E., Kovar, D.R., Hussey, P.J. & Staiger, C.J. (1998) Pollen profilin function depends on interaction with proline-rich motifs. *Plant Cell*, **10**, 981–993.

Gibbon, B. C., Kovar, D. R. & Staiger, C.J. (1999) Latrunculin B has different effects on pollen germination and tube growth, *Plant Cell*, **11**, 2349–2363.

Gilliland, L.U., Kandasamy, M.K., Pawloski, L.C. & Meagher, R.B. (2002) Both vegetative and reproductive actin isovariants complement the stunted root hair phenotype of the *Arabidopsis* act2-1 mutation. *Plant Physiol.*, **130**, 2199–2209.

Glenney, J.R.J., Bretscher, A. & Weber, K. (1980) Calcium control of the intestinal microvillus cytoskeleton: Its implications for the regulation of microfilament organizations. *Proc. Natl. Acad. Sci. USA,* **77**, 6458–6462.

Glotzer, M., Murray, A.W. & Kirschner, M.W. (1991) Cyclin is degraded by the ubiquitin pathway. *Nature*, **349**, 132–138.

Gohla, A. & Bokoch, G.M. (2002) 14–3–3 regulates actin dynamics by stabilizing phosphorylated cofilin. *Curr. Bio*l., **12**, 1704–1710.

Gottswald, U., Brokamp, R., Karakesisoglou, I., Schleicher, M. & Noegel, A.A. (1996) Identification of a cyclase-associated protein (CAP) homologue in Dicytostelium discoideum and characterization of its interaction with actin. *Mol. Biol Cell*, **7**, 261–272.

Graef, R., Daunderer, C. & Schliwa, M. (2000) Dictyostelium DdCP224 is a microtubule-associated protein and a permanent centrosomal resident involved in centrosome duplication. *J. Cell Sci.*, **113**, 1747–1758.

Green, P.B. (1962) Mechanism for plant cellular morphogenesis. *Science*, **138**, 1404–1405.

Grierson, C. & Ketelaar, T. (2003) Development of root hairs, in *The Plant Cytoskeleton in Cell Differentiation and Development* (ed. P.J. Hussey). Annual Plant Reviews Series, Blackwell Publishers: Oxford.

Gu, Y., Vernoud, V., Fu, Y. & Yang, Z.B. (2003) ROP GTPase regulation of pollen tube growth through the dynamics of tip-localized F-actin. *J. Exp. Bot.*, **54**, 93–101.

Gungabissoon, R.A., Jiang, C.J., Drøbak, B.J., Maciver, S.K. & Hussey, P.J. (1998) Interaction of maize actin-depolymerizing factor with F-actin and phosphoinositides and its inhibition of plant phospholipase C. *Plant J.*, **16**, 689–696.

Gungabissoon, R.A., Khan, S., Hussey, P.J. & Maciver, S.K. (2001) Interaction of elongation factor 1α from *Zea mays* (ZmEF-1α) with F-actin and interplay with the maize actin severing protein, ZmADF3. *Cell Motil. Cytoskel.*, **49**, 104–111.

Gutsche-Perelroizen, I., Lepault, J., Ott, A. & Carlier, M.F. (1999) Filament assembly from profilin-actin. *J. Biol. Chem.*, **274**, 6234–6243.

Hartman, J.J., Mahr, J., McNally, K., Okawa, K., Iwamatsu, A., Thomas, S., Cheesman, S., Heuser, J., Vale, R.D. & McNally, F.J. (1998) Katanin, a microtubule-severing protein, is a novel AAA ATPase that targets to the centrosome using a WD40–containing subunit, *Cell.* **93**, 277–287.

Hawkins, M., Pope, B., MacIver, S.K. & Weeds, A.G. (1993) Human actin depolymerizing factor mediates a pH-sensitive destruction of actin-filaments. *Biochem.*, **32**, 9985–9993.

Heath, I.B. (1974) A unified hypothesis for the role of membrane bound enzyme complexes and microtubules in plant cell wall synthesis. *J. Theor. Biol.*, **48**, 445–449.

Hepler, P.K., Vidali, L. & Cheung, A.Y. (2001) Polarized cell growth in higher plants. *Ann. Rev.*

Cell Devel. Biol., **17**, 159–187.

Herrman, A. & Felle, H. (1995) Tip growth in root hair cells of *Sinapis alba* L.: significance of internal and external Ca^{2+} and pH. *New Phytol.*, **129**, 523–533.

Higashijima, T. & Ross, E.M. (1991) Mapping of the mastoparan-binding site on G-proteins – cross-linking of [I-125] Tyr3, Cys11 mastoparan to G0. *J. Biol. Chem.*, **266**, 12655–12661.

Higgs, H.N. & Pollard, T.D. (2001) Regulation of actin filament network formation through ARP2/3 complex: activation by a diverse array of proteins. *Ann. Rev. Biochem.* **70**, 649–676.

Higgs, H.N., Blanchoin, L. & Pollard, T.D. (1999) Influence of the C terminus of Wiskott-Aldrich syndrome protein (WASp) and the Arp2/3 complex on actin polymerization. *Biochemistry*, **38**, 15212–15222.

Hirsh, D. & Vanderslice R. (1976) Temperature-sensitive developmental mutants of *Caenorhabditis elegans*. *Dev. Biol.* **49**, 220–235.

Holt, M.R. & Koffer, A. (2001) Cell motility: proline-rich proteins promote protrusions. *Trends Cell Biol.*, **11**, 38–46.

Hopmann, R. & Miller, K.G. (2003) A balance of capping protein and profilin functions is required to regulate actin polymerization in Drosophila bristle. *Mol Biol Cell.*, **14**, 118–128.

Huang, S.R., McDowell, J.M., Weise, M.J. & Meagher, R.B. (1996) The *Arabidopsis* profilin gene family – Evidence for an ancient split between constitutive and pollen-specific profilin genes *Plant Physiol.*, **111**, 115–126.

Hudson, A.M. & Cooley, L.A. (2002) Subset of dynamic actin rearrangements in Drosophila requires the Arp2/3 complex. *J. Cell Biol.*, **156,** 677–687.

Hülskamp, M., Misera, S. & Jürgens, G. (1994) Genetic dissection of trichome cell development in *Arabidopsis*. *Cell*, **76**, 555–566.

Hussey, P.J., Allwood, E.G. & Smertenko, A.P. (2002) Actin-binding proteins in the *Arabidopsis* genome database: properties of functionally distinct plant actin-depolymerizing factors/cofilins. *Phil. Trans. R. Soc. Lond. B.*,, **357**, 791–798.

Hussey, P.J., Hawkins, T.J., Igarashi, H., Kaloriti, D. & Smertenko, A. (2002) The plant cytoskeleton: recent advances in the study of the plant microtubule-associated proteins MAP-65, MAP-190 and *Xenopus* MAP215-like protein, MOR1. *Plant Molec. Biol.*, **50**, 915–924.

Igarashi, H., Orii, H., Mori, H., Shimmen, T. & Sonobe, S. (2000) Isolation of a novel 190 kDA protein from tobacco BY-2 cells: possible involvement in the interaction between actin filaments and microtubules. *Plant Cell Physiol.*, **41**, 920–931.

Irving, H.R. (1998) Abscisic acid induction of GTP hydrolysis in maize coleoptile plasma membranes. *Aus. J Plant Physiol.*, **25**, 539–546.

Ishizaki, T., Morishima, Y., Okamoto, M., Furuyashiki, T., Kato, T. & Narumiya, S. (2001) Coordination of microtubules and the actin cytoskeleton by the Rho effector mDia1. *Nature Cell Biol.*, **3**, 8–14.

Iwasaki, Y., Kato, T., Kaidoh, T., Ishikawa, A. & Asahi, T. (1997) Characterization of the putative alpha subunit of a heterotrimeric G protein in rice. *Plant Molec. Biol.*, **34**, 563–572.

Janmey, P.A., Matsudaira, P.T. (1988) Functional comparison of villin and gelsolin. Effects of Ca^{2+}, KCl, and polyphosphoinositides, *J. Biol. Chem.*, **263**, 16738–16743.

Jiang, C.J. & Sonobe, S. (1993) Isolation of a group of approximately 65-kDa tobacco proteins that bundle microtubules *in vitro, J. Cell Sci.*, **105**, 891–901.

Jiang, C.J., Weeds, A.G. & Hussey, P.J. (1997) The maize actin depolymerizing factor, ZmADF3, redistributes to the growing tip of elongating root hairs and can be induced to translocate into the nucleus with actin. *Plant J.*, **12**, 1035–1043.

Jones, M.A., Shen, J.J., Fu, Y., Li, H., Yang, Z.B. & Grierson, C.S. (2002) The *Arabidopsis* Rop2 GTPase is a positive regulator of both root hair initiation and tip growth. *Plant Cell*, **14**, 763–776.

Kang, F., Purich, D.L. & Southwick, F.S. (1999) Profilin promotes barbed-end actin filament assembly without lowering the critical concentration. *J. biol. Chem.*, **274**, 36963–36972.

Katoh, H. & Negishi, M. (2003) RhoG activates Rac1 by direct interaction with the Dock180-binding protein Elmo. *Nature*, **424**, 461–464.

Katsuta,, J. & Shibaoka, H. (1998) The roles of the cytoskeleton and the cell wall in nuclear positioning in tobacco BY-2 cell, *Plant Cell Physiol.*, **29**, 403–413.

Katsuta, J., Hashiguchi, Y. & Shibaoka, H. (1990) The role of the cytoskeleton in positioning of the nucleus in premitotic cells, *J. Cell Sci.*, **95**, 413–422.

Ketelaar, T (2002). Spatial organisation of cell expansion by the cytoskeleton. PhD thesis Wageningen University, The Netherlands.

Ketelaar, T., Allwood, E.G., Anthony, R., Voigt, B., Menzel, D. & Hussey, P.J. (2004) The actin-interacting protein AIP1 is essential for actin organization and plant development. *Current Biol.*, **14**, 145–149.

Ketelaar, T., de Ruijter, N.C.A. & Emons, A.M.C. (2003) Unstable F-actin specifies the area and microtubule direction of cell expansion in *Arabidopsis*, root hairs. *Plant Cell*, **15**, 285–292.

Ketelaar, T. & Emons, A.M.C. (2000) The role of microtubules in root hair growth and cellulose microfibril deposition. In: Ridge, R.W. & Emons, A.M.C., eds. *Root hairs. Cell and molecular biology*. Tokyo, Berlin, Heidelberg, New York: Springer-Verlag, pp. 17–28.

Ketelaar, T. & Emons, A.M.C. (2001) The cytoskeleton in plant cell growth: lessons from root hairs. *New Phytol.*, **152**, 409–418.

Ketelaar, T., Faivre-Moskalenko, C., Esseling, J.J., de Ruijter, N.C., Grierson, C.S., Dogterom, M. & Emons, A.M. (2002) Positioning of nuclei in *Arabidopsis* root hairs: an actin-regulated process of tip growth. *Plant Cell*, **14**, 2941–2955.

Kinley, A.W., Weed, S.A., Weaver, A.M., Karginov, A.V., Bissonette, E., Cooper, J.A. & Parsons, J.T. (2003) Cortactin interacts with WIP in regulating Arp2/3 activation and membrane protrusion. *Curr. Biol.*, **13**, 384–393.

Kirik, V., Grini, P.E., Mathur, J., Klinkhammer, I., Adler, K., Bechtold, N., Herzog, M., Bonneville, J.M. & Hülskamp, M. (2002) The *Arabidopsis* tubulin-folding cofactor A gene is involved in the control of the α-/β- tubulin monomer balance. *Plant Cell*, **14**, 2265–2276.

Kiyokawa, E., Hashimoto, Y., Kobayashi, S., Sugimura, H., Kurata, T. & Matsuda, M. (1998) Activation of Rac1 by a Crk SH3–binding protein, DOCK180. *Genes Dev.*, **12**, 3331–3336.

Klahre, U., Friederich, E., Kost, B., Louvard, D. & Chua, N.H. (2000) Villin-like actin-binding proteins are expressed ubiquitously in *Arabidopsis*. *Plant Physiol.*, **122**, 35–48.

Koch, G., Haberman, B., Mohr, C,. Just, I. & Aktories, K. (1991) Interaction of mastoparan with the low-molecular mass GTP-binding protein-Rho protein-Rac. *FEBS Lett.*, **291**, 336–340.

Konzok, A., Weber, I., Simmeth, E., Hacker, U., Maniak, M. & Müller-Taubenberger, A. (1999) DAip1, a *Dictyostelium* homologue of the yeast actin-interacting protein 1, is involved in endocytosis, cytokinesis, and motility. *J. Cell Biol.*, **146**, 453–464.

Kost, B., Lemichez, E., Spielhofer, P., Hong, Y., Tolias, K., Carpenter, C. & Chua, N.H. (1999) Rac homologues and compartmentalized phosphatidylinositol 4,5-bisphosphate act in a common pathway to regulate polar pollen tube growth. *J. Cell Biol.*, **145**, 317–330.

Kovar, D.R., Drøbak, B.K. & Staiger, C.J. (2000) Maize profilin isoforms are functionally distinct. *Plant Cell*, **12**, 583–598.

Kovar, D.R., Kuhn, J.R., Tichy, A.L. & Pollard, T.D. (2003) The fission yeast cytokinesis formin Cdc12p is a barbed end actin filament capping protein gated by profilin. *J. Cell Biol.*, **161**, 875–887.

Kovar, D.R., Yang, P., Sale, W.S., Drøbak, B.K. & Staiger, C.J. (2001b) *Chlamydomonas reinhardtii* produces a profilin with unique biochemical properties *J. Cell Sci.*, **114**,, 4293–4305.

Lam, B.C.H., Sage, T.L., Bianchi, F. & Blumwald, E. Role of SH3 domain-containing proteins in clathrin-mediated vesicle trafficking in *Arabidopsis*. *Plant Cell*, **13**, 2499–2512.

Lambrechts, A., Verschelde, J.L., Jonckheere, V., Goethals, M., Vandekerckhove, J. & Ampe, C. (1997) The mammalian profilin isoforms display complementary affinities for PIP2 and proline-rich sequences. *EMBO J*, **16**, 484–494.

Lancelle, S. A., Callaham, D.A. & Hepler, P.K. (1986) A method for rapid freeze fixation of plants. *Protoplasma*, **131**, 153–165.

Larkin, J.C., Young, N., Prigge, M. & Marks, M.D. (1996) The control of trichome spacing and number in *Arabidopsis*. *Development*, **122**, 997–1005.

Le, J., El-Assal, Sel-D., Basu, D., Saad, M.E. & Szymanski, D.B. (2003) Requirements for *Arabidopsis* ATARP2 and ATARP3 during epidermal development. *Curr. Biol.*, **13**, 1341–1347.

Ledbetter, M.C. & Porter, K.R. (1963) A 'microtubule' in plant fine structure. *J. Cell Biol.*, **19**, 239–250.

Lee, Y.-R.J. & Liu, B. (2000) Identification of a phragmoplast associated kinesin-related protein in higher plants, *Curr. Biol.*, **10**, 797–800.

Lemichez, E., Wu, Y., Sanchez, J.P., Mettouchi, A., Mathur, J. & Chua, N.H. (2001) Inactivation of AtRac1 by abscisic acid is essential for stomatal closure. *Genes Dev.*, **15**, 1808–1816.

Lew, R.R. (1996) Pressure regulation of the electrical properties of growing *Arabidopsis thaliana* L root hairs. *Plant Physiol.*, **112**, 1089–1100.

Lew, R.R. (2000) Electrobiology of Root Hairs, in *Root Hairs: Cell and Molecular Biology* (eds R.W. Ridge and A.M.C. Emons) Springer-Verlag, Tokyo: pp. 115–139.

Lewis, S.A., Tian, G. & Cowan, N.J. (1997) The α- and β-tubulin folding pathways, *Trends Cell Biol.*, **7**, 479–484.

Li, F. & Higgs H.N. (2003) The mouse Formin mDia1 is a potent actin nucleation factor regulated by autoinhibition. *Curr. Biol.*, **13**, 1335–1340.

Li, S., Blanchoin, L., Yang, Z. & Lord, E.M. (2003) The putative *Arabidopsis* Arp2/3 complex controls leaf cell morphogenesis. *Plant Physiol.*, **132**, 2034–2044.

Lin, Y.K., Yang, Z.B. (1997) Inhibition of pollen tube elongation by microinjected anti-Rop1Ps antibodies suggests a crucial role for Rho-type GTPases in the control of tip growth. *Plant Cell*, **9**, 1647–1659.

Liu, B., Cyr, R.J. & Palevitz, B.A. (1996) A kinesin-like protein, KatAp, in the cells of *Arabidopsis* and other plants. *Plant Cell*, **8**, 119–132.

Lloyd, A.M., Schena, M., Walbot, V. & Davis, R.W. (1994) Epidermal cell fate determination in *Arabidopsis*: Patterns defined by a steroid-inducible regulator. *Science*, **266**, 436–439.

Lloyd, C.W. (1987) The plant cytoskeleton: the impact of fluorescence microscopy, *Annu. Rev. Plant Physiol.*, **38**, 119–139.

Lloyd, C., Chan, J. & Hussey, P.J. (2004) Microtubules and microtubule-associated proteins, in *The Plant Cytoskeleton in Cell Differentiation and Development* (P.J. Hussey). Blackwell: Oxford, pp. 3–30.

Lloyd, C.W., Pearce, K.J., Rawlins, D.J., Ridge, R.W. & Shaw, P.J. (1987) Endoplasmic microtubules connect the advancing nucleus to the tip of legume root hairs, but F-actin is involved in basipetal migration. *Cell Motil. Cytoskel.*, **8**, 27–36.

Lu, J. & T. D. Pollard ((2001)). Profilin binding to poly-L-proline and actin monomers along with ability to catalyze actin nucleotide exchange is required for viability of fission yeast. *Mol. Biol. Cell*, **12**, 1161–1175.

McClinton, R.S. & Sung, Z.R. (1997) Organization of cortical microtubules at the plasma membrane in *Arabidopsis*. *Planta*, **201**, 252–260.

Machesky, L.M. & Insall, R.H. S (1998) Scar1 and the related Wiskott-Aldrich syndrome protein, WASP, regulate the actin cytoskeleton through the Arp2/3 complex. *Curr. Biol.*, **8**, 1347–1356.

Machesky, L.M., Mullins, R.D., Higgs, H.N., Kaiser, D.A., Blanchoin, L., May, R.C., Hall, M.E. & Pollard, T.D. (1999) Scar, a WASp-related protein, activates nucleation of actin filaments by the Arp2/3 complex. *Proc. Natl. Acad. Sci. USA*, **96**, 3739–3744.

McKinney, E.C., Kandasamy, M.K. & Meagher, R.B. (2002) *Arabidopsis*, contains ancient classes of differentially expressed actin-related protein genes. *Plant Physiol.* **128**, 997–1007.

McNally, F.J. & Thomas, S. (1998) Katanin is responsible for the M-phase microtubule-severing activity in Xenopus eggs. *Mol Biol Cell*, **9**, 1847–1861.

McNally, F.J. & Vale, R.D. (1993) Identification of katanin, ATPase that severs and disassembles stable microtubules *Cell*, **75**, 419–429.

McNally, F.J. (1996) Modulation of microtubule dynamics during the cell cycle, *Curr. Opin. Cell Biol.*, **8**, 23–29.

Malhó, R. (1998) The role of inositol(1,4,5)triphosphate in pollen tube growth and orientation. *Sex. Plant Reprod.*, **11**, 231–235.

Marc, J., Granger, C.L., Brincat, J., Fisher, D.D., Kao, T., McCubbin, A.G. & Cyr, R.J. (1998) A GFP-MAP4 reporter gene for visualizing cortical microtubule rearrangements in living epidermal cells. *Plant Cell*, **10**, 1927–1940.

Mathur, J. & Chua, N.H. (2000) Microtubule stabilization leads to growth reorientation in *Arabidopsis* trichomes. *Plant Cell*, **12**, 465–477.

Mathur, J., Mathur, N., Kernebeck, B. & Hülskamp, M. (2003a). Mutations in actin-related proteins 2 and 3 affect cell shape development in *Arabidopsis*. *Plant Cell*, **15**, 1632–1645.

Mathur, J., Mathur, N., Kirik, V., Kernebeck, B., Srinivas, B.P. & Hülskamp M (2003b). *Arabidopsis*, *CROOKED* encodes for the smallest subunit of the ARP2/3 complex and controls cell shape by region specific fine F-actin formation. *Development*, **130**, 3137–3146.

Mathur, J., Spielhofer, P., Kost, B. & Chua, N.H. (1999) The actin cytoskeleton is required to elaborate and maintain spatial patterning during trichome cell morphogenesis in *Arabidopsis thaliana. Development*, **126**, 5559–5568.

Melaragno, J.E., Mehrota, B. & Coleman, A.W. (1993) Relationship between endoploidy and cell size in epidermal tissue of *Arabidopsis. Plant Cell*, **5**, 1661–1668.

Metzler, W.J., Bel,l A.J., Ernst, E., Lavoie, T.B. & Mueller, L. (1994) Identification of the poly-L-proline-binding site on human profilin. *J. Biol. Chem.*, **269**, 4620–4625.

Miki, H., Miura, K. & Takenawa, T. (1996) N-WASP, a novel actin-depolymerizing protein, regulates the cortical cytoskeletal rearrangement in a PIP2-dependent manner downstream of tyrosine kinases. *EMBO J.*, **15**, 5326–5335.

Miller, D.D., De Ruijter, N.C.A., Bisseling, T. & Emons, A.M.C. (1999) The role of actin in root hair morphogenesis: Studies with lipochito-oligosaccharide as a growth stimulator and cytochalasin as an actin perturbing drug. *Plant J.*, **17**, 141–154.

Miller, D.D., De Ruijter, N.C.A. & Emons, A.M.C. (1997) From signal to form: aspects of the cytoskeleton-plasma membrane-cell wall continuum in root hair tips. *J. Exp. Bot.*, **48**, 1881–1896.

Miller, D.D., Leferink-ten Klooster, H.B. & Emons, A.M.C. (2000) Lipochito-oligosaccharide nodulation factors stimulate cytoplasmic polarity with longitudinal endoplasmic reticulum and vesicles at the tip in vetch root hairs. *Molec. Plant Microbe Inter.*, **13**, 1385–1390.

Mittermann, I., Swoboda, I., Pierson, E., Eller, N., Kraft, D., Valenta, R. & Heberlebors, E. (1995) Molecular-cloning and characterization of profilin from tobacco (*Nicotiana tabacum*) – increased profilin expression during pollen maturation. *Plant Molec. Biol*, **27**, 137–146.

Molendijk, A.J., Bischoff, F., Rajendrakumar, C.S.V., Friml, J., Braun, M., Gilroy, S. & Palme, K. (2001) *Arabidopsis thaliana* Rop GTPases are localized to tips of root hairs and control polar growth. *EMBO J.*, **20**, 2779–2788.

Moreau, V., Galan, J.M., Devilliers, G., Haguenauer-Tsapis, R. & Winsor, B. (1997) The yeast actin-related protein Arp2p is required for the internalization step of endocytosis. *Mol. Biol. Cell*, **8**, 1361–1375.

Moriyama, K. & Yahara, I. (2002) Human CAP1 is a key factor in the recycling of cofilin and actin for rapid actin turnover. *J. Cell Sci.,*, **115**, 1591–1601.

Moutinho, A., Hussey, P.J., Trewavas, A.J. & Malhó, R. (2001) cAMP acts as a second messenger in pollen tube growth and reorientation. *Proc. Natl. Acad, Sci. USA*, **98**, 10481–10486.

Mullins, R.D., Heuser, J.A. & Pollard, T.D. (1998) The interaction of Arp2/3 complex with actin: Nucleation, high affinity pointed end capping, and formation of branching networks of filaments. *Proc. Natl. Acad, Sci. USA*, **95**, 6181–6186.

Mullins, R.D., Stafford, W.F. & Pollard, T.D. (1997) Structure, subunit topology, and actin-binding activity of the Arp2/3 complex from Acanthamoeba. *J. Cell Biol.*, **136**, 331–343.

Nishihama, R., Soyano, T., Ishikawa, M., Araki, S., Tanaka, H., Asada, T., Irie, K., Ito, M., Terada, M., Banno, H., Yamazaki, Y. & Machida ,Y. (2002) Expansion of the cell plate in plant cytokinesis requires a kinesin-like protein/MAPKKK complex. *Cell* **109**, 87–99.

Noegel, A.A., Rivero, F., Albrecht, R., Janssen, K.P., Kohler, J., Parent, C.A. & Schleicher, M. (1999) Assessing the role of the ASP56/CAP homologue of *Dictyostelium discoideum*, and the requirements for subcellular localization. *J. Cell Sci.*, **112**, 3195–3203.

Nolan, K.M., Barrett, K., Lu, Y., Hu, K.Q., Vincent, S. & Settleman J. (1998) Myoblast city, the Drosophila homolog of DOCK180/CED-5, is required in a Rac signaling pathway utilized for multiple developmental processes. *Genes Dev.*, **12**, 3337–3342.

Ohkura, H., Adachi, Y., Kinoshita, N., Niwa, O., Toda, T. & Yanagida, M. (1988) Cold-sensitive and caffeine-supersensitive mutants of the *Schizosaccharomyces pombe dis*, genes implicated in sister chromatid separation during mitosis. *EMBO J.*, **7**, 1465–1473.

Okada, K., Blanchoin, L., Abe, H., Chen, H., Pollard, T.D. & Bamburg, J.R.. (2002) Xenopus actin-interacting protein 1 (XAip1) enhances cofilin fragmentation of filaments by capping filament ends. *J. Biol. Chem.*, **277**, 43011–43016.

Ono, S. (2003) Regulation of actin filament dynamics by actin depolymerizing factor/cofilin and the actin-interacting protein 1: new blades for twisted filaments. *Biochemistry*, **42**, 13363–13370.

Oppenheimer, D.G., Pollock, M.A., Vacik, J., Szymanski, D.B., Ericson, B., Feldmann, K. & Marks, M.D. (1997) Essential role of kinesin-like protein in *Arabidopsis* trichome morphogenesis. *Proc. Natl, Acad. Sci. USA*, **94**, 6261–6266.

Ovecka, M., Nadubinska, M., Volkmann, D. & Baluška, F. (2000) Actomyosin and exocytosis inhibitors alter root hair morphology in *Poa annua.*, *Biologia*, **55**, 105–114.

Panchal, S.C., Kaiser, D.A., Torres, E., Pollard, T.D. & Rosen, M.K. (2003) A conserved amphipathic helix in WASP/Scar proteins is essential for activation of Arp2/3 complex. *Nature Struct. Biol.*, **10**, 591–598.

Perelroizen, I., Didry, D., Christensen, H., Chua, N.H. & Carlier, M.F. (1996) Role of nucleotide exchange and hydrolysis in the function of profilin in actin assembly. *J. Biol. Chem.*, **271**, 12302–12309.

Petersen, J., Nielsen, O., Egel, R. & Hagan, I.M. (1998a) F-actin distribution and function during sexual differentiation in Schizosaccharomyces pombe. *J Cell Sci.*. **111**, 867–876.

Petersen, J., Nielsen, O., Egel, R. & Hagan, I.M. (1998b) FH3, a domain found in formins, targets the fission yeast formin Fus1 to the projection tip during conjugation. *J. Cell Biol.*, **141**, 1217–1228.

Pollard, T.D., Blanchoin, L. & Mullins, R.D. (2000) Molecular mechanisms controlling actin filament dynamics in nonmuscle cells. *Ann. Rev. Bioph. Biom.*, **29**, 545–576.

Pope, B., Way, M., Matsudaira, P.T. & Weeds, A. (1994) Characterisation of the F-Actin binding domains of villin: classification of F-Actin binding proteins into two groups according to their binding sites on actin. *FEBS Lett.*, **338**, 58–62.

Pring, M., Evangelista, M., Boone, C., Yang, C. & Zigmond, S.H. (2003) Mechanism of formin-induced nucleation of actin filaments. *Biochemistry*, **42**, 486–96.

Pruyne, D. & Bretscher, A. (2000) Polarization of cell growth in yeast. *J. Cell Sci.*, **113**, 571–85.

Pruyne, D., Evangelista, M., Yang, C., Bi, E., Zigmond, S., Bretscher, A. & Boone, C. (2002) Role of formins in actin assembly: nucleation and barbed-end association. *Science*, **297**, 612–615.

Qiu, J.L., Jilk, R., Marks, M.D. & Szymanski, D.B. (2002) The *Arabidopsis*, *SPIKE1* gene is required for normal cell shape control and tissue development. *Plant Cell*, **14**, 101–118.

Ramachandran, S., Christensen, H.E., Ishimaru, Y., Dong, C.H., Chao-Ming, W., Cleary, A.L. & Chua, N.H. (2000) Profilin plays a role in cell elongation, cell shape maintenance, and flowering in *Arabidopsis*. *Plant Physiol.*, **124**, 1637–1647.

Reddy, A.S. & Day, I.S. (2001) Kinesins in the *Arabidopsis* genome: a comparative analysis among eukaryotes, *BMC Genomics*, **2**, 2 (http://www.biomedcentral.com/1471-2164/2/2).

Reddy, A.S.N, Narasimhulu, S.B., Safadi, F. & Golovkin, M. (1996a) A plant kinesin heavy chain-like protein is a calmodulin-binding protein. *Plant J.*, **10**, 9–21.

Reddy, A.S.N., Safadi, F., Narasimhulu, S.B., Golovkin, M. & Hu, X. (1996b) A novel plant calmodulin-binding protein with a kinesin heavy chain motor domain. *J. Biol. Chem.*, **271**, 7052–7060.

Ringli, C., Baumberger, N., Diet, A., Frey, B. & Keller, B. (2002) ACTIN2 is essential for bulge site selection and tip growth during root hair development of *Arabidopsis*. *Plant Physiol.*, **129**, 1464–1472.

Roberts, K. (1994) The plant extracellular matrix: in a new expansive mood. *Curr. Opin. Cell Biol.*, **6**, 688–694.

Roberts, M.R. (2003) 14-3-3 Proteins find new partners in plant cell signalling. *Trends Plant Sci.*, **8**, 218–223.

Robinson, R.C., Turbedsky, K., Kaiser, D.A., Marchand, J.B., Higgs, H.N., Choe, S. & Pollard, T.D. (2001) Crystal structure of Arp2/3 complex. *Science*, **294**, 1679–1684.

Rodriguez, O.C., Schaefer, A.W., Mandato, C.A., Forscher, P., Bement, W.M. & Waterman-Storer, C.M. (2003) Conserved Microtubule-Actin Interactions in Cell Movement and Morphogenesis. *Nature Cell. Biol.*, **5**, 599–609.

Rohatgi, R., Ma, L., Miki, H., Lopez, M., Kirchhausen, T., Takenawa, T. & Kirschner, M.W. (1999) The interaction between N-WASP and the Arp2/3 complex links Cdc42–dependent signals to actin assembly. *Cell*, **97**, 221–231.

Sagot, I., Klee, S.K. & Pellman, D. (2002a) Yeast formins regulate cell polarity by controlling the assembly of actin cables. *Nature Cell Biol.*, **4**, 42–50.

Sagot, I., Rodal, A.A., Moseley, J., Goode, B.L. & Pellman, D. (2002b) An actin nucleation mechanism mediated by Bni1 and profilin. *Nature Cell Biol.*, **4**, 626–631.

Schiefelbein, J.W., Shipley, A. & Rowse P. (1992) Calcium influx at the tip of growing root-hair cells of *Arabidopsis*, *thaliana*. *Planta*, **187**, 455–459.

Schindelman, G., Morikami, A., Jung, J., Baskin, T.I., Carpita, N.C., Derbyshire, P., McCann, M.C. & Benfey, P.N. (2001) *COBRA* encodes a putative GPI-anchored protein, which is polarly localized and necessary for oriented cell expansion in *Arabidopsis*. *Genes Dev.*, **15**, 1115–1127.

Schmidt, A. (2002) Actin During Mitosis and Cytokinesis, in *Actin: A Dynamic Framework for Multiple Plant Cell Functions* (eds. Staiger, C.J., Baluška, F., Volkmann, D. & Barlow) Kluwer Academic Publishers: Dordrecht, Boston, London, pp. 437–456.

Schuyler, S.C., Liu, J.Y. & Pellman, D. (2003) The molecular function of Ase1p: evidence for a MAP-dependant midzone-specific spindle matrix, *J. Cell. Biol.*, **160**, 517–528.

Schwab, B., Mathur, J., Saedler, R., Schwarz, H., Frey, B., Scheidegger, C. & Hülskamp, M. (2003) Regulation of cell expansion by the *DISTORTED* genes in *Arabidopsis thaliana*: actin controls the spatial organisation of microtubules. *Molec. Gen. Genomics*, **269**, 350–360.

Scott, M.P., Zappacosta, F., Kim, E.Y., Annan, R.S. & Miller, W.T. (2002) Identification of novel SH3 domain ligands for the Src family kinase Hck. Wiskott-Aldrich syndrome protein (WASP), WASP-interacting protein (WIP), and ELMO1. *J. Biol Chem.*, **277**, 28238–28246.

Shaw, S.L., Dumais, J. & Long, S.R. (2000) Cell surface expansion in polarly growing root hairs of *Medicago truncatula. Plant Physiol.*, **123**, 959–969.

Sieberer, B. & Emons, A.M.C. (2000) Cytoarchitecture and pattern of cytoplasmic streaming in root hairs of *Medicago truncatula* during development and deformation by nodulation factors. *Protoplasma*, **214**, 118–127.

Sieberer, B.J., Timmers, A.C.J., Lhuissier, F.G.P. & Emons, A.M.C. (2002) Endoplasmic Microtubules configure the subapical cytoplasm and are required for fast growth of *Medicago truncatula* root hairs. *Plant Physiol.*, **130**, 977–988.

Singh, S., Powell, D.W., Rane, M.J., Millard, T.H., Trent, J.O., Pierce, W.M., Klein, J.B., Machesky, L.M. & McLeish, K.R. (2003) Identification of the p16-Arc subunit of the Arp2/3 complex as a substrate of MAPK-activated protein kinase-2 by proteomic analysis. *J. Biol. Chem.*, 10.1074/jbc.M306428200 [Epub ahead of print].

Smertenko, A.P., Allwood, E.G., Khan, S., Jiang, C.J., Maciver, S.K., Weeds, A.G. & Hussey, P.J. (2001) Interaction of pollen-specific actin-depolymerizing factor with actin. *Plant J.*, **25**, 203–12.

Smertenko, A.P., Jiang, C-J., Simmons, N.J., Weeds, A.G., Davies, D.R. & Hussey, P.J. (1998) Ser6 in the maize actin-depolymerizing factor, ZmADF3, is phosphorylated by a calcium-stimulated protein kinase and is essential for the control of functional activity. *Plant J.*, **14**, 187–193.

Smertenko, A., Saleh, N., Igarashi, H., Mori, H., Hauser-Hahn, I., Jiang, C.J., Sonobe, S., Lloyd, C.W. & Hussey, P.J. (2000) Sequencing a plant MAP-65 cDNA reveals a novel class of microtubule-associated protein. *Nature Cell Biol.*, **2**, 750–753.

Smirnova, E.A., Reddy, A.S.N., Bowser, J. & Bajer, A.S. (1998) Minus end-directed kinesin-like motor protein, Kcbp, localizes to anaphase spindle poles in *Haemanthus* endosperm. *Cell Motil. Cytoskel.*, **41**, 271–280.

Smith, L.G. (2003) Cytoskeletal control of plant cell shape: getting the fine points. *Curr. Opin. Plant Biol.*, **6**, 63–73.

Sohn, R. H. & Goldschmidt-Clermont, P.J. (1994) Profilin: At the Crossroads of Signal Transduction and the Actin Cytoskeleton. *Bioessays*, **16**, 465–472.

Song, H., Golovkin, M., Reddy, A.S.N. & Endow, S.A. (1997) *In vitro* motility of AtKCBP, a calmodulin-binding kinesin protein of *Arabidopsis*, *Proc. Natl. Acad. Sci. USA*, **94**, 322–327.

Staiger, C.J. & Hussey P.J. (2004) Actin and actin modulating proteins, in *The Plant Cytoskeleton* (ed. P.J. Hussey) Blackwell: Oxford, pp. 31–80.

Staiger, C.J., Baluška, F., Volkmann, D. & Barlow, P. eds, *Actin – A Dynamic Framework for Multiple Plant Functions*. Dordrecht, Boston, London, Kluwer Academic Publishers.

Steinborn, K., Maulbetsch, C., Priester, B., Trautmann, S., Pacher, T., Geiges, B., Kuttner, F., Lepiniec, L., Stierhof, Y.D., Schwarz, H., Jürgens, G. & Mayer, U. (2002) The *Arabidopsis* PILZ group genes encode tubulin-folding cofactor orthologs required for cell division but not cell growth, *Gene Dev.*, **16**, 959–971.

Stevenson, V., Hudson, A., Cooley, L. & Theurkauf, W.E. (2002) Arp2/3-dependent pseudo-cleavage furrow assembly in syncytial Drosophila embryos. *Curr. Biol.*, **12**, 705–711.

Stoppin-Mellet, V., Gaillard, J. & Vantard, M. (2002) Functional evidence for in vitro microtubule severing by the plant katanin homologue. *Biochem. J.*, **365**, 337–342.

Strompen, G., El Kasmi, F., Richter, S., Lukowitz, W., Assaad, F.F., Jürgens, G. & Mayer U. (2002) The *Arabidopsis* HINKEL gene encodes a kinesin-related protein involved in cytokinesis and is expressed in a cell cycle-dependent manner. *Curr. Biol.*, **12**, 153–159.

Sugimoto, K., Himmelspach, R., Williamson, R.E. & Wasteneys, G.O. (2003) Mutation or drug-dependent microtubule disruption causes radial swelling without altering parallel cellulose microfibril deposition in *Arabidopsis* root cells. *Plant Cell*, **15**, 1414–1429.

Sumi, T., Matsumoto, K., Takai, Y. & Nakamura, T. (1999) Cofilin phosphorylation and actin cytoskeletal dynamics regulated by rho- and Cdc42-activated LIM-kinase 2. *J. Cell Biol.*, **147**, 1519–1532.

Szymanski, D.B. & Marks, M.D. (1998) GLABROUS1 overexpression and TRIPTYCHON alter the cell cycle and trichome cell fate in *Arabidopsis*. *Plant Cell*, **10**, 2047–2062.

Szymanski, D.B., Jilk, R.A., Pollock, S.M. & Marks, M.D. (1998) Control of GL2 expression in *Arabidopsis* leaves and trichomes. *Development*, **125**, 1161–1171.

Szymanski, D.B., Marks, M.D. & Wick, S.M. (1999) Organized F-actin is essential for normal trichome morphogenesis in *Arabidopsis*. *Plant Cell*, **11**, 2331–2347.

Tao, L.Z., Cheung, A.Y. & Wu, H.M. (2002) Plant Rac-like GTPases are activated by auxin and mediate auxin-responsive gene expression. *Plant Cell*, **14**, 2745–2760.

Taunton, J., Rowning, B.A., Coughlin, M.L., Wu, M., Moon, R.T., Mitchison, T.J. & Larabell, C.A. (2000) Actin-dependent propulsion of endosomes and lysosomes by recruitment of N-WASP. *J. Cell Biol.*, **148**, 519–530.

Tian, G.L., Huang, Y., Rommelaere, H., Vandekerckhove, J., Ampe, C. & Cowan, N.J. (1996) Pathway leading to correctly folded tubulin. *Cell*, **86**, 287–296.

Tian, G.L., Lewis, S.A., Feierbach, B., Stearns, T., Rommelaere, H., Ampe, C. & Cowan, N.J. (1997) Tubulin subunits exist in an activated confirmational state generated and maintained by protein cofactors. *J. Cell Biol.*, **138**, 821–832.

Tominaga, M., Morita, K., Sonobe, S., Yokota, E. & Shimmen, T. (1997) Microtubules regulate the organization of actin filaments at the cortical region in root hair cells of *Hydrocharis*. *Protoplasma*, **199**, 83–92.

Tominaga, M., Yokota, E., Vidali, L., Sonobe, S., Hepler, P.K. & Shimmen, T. (2000) The role of plant villin in the organization of the actin cytoskeleton, cytoplasmic streaming and the architecture of the transvacuolar strand in root hair cells of *Hydrocharis*. *Planta*, **210**, 836–843.

Toshima, J., Toshima, J.Y., Amano, T., Yang, N., Narumiya, S. & Mizuno, K. (2001) Cofilin phosphorylation by protein kinase testicular protein kinase 1 and its role in integrin-mediated actin reorganization and focal adhesion formation. *Mol. Biol. Cell*, **12**, 1131–1145.

Tournebize, R., Popov, A., Kinoshita, K., Ashford, A.J., Rybina, S., Pozniakovsky, A., Mayer, T.U., Walczak, C.E., Karsenti, E. & Hyman, A.A. (2000) Control of microtubule dynamics by the antagonistic activities of XMAP215 and XKCM1 in Xenopus egg extracts. *Nature Cell Biol.*, **2**, 13–19.

Traas, J.A., Bellini, C., Nacry, P., Kronenberger, J., Bouchez, D. & Caboche, M. (1995) Normal differentiation patterns in plants lacking microtubular preprophase bands. *Nature*, **375**, 676–677.

Trotochaud, A.E., Jeong, S. & Clark, S.E. (1999) CLAVATA3, a multimeric ligand for the CLAVATA1 receptor-kinase. *Science*, **289**, 613–617.

Twell, D., Park, S.K., Hawkins, T.J., Schubert, D., Schmidt, R., Smertenko, A. & Hussey, P.J. (2002) MOR1/GEM1 has an essential role in the plant-specific cytokinetic phramoplast. *Nature Cell Biol.*, **4**, 711–714.

Tzafrir, I. McElver, J.A. Liu, C.M. Yang L.J., Wu J.Q., Martinez, A.M., Patton, D.A. & Meinke, D.W. (2002) Diversity of TITAN Functions in *Arabidopsis* Seed Development. *Plant Physiol.*, **128**, 38–51.

Ueda, K., Matsuyama, T. & Hashimoto, T. (1999) Visualization of microtubules in living cells of transgenic *Arabidopsis thaliana*. *Protoplasma*, **206**, 201–206.

Vale, R.D. (1991) Severing of stable microtubules by a mitotically activated protein in *Xenopus* egg extracts, *Cell*, **64**, 827–839.

Valster, A.H., Pierson, E.S., Valenta, R., Hepler, P.K. & Emons, A.M.C. (1997) Probing the plant actin cytoskeleton during cytokinesis and interphase by profilin microinjection. *Plant Cell*, **9**, 1815–1824.

Vartiainen, M.K., Mustonen, T., Mattila, P.K., Ojala, P.J., Thesleff, I., Partanen, J. & Lappalainen, P. (2002) The three mouse actin-depolymerizing factor/cofilins evolved to fulfill cell-type-specific requirements for actin dynamics. *Molec. Biol. Cell*, **13**, 183–194.

Very, A.A. & Davies, J.M. (2000) Hyperpolarization-activated calcium channels at the tip of *Arabidopsis* root hairs. *Proc. Natl. Acad. Sci. USA*, **97**, 9801–9806.

Vidali, L. & Hepler, P.K. (1997) Characterization and localization of profilin in pollen grains and

tubes of *Lilium longiflorum*. *Cell Motil. Cytoskel.*, **36**, 323–338.
Vidali, L., Yokota, E., Cheung, A.Y., Shimmen, T. & Hepler, P.K. (1999) The 135 kDa actin-bundling protein from *Lilium longiflorum* pollen is the plant homologue of villin. *Protoplasma*, **209**, 283–291.
von Witsch, M., Baluška, F., Staiger, C.J. & Volkmann, D. (1998) Profilin is associated with the plasma membrane in microspores and pollen. *Eur. J. Cell Biol.*, **77**, 303–312.
Vos, J.W., Safadi, F., Reddy, A.S.N. & Hepler, P.K. (2000) The kinesin-like calmodulin binding protein is differentially involved in cell division. *Plant Cell*, **12**, 979–990.
Wallar, B.J. & Alberts, A.S. (2003) The formins: active scaffolds that remodel the cytoskeleton. *Trends Cell Biol.*, **13**, 435–46.
Wang, P.J. & Huffaker, T.C. (1997) Stu2p: A microtubule-binding protein that is an essential component of the yeast spindle pole body. *J. Cell. Biol.*, **139**, 1271–1280.
Wasteneys, G.O. (2000) The cytoskeleton and growth polarity. *Curr. Opin. Plant Biol.*, **3**, 503–511.
Watanabe, N., Kato, T., Fujita, A., Ishizaki, T. & Narumiya, S. (1999) Cooperation between mDia1 and ROCK in Rho-induced actin reorganization. *Nature Cell Biol.*, **1**, 136–143.
Webb, M. (2002) Cell specification in *Arabidopsis* root eperdermis requires the activity of ECTOPIC ROOT HAIR 3 a katanin-p60 protein. *Development*, **129**, 123–131.
Welch, M.D., Iwamatsu, A. & Mitchison, T.J. (1997) Actin polymerization is induced by Arp2/3 protein complex at the surface of Listeria monocytogenes. *Nature*, **385**, 265–269.
Whittington, A.T., Vugrek, O., Wei, K.J., Hasenbein, N.G., Sugimoto, K., Rashbrooke, M.C. & Wasteneys, G.O. (2001) MOR1 is essential for organizing cortical microtubules in plants. *Nature*, **411**, 610–613.
Wiedemeier, A.M., Judy-March, J.E., Hocart, C.H., Wasteneys, G.O., Williamson, R.E. & Baskin, T.I. (2002) Mutant alleles of *Arabidopsis*, *RADIALLY SWOLLEN 4* and *7* reduce growth anisotropy without altering the transverse orientation of cortical microtubules or cellulose microfibrils. *Development*, **129**, 4821–4830.
Wills, Z., Emerson, M., Rusch, J., Bikoff, J., Baum, B., Perrimon, N. & Van Vactor, D. (2002) A Drosophila homolog of cyclase-associated proteins collaborates with the Abl tyrosine kinase to control midline axon pathfinding. *Neuron,* **36**, 611–622.
Winter, D.C., Choe, E.Y. & Li, R. (1999) Genetic dissection of the budding yeast Arp2/3 complex: a comparison of the *in vivo* and structural roles of individual subunits. *Proc. Natl. Acad. Sci. USA*, **96**, 7288–7293.
Winter, D., Podtelejnikov, A.V., Mann, M. & Li, R. (1997) The complex containing actin-related proteins Arp2 and Arp3 is required for the motility and integrity of yeast actin patches. *Curr. Biol.* **7**, 519–529.
Wolfe, J. & Steponkus, P.L. (1981) The stress-strain relation of the plasma membrane of isolated protoplasts. *Biochim. Biophys. Acta*, **643**, 663–668.
Wolfe, J. & Steponkus, P.L. (1983) Mechanical properties of the plasma membrane of isolated plant protoplasts. *Plant Physiol.*, **71**, 276–285.
Woychik, R.P., Maas, R.L., Zeller, R., Vogt, T.F. & Leder, P. (1990) 'Formins': proteins deduced from the alternative transcripts of the limb deformity gene. *Nature*, **346**, 850–853.
Wu, G., Gu, Y., Li, S. & Yang, Z. (2001) A genome-wide analysis of *Arabidopsis* Rop-interactive CRIB motif-containing proteins that act as Rop GTPase targets. *Plant Cell*, **13**, 2841–56. Erratum in: *Plant Cell*, **14**, 741 (2002).
Wymer, C.L., Bibikova, T.N. & Gilroy, S. (1997) Cytoplasmic free calcium distributions during the development of root hairs of *Arabidopsis*, *thaliana*. *Plant J.*, **12**, 427–439.
Yokota, E., Muto, S. & Shimmen, T. (2000) Calcium-calmodulin suppresses the filamentous actin-binding activity of a 135-kilodalton actin-bundling protein isolated from lily pollen tubes. *Plant Physiol.*, **123**, 645–654.
Yokota, E., Takahara, K.-I. & Shimmen, T. (1998) Actin-bundling protein isolated from pollen tubes of lily, *Plant Physiol.*, **116**, 1421–1429.
Zhou, G.L., Yamamoto, T., Ozoe, F., Yano, D., Tanaka, K., Matsuda, H. & Kawamukai, M. (2000) Identification of a 14-3-3 protein from *Lentinus*, edodes that interacts with CAP (adenylyl cyclase-associated protein), and conservation of this interaction in fission yeast. *Biosci. Biotech. Biochem.*, **64**, 149–159.

2 Mechanisms of cell polarity establishment and polar auxin transport

Arthur J. Molendijk, Olaf Tietz, Benedetto Ruperti,
Ivan A. Paponov and Klaus Palme

2.1 Introduction

The mechanism by which cells become polarised in response to endogenous and exogenous cues is a fundamental question in the study of plant development. Polarity has been defined as a 'persistent asymmetrical and ordered distribution of structures along a [spatial] axis' (Cove *et al.*, 1999) and is an intrinsic feature of all life-forms. It occurs at all hierarchical levels, from bacteriophage to complex multicellular organism. At each level, polarity is essential for the proper functioning of cells and organisms. Well established examples are the polar transport of nutrients across epithelia, segmentation during embryogenesis, and the development of plant roots and their growth, either in the ground or away from light. Besides polarity, the polar axis can be described by two additional characteristics: axis alignment and axis orientation. Axis alignment specifies the relationship between the axis and external references, whereas axis orientation defines the direction in which the 'asymmetrical and ordered distribution' occurs (Cove, 2000).

At present, research is focused on establishing the molecular mechanisms by which polarity originates and is maintained (Dubrin & Nelson, 1996; Shulman & Johnston, 1999). Cellular polarity is based on different, highly specialised domains both within the cytoplasm and the plasma membrane, and also, in the case of plants and bacteria, within the cell wall. These domains may be based on factors such as asymmetric distribution of cell organelles and messenger RNA, regulated targeting of proteins, uneven localisation of membrane proteins, intracellular Ca^{2+} gradients, polar orientation of cytoskeletal elements, or the formation of cilia. Although studies on polarity started more than 100 years ago (Vöchting, 1878), insights into the molecular mechanisms and contributing factors underlying cell polarity are just arising (Dubrin, 2000). Many genome sequencing projects have been completed. The recent publication of a vast amount of sequence information and the availability of several model organisms that are conducive to genetic analysis enables us now to address questions about the molecular mechanisms of polarity establishment and maintenance. We can also now gain insights into the relationship between specific genetic elements during the evolution of different organisms and can infer specific molecular adaptations according to the particular selection pressures on the respective organisms. The repertoire of genetic and data mining tools, as well as the many new and advanced microscopic tools (Godde *et al.*, 1999), will open various windows for an unprecedented understanding of the molecular basis of cell polarity in different model organisms.

Model organisms to study cell polarity are *Drosophila* and *Caenorhabditis*, yeast and brown algae (see Chapter 5, for polarity in *Fucus*). Only recently has the model plant *Arabidopsis thaliana* become a favourite for researchers (Fowler & Quatrano, 1997; Scheres & Benfey, 1999). This article will examine different areas of cell polarity of *Arabidopsis* and summarise relevant research to date.

2.2 Root hairs

A good example of cell polarity in plants is seen in tip growth. Tip growing cells, as the name implies, grow by secretory growth at the tip, without elongation growth in the trailing cell tube (Shaw *et al.*, 2000). Tip growth is defined by three characteristics: a clear zonation of cell organelles towards the growing pole, tip-located high Ca^{2+} levels, and a reverse fountain type of cytoplasmic streaming (Sherrier & VandenBosch, 1994; Geitmann & Emons, 2000; Hepler *et al.*, 2001). This growth type contrasts with polar diffuse growth of elongating cells, such as in young hypocotyls, where the lateral cell walls expand unidirectionally and no growth occurs at the cross walls (for a review see Kropf *et al.*, 1998; Baluška *et al.*, 2003b; Smith, 2003; see Chapter 1). Tip growth is also seen in fungal hyphae, and might represent a favourable type of growth for cells through a solid substrate. Well known examples of tip-growing cells in plants are pollen tubes, fern and moss protonemata, root hairs, and the rhizoid of the green alga *Chara*.

Root hairs are increasingly popular model plant cells, allowing easy experimental observation and manipulation, initiating as single buds on the outer surface of root epidermal cells, termed trichoblasts. The development from budding to tip growth is best characterised by the appearance of the tip-high Ca^{2+} gradient (Pierson *et al.*, 1996; Miller *et al.*, 1997; Wymer *et al.*, 1997; Bibikova *et al.*, 1997; Bibikova & Gilroy, 1998). The process of root hair formation can be followed from beginning to end in epidermal cell files which consist of an orderly sequence of cells of increasing age away from the root apical meristem. In contrast to pollen tubes, root hairs are not essential for plant viability, and a large number of mutants in root hair formation have been described in *Arabidopsis*. A recent publication deals specifically with the different aspects of root hair biology (Ridge & Emons, 2000).

2.3 Rho GTPases

RHO GTPases (Rac, CDC42, and Rho subfamilies) are conserved switches controlling actin organisation among other processes which include gene expression, cell wall synthesis, and cell cycle progression both in yeast and animals (Van Aelst & d'Souza-Schorey, 1997; Schmidt & Hall, 1998; Hall, 1998; Bishop & Hall, 2000). Plants are unique in their lack of Ras, Rac/CDC42, and Rho GTPases. Instead, they possess a plant-specific class of small GTPases belonging to the RHO family termed Rop for RHO of plant (Yang & Watson, 1993). In *Arabidopsis*, the Rop multigene family has 11 members (Winge *et al.*, 2000). RHO GTPases are usually isoprenylated at the C-terminus; this lipid modification allows membrane association. RHO proteins can be stripped from membranes *in vivo* by cytosolic GDI factors (guanosine nucleotide dissociation inhibitor), and thus are able to cycle between membranes and cytosol where they are present in a binary complex with RHOGDI (Olofsson, 1999). Only the membrane-localised, GTP-bound forms are active and interact with effector molecules. The GTPase cycle of RHOs is regulated by GTP/GDP exchange factors (GEF), and GTPase activating proteins (GAP). The effector proteins of RHO GTPases and their GTPase binding motifs have recently been reviewed (Bishop & Hall, 2000).

The most frequently occurring binding motif for Rac/CDC42 GTPases is the CRIB (CDC42/Rac interactive binding) motif (for a review see Bishop & Hall, 2000; Hoffman & Cerione, 2000), which has also been identified in *Arabidopsis* proteins (Borg *et al.*, 1999; Wu *et al.*, 2000, 2001). The *Arabidopsis* genome contains 3 RHOGDI genes (Bischoff *et al.*, 2000) and 3 different classes of RHOGAPs comprising 12 genes (AGI, 2000). Remarkably, *Arabidopsis* lacks typical RHOGEFs with the catalytic Dbl homology domain, but contains a single copy homolog (*SPIKE1*; Qiu *et al.*, 2002) of the unconventional RHOGEFs Zizimin/Dock180 (Meller *et al.*, 2002; Brugnera *et al.*, 2002).

2.3.1 Rho GTPases and cell polarity

CDC42 is crucial to cell polarity in yeast and other types of eukaryotic cells as recently illustrated by protein interaction mapping (Johnson, 1999; Drees *et al.*, 2001). It localises to buds (Ziman *et al.*, 1993) and to growing hyphal tips and controls cell polarity in both locations. The role of GTPases can be investigated by using two classes of mutants. Those of the first class are locked in the GTP-bound form and mutants are therefore constitutively active (CA). The second contains dominant negative (DN) mutants that inhibit the GDP/GTP exchange activity of a corresponding GEF, resulting in all target GTPases remaining in the GDP-bound form (off conformation). Expression of CA CDC42 typically results in apolar growth in both yeast and fungal hyphae.

Rop GTPases are also localised to growing tips of pollen tubes (Lin *et al.*, 1996; Kost *et al.*, 1999) and root hairs, and are polarly localised before and during root hair budding (Molendijk *et al.*, 2001; Jones *et al.*, 2002). That Rop GTPases control cell polarity is evident from the apolar growth phenotypes (branching and swelling) observed in root hairs and pollen tubes either expressing CA Rops or overexpressing wildtype Rop (OX). It is also indicated by the growth inhibitory effects of Rop antibody injection (Lin & Yang, 1997), and DN mutant Rops (Kost *et al.*, 1999; Li *et al.*, 1999; Molendijk *et al.*, 2001; Gu *et al.*, 2003; Jones *et al.*, 2002; Tao *et al.*, 2002; Chen *et al.*, 2003: Cheung *et al.*, 2003). Apolar growth phenotypes are also observed in leaf epidermal cells which fail to produce the characteristic lobed cell morphology (Fu *et al.*, 2002), as well as in hypocotyl epidermal cells which normally expand by polar diffuse growth, but are swollen when expressing CA Rop (Molendijk *et al.*, 2001). This indicates a more general role of Rops in plant cell polarity. Not all CA or OX Rops produce apolar growth phenotypes, and specific phenotypic differences also exist between the different Rops, indicating a functional diversity among the Rop GTPases (Lavy, *et al.*, 2002; Cheung *et al.*, 2003). *Arabidopsis* CA Rops cause root hair swelling in the case of AtRop4 and AtRop6 and branching in the case of AtRop2. Both CA and OX AtRop2 may result in multiple root hairs from a single trichoblast, further indicating that Rop GTPases control cell polarity during root hair initiation as well as during tip growth (Jones *et al.*, 2002). Similar apolar growth phenotypes can also be observed in many root hair morphology mutants (for a review see Hülskamp *et al.*, 1998; Cavell & Grierson, 2000; Mathur & Hülskamp, 2001). However, it cannot be argued convincingly that the identified genes control cell polarity because the corresponding proteins do not function as comparibly versatile molecular switches in signal transduction (for a review see Yang, 1998; Valster *et al.*, 2000; Zheng & Yang, 2000a,b; Fu & Yang, 2001). A possible exception is *RHD3* which encodes an evolutionarily conserved GTP-binding protein involved in cell expansion (Wang *et al.*, 1997, 2002). Several such mutants represent gene products that might be direct targets of the Rop GTPase switch, such as *RHD2*, encoding an NADPH oxidase (Foreman *et al.*, 2003).

2.4 Specific gene expression in root hair formation

Several *Arabidopsis* genes controlling trichoblast/atrichoblast differentiation have been identified (for a review see Schiefelbein, 2000; Dolan, 2001). Recently, the role of GL2 (a homeodomain transcription factor) in inhibiting ectopic, root hair-specific gene expression has been illustrated in case of the expansin gene *ATEXP7*. Remarkably, GL2 modified to include a trans-activating function caused leaf epidermal cells to develop ectopic root hairs or root hair-like structures (Ohashi *et al.*, 2003). The morphological development of cells specified as trichoblasts (root hair-forming root epidermal cells) requires root hair-specific gene expression. Gene expression profiling is not yet complete, but a large number of genes involved in root hair morphogenesis has already been shown to be expressed in trichoblast cell files but not in atrichoblast cell files (non-root hair forming root epidermal cells adjacent to trichoblast cell files in *Arabidopsis*). Interestingly, many of the corresponding gene products are involved in cell wall synthesis; the leucine-rich repeat/extensin cell wall protein LRX1 (Baumberger *et al.*, 2001), the expansins AtEXP7 and AtEXP18 (Cho & Cosgrove, 2002), the proline-rich cell wall protein AtPRP3 (Bernhardt & Tierney, 2000), the cellulose synthase-like protein Kojak/AtCSLD3 (Favery *et al.*, 2001; Wang *et al.*, 2001), and the membrane anchored endo-1,4-β-D glucanase KOR2 (Mølhøj *et al.*, 2001).

2.4.1 Plasma membrane trafficking is required for establishment of cell polarity in root hair formation

Trichoblasts can be recognised by their unusually high ER content, among other cytological characteristics, before a visible outgrowth of root hairs appears, indicating an early start of root hair morphogenesis-specific gene expression and cell differentiation (Ridge & Emons, 2000; Baluška *et al.*, 2000b; Bibikova & Gilroy, 2002; see also Chapter 7). Root hair initiation, as defined by polar Rop localisation, occurs during the polar diffuse growth phase, when secretory growth is occurring uniformly in the four lateral cell walls of the trichoblast. In the same cell, a membrane/cell wall domain is established with distinct characteristics. A local acidification of the cell wall accompanies bulging (Bibikova *et al.*, 1998). This bulging phase is distinct from the subsequent tip growth phase, as it is insensitive not only to Ca^{2+} influx, but also actin and tubulin inhibitors (Wymer *et al.*, 1997; Miller *et al.*, 1999; Ovecka *et al.*, 2000; Bibikova & Gilroy, 2000; Baluška *et al.*, 2000b). Before and during visible bulging, the cell wall is locally modified by xyloglucan endotransglycosylase (XET) action and expansins (Vissenberg *et al.*, 2001; Baluška *et al.*, 2000b). XET action and expansin localisation in the cell wall are also insensitive to the actin inhibitor latrunculin D. Bulges contain new actin structure and profilin (Braun *et al.*, 1999; Baluška *et al.*, 2000), but the intracellular Ca^{2+} concentration in the bulge is not elevated (Bibikova & Gilroy, 2000). It seems likely that bulging is not merely a local expansion of the cell wall, but rather an example of localised diffuse growth, involving secretion as suggested in the case of expansins (Mathur & Hülskamp, 2001).

The occurrence of early plasma membrane trafficking in root hair formation is indicated by the fact that early polar localisation of Rop is sensitive to the secretion inhibitor Brefeldin A (BFA) but not to actin or tubulin inhibitors (Molendijk *et al.*, 2001). BFA targets the GTP/GDP exchange factors of the

Arf GTPases (Arf GEF) of the Gea/GNOM class. This class is represented by three proteins in *Arabidopsis*. Indeed, GNOM has been demonstrated to be a target for BFA action in plants (Steinmann *et al*., 1999; Jürgens & Geldner, 2002; Geldner *et al*., 2003). The presence of 'BFA' or 'fusiform' compartments in plant cells after treatment with the drug has indicated both the presence of a recycling endosome, and its functional importance in establishing cell polarity. The plant recycling endosome is thought to be composed of different subdomains and to be an important sorting site for proteins targeted to different plasma membrane destinations (Steinmann *et al*., 1999; Geldner *et al*., 2001; Grebe *et al*., 2002; Jürgens & Geldner, 2002, Geldner *et al*., 2003; Baluška *et al*., 2003b). Not all cells have BFA compartments as revealed by pectin staining, but trichoblasts do, indicating a recycling endosome is active during root hair formation (Baluška *et al*., 2002). BFA-sensitive membrane recyling was originally demonstrated at the cross wall membranes of root meristematic cells (Geldner *et al*., 2001; Baluška *et al*., 2002, 2003a). Polar localisation of Rop to these membranes is also sensitive to BFA (Molendijk *et al*., 2001), suggesting a more general connection between Rop localisation and endosomal recycling. Recovery from BFA treatments is latrunculin D-sensitive for vesicle markers, but not for cytosolic Rop GTPase in root hair formation, perhaps indicating a less efficient process of Arf1 dependent vesicle trafficking in the presence of the inhibitor.

2.5 Does Rop localise secretion at the future bud site?

Since new plasma membrane is continuously inserted at the tip of the cell during secretory growth, Rop must be retrieved from the flanks by RopGDI and relocalised to the tip. Rop is probably activated at the tip and progressively inactivated away from the tip by RopGAP activity. In fact, a tip-high Rop gradient has been shown to exist along the plasma membrane of a tip-growing cell (Gu *et al*., 2003). Such a spatio-temporal cycle of Rop activation, together with RopGAP and RopGDI action might also explain the polar localisation of Rop GTPase to a relatively small membrane patch during the earlier phase of root hair initiation. From the reversible inhibitory effect of BFA on Rop localisation, it is clear that this Rop activation/localisation cycle functions in concert with Arf1–dependent vesicle trafficking. It might be that a Rop-containing plasma membrane patch marks a site of increased Arf1–dependent vesicle trafficking. Many different root hair mutants exist with multiple hairs on one trichoblast or a changed position of the root hair along the apical-basal axis of the cell, showing that the number and position of root hairs are not stable characteristics (Bibikova & Gilroy, 2002). Auxin analogues and low concentrations of BFA affect the apical-basal position of the root hair (Grebe *et al*., 2002). The absence of a fixed cue in the plasma membrane or cell wall is also suggested by the fact that BFA treatment may leave multiple Rop-positive membrane patches on the trichoblast plasma membrane (Molendijk *et al.*, 2001), and by the induction of double root hair formation during growth in low concentrations of BFA (Grebe *et al.*, 2002). Thus, in the absence of a fixed cue, Rop GTPase might focus Arf1-dependent endosomal recycling and determine a future bud site, with activation of Rop at the site where vesicles are inserted. How Rop would function in directing vesicle fusion with the plasma membrane is not completely understood. Mathur *et al.* (2003a,b) show an important role for regulation of actin dynamics by the Arp2/3 complex in determining where and when cell expansion can occur. The Arp2/3 complex is regulated by Rho GTPase signalling in animal systems, and a likely downstream target for the Rop GTPase in plants (Mathur *et al.*, 2003a,b; see Chapter 1, for more detail).

2.5.1 How are Rops activated/deactivated?

Several studies have shown a role for Rop GTPases in regulating actin dynamics (Fu *et al.*, 2001a, 2002; Gu *et al.*, 2003; Chen *et al.*, 2003), but information about a possible role for signal transduction from membrane receptors to actin is scarce (Miller *et al.*, 1997; Staiger, 2000; Baluška *et al.*, 2000a; Schmelzer, 2002; Vantard & Blanchoin, 2002; Schultheiss *et al.*, 2002). However, Rho GTPases generally link membrane receptor signalling to the actin skeleton (Hall, 1998; Schmidt & Hall, 1998; Kjøller & Hall, 1999). If this property of Rho GTPases is conserved, it might be that Rop functions in signal transduction via the most widely present class of plant receptors, the receptor linked kinases (RLKs) (Shiu & Bleecker, 2001). Importantly, Rop is a component of the active RLK CLV1 signalling complex (Trotochaud *et al.*, 1999). Yeast 2-hybrid screening using Rop GTPase as a bait yields several RLKs (our unpublished results), which would agree with a link between Rops and RLKs. Whether and how Rop is activated in RLK complexes demands further investigation.

Many developmental processes modulated by phytohormones such as auxin, brassinosteroid and ABA are affected in CA or DN Rop transgenics, suggesting Rops might mediate hormone responses (Li *et al.*, 2001). Auxin activates Rop (Tao *et al.*, 2002) and ABA deactivates Rop (Lemichez *et al.*, 2001; Zheng *et al.*, 2002), but again the mechanisms by which it does this remain unclear. A comparison with Rho GTPase signalling mechanisms in yeast or animals here is of limited use, especially because the *Arabidopsis* genome does not contain conventional RHOGEFs with Dbl homology domains. The only gene likely to encode a RopGEF identified so far is the single copy *SPIKE1* gene (Qiu *et al.*, 2002). It remains to be seen if the absence of conventional RHOGEFs might be compensated by alternative regulation of GTP/GDP exchange on Rops. Since BFA reversibly inhibits polar localisation/activation of Rop GTPase during early root hair formation, it might be that Arf1–dependent plasma membrane/vesicle trafficking inserts a Rop activator into the plasma membrane. Recently, it has been shown that AtSERK (an RLK) plant receptors can undergo internalisation/recycling (Shah *et al*., 2002); a process which is likely to be sensitive to BFA.

Deactivation by GAPs is a conserved feature in the regulation of Rop GTPase switches. Three different classes of RopGAPs are present in the *Arabidopsis* genome. One class containing a CRIB domain (Borg *et al*., 1999; Wu *et al*., 2000), a class containing a PH (Pleckstrin Homology) domain, and a third class. The presence of a CRIB domain might result in negative feedback by GTP-bound Rop at the plasma membrane, and the PH domain might give negative feedback via its ligand, perhaps plasma membrane localised PIP_2, but regulation of GAP activity has not been studied yet.

2.5.2 How do Rops control cell polarity?

The importance of actin (Jiang *et al*., 1997; Miller *et al*., 1999; Ramachandran *et al*., 2000; Esseling *et al*., 2000; Swoboda *et al*., 2001; Gilliland *et al*., 2002; Ketelaar *et al*., 2002; Ringli *et al*., 2002; Ketelaar *et al*., 2003), tubulin (Bibikova *et al*., 1999; Ketelaar & Emons, 2000; Bao *et al*., 2001), and Ca^{2+} (Véry & Davies, 2000) in tip growth has been well documented (Geitmann & Emons, 2000; Ketelaar & Emons, 2001; Hepler *et al*., 2001; Mathur & Hülskamp, 2002a; see Chapter 1 for a detailed review). Expression of CA Rop results in abolition of tip growth; in this case, not surprisingly, actin and tubulin cytoskeletons, and the tip-high Ca^{2+} gradient are all affected (Li *et al*., 1999; Molendijk *et al*., 2001; Mathur & Hülskamp, 2002b). The molecular mechanisms underlying regulation of actin dynamics (and possibly Ca^{2+} concentration) by Rop GTPase in its control of polar growth are being unraveled.

Several plant homologues of Rac effectors in mammalian signalling pathways to actin are notably absent in the plant genome (e.g. PAK and WASP), but homologues of other downstream factors in animal cells such as HSPC300 (Maize BRK1) and Arp2/3 do exist. BRK1 is probably involved in Arp2/3 dependent actin polymerisation (Frank *et al*., 2003). Knock-outs in *BRK1*, *ARP2*, and *ARP3* genes result in apolar growth phenotypes of leaf epidermal cells that lack the characteristic lobed morphology (Frank & Smith, 2002; Mathur *et al*., 2003a,b). In this respect, the knockouts resemble both *spike1* (Qiu *et al*., 2002) and *rop CA* mutant phenotypes (Fu *et al*., 2002), and strongly suggest a role for Rop in regulating actin polymerisation through BRK1 and Arp2/3 (Frank *et al*., 2003; Mathur *et al*., 2003a,b). Formins are intermediates in Rho GTPase signalling to actin in animal cells and are also found in *Arabidopsis*. They are possible candidates for Rop-dependent activation by a different mechanism from that of Rho GTPases in animal

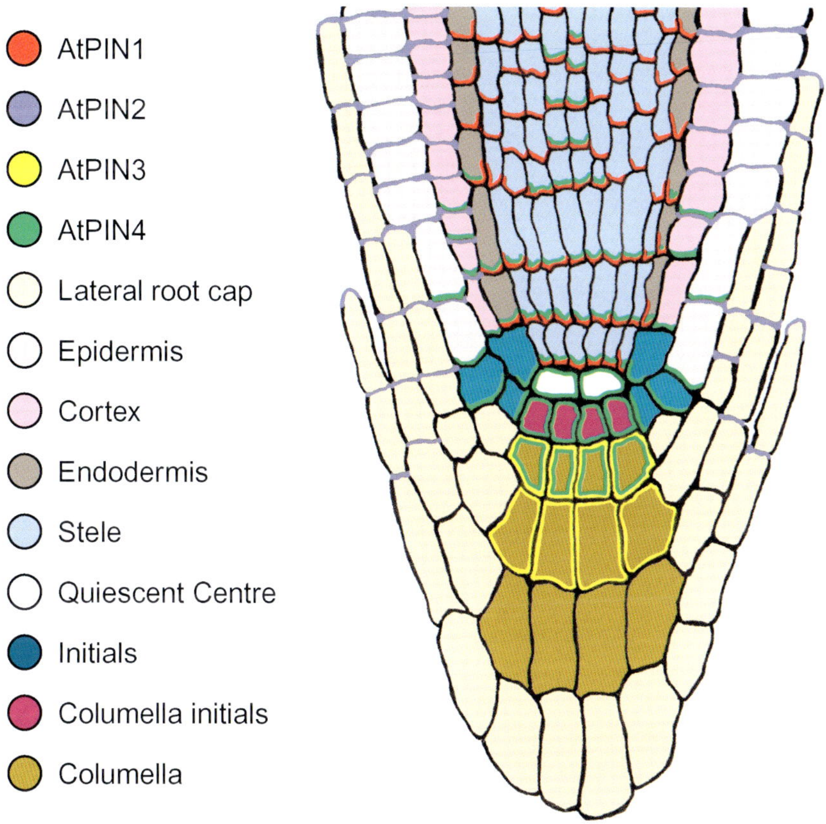

Figure 2.1 PIN localisation in *Arabidopsis* root tip.

systems (Deeks *et al.*, 2002). Rop may also act through ADF (Actin Depolymerising Factor) to regulate pollen tube growth, perhaps involving an ADF protein kinase (Chen *et al.*, 2003). Another important signalling molecule PIP_2 is localised in the plasma membrane at the tip of pollen tubes (see Chapter 1). Rop associates with PIP kinase (Kost *et al.*, 1999), providing another possible link between Rop and regulation of actin dynamics.

Reactive oxygen species (ROS) production and, consequently, pathogen defence responses are enhanced in transgenic plants and plant cell cultures expressing CA Rop, and decreased in plants expressing DN Rop or antisense *Rop* genes (Potikha *et al.*, 1999; Kawasaki *et al.*, 1999; Park *et al.*, 2000; Schiene *et al.*, 2000; Ono *et al.*, 2001; Suharsono *et al.*, 2002; Baxter-Burrell, 2002). ROS, especially H_2O_2, are important signalling molecules in plants (for a review, see Neill *et al.*, 2002). Recently ROS production at the plasma membrane by NADPH oxidase has been shown to regulate root hair tip growth (Foreman *et al.*, 2003). The root hair-defective *rhD2* mutant, recently identified as a root NADPH oxidase mutant, shows normal bulging, but is inhibited in tip growth. A tip-focused ROS gradient was reported in root hair tip growth and ROS were shown to activate root hair hyperpolarisation-activated Ca^{2+} channels. Foreman *et al.* (2003) propose that Rop might activate ROS production in tip growth, resulting in Ca^{2+} channel activation and, consequently, in the tip-high Ca^{2+} gradient (Foreman *et al.*, 2003). It is possible that Rop GTPases might co-ordinate the seemingly distinct processes of ROS production/pathogen defence and regulation of cytoskeletal dynamics/cell polarity through different signalling pathways (e.g. Agrawal *et al.*, 2003).

The first Rop molecular targets to be identified were the RICS, small CRIB domain-containing proteins that share little sequence homology outside the CRIB domain. Overexpression of RIC3 and RIC4 causes depolarised growth in pollen, as might be expected for Rop effectors (Wu *et al.*, 2001). The role of RICs in the regulation of intracellular Ca^{2+} levels and cytoskeletal dynamics is being investigated (Gu *et al.*, 2003). To fully appreciate the role of Rop in plant cell polarity it will be necessary to identify the relevant molecular interaction networks using yeast 2-hybrid analysis, affinity chromatography, or *in vivo* tagging of Rop-containing protein complexes.

2.6 PIN proteins and cell polarity

PIN proteins were the first polarly localised plasma membrane proteins in plants to be discovered. They are encoded by the so called *PIN* gene family, a family which in *Arabidopsis* comprises eight members, sharing a high degree of amino acid homology. They are plasma membrane proteins with similarity to prokaryotic and eukaryotic carrier proteins, indicating a function in trans-membrane transport processes. Their structure is based on two hydrophobic, highly conserved domains, each comprising four to six membrane-spanning segments. Major implications on the function of PIN proteins come from the analysis of corresponding knockout mutants. These mutants display defects in cellular efflux of the plant hormone auxin, suggesting that the PIN proteins are components of the auxin efflux carrier complex (for a review see Palme & Gälweiler, 1999; Friml & Palme, 2002).

Auxin efflux carrier proteins (AEC) have been hypothesised as key players in polar auxin transport in plants since the early 1970s. This polar transport through plant tissues, which is a unique feature of the plant hormone auxin, is almost certainly based on a cell-to-cell transport. Unequally distributed auxin efflux carrier proteins in the plasma membrane are thought to generate directional auxin efflux from the cells.

Consistent with the proposed location of AECs, AtPIN1 is localised in the basal plasma membrane of xylem parenchyma cells in the inflorescence axis of *Arabidopsis* (Gälweiler *et al.*, 1998). AtPIN1 was also found in the apical plasma

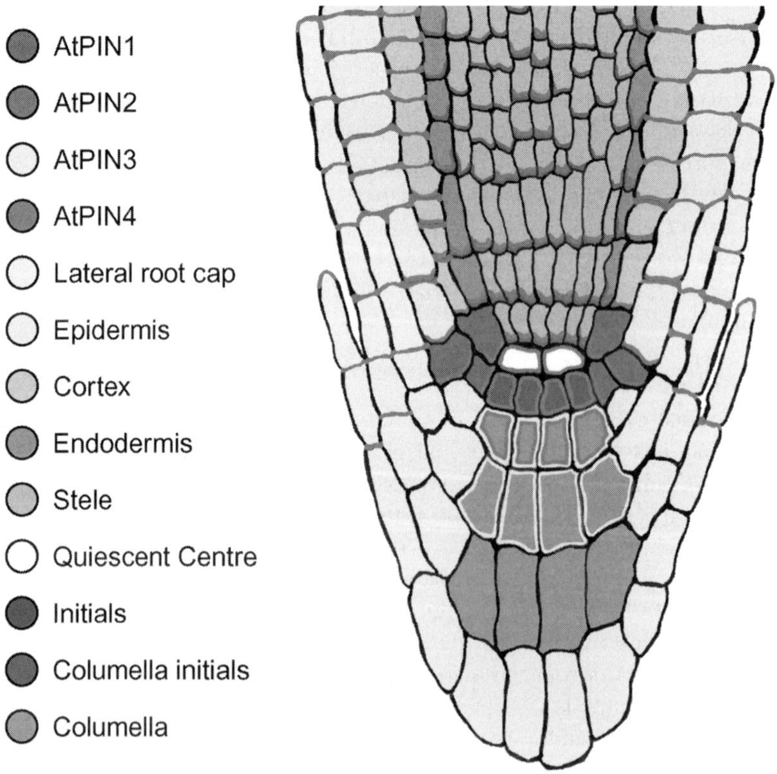

Figure 2.1 PIN localisation in *Arabidopsis* root tip. A coloured representation of this figure can be found in the colour plate section.

membrane of root stele cells (Fig. 2.1; Geldner *et al.*, 2001), where it might facilitate polar auxin transport through the root central cylinder towards the root tip (Rashotte *et al.*, 2000). Other PIN proteins like AtPIN2, AtPIN3 and AtPIN4 are also found in root cells (Fig. 2.1). In contrast to the apical localisation of AtPIN1 in stele cells, AtPIN2 is localised at the basal plasma membrane of root epidermal cells (Müller *et al.*, 1998), where it is involved in basal auxin transport (Rashotte *et al.*, 2001). In columella cells, where lateral relocation of auxin is thought to occur, AtPIN3 and AtPIN4 are not polarly localised, when plants grow vertically (Friml *et al.*, 2002a,b; Ottenschläger *et al.*, 2003). Gravi-stimulation leads to unequal distribution of auxin in root tips and to relocation of AtPIN3 to the new side of auxin efflux (Friml *et al.*, 2002b; Ottenschläger *et al.*, 2003). These data indicate that intracellular PIN localisation is generally consistent with a role for these proteins in polar auxin transport.

A relocalisation of PIN proteins, as in the case of AtPIN3, is also observed during embryonic development and lateral root formation. During the earlier stages of embryogenesis, intracellular AtPIN1 localisation does not yet display a pronounced basal polarity. A basal plasma membrane localisation of AtPIN1 is established during the mid-globular stage of *Arabidopsis* embryo development, where AtPIN1 is gradually confined to the basal plasma membrane of elongated cells (Steinmann *et al.*, 1999). In lateral root formation, AtPIN1 is expressed in pericycle cells that give rise to the new side-root primordia. At first it is exclusively localised to the basal anticlinal sides, but during stages V–VI (according to Malamy & Benfey, 1997), AtPIN1 relocalises towards the periclinal sides of these cells (our unpublished results). These data indicate that AtPIN1 localisation during development is dependent on the differentiation status of the cell.

2.6.1 Molecular mechanisms that control PIN polarity – current models

Spatial control of exocytosis is of fundamental importance in achieving cell polarity and has been well studied in yeast and animal cells (Spiliotis & Nelson, 2003). In contrast, the mechanisms by which polar secretion in plants is ensured have been poorly understood until recently (Jürgens & Geldner, 2002). Clearly, the cytoskeleton should play a crucial role in the establishment and maintenance of cellular growth and polarity. In plants, microtubules (MT) seem to provide the direction of the growth axis, whereas actin filaments are predominantly involved in the delivery of cell wall and plasma membrane components. A major route for exocytotic vesicles from the Golgi apparatus to the plasma membrane is via the actin cytoskeleton (for a review see Mathur & Hülskamp, 2002a; Chapter 7). Interestingly, AtPIN1-containing vesicles seem to utilise both actin and MTs depending on the phase of the cell cycle. Disruption of MT leads to aggregation of AtPIN1 in patches together with KNOLLE, a cell plate-localised cytokinesis-specific syntaxin (Lauber *et al.*,

1997; Geldner *et al.*, 2001). In contrast, during interphase, AtPIN1 trafficking proceeds along actin filaments, as indicated by an altered localisation of AtPIN1 upon disruption of the actin cytoskeleton (Geldner *et al.*, 2001). The underlying mechanisms which provide the switch from one route to the other are not yet known, but demonstrate the complexity of exocytotic protein trafficking. The secretion inhibitor BFA specifically inhibits Sec7-domain-containing ArfGEFs, thereby interfering with Arf-dependent vesicle trafficking (Peyroche *et al.*, 1999; Sata *et al.*, 1999). BFA treatment of *Arabidopsis* seedlings results in rapid accumulation of AtPIN1 in intracellular fusiform membrane compartments, even in the presence of cycloheximide, a compound that inhibits protein biosynthesis (Geldner *et al.*, 2001). The effect of BFA on PIN localisation is reversible and is also observed for AtPIN2 and AtPIN3 (Geldner *et al.*, 2003; Friml *et al.*, 2002b). From these experiments it can be concluded that AtPIN proteins are cycling rapidly between the plasma membrane and endomembrane compartments. The physiological relevance of such cycling might be to control tightly the number of AtPIN molecules at the plasma membrane, thus ensuring fine-tuning of auxin efflux. This would be analogous to mammalian GLUT4 and its substrate glucose, controlling intracellular sugar levels (for a review see Muday & Murphy, 2002). A crucial factor in distributing AtPIN1 between endosomes and the plasma membrane is the ArfGEF GNOM protein. AtPIN1 is localised nonpolarly in *gnom*-embryos, which show severely disturbed apical-basal axis formation (Steinmann *et al.*, 1999). GNOM action seems to be specific for AtPIN proteins, as *gnom*-mutants show no general defects in cytokinesis or pollen tube growth (Bednarek & Fabel, 2002; Parton *et al.*, 2001). A mutation engineered in the Sec7 domain of GNOM confers resistance of polar auxin transport to BFA in transgenic *Arabidopsis*, and also abolishes cycling of AtPIN1, but not of other marker proteins. This experiment demonstrates the specific dependency of polar AtPIN1 localisation on GNOM-mediated vesicle trafficking (Geldner *et al.*, 2003).

It is not yet clear where in the cell the decision is made as to which part of the plasma membrane AtPINs are targeted, but a rather specific sorting mechanism is expected, because different plasma membrane proteins may show different localisations in the same root meristem cell. The protein COBRA, for example, is localised at the lateral plasma membranes, the putative Auxin Influx Carrier AUX1 at the basal PM, and AtPIN1 predominantly at the apical plasma membrane (Schindelmann *et al.*, 2001; Swarup *et al.*, 2001). It is not clear yet to which extent specific proteins are targeted directly from the *trans*-Golgi network (TGN) to the plasma membrane, or via endosomes. For AtPIN1 the latter route is supposed to be more relevant, as indicated by the BFA-induced intracellular PIN localisation in the presence of cycloheximide (Geldner *et al.*, 2001). Targeted secretion to the plasma membrane in plants is probably mediated by clathrin-coated vesicles (Lam *et al.*, 2001). Membrane docking of clathrin-coated vesicles in mammalian cells is mediated by synaptobrevin (v-SNARE) and syntaxin (t-SNARE) interactions, leading to the fusion of vesicles with the plasma membrane. Matching v-SNARE/t-SNARE pairs determine the place of fusion events (for a review see Chen *et al.*, 2001). In polarised mammalian cells, either basolateral or apical localisation of specific syntaxins is required for correct sorting of marker proteins (Low *et al.*, 1998). The *Arabidopsis* genome encodes 24 putative syntaxins and a comprehensive analysis of their intracellular localisation will be interesting (Sanderfoot *et al.*, 2000). Plant cells lack a plasma membrane diffusion barrier, analogous to the tight junctions in polarised epithelial mammalian cells, which prevent mixing of apical and baso-lateral proteins. This might indicate the existence of a plant-specific mechanism to avoid mixing of polarly localised membrane components.

Localisation of plasma membrane proteins in mammals can be ensured by lipid-rafts. These are sphingolipid-cholesterol enriched microdomains whose protein content does not diffuse into the plasma membrane. The concentration of sterols determines the integrity of the raft (for a review see Maxfield, 2002). Cholesterol is important in vesicle transport and the establishment of cell polarity in *Caenorhabditis* (Micheaux *et al.*, 2000). Similar mechanisms probably also exist in plants: mutations in the *Arabidopsis* gene *SMT1*, which is required for sterol biosynthesis, result in an altered membrane sterol composition as well as a reduced rate of polar auxin transport in shoots (Diener *et al.*, 2000; Willemsen *et al.*, 2003). Consistent with an effect on polar auxin transport, AtPIN1 and AtPIN3 are mislocalised. However, AUX1 localisation is unchanged in the *smt1* mutant, suggesting different sorting mechanisms operate for PIN and AUX1, respectively (Willemsen *et al.*, 2003). Since membrane fluidity in *smt1* mutants is unaltered, polarity defects might involve changes in lipid-rafts. Although lipid rafts have not yet been shown to exist in plants, the COBRA protein contains a GPI-anchor, which is a widespread mechanism to keep proteins in lipid rafts (Willemsen *et al.*, 2003; Schindelmann *et al.*, 2001; Keller & Simons, 1997). Also, AtPIN2 co-localises with sterols in 'BFA-patches', underlining the importance of sterols for AtPIN protein sorting (Grebe *et al.*, 2003).

There is also evidence for intrinsic protein sorting mechanisms of AtPINs. Ectopically expressed AtPIN1 in root epidermal cells is located to the apical side of the cell (our unpublished results), whereas AtPIN2 is at the basal end (Müller *et al.*, 1998). C-Terminal sorting signals are known that determine apical or basolateral localisation in polarised mammalian cells. These sequences are not highly conserved and do not work ubiquitously in different polarised cell types (for a review see Keller & Simons, 1997). Confusingly, AtPIN1 and AtPIN2 co-localise in neighbouring cortex cells at the basal end, suggesting cell type-dependent AtPIN sorting may also occur (our unpublished results).

Proteins that directly mediate AtPIN targeting have not yet been discovered, but there is evidence that ABC transporters are needed for AtPIN1 trafficking: *Arabidopsis* knockouts in *MDR1* and double knockouts in *MDR1* and *PGP1* show pleiotropic defects and reduced polar auxin transport in shoots (Noh *et al.*, 2001). In addition, AtPIN1 is mislocalised. Instead of a polar, basal AtPIN1 localisation as in wildtype, it is distributed around the whole PM and diffusely in the cytosol (Noh *et al.*, 2003). Whether MDR1 and PGP1 have a direct and specific effect on AtPIN1 trafficking is not yet clear. Clearly, more functional studies will be required to understand membrane trafficking and establishment of cell polarity in plants.

Acknowledgements

We are grateful to Dr William Teale for comments and careful reading of the manuscript. We also gratefully acknowledge the support of the DFG as well as The Fifth Framework Programme of the European Community – 'Quality of Life and Management of Living Resources'.

References

Agrawal, G.K., Iwahashi, H. & Rakwal, R. (2003) Small GTPase Rop: molecular switch for plant defense responses. *FEBS Lett.*, **546**, 173–180.

Arabidopsis Genome Initiative (AGI) (2000) Analysis of the genome sequence of the flowering plant *Arabidopsis thaliana*. *Nature*, **408**, 796–815.

Baluška, F., Barlow, P.W. & Volkmann, D. (2000a) Actin and myosin VII in developing root cells, in *Actin: A dynamic framework for multiple plant cell functions* (eds. C.J. Staiger, F. Baluška, D. Volkmann & P.W. Barlow). Kluwer Academic: Dordrecht, pp. 457–476.

Baluška, F., Salaj, J., Mathur, J., Braun, M., Jasper, F., Samaj, J., Chua, N.-H., Barlow, P.W. & Volkmann, D. (2000b) Root hair formation: F-actin dependent tip growth is initiated by a local assembly of profilin-supported F-actin meshworks accumulated within expansin-enriched bulges. *Dev. Biol.*, **227**, 618–632.

Baluška, F., Hlavacka, A., Samaj, J., Palme, K., Robinson, D.G., Matoh, T., McCurdy, D. W., Menzel, D. & Volkmann, D. (2002) F-actin-dependent endocytosis of cell wall pectins in meristematic root cells. Insights from Brefeldin A-induced compartments. *Plant Physiol.*, **130**, 422–431.

Baluška, F, Samaj, J. & Menzel, D. (2003a) Polar transport of auxin: carrier-mediated flux across the plasma membrane or neurotransmitter-like secretion. *Trends Cell Biol.*, **13**, 282–285.

Baluška F., Woytaszek, P., Volkmann, D. & Barlow, P. (2003b) The architecture of polarized cell growth: the unique status of elongating plant cells. *Bioessays*, **25**, 569–576.

Bao, Y, Kost, B. & Chua, N-H. (2001) Reduced expression of α-tubulin genes in *Arabidopsis thaliana* specifically affects root growth and morphology, root hair development and root gravitropism. *Plant J.*, **28**, 145–157.

Baumberger, N., Ringli, C. & Keller, B. (2001) The chimeric leucine-rich repeat/extensin cell wall protein LRX1 is required for root hair morphogenesis in *Arabidopsis thaliana*. *Genes Dev.*, **15**, 1128–1139.

Baxter-Burrell, A., Yang, Z., Springer, P.S. & Bailey-Serres, J. (2002) RopGap4-dependent Rop GTPase rheostat control of *Arabidopsis* oxygen deprivation tolerance. *Science*, **296**, 2026–2028.

Bednarek, S. Y. & Falbel, T. G. (2002) Membrane trafficing during plant cytokinesis. *Traffic*, **3**, 621–629.

Bernhardt, C. & Tierney, M.L. (2000) Expression of AtPRP3, a proline-rich structural cell wall protein from *Arabidopsis,* is regulated by cell-type-specific developmental pathways involved in root hair formation. *Plant Physiol.*, **122**, 705–714.

Bibikova T. & Gilroy S. (2002) Root hair development. *J. Plant Growth Reg.*, **21**, 383–415.

Bibikova, T.N. & Gilroy, S. (1998) Expanding tip-growth theory – reply. *Trends Plant Sci.*, **3**, 42.

Bibikova, T. & Gilroy, S. (2000) Calcium in root hair growth, in *Root hairs: Cell and Molecular Biology* (R.W. Ridge & A.M.C. Emons). Springer-Verlag: Tokyo, pp. 141–164.

Bibikova, T.N., Zhigilei, A. & Gilroy, S. (1997) Root hair growth in *Arabidopsis thaliana* is directed by calcium and an endogenous polarity. *Planta*, **203**, 495–505.

Bibikova, T. N., Jacob, T., Dahse, I. & Gilroy, S. (1998) Localized changes in apoplastic and cytoplasmic pH associated with root hair development in *Arabidopsis thaliana. Development*, **125**, 2925–2934.

Bibikova, T. N., Blancaflor, E.B. & Gilroy, S. (1999) Microtubules regulate tip growth in *Arabidopsis* root hairs. *Plant J.*, **17**, 657–665.

Bischoff, F., Vahlkamp, L., Molendijk, A. & Palme, K. (2000) Localization of AtRop4 and AtRop6 and interaction with the guanine nucleotide dissociation inhibitor AtRhoGDI1 from *Arabidopsis*. *Plant Molec. Biol.*, **42**, 515–530.

Bishop, A.L. & Hall, A. (2000) Rho GTPases and their effector proteins. *Biochem. J.*, **348**, 241–255.

Borg, S., Podenphant, L., Jensen, T.J. & Poulsen, C. (1999) Plant cell growth and differentiation may involve GAP regulation of Rac activity. *FEBS Lett.*, **453**, 341–345.

Braun, M., Baluška, F., von Witsch, M. & Menzel, D. (1999) Redistribution of actin, profilin and phosphatidylinositol-4,5-bisphosphate in growing and maturing root hairs. *Planta*, **209**, 435–443.

Brugnera, E., Haney, L., Grimsley, C., Lu, M.J., Walk, S.F., Tosello-Trampont, A.C., Macara, I.G., Madhani, H., Fink. G.R. & Ravichandran, K.S. (2002) Unconventional Rac-GEF activity is mediated through the Dock180-ELMO complex. *Nature Cell Biol.*, **4**, 574–582.

Cavell, A. & Grierson, C.S. (2000) Genetics of root hairs development, in *Root Hairs: Cell and Molecular Biology* (eds. R.W. Ridge & A.M.C. Emons). Springer-Verlag: Tokyo, pp. 211–222.

Chen, C.Y., Cheung, A.Y. & Wu, H. (2003) Actin-Depolymerizing Factor mediates Rac/Rop GTPase-regulated pollen tube growth. *Plant Cell*, **15**, 237–249.

Chen, Y.A. & Scheller, R.H. (2001) SNARE-mediated membrane fusion. *Nature Rev. Mol. Cell Biol.*, **2**, 98–106.

Cheung, A.Y., Chen, C., Y., Tao, L., Andreyeva, T., Twell, D. & Wu, H. (2003) Regulation of pollen tube growth by Rac-like GTPases. *J. Exp. Bot.*, **54**, 73–81.

Cho, H-T. & Cosgrove, D.J. (2002) Regulation of root hair initiation and expansin gene expression in *Arabidopsis. Plant Cell*, **14**, 3237–3253.

Cove, D.J. (2000) The generation and modification of cell polarity. *J. Exp. Bot.*, **51**, 831–838.

Cove, D.J., Hope, I.A. & Quatrano, R.S. (1999) Polarity in biological systems, in *Development. Genetics, Epigenetics and Environmental Regulation* (eds. Russo, V.E.A., Cove, D.J., Edgar, L.G., Jaenisch, R. & Salamini, F.), Springer-Verlag: Berlin, pp. 507–524.

Deeks, M.J., Hussey, P.J. & Davies, B (2002) Formins: intermediates in signal-transduction cascades that affect cytoskeletal reorganisation. *Trends Plant Sci.*. **7**, 492–498.

Diener, A.C., Li, H., Zhou, W., Whoriskey, W.J., Nes, W.D. & Fink G.R. (2000) Sterol methyltransferase 1 controls the level of cholesterol in plants. *Plant Cell*, **12**, 853–870.

Dolan, L. (2001) How and where to build a root hair. *Curr. Opin. Plant Biol.*, **4**, 550–554.

Drees, B.L., Sundin, B., Brazeau, E., Caviston, J.P., Chen, G.C., Guo, W., Kozminski, K.G., Lau, M.W., Moskow, J.J., Tong, A., Schenkman, L.R., McKenzie, A., Brennwald, P., Longtine, M., Bi, E., Chan, C., Novick, P., Boone, C., Pringle, J.R., Davis, T.N., Fields, S. & Drubin, D.G. (2001) A protein interaction map for cell polarity development. *J. Cell Biol.*, **154**, 549–571.

Dubrin, D. (2000) *Cell Polarity*. Oxford University Press, Oxford.

Dubrin, D.G. & Nelson, W.J. (1996) Origins of cell polarity. *Cell*, **84**, 335–344.

Esseling J., de Ruijter N. & Emons, A.M.C. (2000) The root hair actin cytoskeleton as backbone, highway, morphogenetic instrument and target for signalling. In *Root Hairs: Cell and Molecular Biology* (eds. R.W. Ridge & A.M.C. Emons). Springer-Verlag: Tokyo, pp. 29–52.

Favery, B., Ryan, E., Foreman, J., Linstead, P., Boudonck, K., Steer, M., Shaw, P. & Dolan, L. (2001) *KOJAK* encodes a cellulose synthase-like protein required for root hair cell morphogenesis in *Arabidopsis*. *Genes Dev*., **15**, 79–89.

Foreman, J., Demidchik, V., Bothwell, J.H.F., Mylona, P., Miedema, H., Torres, M.A., Linstead, P., Costa, S., Brownlee, C., Jones, J.D.G., Davies, J.M. & Dolan, L. (2003) Reactive oxygen species produced by NADPH oxidase regulate plant cell growth. *Nature*, **422**, 442–446.

Fowler, J.E. & Quatrano, R.S. (1997) Plant cell morphogenesis. *Ann. Rev. Cell Dev. Biol.*, **13**, 697–743

Frank M. J. & Smith, LG. (2002) A small, novel protein highly conserved in plants and animals promotes the polarized growth and division of maize leaf epidermal cells. *Curr. Biol.*, **12**, 849–853.

Frank, M.J., Cartwright, H.N. & Smith, L.G. (2003) Three Brick genes have distinct functions in a common pathway promoting polarized cell division and cell morphogenesis in the maize leaf epidermis. *Development*, **130**, 753–762.

Friml, J. & Palme, K. (2002) Polar auxin transport old questions and new concepts? *Plant Molec. Biol.*, **49**, 273–284.

Friml, J., Benkova, E., Ikram, B., I.,Wisniewska J., Hamann, T., Ljung, K., Wood, S., Sandberg, G., Scheres, B. & Palme K. (2002a) AtPIN4 mediates sink-driven gradients and root patterning in *Arabidopsis*. *Cell*, **108**, 661–673.

Friml, J., Wisniewska J., Benkova, E., Mendgen, K. & Palme K. (2002b) Lateral relocation of auxin efflux regulator PIN3 mediates tropism in *Arabidopsis*. *Nature*, **415**, 806–809.

Fu, Y. & Yang, Z. (2001) Rop GTPase: a master switch of cell polarity development in plants. *Trends Plant Sci.*, **6**, 545–547.

Fu, Y., Wu, G. & Yang, Z. (2001) Rop GTPase-dependent dynamics of tip-localized F-actin controls tip growth in pollen tubes. *J. Cell Biol.*, **152**, 1019–1032.

Fu, Y., Li, H. & Yang, Z. (2002) The Rop2 GTPase controls the formation of cortical fine F-actin and the early phase of directional cell expansion during *Arabidopsis* organogenesis. *Plant Cell*, **14**, 777–794.

Gälweiler, L., Guan, C., Müller, A., Wisman, E., Mendgen, K., Yephremov, A. & Palme, K. (1998) Regulation of polar auxin transport by AtPIN1 in *Arabidopsis* vascular tissue. *Science*, **282**, 2226–2230.

Geitmann A. & Emons, A.M.C. (2000) The cytoskeleton in plant and fungal cell tip growth. *J. Microscopy* 198, 218–245.

Geldner, N., Friml, J., Stierhof, Y.D., Jürgens, G. & Palme, K. (2001) Auxin transport inhibitors block PIN1 cycling and vesicle traficking. *Nature*, **413**, 425–428.

Geldner, N., Anders, N. Wolters, H., Keicher, J., Komberger, W., Muller, P., Delbarre, A., Ueda, T., Nakano, A. & Jürgens, G. (2003) The *Arabidopsis* GNOM ARF-GEF mediates endosomal recycling, auxin transport, and auxin-dependent plant growth. *Cell*, **112**, 219–230.

Gilliland, L.U., Kandasamy, M.K., Pawloski, L.C. & Meagher, R.B. (2002) Both vegetative and reproductive actin isovariants complement the stunted root hair phenotype of the *Arabidopsis* act2-1 mutation. *Plant Physiol*., **130**, 2199–2209.

Godde, M., Tretyn, A., Uhl, R. & Palme, K. (1999) Novel tools for elucidation of gene functions – An array of genetic and microscopic possibilities in plant physiology. In *Advances in Regulation of Plant Growth and Development* (eds. Strnad, M., Pec, P. & Breck, E.), Peres Publishers: Prague, pp. 7–22.

Grebe, M., Friml, J., Swarup, R., Ljung, K., Sandberg, G., Terlou, M., Palme, K., Bennett, M.J. & Scheres, B. (2002) Cell polarity signaling in *Arabidopsis* involves a BFA-sensitive auxin influx pathway. *Curr. Biol*., **12**, 329–334.

Grebe, M., Xu, J., Möbius, W., Ueda, T., Nakano, A., Geuze, H.J., Rook, M.B. & Scheres, B. (2003) *Arabidopsis* sterol endocytosis involves actin-mediated trafficking via ARA6-positive early endosomes. *Curr. Biol.*, **13**, 1378–1387.

Gu, Y., Vernoud, V., Fu, Y. & Yang, Z. (2003) Rop GTPase regulation of pollen tube growth through the dynamics of tip-localized F-actin. *J. Exp. Bot.*, **54**, 93–101.

Hall, A. (1998) Rho GTPases and the actin cytoskeleton. *Science*, **279**, 509–514.

Hepler, P.K., Vidali, L. & Cheung, A.Y. (2001) Polarized cell growth in higher plants. *Ann. Rev. Cell Dev. Biol.*, **17**, 159–187.

Hoffman, G.R. & Cerione, R.A. (2000) Flipping the switch: The structural basis for signalling through the CRIB motif. *Cell*, **102**, 403–406.

Hülskamp, M., Folkerts, U. & Grini, P.E. (1998) Cell morphogenesis in *Arabidopsis. BioEssays*, **20**, 20–29.

Jiang, C-J., Weeds, A.G. & Hussey, P.J. (1997) The maize actin-depolymerizing factor ZmADF3 redistributes to the growing tip of elongating root hairs and can be induced to translocate into the nucleus with actin inhibitors. *Plant J.*, **12**, 1035–1043.

Johnson, D.I. (1999) Cdc42: An essential Rho-type GTPase controlling eukaryotic cell polarity. *Microbiol. Mol. Biol. Rev.*, **63**, 54–105.

Jones, M.A., Shen, J., Fu, Y., Li, H., Yang, Z. & Grierson, C.S. (2002) The *Arabidopsis* Rop2 GTPase is a positive regulator of both root hair initiation and tip growth. *Plant Cell*, **14**, 763–776.

Jürgens, G. & Geldner, N. (2002) Protein secretion in plants: From the trans-Golgi network to the outer space. *Traffic*, **3**, 605–613.

Kawasaki, T., Henmi, K., Ono, E., Hatakeyama, S., Iwano M., Satoh, H. & Shimamoto, K. (1999) The small GTP-binding protein Rac is a regulator of cell death in plants. *Proc. Natl. Acad. Sci. USA*, **96**, 10922–10926.

Keller, P. & Simons, K. (1997) Post-Golgi biosynthetic trafficing. *J. Cell Sci.*, **110**, 3001–3009.

Ketelaar, T. & Emons, A.M.C. (2000) The role of microtubules in root hair growth and cellulose microfibril deposition. in *Root Hairs: Cell and Molecular Biology* (eds. R.W. Ridge & A.M.C. Emons). Springer-Verlag: Tokyo, pp. 17–18.

Ketelaar, T. & Emons, A.M.C. (2001) The cytoskeleton in plant cell growth: lessons from root hairs. *New Phytol.*, **152**, 409–418.

Ketelaar, T., Faivre-Moskalenko, C., Esseling, J.J., de Ruijter, N.C.A., Grierson, C.S., Dogterom, M. & Emons, A.M.C. (2002) Position of nuclei in *Arabidopsis* root hairs: An actin-regulated process of tip growth. *Plant Cell*, **14**, 2941–2955.

Ketelaar, T., de Ruijter, N.C.A. & Emons, A.M.C. (2003) Unstable F-actin specifies the area and microtubule direction of cell expansion in *Arabidopsis* root hairs. *Plant Cell*, **15**, 285–292.

Kjøller, L. & Hall, A. (1999) Signaling to Rho GTPases. *Exp. Cell Res.*, **253**, 166–179.

Kost, B., Lemichez, E., Spielhofer, P., Hong, Y., Tolias, K., Carpenter, C. & Chua, N-H. (1999) Rac homologues and compartmentalized phospatidylinositol 4,5-bisphospate act in a common pathway to regulate polar pollen tube growth. *J. Cell Biol.*, **145**, 317–330.

Kropf, D.L., Bisgrove, S.R. & Hable, W.E. (1998) Cytoskeletal control of polar growth in plant cells. *Curr. Opin. Cell Biol.*, **10**, 117–122.

Lam, B. C., Sage, T. L., Bianchi, F. & Blumwald, E. (2001) Role of SH3 domain-containing proteins in clathrin-mediated vesicle trafficking in *Arabidopsis. Plant Cell,* **13**, 2033–2051.

Lauber, M.H., Wiaznegger, I., Steinmann, T., Schwartz, H., Mayer, U., Hwang, I., Lukowitz, W. & Jürgens, G. (1997) The *Arabidopsis* KNOLLE protein is a cytokinesis specific-syntaxin. *J. Cell. Biol.*, **139**, 1485–1493.

Lavy, M., Bracha-Drori, K., Sternberg, H. & Yalovsky, S. (2002) A cell-specific, prenylation-independent mechanism regulates targeting of type II RACs. *Plant Cell*, **14**, 2431–2450.

Lemichez, E., Wu, Y., Sanchez, J-P., Mettouchi, A., Mathur, J. & Chua, N-H. (2001) Inactivation of AtRac1 by abscisic acid is essential for stomatal closure. *Genes Dev.*, **15**, 1808–1816.

Li, H., Lin, Y.K., Heath, R.M., Zhu, M.X. & Yang, Z.B. (1999) Control of pollen tube tip growth by a Rop GTPase dependent pathway that leads to tip-localized calcium influx. *Plant Cell*, **11**, 1731–1742.

Li, H., Shen, J-J., Zheng, Z-L., Lin, Y. & Yang, Z. (2001) The Rop GTPase switch controls multiple developmental processes in *Arabidopsis. Plant Physiol.*, **126**, 670–684.

Lin, Y. & Yang, Z. (1997) Inhibition of pollen tube elongation by microinjected anti-Rop1Ps antibodies suggests a crucial role for Rho-type GTPases in the control of tip growth. *Plant Cell*, **9**, 1674–1659.

Lin, Y., Wang, Y., Zhu, J. & Yang, Z. (1996) Localization of Rho GTPase implies a role in tip growth and movement of the generative cell in pollen tubes. *Plant Cell*, **8**, 293–303.

Low, S.H., Roche P.A., Anderson, H.A., van Ijzendoorn S.C., Zhang, M., Mostov K.E. &

Weimbs T. (1998) Targeting of SNAP-23 and SNAP-25 in polarized epithelial cells. *J. Biol. Chem.*, **273**, 3422–3430.

Malamy, J.E. & Benfey, P.N. (1997) Organization and cell differentiation in lateral roots of *Arabidopsis thaliana*. *Development*, **124**, 33–44.

Mathur, J. & Hülskamp, M. (2001) Cell growth: How to grow and where to grow. *Curr. Biol.*, **11**, R402–R404.

Mathur, J. & Hülskamp, M. (2002a) Microtubules and microfilaments in cell morphogenesis in higher plants. *Curr. Biol.*, **12**, R669–R676.

Mathur, J. & Hülskamp, M. (2002b) Signal transduction: Rho-like proteins in plants. *Curr. Biol.*,**12**, R526–R528.

Mathur, J., Mathur, N., Kernebeck, B. & Hülskamp, M. (2003a) Mutations in actin-related proteins 2 and 3 affect cell shape development in *Arabidopsis*. *Plant Cell*, **15**, 1632–1645.

Mathur, J., Mathur, N., Kirik, V., Kernebeck, B., Srinivas, B. & Hülskamp, M. (2003b) *Arabidopsis CROOKED* encodes for the smallest subunit of the ARP2/3 complex and controls cell shape by region specific F-actin formation. *Development*, **130**, 3137–3146.

Maxfield, F.R. (2002) Plasma membrane microdomains. *Curr. Opin. Cell Biol.*, **14**, 483–487.

Meller, N., Irani-Tehrani, M., Kiosses, W.B., Del Pozo, M.A. & Schwartz, M.A. (2002) Zizimin1, a novel CDC42 activator, reveals a new GEF domain for Rho proteins. *Nature Cell Biol.*, **4**, 639–647.

Micheaux, G., Gansmüller, A., Hindelang, C. & Labouesse, M. (2000) CHE-14, a protein with a sterol-sensing domain is required for apical sorting in *C. elegans* ectodermal epithelial cells. *Curr. Biol.*, **10**, 1098–1107.

Miller, D.D., de Ruijter, N.C.A. & Emons, A.M.C. (1997) From signal to form: aspects of the cytoskeleton-plasma membrane-cell wall continuum in root hair tips. *J. Exp. Bot.*, **316**, 1881–1896.

Miller, D.D., de Ruijter, N.C.A., Bisseling, T. & Emons, A.M.C. (1999) The role of actin in root hair morphogenesis: studies with lipochito-oligosaccharide as a growth stimulator and cytochalasin as an actin perturbing drug. *Plant J.*, **17**, 141–154.

Molendijk, A.J., Bischoff, F., Rajendrakumar, C.S.V., Friml, J., Braun, M., Gilroy, S. & Palme, K. (2001) *Arabidopsis thaliana* Rop GTPases are localized to tips of root hairs and control polar growth. *EMBO J.*, **20**, 2779–2788.

Mølhøj, M., Jørgensen, B., Ulvskov, P. & Borkhardt, B. (2001) Two *Arabidopsis thaliana* genes, *KOR2* and *KOR3*, which encode membrane-anchored endo-1,4-β-D-glucanases, are differentially expressed in developing leaf trichomes and their supprt cells. *Plant Molec. Biol.* **46**, 263–275.

Müller, A., Guan, C., Gälweiler, L., Tänzler, P., Huijser, P., Marchant, A., Parry, G., Bennett, M., Wisman, E. & Palme, K. (1998) AtPIN2 defines a locus of *Arabidopsis* for root gravitropism control. *EMBO J.*, **17**, 6903–6911.

Muday, G. & Murphy, A. (2002) An emerging model for auxin transport regulation. *Plant Cell*, **14**, 293–297.

Neill, S., Desikan, R. & Hancock, J. (2002) Hydrogen peroxide signalling. *Curr. Opin. Plant Biol.*, **5**, 398–395.

Noh, B., Murphy, A.S. & Spalding, E.P. (2001) Multidrug resistance-like genes of *Arabidopsis* required for auxin transport and auxin-mediated development. *Plant Cell*, **13**, 2441–2454.

Noh, B., Bandyopdhyay, A., Peer, W.A., Spalding, E.P. & Murphy, A.S. (2003) Enhanced gravi- and phototropism in plant *mdr* mutants mislocalizing the auxin efflux protein PIN1. *Nature*, **423**, 999–1002.

Ohashi, Y., Oka, A., Rodriguez-Pousada, R., Possenti, M., Ruberti, I., Morelli, G. & Aoyama, T. (2003) Modulation of phospholipid signaling by GLABRA2 in root-hair pattern formation. *Science*, **300**, 1427–1430.

Olofsson, B. (1999) Rho guanine dissociation inhibitors: Pivotal molecules in cellular signalling. *Cell. Signal.*, **11**, 545–554.

Ono, E., Wong, H.L., Kawasaki, T., Hasegawa, M., Kodama, O. & Shimamoto, K. (2001) Essential role of the small GTPase Rac in disease resistance of rice. *Proc. Natl. Acad. Sci. USA*, **98**, 759–764.

Ottenschläger, I., Wolff, P., Wolverton, C., Bhalerao, R.P., Sandberg, G., Ishikawa, H., Evans, M. & Palme K. (2003) Gravity-regulated differential auxin transport from columella to lateral root cap cells. *Proc. Natl. Acad. Sci. USA*, **100**, 2987–2991.

Ovečka M., Nadubinska, M., Volkmann, D. & Baluška, F. (2000) Actomyosin and exocytosis inhibitors alter root hair morphology in *Poa annua*. *Biologia*, **55**, 105–114.

Palme, K. & Gälweiler, L. (1999) PIN-pointing the molecular basis of auxin transport. *Curr. Opin. Plant Biol.*, **2,** 375–381.

Park, J., Choi, H-J., Lee, S., Lee, T., Yang, Z. & Lee, Y. (2000) Rac-related GTP-binding protein in elicitor-induced reactive oxygen generation by suspension-cultured soybean cells. *Plant Physiol.*, **124**, 725–732.

Parton, R.M., Fischer-Parton, S., Watahiki, M.K. & Trewavas, A.J. (2001) Dynamics of the apical vesicle accumulation and the rate of growth are related in individual pollen tubes. *J. Cell Sci.*, **11**, 2685–2695.

Peyroche, A., Antonny, B., Robineau, S., Acker, J., Cherfils, J. & Jackson, C.L. (1999) Brefeldin A acts to stabilise an abortive ARF-GDP-Sec7 domain protein complex: involvement of specific residues to the Sec7 domain. *Mol. Cell*, **3**, 275–285.

Pierson, E.S, Miller, D.D., Callaham, D.A., Van Aken, J., Hackett, G. & Hepler, P.K. (1996) Tip-localized calcium entry fluctuates during pollen tube growth. *Dev. Biol.*, **174**, 160–173.

Potikha, T., Collins, C.C., Johnson, D.I., Delmer, D.P. & Levine, A. (1999) The involvement of hydrogen peroxide in the differentiation of secondary walls in cotton fiber. *Plant Physiol.*, **119**, 849–858.

Qiu, J.L., Jilk, R., Marks, M.D. & Szymanski, D.B. (2002) The *Arabidopsis SPIKE1* gene is required for normal cell shape control and tissue development. *Plant Cell*, **14**, 101–118.

Ramachandran, S., Christensen, H.E.M., Ishimaru, Y., Dong, C-H., Chao-Ming, W., Cleary, A.L. & Chua, N-H. (2000) Profilin plays a role in cell elongation, cell shape maintenance, and flowering in *Arabidopsis*. *Plant Physiol.*, **124**, 1637–1674.

Rashotte, A.M., Brady, S.R., Reed, R.C., Ante, S.J. & Muday, G.K. (2000) Basipetal auxin transport is required for gravitropism in roots of *Arabidopsis*. *Plant Physiol.*, **122**, 481–490.

Rashotte, A.M., DeLong, A. & Muday, G.K. (2001) Genetic and chemical reductions in protein phosphatase activity alter auxin transport, gravity response and lateral root growth. *Plant Cell*, **13**, 1683–1697.

Ridge, R.W. & Emons, A.M.C. (2000) *Root hairs: Cell and Molecular Biology*. Springer-Verlag: Tokyo.

Ringli, C., Baumberger, N., Diet, A., Frey, B. & Keller, B. (2002) Actin2 is essential for bulge site selection and tip growth during root hair development of *Arabidopsis*. *Plant Physiol.*, **129**, 1464–1472.

Sanderfoot, A.A., Assaad, F.F. & Raikhel, N.V. (2000) An abundance of soluble N-ethylmaleimide-sensitive factor adaptor protein receptor. *Plant Physiol.*, **124**, 1558–1569.

Sata, M., Moss, J. & Vaughan, M. (1999) Structural basis for the inhibitory effect of brefeldin A on guanine nucleotide-exchange proteins for ADP-ribosylation factors. *Proc. Natl. Acad. Sci. USA*, **96**, 2752–2757.

Scheres, B. & Benfey, P.N. (1999) Asymmetric cell division. *Annu. Rev. Plant Physiol. Plant Mol.Biol.*, **50**, 505–537.

Schiefelbein, J.W. (2000) Constructing a plant cell. The genetic control of root hair development. *Plant Physiol.*, **124**, 1525–1531.

Schiene, K., Pühler, A. & Niehaus, K. (2000) Transgenic tobacco plants that express an antisense construct derived from a *Medicago sativa* cDNA encoding a Rac-related small GTP-binding protein fail to develop necrotic lesions upon elicitor infiltration. *Mol. Gen. Genet.*, **263**, 761–770.

Schindelmann G., Morikarni, A., Jung, J., Baskin, T.L., Carpita, N.C., Derbyshire, P., McCann, M.C. & Benfey, P.N. (2001) COBRA encodes a putative GPI-anchored protein, which is polarly localized and necessary for oriented cell expansion in *Arabidopsis*. *Genes Dev.*, **15**, 1115–1127.

Schmelzer, E. (2002) Cell polarization, a crucial process in fungal defence. *Trends Plant Sci.*, **7**, 411–415.

Schmidt, A. & Hall, M.N. (1998) Signaling to the actin cytoskeleton. *Ann. Rev. Cell Dev. Biol.*, **14**, 305–338.

Schultheiss, H., Dechert, C., Kogel, K. & Hückelhoven, R. (2002) A small GTP-binding host protein is required for entry of powdery mildew fungus into epidermal cells of Barley. *Plant*

Physiol., **128**, 1447–1454.
Shah, K., Russinova, E., Gadella, T.W.J., Willemse, J. & de Vries, S.C. (2002) The *Arabidopsis* kinase-associated protein phosphatase controls internalization of the somatic embryogenesis receptor kinase 1. *Genes Dev.*, **16**, 1707–1720.
Shaw, S.L., Dumais, J. & Long, S.R. (2000) Cell surface expansion in polarly growing root hairs of *Medicago trunculata. Plant Physiol.*, **124**, 959–969.
Sherrier, D.J. & Vandenbosch, K.A. (1994) Secretion of cell wall polysaccharides in *Vicia* root hairs. *Plant J.*, **5**, 185–195.
Shiu, S-H. & Bleecker, A.B. (2001) Receptor-like kinases from *Arabidopsis* form a monophyletic gene family related to animal receptor kinases. *Proc. Natl. Acad. Sci. USA*, **98**, 10763–10768.
Shulman, J.M. & St. Johnston, D. (1999) Pattern formation in single cells. *Trends Cell Biol.,* **9**, M60–M64.
Smith, L.G. (2003) Cytoskeletal control of plant shape: getting the fine points. *Curr. Opin. Plant Biol.*, **6**, 63–73.
Spiliotis, E.T. & Nelson, J. (2003) Spatial control of exocytosis. *Curr. Opin. Cell Biol.*, **15**, 430–437.
Steinmann, T., Geldner, N., Grebe, M., Mangold, S., Jackson C. L., Paris S., Gälweiler, L., Palme, K. & Jürgens, G. (1999) Coordinated polar localization of auxin efflux carrier PIN1 by GNOM ARF GEF. *Science,* **286**, 316–318.
Staiger, C.J. (2000) Signaling to the actin cytoskeleton in plants. *Ann. Rev. Plant Mol. Biol.*, **51**, 257–288.
Suharsono, U., Fujisawa, Y., Kawasaki, T., Iwasaki, Y., Satoh, H. & Shimamoto, K. (2002) The heterotrimeric G protein alpha subunit acts upstream of the small GTPase Rac in disease resistance of rice. *Proc. Natl. Acad. Sci. USA*, **99**, 13307–13312.
Swarup, R., Friml, J., Marchant, A., Ljung, K., Sandberg, G., Palme K. & Bennett, M. (2001) Localisation of the auxin permease AUX1 suggests two functionally distinct hormone transport pathways operate in *Arabidopis* root apex. *Genes Dev.*, **15**, 2648–2653.
Swoboda, I., Bhalla, P.L., Xu, H., Zhang, Y., Mittermann, I., Valenta, R. & Singh, M.B. (2001) Identification of *pronp1*, a tobacco profilin gene activated in tip-growing cells. *Plant Molec. Biol.*, **46**, 531–538.
Tao, L., Cheung, A.Y. & Wu, H. (2002) Plant rac-like GTPases are activated by auxin and mediate auxin-responsive gene expression. *Plant Cell*, **14**, 2745–2760.
Trotochaud, A.E., Hao, T., Wu, G., Yang, Z. & Clark, S.E. (1999) The CLAVATA1 receptor-like kinase requires CLAVATA3 for the asembly into a signalling complex that includes KAPP and a Rho-related protein. *Plant Cell*, **11**, 393–405.
Valster, A.H., Hepler, P.K. & Chernoff, J. (2000) Plant GTPases: the Rhos in bloom. *Trends Cell Biol.*, **10**, 141–146.
Van Aelst, L. & D'Souza-Schorey, C. (1997) Rho GTPases and signaling networks. *Genes Dev.*, **11**, 2295–2322.
Vantard, M. & Blanchoin, L. (2002) Actin polymerization processes in plants. *Curr. Opin. Biol.*, **5**, 502–506.
Véry, A. & Davies, J.M. (2000) Hyperpolarization-activated calcium channels at the tip of *Arabidopsis* root hairs. *Proc. Natl. Acad. Sci. USA*, **97**, 9801–9806.
Vissenberg, K., Fry, S.C. & Verbelen, J. (2001) Root hair initiation is coupled to a highly localized increase of xyloglucan endotransglycosylase action in *Arabidopsis* roots. *Plant Physiol.*, **127**, 1125–1135.
Vöchting, H. (1878) *Über Organbildung im Pflanzenreich*. Cohen: Bonn.
Wang, H.Y., Lockwood, S.K., Hoeltzel, M.F. & Schiefelbein, J.W. (1997) The root hair defective3 gene encodes an evolutionary conserved protein with GTP-binding motifs and is required for regulated cell enlargement in *Arabidopsis*. *Genes Dev.*, **11**, 799–811.
Wang, H., Lee, M.M. & Schiefelbein, J.W. (2002) Regulation of the cell expansion gene *RHD3* during *Arabidopsis* development. *Plant Physio*l., **129**, 638–649.
Wang, X., Cnops, G., Vanderhaegen, R., De Block, S., Van Montagu, M. & Van Lijsebettens, M. (2001) *AtCSLD3*, A cellulose synthase-like gene important for root hair growth in *Arabidopsis*. *Plant Physiol.*, **126**, 575–586.
Willemsen, V., Friml, J., Grebe M., van den Toorn, A., Palme, K. & Scheres, B. (2003) Cell

polarity and PIN positioning in *Arabidopsis* require *STEROL METHYTRANSFERASE1* function. *Plant Cell*, **15**, 612–625.

Winge, P., Brembu, T., Kristensen, R. & Bones, A.M. (2000) Genetic structure and evolution of RAC-GTPases in *Arabidopsis thaliana*. *Genetics*, **156**, 1959–1971.

Wu, G., Li, H. & Yang, Z. (2000) Arabidopsis RopGAPs are a novel family of Rho GTPase-activating proteins that require the Cdc42/Rac-interactive binding motif for Rop-specific GTPase stimulation. *Plant Physiol.*, **124**, 1625–1636.

Wu, G., Gu, Y., Li, S. & Yang, Z. (2001) A genome-wide analysis of *Arabidopsis* Rop-interactive CRIB motif-containing proteins that act as Rop GTPase targets. *Plant Cell*, **13**, 2841–2856.

Wymer, C.L., Bibikova, T.N. & Gilroy, S. (1997) Cytoplasmic free calcium distributions during the development of root hairs of *Arabidopsis thaliana*. *Plant J.*, **12**, 427–439.

Yang, Z. (1998) Signaling tip growth in plants. *Curr. Opin. Plant Biol.*, **1**, 525–530.

Yang, Z. & Watson, J.C. (1993) Molecular cloning and characterization of rho, a ras-related small GTP-binding protein from the garden pea. *Proc. Natl. Acad. Sci. USA*, **90**, 8732–8736.

Zheng, Z-L. & Yang Z. (2000a) The Rop GTPase switch turns on polar growth in pollen. *Trends Plant Sci.*, **5**, 298–303.

Zheng, Z-L. & Yang, Z. (2000b) The Rop GTPase: an emerging signaling switch in plants. *Plant Molec. Biol.*, **44**, 1–9.

Zheng, Z-L., Nafisi, M., Tam, A., Li, H., Crowell, D.N., Chary, S.N., Schroeder, J.I., Shen, J. & Yang, Z. (2002) Plasma membrane-associated Rop10 small GTPase is a specific negative regulator of abscisic acid responses in *Arabidopsis*. *Plant Cell*, **14**, 2787–2797.

Ziman, M., Preuss, D., Mulholland, J., O'Brien, J.M., Botstein, D. & Johnson, D.I. (1993) Subcellular localization of Cdc42p, a *Saccharomyces cerevisiae* GTP-binding protein involved in the control of cell polarity. *Mol. Biol. Cell*, **4**, 1307–1316.

3 Polarity and cell walls

Przemysław Wojtaszek, Dieter Volkmann and František Baluška

3.1 Introduction

Polarity is defined as a general asymmetry, actively established and maintained by the cells. It can vary from differential distribution of cellular components, localisation of mRNAs, distribution and functioning of translated proteins and products of their activity through vesicle trafficking and directional relocation of cellular organelles, to differences in the shape, growth, and developmental patterns. Although it is predominantly a feature of individual cells, both pro- and eukaryotic, the polarity phenomena are also observed at higher levels of organisation in multicellular organisms (Drubin & Nelson, 1996; Nelson, 2003).

Plant cells are different from animal cells in that they are usually encased permanently within surrounding walls. From the structural and physiological points of view, they, together with fungal and many bacterial cells, can be classified as 'walled' cells as opposed to 'naked' cells of other organisms (Peters *et al.*, 2000; Baluška *et al.*, 2003c). This creates a rather specific situation, also with respect to cell polarity. In animals, polarity is related to the events occurring at their boundary, and the cytoskeleton seems to be a key player. In plants, although polar phenomena are occurring inside the cells, the visible, morphogenetic output of those is ultimately related to and maintained by the cell walls. Moreover, cell walls are not only metabolically active, dynamic cellular organelles. They are also at the same time embedded within a supracellular structure – the apoplast, and this makes the cell wall a unique linker between different levels of biological organisation (Wojtaszek, 2000, 2001). In other words, cell walls are located physically and metaphorically at the 'transition zone' from uni- to multicellularity and from cellular to organismal level.

Studies on the role of cell walls in establishment and maintenance of cellular polarity in plants seem generally not to be an easy topic. The formation of the walls involves the whole biochemical machinery of the protoplast (Wojtaszek, 2000). It is thus often observed that alterations at earlier stages of wall biosynthesis and assembly, not even in the walls itself, can lead to modified properties and malfunctions of the walls. Clear separation of events is not always possible. Molecular genetic approaches are starting to give some initial answers, although the analyses of *Arabidopsis* (Arabidopsis Genome Initiative, 2000) and rice (Goff *et al.*, 2002; Yu *et al.*, 2002a) genomes are showing that there is a relatively huge number of genes which might be related to wall functioning. However, their identification and proper annotation is hampered by the lack of known homologues from the other organisms. It seems thus, that at present, the approaches of functional genomics aimed at the full characterisation of individual genes would provide much better understanding of wall assembly and functioning.

3.2 Cell walls and turgor – the physiology of 'walled' cells

Living entities, living cells in particular, could be defined as autopoietic systems, i.e. systems able to maintain themselves through the network of interactions of their own internal components (Varela *et al.*, 1974; Fleischaker, 1988; Harold, 2001b). Two of the crucial criteria of autopoiesis are related to the boundaries of the unity thought to be 'autopoietic'. All living cells, both in uni- and multicellular organisms, have identifiable boundaries giving them identity. Moreover, these boundaries are constituted by the relations among cellular components, and not imposed from the outside which fulfils the criterion of self-boundingness (Fleischaker, 1988). This view coincides with that of Lintilhac (1999) who sees the living cell not as a histologically determined entity but reduces it to an 'event surface' separating externum from internum. This points to the most important event in the evolution of Life – the generation of the border, separating living units (cells) from their environment (Peters *et al.*, 2000) as a culmination of earlier prebiotic chemical events leading to vectorial biochemistry powered by sustainable sources of energy, and susceptible to natural selection (Harold, 2001a). It can be assumed that these borders were initially of a different, although chemically plausible, nature as long as they fulfilled their crucial function, i.e. maintenance of relatively constant internal environment with respect to cellular components irrespectively of chemical fluctuations in the external environment. Finally, amphiphilic lipid bilayers have been selected as natural cellular borders. Such a permeability barrier created, however, an osmotic gradient causing water to flow along its chemical potential into the cells (Money, 1997).

To be stabilised structurally, cells make use of the simplest mechanical design where the bordering membranes bear tensile stresses while the internal liquid – the cytosol – is under hydrostatic compression. However, lipid membranes are not very elastic and can be stretched by only about 2% without rupturing, which corresponds to the pressure difference of about 8000 Pa when cell radius is 5 μm (Boal, 2002). As the water influx creates high hydrostatic pressure this has to be somehow counterbalanced to prevent cell rupture. Two strategies for coping with the problem of 'death-by-flooding' (Stein, 1995) leading to cell rupture have emerged during the evolution. One of them is based on the active generation and maintenance of chemical gradients of osmotically active solutes leading to effective separation of various substances across cell membranes. This is exemplified by Na^{+}/K^{+}-ATPase commonly found in animal cells (Peters *et al.*, 2000). Another approach has found its fulfilment in plant, fungal, and bacterial cells in the form of walled cells where living protoplasts are surrounded by a more or less hermetical extracellular matrix (ECM) in the form of cell walls, counteracting the high hydrostatic pressure coming from the inside the cells (Peters *et al.*, 2000; Harold, 2002; Hohmann, 2002).

Both evolutionary designs have an immediate bearing on the developmental and morphogenetic processes, among them those related to polarity, especially to polar growth of cells (Baluška *et al.*, 2003c). Naked animal cells depend on their tensegrally organised cytoskeletal elements when controlling their shape (Ingber, 1993a, 2003a,b). Although such a way of osmotic regulation seems to be less energetically favourable, it offers a great advantage in the form of cellular motility also powered and controlled through cytoskeletal rearrangements (Pollard, 2003). Some positional cues, affecting more or less directly the cell shape (Brock *et al.*, 2003) or cell fate (Huang & Ingber, 1999), come from the spatially limited contacts with ECM (focal adhesions – Zamir & Geiger, 2001; Wehrle-Haller & Imhof, 2002) or with neighbouring cells (adherens junctions – Perez-Moreno *et al.*, 2003), and are transduced inside the cell through the cytoskeleton-plasma membrane cross-talk (Geiger *et al.*, 2001). This enables generation of complex cellular shapes and coordinated developmental patterns both in unicellular as well as multicellular organisms.

In walled cells, the cytoskeleton also plays an important role. However, the control and maintenance of cellular shapes results mostly from an interplay between turgor and mechanical properties of the surrounding walls. As concluded by Peters *et al.* (2000) such a situation has important functional implications, among which three seem to be of crucial importance. First, walled cells are incapable of fast movements. Thus, they are either equipped with motor apparatuses, like bacteria, or they are sessile organisms. Second, in walled cells a steep gradient of hydrostatic pressure (exceeding 2 MPa) might occur and it is used mechanically to stabilise the multicellular body of higher plants. Third, the presence of the walls fixes the position of the cells in relation to their neighbours. Thus, as turgor is omnidirectional, any morphogenic events have to be related to wall remodelling resulting from differential and vectorial secretion of wall elements and their assembly into functional walls with required mechanical properties.

Two mechanical situations mark the opposite poles of turgor regulation. On the one side, walled cells are capable of generating extremely high turgor pressures. The pressure difference needed to transport water through a 100 m tall tree is usually calculated as being up to 3 MPa (Taiz & Zeiger, 1998), and the walls of tracheary elements are constructed in such a way as to withstand pressure difference reaching 6 MPa. A real extreme, however, could be found in some appressorium-forming pathogenic fungi. These single-cell infection structures accumulate osmotically active substances to a concentration of more than 3 M (de Jong *et al.*, 1997) creating turgor pressures exceeding 8 MPa (Howard *et al.*, 1991). This pressure is used to generate invasive forces in the range of 15–20 μN for directional mechanical penetration of host plant walls (Bechinger *et al.*, 1999). On the other side, living plant and fungal cells are able to dynamically change the volume of their protoplasts in response to changes in their environment. Sometimes, the retraction of protoplasts from the walls could be seen in cells adjoining the cells infected by the fungus (Bolwell *et al.*, 2001). In most cases, however, this is the type of reaction – the plasmolytic cycle – commonly found in cells responding to changes in osmoticum (Oparka, 1994; Lang-Pauluzzi & Gunning, 2000; Komis *et al.*, 2003). Interestingly, detachment of the protoplast from the walls during plasmolysis is never complete, and the plasma membrane remains attached to the walls through Hechtian strands, indicating the need for the maintenance of functional connections between various cellular compartments (see below). Plasmolysis is usually a relatively rapid event (Lang-Pauluzzi & Gunning, 2000), but in longer terms its extent could be actively regulated by for example synthesis of osmotically active substances, like proline (Nanjo *et al.*, 1999). On the other hand, permanent modification of internal turgor finds its counterbalance in modulated synthesis of wall components and assembly of the walls finely attuned to such new situation (Hoffmann-Benning *et al.*, 1997).

A final point which should be made here is the relation between turgor, cell walls and the growth of walled cells (Niklas, 1992; Brett & Waldron, 1996). It is usually assumed that turgor is the driving force for growth, and that cellular expansion is achieved through the cycles of localised wall yielding to turgor with subsequent hardening of the walls leading to changes in cell shape. Recent discoveries point to some other important participants of this process. In some cases, like in growing cotton fibres or, to some extent, in developing root hairs, symplastic isolation of individual expanding cells through the selective gating of plasmodesmata comes into play (Ehlers *et al.*, 1999; Blackman & Overall, 2001; Ruan *et al.*, 2001). On the other hand, observations from oomycetes, like *Achlya* or *Saprolegnia*, show them as exceptional organisms, growing despite almost complete lack of turgor pressure (reviewed in Harold, 2002). This leads to the suggestion that although turgor is required for growth, it is not a decisive factor determining the growth rate (Money, 1997; Harold, 2002). This might be true not only for tip-growing fungal cells, but also, to some extent, for some types of plant cells, for example root hairs and pollen tubes showing the same pattern of cellular growth (Hepler *et al.*, 2001; see also Chapter 1). In such cases, the role of other 'partners in secretion', especially actin-driven movement of secretory vesicles should not be overlooked.

As discussed below, the cell wall is thin and extremely loosened at the tips of root hairs and pollen tubes. Therefore, the presence of the force-generating actin polymerisation machinery in plant cells, which is able to literally push the plasma membrane (Borisy & Svitkina, 2000), may be proposed to drive the plant tip growth (Baluška & Volkmann, 2002; Baluška *et al.*, 2000d; Vantard & Blanchoin, 2002). In support of this notion, the tip growth machinery is sensitive to the extremely low levels of latrunculin B which do not affect cytoplasmic streaming based on longer actin filaments, but block the growth at the tip (Gibbon *et al.*, 1999; Vidali *et al.*, 2001) which is enriched by dense meshworks (Baluška *et al.*, 2000c; Baluška & Volkmann, 2002; Foissner *et al.*, 2002). These meshworks consist probably of short and ephemeral, but extremely dynamic, actin filaments or even of actin oligomers. As expected, all actin-binding molecules required for the force-generating actin polymerisation machinery have been found within the *A. thaliana* genome (Hussey *et al.*, 2002; Meagher & Fechheimer, 2003). Moreover, mutants in actin and individual actin-binding proteins of the force-generating actin polymerisation machinery compromise the tip growth and also have dramatic effects of growth and polarisation of other plant cells (Ramachandran *et al.*, 2000; Dong *et al.*, 2001; Gilliland *et al.*, 2003; Mathur *et al.*, 2003a,b; Chapter 1).

3.3 Cell wall-plasma membrane-cytoskeleton continuum in plant cells

It seems that tensegral organisation of cytoskeletal elements is becoming a prevailing view (Ingber, 1993a; Pickett-Heaps *et al.*, 1999). In animal cells, interactions of the cytoskeleton with the ECM through the plasma membrane at localised contact points are used to converge physical and chemical stimuli into cellular responses (Ingber, 2003a,b). Moreover, the mechanical environment of animal cells not only regulates the extent of those reactions (Zhu *et al.*, 2000; Wehrle-Haller & Imhof, 2002), but also affects the structure and composition of focal adhesions (Katz *et al.*, 2000). It is worth noting here that in an animal's body the ECM is produced as a cooperative effort by groups of cells. As a result, different cell types, sharing and managing the same matrix, can transmit, for example, mechanical stress, and induce matrix remodelling in response to it (Swartz *et al.*, 2001).

Although plant and animal ECMs are based on similar mechanical designs (Wojtaszek, 2000), they also show some telling differences which affect the type of growth and developmental patterns. Most importantly, plant walls are constructed and controlled by individual protoplasts. As a consequence, when protoplasts are isolated, they are unable to divide mitotically before their functional walls are regenerated (Meyer & Abel, 1975; Meyer & Herth, 1978). Also at later stages the existence of a structural framework, consisting of the walls and the cytoskeleton (Wyatt & Carpita, 1993; Reuzeau & Pont-Lezica, 1995) permeating the whole cells, is crucial for their proper functioning (Fowler & Quatrano, 1997). Plant cells monitor wall integrity, and respond accordingly to any changes in wall architecture (Nakagawa & Sakurai, 2001). In addition, the mechanically stable walls are used by the tensegrally organised cytoskeleton as a 'spatial anchor' to orient and position cellular compartments (Katsuta & Shibaoka, 1988; Grolig, 1998) and/or processes, like cytoplasmic streaming (Masuda *et al.*, 1991; Ryu *et al.*, 1995) or mRNA localisation (Okita & Choi, 2002).

In order to maintain the integrity of the cell wall-plasma membrane-cytoskeleton continuum, there have to exist various complements of interacting proteins, glycoproteins, and carbohydrates (Pont-Lezica *et al.*, 1993; Reuzeau & Pont-Lezica, 1995; Heath, 2001). As these aspects have been recently thoroughly reviewed (Kohorn, 2000; Baluška *et al.*, 2003b), only a few points will be highlighted here.

Earlier attempts aimed to identify interacting components, and concentrated on the demonstration (most often by immunocytochemical methods) of animal-like proteins, such as integrins (e.g. Katembe *et al.*, 1997; Reuzeau *et al.*, 1997) or vitronectin and fibronectin (Sanders *et al.*, 1991; Zhu *et al.*, 1993b, 1994). However, the differences between plant and animal cells are the most pronounced at their peripheries (Meyerowitz, 2002), and it is suggested that different complements of the building blocks in both exocellular matrices dictate different types of intermolecular interactions (Baluška *et al.*, 2003b). It thus seems that in these cases the functional approaches should be much more successful. Indeed, plasma membrane-cell wall continuity has been identified as being crucial for the expression of plant defence responses (Kiba *et al.*, 1998; Mellersh & Heath, 2001). Interestingly, however, arginyl-glycyl-aspartic (RGD)-containing polypeptides, involved in interactions of ECM molecules with animal integrins, seem to interfere with various cellular processes in plants, although the cooperating partners have not been identified. Suspension-cultured cells, grown in the presence of RGD peptides, show aberrant wall-plasma membrane interactions and organisation (Schindler *et al.*, 1989; Faik *et al.*, 1998). RGD peptides change the pattern of plasmolysis in plant tissues (Canut *et al.*, 1998) and inhibit gravisensing in internodal cells of *Chara corallina* (Wayne *et al.*, 1992). Addition of RGD peptides to the growth medium induces the transition of normally-growing maize calluses into embryogenic ones (Labouré *et al.*, 1999), as occurs in agarose-embedded sunflower protoplasts (Barthou *et al.*, 1999). Significantly, flax protoplasts were also shown to be able to recognise and to react differently to the type of surrounding matrix, such as by distinct morphogenic responses (Roger *et al.*, 1996).

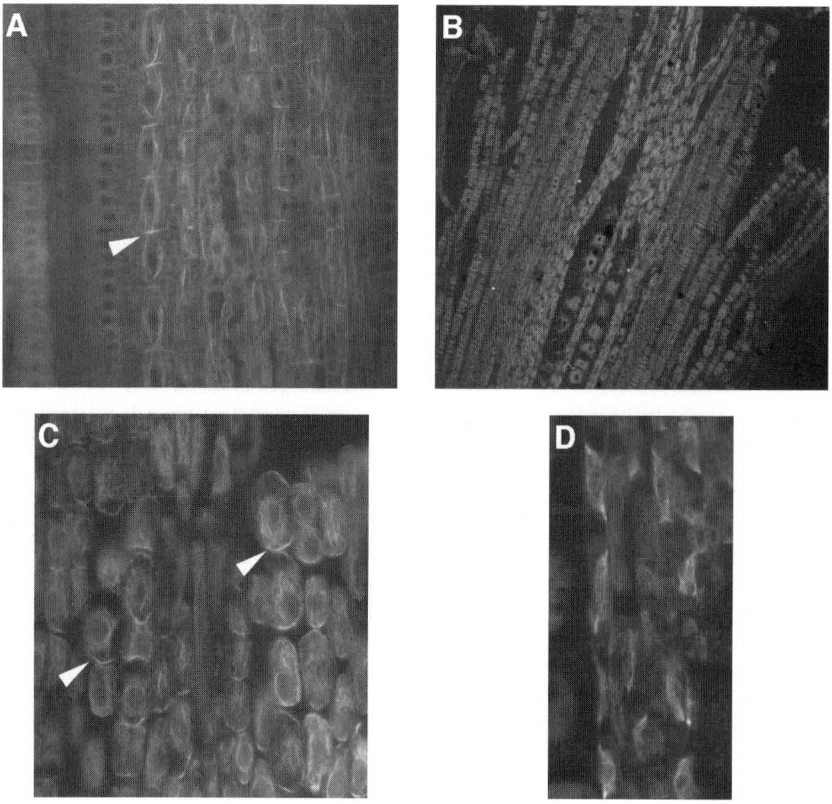

Figure 3.1 The influence of mechanical stability of plant cell walls on the arrangement of actin cytoskeleton in cells of maize root tips. A. Under normal conditions, with cell walls intact, the arrangement of actin is characteristic of the cell type and its developmental stage. Note the prominent and even labelling of cell peripheries near cross-walls (arrowhead). B. At 90 minutes, root tips treated with standard protoplasting solution (cellulase Onozuka RS + Pectolyase + osmoticum) are mostly digested and start to separate along longitudinal walls into individual cell files. Cross-walls still seem to be unaffected. Again, arrangement of actin in various cells is cell type-specific. Note strong labelling of actin in parenchymatic cells of the stele. C. Detailed view of the root treated the same way as in B. In cells with digested walls, protoplasts are largely plasmolysed and ready for release from the surrounding walls. However, they still show strong labelling of protoplasts' peripheries at places where they were in contact with cross-walls (arrowheads). D-F. When different enzymes are used for treatments in the presence of osmoticum, different patterns of plasmolysis and shaping of resulting protoplasts could be seen, especially in parenchymatous cells of the stele. Actin filaments are mostly

Several other approaches for revealing the macromolecular organisation of cytoskeleton-plasma membrane-cell wall contact points could be envisaged, and three of them can be considered here. First, suspension-cultured, NaCl-adapted tobacco cells develop different wall architectures (McCann *et al.*, 1994) and tight zones of wall-membrane adhesion. These strong adhesive

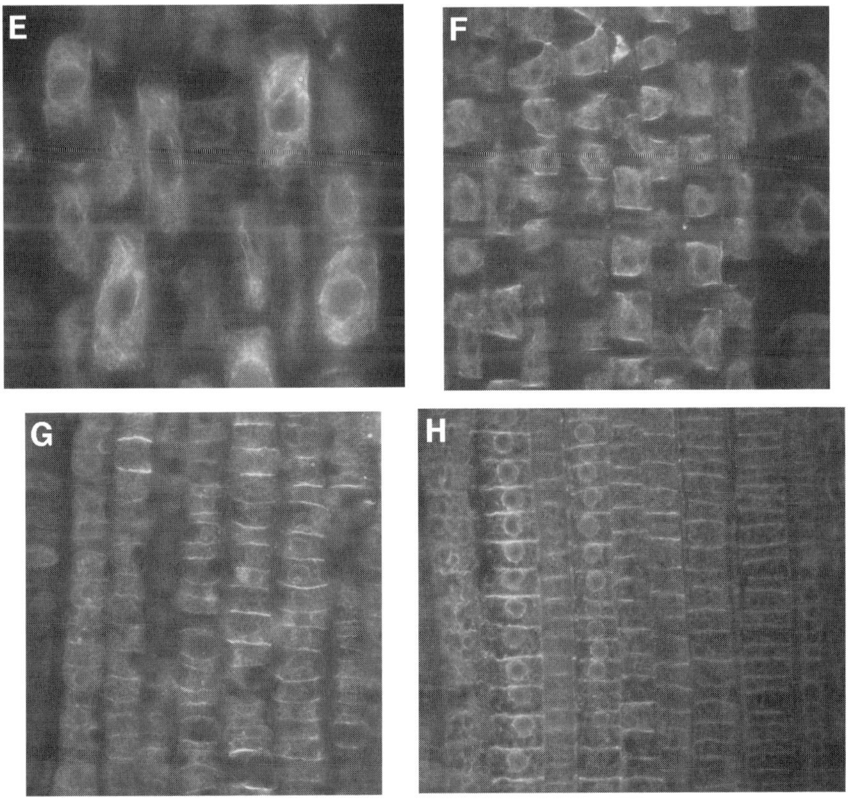

arranged longitudinally under standard protoplasting conditions (E) with some Hechtian strands still visible. Treatment of the root tips with collagenase (D) induces complete rearrangement of actin which becomes focused at the opposite ends of the cells, reflecting probably the loss of mechanically stable connections with surrounding walls. On the other hand, pectinase (F) induces the formation of cubic protoplasts with membrane domains facing cross-walls completely detached from the walls, implicating pectins in membrane-wall interactions. However, actin is strongly labelled at those domains. G-H. Protoplasting induces strong labelling of cross-walls in the cortex (G). Similar pattern of labelling could also be induced by chitinase acting even in the absence of osmoticum. Note the ring-like labelling of actin around nuclei (H).

Maize root tips were subjected to enzymatic treatments for 90 minutes in the absence or presence of osmoticum (0.6 M sorbitol). Following fixation and embedding in Steedman wax, sections were labelled immunofluorescently using anti-actin antibodies. Control treatment was done in buffer only.

properties are also retained in wall-less protoplasts (Zhu *et al.*, 1993b). At the same time, cell wall extensibility is reduced and arabinogalactan proteins (AGPs) are lost from the membranes (Zhu *et al.*, 1993a). Analysis of plants responding to water deficit enabled the identification of plasma membrane- and cell wall-derived interacting partner proteins (Garciá-Gomez *et al.*, 2000). More indirectly, a second approach is the analysis of the plasmolytic cycle

(Cleary, 2001) and the formation of Hechtian strands. As destabilisation of the cytoskeleton has only slight effect on plasmolysis, it has been suggested that in Hechtian strands membrane-wall attachments might be independent of cytoskeletal elements (Lang-Pauluzzi, 2000; Lang-Pauluzzi & Gunning, 2000) or of intracellular cytoskeletal organisation. However, in some plants, like *Chlorophytum comosum*, the cytoskeleton is absolutely essential for the volume regulation of the plasmolysed protoplast (Komis *et al.*, 2003). On the other hand, cold-hardened cells, when plasmolysed, show an increased number of Hechtian strands, and the binding strength within them is also higher (Buer *et al.*, 2000). Finally, cell wall-cytoskeleton interactions at the plasma membrane could be analysed enzymatically. Cell wall and/or contact point components could be bound via both covalent bridges and non-covalent interactions, and the extent of those could be regulated developmentally or in response to stress (Carpita & Gibeaut, 1993; Wojtaszek, 2000). Biochemical extraction experiments and proteomic analyses revealed that there exists a kind of higher-order, hierarchical organisation of macromolecules which can be disrupted by sequential extraction with various solvents (Robertson *et al.*, 1997; Blee *et al.*, 2001; Chivasa *et al.*, 2002). However, when plant cells or tissues are treated exogenously with enzymes, various intracellular rearrangements could be observed, allowing conclusions to be drawn on the nature of extracellular interactions. Through this type of study, functional associations, sensitive to proteolytic digestion, between cortical microtubules and cell walls of tobacco cells have been identified (Akashi *et al.*, 1990; Akashi & Shibaoka, 1991). Similarly, actin-dependent cytoplasmic streaming in *Vallisneria* has been shown to be regulated by protease-sensitive anchorage of actin at the end walls (Masuda *et al.*, 1991; Ryu *et al.*, 1995). Involvement of pectins and/or chitinase-sensitive components, presumably originating from oligosaccharide side chains of AGPs (Showalter, 2001; van Hengel *et al.*, 2001), in the stabilisation of the actin cytoskeleton has been recently identified in maize root cells (see Fig. 3.1).

Wall-associated kinases (WAKs) and interacting proteins/carbohydrates are a very good example of structural hierarchy governing the organisation of membrane-wall contact points (Anderson *et al.*, 2001). Typical WAKs have a cytoplasmic kinase domain, a single transmembrane domain, and an extracellular domain (He *et al.*, 1996). In *Arabidopsis*, the latter is the most diverse part of WAK isoforms, showing the presence of EGF-like repeats, and sequences resembling tenascin, collagen, and neurexin, which most possibly confer some kind of specificity (He *et al.*, 1999). The presence of WAKs at the wall-membrane interface is essential for normal cell elongation and plant development (Lally *et al.*, 2001; Wagner & Kohorn, 2001). Pectins have been suggested as covalent linkers for exocellular WAK domains (Wagner & Kohorn, 2001). However, some protein counterparts have also been identified, and these include the glycine-rich protein AtGRP-3, and KAPP protein kinase

(Park *et al.*, 2001). Less obvious are interactions between WAKs and arabinogalactan proteins located at the external face of plasma membrane (Gens *et al.*, 2000). In maize root cells treated exogenously with collagenase, we have observed relocations of WAKs (Wojtaszek, Baluška and Volkmann, unpublished). Since in *Arabidopsis*, WAK2 and 4 isoforms contain collagen-like domains, this might imply direct digestion and release of WAKs' cytoplasmic domains. However, glycine-rich proteins could also be digested with collagenases (Ringli *et al.*, 2001). In addition, WAKs could also be released by wall-degrading enzymes (He *et al.*, 1996). This indicates a high level of complexity in this type of physical attachment between plasma membrane and cell walls. Another kind of interaction has been predicted for plant receptor kinase with an extracellular domain revealing similarity to carbohydrate-binding proteins – lectins (Herve *et al.*, 1996) or for chimeric LRX1 and LRX2 cell wall proteins of the LRR/extensin family with receptor-like domains (Baumberger *et al.*, 2001, 2003).

3.4 Cell walls and polarity at the cellular level: targeted exocytosis and local endocytosis establish and maintain specialised wall domains

In all groups of eukaryotic organisms, interactions of cells with their immediate surroundings are necessary for their growth, development and survival. Importantly, cell walls of plant cells fullfil not only protective roles but also exert morphogenic roles via their direct impacts on growth domains, cell shape and morphogenesis (Kohorn, 2000; Wojtaszek, 2000; Baluška *et al.*, 2003c). One can presume that the differential composition of cell walls is driving the differential shaping of plant cells. Moreover, besides having a direct mechanical impact, evidence is emerging that secreted and diffusible cell wall proteins have signalling roles in plant embryogenesis (Pennell *et al.*, 1991, 1992; van Engelen & de Vries, 1992; McCabe *et al.*, 1997). At the cellular level, plant cells are expected to generate and maintain subcellular cell periphery domains differing in their mechanical and informational properties. However, this important aspect of plant cell walls is still poorly recognised and has not been well analysed by the mainstream cell wall community.

Here, we survey emerging data on the two types of cell wall domains in growing plant cells. First, we focus on loosened cell walls at growing apices of tip-growing root hairs and pollen tubes. With respect to root hairs, the mechanical properties of the walls seem to be essential both for the onset and for the maintenance of tip growth. This is because root hairs are formed from root epidermis cells, elongating along apical-basal axis, which are characterised by thick tangential cell wall facing the outside world of roots. Second, we highlight fundamental differences between the non-growing cross-walls and diffusely growing side-walls which have an immediate bearing on the polarity of axially growing plant cells. As cell walls are expected to be hard-wired, via transmembrane linker molecules to the cytoskeleton (Baluška *et al.*, 2003b) which again interacts with the vesicle-based membrane trafficking machinery, cell walls can be considered to represent one of the most critical components of plant cells with respect of their polarity.

3.4.1 Unique properties of cell wall domains are essential for the establishment and maintenance of polarity in tip-growing cells

In growing plant cells, robust cell walls are dynamic structures, continuously remodelled via both targeted exocytosis and local endocytosis, which directly underpin the differential growth that drives polarity in individual cells. Currently available models and concepts consider targeted secretion as the major cellular tool to establish various wall domains differing in their mechanical properties. For instance, plant cells performing intrusive types of growth (Lev-Yadun, 2001) and growing at their tips, like pollen tubes and root hairs (Hepler *et al.*, 2001), maintain loosened cell walls only at their extending apices and undergo progressive stiffening of the walls along the apical-basal axis of these tubular plant cells (for pollen tubes see Li *et al.*, 1997; Franklin-Tong, 1999; Feijó *et al.*, 2001). At the very tips of such cells, pectins are the major wall component and their mechanical properties are controlled via *in muro* de-esterification and subsequent cross-linking with calcium and boron (Matoh & Kobayashi, 1998; Kobayashi *et al.*, 1999). Until recently, integrated cell wall molecules were considered not to be available for endocytosis in order to be internalised and/or degraded. Therefore, it was quite a surprise for us to find out that wall pectins cross-linked with boron and calcium are vigorously internalised by root cells (Baluška *et al.*, 2002; Yu *et al.*, 2002b). This exciting twist in our viewing of plant cell walls, having immediate impacts on their polarity and morphogenesis, implies that plant cells are in a position to control locally the composition of wall domains via both targeted exocytosis and local endocytosis.

Root hair initiation provides a paradigm of local thinning of cell walls leading to initiation of new polarity in specialised root epidermis cells, so called trichoblasts, which continue in their axial growth. The outer tangential cell wall of root epidermis is one of the thickest walls in this portion of the root apex. During root hair initiation, the very first cytological event is a local thinning of this wall during which components released locally from weakened cell walls are apparently internalised and seem to be degraded within vacuoles (Čiamporová *et al.*, 2003). Moreover, wall-remodelling proteins acting *in muro* such as expansins (Baluška *et al.*, 2000c; Cho & Cosgrove, 2002) and xyloglucan endotransglycosylases (Vissenberg *et al.*,

2001, 2003) are targeted and activated locally, allowing this restricted wall thinning to be executed only at the outgrowing bulge. Other less well-known wall-loosening molecules, such as AGPs and elusive osmiophilic particles, also accumulate at bulging sites (Šamajová *et al.*, 1998; Šamaj *et al.*, 1999). Interestingly in this respect, the absence of fucosylated AGPs in the *mur1* mutant of *Arabidopsis* results in an excessive bulging of root hair bases (van Hengel & Roberts, 2002). A similar phenomenon was reported also in trichoblasts of the *reb1 Arabidopsis* mutant, having depleted AGPs in its cell walls (Ding & Zhu, 1997; Andème-Onzighi *et al.*, 2002); and in trichoblasts of the *der1* mutant of *Arabidopsis* which is depleted in or lacks ACTIN2 protein (Ringli *et al.*, 2002). The crucial role of AGPs in shaping and polarity of trichoblasts was confirmed using Yariv reagent. This compound binds selectively to AGPs affecting their functioning. Thus, Yariv reagent induces dramatic bulging of the trichoblast outer cell wall (Willats & Knox, 1996; Ding & Zhu, 1997), and also perturbs polarity and tip growth of pollen tubes (Roy *et al.*, 1998, 1999; Majewska-Sawka & Nothnagel, 2000; Mollet *et al.*, 2002). As discussed later, like the potent anti-actin drug latrunculin B (Baluška *et al.*, 2001b), Yariv reagent inhibits rapid cell elongation making interactions between AGPs and the actin cytoskeleton even more plausible.

Wall-loosening activity of expansins is known to be enhanced by slightly acidic pH in the walls (Cosgrove, 2000). A local decrease in pH precedes and accompanies bulging of trichoblasts (Bibikova *et al.*, 1998). Moreover, local application of exogenous α-expansin on root hairs of cucumber induced bulging and bursting of their apices, confirming that expansins have morphogenetic effects mediated through their wall-loosening actions (Cosgrove *et al.*, 2002). Similar bursting of hair apices was observed also in *KOJAK* and LRX mutants of *Arabidopsis*, implicating cellulose synthase-like proteins and LRR/extensin structural wall proteins in root hair morphogenesis (Favery *et al.*, 2001; Wang *et al.*, 2001; Baumberger *et al.*, 2001, 2003). Interestingly, cellulose synthase-like proteins are important also for the tip growth of pollen tubes (Goubet *et al.*, 2003; Lazzaro *et al.*, 2003). Moreover, bursting of pollen tube apices was observed under boron deficiency (Geitmann, 1997) while boron was shown to be accumulated in high amounts at hair apices (Glüsenkamp *et al.*, 1997). These local changes of cell wall mechanics are immediately confronted with global forces of turgor pressure, inducing local mechanical stress resulting in local bulging during root hair initiation. Not surprisingly, stress-induced and mitogen-activated protein kinases (MAPKs) such as SIMK were found to be both recruited and locally activated at these cell periphery domains (Šamaj *et al.*, 2002, 2004). SIMK seems to be a part of a safeguard apparatus for monitoring and maintaining cell wall integrity via both sensory and protective activities. One plausible scenario would be that MAPKs are involved in the local recruitment of the actin cytoskeleton which is essential not only for the protection of structurally weakened cell walls, but also for both the onset and performance of tip growth (Baluška *et al.*, 2000c,d; Vidali *et al.*, 2001; Hepler *et al.*, 2001; Ringli *et al.*, 2002; Gilliland *et al.*, 2002; Šamaj *et al.*, 2002). However, activated MAPK modules may also fine tune the delicate balance between wall loosening and stiffening activities.

A recent breakthrough paper by Foreman *et al.* (2003) implicates reactive oxygen species (ROS) in the structural maintenance of delicate cell periphery complex of emerging root hairs. RHD2 protein, identified as NADPH oxidase, is essential for the onset of hair tip growth and for the assembly of a tip-focused calcium gradient. As ROS are known to be involved in cross-linking of wall molecules (Wojtaszek, 1997; Bolwell *et al.*, 2001), a plausible scenario would be that lack of control over cell wall mechanics prevents onset of the tip-growth in *rhd2* mutant. But ROS are also known, especially in animal cells, to impinge directly on the cytoskeleton and vesicle-mediated membrane trafficking (Yan & Smith, 2000; Nimnual *et al.*, 2003). Root hairs, with a whole complement of available hair mutants (Table 3.1), will provide us with an excellent model system to study putatively diverse ROS roles in plant polarity.

In order to prevent bursting of root hair and pollen tube apices, structural wall proteins (Bernhardt & Tierney, 2000; Baumberger *et al.*, 2001, 2003; Bucher *et al.*, 2002), wall-associated kinases (Sivaguru *et al.*, 2003), and flavonols (Derksen *et al.*, 1999) have been proposed to support the already mentioned MAPKs and ROS-driven signalling cascades. Additionally, cross-linking of wall pectins with calcium and boron (Matoh & Kobayashi, 1998; Kobayashi *et al.*, 1999) is also critically involved in the maintenance of the delicate balance between wall loosening and stiffening at the apices of tip-growing cells. Our recent findings suggest that, besides targeted exocytosis, local endocytosis of pectins cross-linked with both calcium and boron is important for the regulation of mechanical properties of cell walls (Baluška *et al.*, 2002; Yu *et al.*, 2002b). One could imagine that the characteristic dome-shaped apex of the tip-growing cell (Shaw *et al.*, 2000) is maintained via distinct zones of exo- and endocytosis (Derksen *et al.*, 1995; Miller *et al.*, 1997), with the very tip performing predominantly exocytosis followed by the progressive increase and then subsequent decrease of endocytosis (Šamaj *et al.*, 2002). The latter process would internalise cross-linked pectins while the very tip would contain only loosely organised esterified pectins which can not be cross-linked with calcium. This pattern of pectin distribution and cross-linking would finely modulate mechanical properties of pectin-based cell walls at the growing tip. In accordance with this model, both exo- and endocytotic events are known to be abundant at tips of root hairs and pollen tubes (Picton & Steer, 1983; Derksen *et al.*, 1995; Miller *et al.*, 1997; Parton *et al.*, 2001; Camacho & Malhó, 2003). Moreover, cell wall pectins were reported to be accumulated within brefeldin A (BFA)-induced compartments

Table 3.1 Selected mutants of *Arabidopsis* with affected cell types showing tip-growing mode.

Mutant	Brief characteristics	References
	Root Hair Development	
deformed root hairs 1 (*der1*)	Root hairs usually shorter and frequently swollen. DER1 is ACTIN2 and is required for correct selection of bulge site and for root hair tip-growth.	Ringli *et al*. (2002)
mur1	Depletion of fucosylated arabinogalactan proteins causing excessive bulging of trichoblasts. MUR1 is GDP-D-mannose-4,6-dehydratase catalysing the first step in fucose biosynthesis.	Bonin *et al*. (1997); Andème-Onzighi *et al*. (2002)
kojak	Depletion of cellulose-synthase-like protein causing bursting of root hair apices.	Favery *et al*. (2001)
AtCLSD3	Depletion of cellulose-synthase-like protein causing bursting of root hair apices.	Wang *et al*. (2001)
AtPRP3	Proline-rich structural cell wall protein expressed in root hairs. Its expression is altered in root hair mutants background.	Bernhardt & Tierney (2000)
lrx1, *lrx2*	Depletion of LRR/extensin class of structural cell wall proteins causing excessive thickening of cell walls and bursting of root hair apices.	Baumberger *et al*. (2001, 2003)
root hair defective (*rhd*; different alleles)	Various defects in root hair development. RHD4 is required for maintenance of a uniform rate of tip growth.	Galway *et al*. (1998)
	Pollen Tube Development	
AtCSLA7	Depletion of cellulose synthase-like protein inhibiting pollen tube tip growth and causing abnormal cell division planes.	Goubet *et al*. (2003)

in apices of pollen tubes (Geitmann *et al.*, 1996; Ueda *et al.*, 1996), as in cells of the root body (Baluška *et al.*, 2002). Importantly, these BFA-induced compartments accumulate endocytic markers like FM dyes and recycling molecules (Geldner *et al.*, 2003; Grebe *et al.*, 2003).

Vigorous tip-based vesicle trafficking appears to be closely related to the high level of cytoplasmic calcium found at apices of root hairs and pollen tubes (Bibikova *et al.*, 1997; Wymer *et al.*, 1997; Brownlee *et al.*, 1999; Camacho & Malhó, 2003; Parton *et al.*, 2003). We propose that this tip-focused calcium gradient apparently regulates the amount of internalised wall pectins because, throughout the eukaryotic superkingdom, vesicle-based membrane trafficking is known to be controlled by the level of cytoplasmic calcium. For instance in neurons, which represent the best understood model for vesicle-based membrane trafficking, there are distinct zones which are specialised in either exo- or endocytosis (Blanpied *et al.*, 2002). Recently, several proteins mediating calcium-regulated vesicle trafficking, originally considered as neuron-specific, were characterised also in non-neuronal cells. For instance, synaptotagmins emerge as important vesicular calcium sensors which regulate both exo- and endocytosis (Chapman, 2002). Intriguingly, there are three synaptotagmin genes in the *Arabidopsis* genome, closely related to human genes (Craxton, 2001; Fukuda, 2003). Moreover, several recent findings implicate that plants also contain specialised synapse-like domains (see below) with auxin emerging as a plant 'neurotransmitter' (Baluška *et al.*, 2003a,c). As discussed below, plant synaptotagmins might be involved in synaptic activity-dependent and calcium-regulated adhesions between vesicles, plasma membrane and cell walls, similar to the case for neuronal synapses (Schivell *et al.*, 1996).

An attractive concept is that endocytosis can be uncoupled from exocytosis, allowing active modulation of cell wall composition independently from exocytosis. In the most extreme view, endocytosis of wall pectins would be essential for the tip-growth as it would guarantee the tip-based gradient of cell-wall rigidification which is crucial for the maintenance of tubular shapes of these cells (Shaw *et al.*, 2000). This tubular shape is essential for the rapid vectorial growth of such exploratory cells. In accordance with this view, several treatments which stop tip-growth still allow exocytosis to continue as evidenced by the thickenned cell walls at cell apices (Schröter & Sievers, 1971; Roy *et al.*, 1999; Yang *et al.*, 1999). Anytime the finely-tuned balance between exocytosis and endocytosis is perturbed in favour of exocytosis, the tip-growth machinery seems to be inhibited and stopped after a short period of isodiametric growth. Short-term F-actin depolymerisation has the same impact (Gibbon *et al.*, 1999; Ovečka *et al.*, 2000; Ketelaar *et al.*, 2003). Importantly, endocytosis is dependent on the intact actin cytoskeleton (for plant cells see Baluška *et al.*, 2002; Geldner *et al.*, 2001, 2003; Grebe *et al.*, 2003) while plant exocytosis can proceed further as evidenced by the continued, albeit slow, growth of root apices devoid of F-actin (Baluška *et al.*, 2001b). All this suggests that F-actin-dependent endocytosis is essential, not for the retrieval of excess plasma membrane provided by abundant exocytosis, as proposed originally (Picton & Steer, 1983), but rather for cytomorphogenesis of tubular shapes of tip-growing cells which enables rapid and explorative motility of their apices.

Stress-induced MAP kinase SIMK (Baluška *et al.*, 2000b; Šamaj *et al.*, 2004) is apparently involved in coupling the actin-based tip-growth machinery to the vesicle-based membrane trafficking, as its inhibition results in bulging of root hair apices and subsequent inhibition of the tip growth (Šamaj *et al.*, 2002, 2004). Importantly, both depolymerisation of F-actin and inhibition of SIMK result in very similar responses when bulging of root hair apices is associated with the disintegration of an actin-based vesicle-rich 'clear zone' and protrusion of tubular vacuoles up to the very tip (Ovečka *et al.*, 2000; Šamaj *et al.*, 2002, 2004). As an intact actin cytoskeleton is essential for plant endocytosis (Baluška *et al.*, 2002; Geldner *et al.*, 2001, 2003), one could propose that a loss of F-actin in tip-growing cells causes cessation of growth due to the inhibition of F-actin-dependent endocytosis. Accordingly, tip-growing cells are very sensitive to actin-disrupting drugs (e.g. latrunculin B) and stop their growth with cytoplasmic streaming still intact (Gibbon *et al.*, 1999; Vidali *et al.*, 2001; Ketelaar *et al.*, 2003). For the future, it will be important to analyse the thickness of the walls in root hair apices inhibited in their tip growth. A recessive mutation in the *RHD4* gene might be instrumental in answering the question: how are thickened cell walls related to the inhibited tip-growth of root hairs? *RHD4* root hairs grow slowly, have increased diameters, and show aberrantly thickened cell walls (Galway *et al.*, 1999).

3.4.2 Cell wall domains of cells elongating axially along the apical-basal polarity axis: cross-walls versus side-walls

Although tip-growing cells show an extreme form of polarity, there are relatively few plant cell types growing via this rather ancient mode of growth, which is found also in unicellular yeast cells and animal neurons (Lev-Yadun, 2001; Baluška *et al.*, 2003c). The majority of plant cells grow via so called diffused (or intercalary) growth in which cell expansion is accomplished axially over the whole lateral surface of cells elongating along the apical-basal polarity axis. Intriguingly, cross-walls attain their non-growing status during the transition from the slow cytoplasmic expansion typical for meristematic cells to the vacuole-based rapid extension characteristic for elongating cells (Baluška *et al.*, 2001c, 2003c). This transition is accomplished in a developmentally unique sensory growth zone (Baluška *et al.*, 2001c). Importantly, rapid axial elongation is a mode of growth specific for plants (Baluška *et al.*, 2003c), and it allows

Table 3.2 Selected mutants of *Arabidopsis* impaired in the regulation of cell shape.

Mutant	Brief characteristics	References
	General	
abp1	Null mutation is embryo lethal; at globular stage cells fail to elongate; cell division also affected. ABP1 is an auxin-binding protein.	Chen *et al.* (2001a,b)
boule (*bul*)	Extreme dwarfism of seedlings resulting from reduced cell elongation, delayed development and reduced fertility. BUL1 protein involved in growth processes regulated by brassinosteroids.	Catterou *et al.* (2001)
elongation defective 1 (*eld1*)	Elongation impaired in all organs, altered development of vascular tissue, ectopic deposition of suberin.	Cheng *et al.* (2000)
stunted plant 1 (*stp1*)	Semi-dwarf mutant. Cells reveal constant elongation rate. STP1 mediates the effects of cytokinin on cell elongation.	Baskin *et al.* (1995)
	Carbohydrates/proteins	
cobra (*cob*)	Orientation of cell expansion affected; reduced amount of crystalline cellulose in the walls. COB is a GPI-anchored protein localized to the longitudinal walls; member of a multigene family.	Schindelman *et al.* (2001); Roudier *et al.* (2002)
emb30/gnom	Abnormal deposition of wall material; affected cell division, elongation and adhesion. EMB30 is similar to Sec7 and is thought to be involved in a secretory pathway.	Mayer *et al.* (1993); Shevell *et al.* (1994); Shevell *et al.* (2000); Geldner *et al.* (2003)
kobito (*kob*)	Plant dwarfism accompanied by cellulose deficiency; mostly specific for elongation zones. *KOB1* is predicted to code for membrane-bound protein.	Pagant *et al.* (2002)
korrigan (*kor*)	Aberrant cell plates, incomplete walls, multinucleated cells; few allelic mutants differing in the extent of aberrations. *KOR1* encodes endo-1,4-β-glucanase required for cellulose synthesis.	Nicol *et al.* (1998); Zuo *et al.* (2000); Sato *et al.* (2001)
procuste1 (*prc1*)	Decreased cell elongation, cellulose deficiency, gapped walls; *PRC1* is a member of cellulose synthase catalytic subunit family.	Fagard *et al.* (2000)
quasimodo1	Mutants dwarfed with reduced cell adhesion resulting probably from reduced pectin content. QUA1 is a putative membrane-bound glucosyltransferase of family 8.	Bouton *et al.* (2002)
radially swollen (*rsw*; different alleles)	Mutants reveal reduced to various extent degree of anisotropic expansion of the roots. RSW1 is a catalytic subunit of cellulose synthase complex.	Arioli *et al.* (1998); Wiedermeier *et al.* (2002)

sku5	Roots deviate from normal direction of growth and exhibit left-handed axial rotation bias. SKU5 is a GPI-anchored protein located both at plasma membrane and cell wall; similar to multiple-copper oxidases with no functional activity detected.	Sedbrook *et al.* (2002)
salt overly sensitive5 (*sos5*)	Root swelling phenotype; root growth arrested under salt stress. SOS5 is a GPI-anchored protein most probably involved in cell adhesion.	Shi *et al.* (2003)
	Lignin deposition	
ectopic lignin deposition (*elp*)	Altered lignin deposition; increased biosynthesis of lignin not accompanied by increased formation of secondary walls; ELP1 is a chitinase-like protein.	Zhong *et al.* (2000); Zhong *et al.* (2002b)
ectopic lignification (*eli*)	Abnormal lignification pattern associated with reduced cell elongation and thicker primary roots.	Caño-Delgado *et al.* (2000)
	Cytoskeletal organization	
fragile fibre (*fra*; different alleles)	Altered orientation of deposition of cellulose microfibrils; FRA1 is a kinesin-like protein while FRA2 is a katanin-like protein involved in severing of microtubules.	Burk *et al.* (2001); Burk & Ye (2002)
microtubule organization 1 (*mor1*)	Disrupted cortical microtubules; left-handed root twisting and radial swelling; orientation of cellulose microfibrils is not disturbed; MOR1 belongs to the family of microtubule-associated proteins.	Whittington *et al.* (2001); Sugimoto *et al.* (2003)
spike1	Impaired polarized growth and assembly of microtubules. SPIKE1 is similar to animal proteins transforming exocellular signals into reorganization of cytoskeleton.	Qiu *et al.* (2002)
wave-dampened (*wvd*)	Suppressed root waving and leftward root slanting; shorter and thicker roots resulting from impaired anisotropic cell expansion. WVD2 is proposed to interact with cortical microtubules.	Yuen *et al.* (2003)
	Cell-to-cell signalling	
clavata3 (clv3)	Acts cell non-autonomously restricting the amount of stem cells of shoot and flower meristems; CLV3 localizes in the apoplast and works as a signalling molecule activating CLV2 receptor.	Rojo *et al.* (2002); Sharma *et al.* (2003)
stomatal density and distribution (sdd1)	In the mutant: clustering of stomata and increased stomatal density; no changes in internal leaf organization. SDD1 is a subtilisin-like serine protease possibly involved in generation of an exocellular signal.	Berger & Altmann (2000); von Groll *et al.* (2002)

otherwise immobile plant organs rapid directional movements. Under stress condition, and in the absence of F-actin, this growth mode can be switched off. Such plant organs exhibit truncated growth zones devoid of an elongation region, and plants develop only dwarfed statures resembling bonsai plants (Baluška *et al.*, 2001b). As the number of *Arabidopsis* mutants lacking rapid cell elongation is very high (Baluška *et al.*, 2001c, 2003c; see also Table 3.2), this topic will surely be a fruitful field of plant cell biology for the future. Interestingly, some plants naturally lack a cell elongation zone at their root apices (Koi & Kato, 2003). It would be rewarding to compare the composition of their cell walls with other plants. Data from suspension-cultured cells already confirmed the differences in wall components of spherical and elongating plant cells (Suzuki *et al.*, 1990).

It seems that axial cell elongation and the apical-basal axis of polarity are already established in the fertilised zygote, as it prepares for its first asymmetric mitotic division. Both processes are inherently linked to polar auxin transport, which occurs across non-growing cross-walls (Friml *et al.*, 2002; Geldner *et al.*, 2003; Baluška *et al.*, 2003a,c). Intriguingly, early polarity of the zygote appears to be related to callose deposition (Williams *et al.*, 1984) and differential properties of wall domains with thicker walls at the micropylar end; and thinner, almost wall-less, cell periphery complex found at the opposite chalazal end (Natesh & Rau, 1984; Williams *et al.*, 1984). JIM8 arabinogalactan epitope marks all cells of *Arabidopsis* early embryos at first, whereas later it becomes restricted only to cell walls of derivatives of the basal cell (Pennell *et al.*, 1991). Loosened and dynamically remodelled cell walls might channel cell expansion towards the apical cell derivatives which form the proembryo while the root pole develops only later. Another arabinogalactan epitope, reactive to JIM4 antibody, seems to be involved in the generation of the shoot pole, as it is expressed only in protodermal cells of this pole during the transition from globular to heart stage and at emerging cotyledonary ridges of early torpedo stage (Stacey *et al.*, 1990). Following this phase, only a few epidermal cells express JIM4 epitope. Interestingly in this respect, Yariv reagent, when applied on early carrot embryos, disturbed selectively morphogenic events at the shoot pole (Thompson & Knox, 1998). The importance of the cell wall for polarity and cell shapes is obvious also from the *rsh* mutant of *Arabidopsis* which lacks a HRGP-type wall protein. RSH is essential for the proper positioning of cytokinetic cell plates, assembly of the adhesive cross-walls instructing alignment of cells into cell files, and for ordered cell shapes (Hall & Cannon, 2002). Similarly, abnormal positioning of cell division planes was reported also for *Arabidopsis* early embryos lacking cellulose synthase-like protein AtCLSA7 (Goubet *et al.*, 2003). RSH was localised to cell walls as well as to *trans*-Golgi networks suggesting its dynamic trafficking. As the closest phenotype to the *rsh* mutant is that of *gnom/emb30* mutants defective in vesicular recycling and pectin distribution (Shevell *et al.*, 2000; Geldner *et al.*, 2003), one could speculate that RSH glycoprotein recycles in a GNOM-dependent manner.

Auxin in plants is transported in a polar fashion, along the apical-basal axis, initially towards the rapidly dividing descendants of the apical cell, while reversal of auxin flow polarity occurs when root pole starts to develop (Friml *et al.*, personal communication; see also Chapter 2). At this stage, apical-basal polarity becomes visible also in derivatives of the apical cell which starts to form distinct cell files. This new polarity of auxin transport, towards the emerging root pole, is maintained up to the end of embryogenesis as well as throughout the whole postembryonic development. An important feature of embryonic cells at this particular stage is that they start to organise into cell files via the establishment and maintenance of dynamic, but non-growing cross-walls which interconnect adjacent cells into discrete and lengthy cell files (Baluška *et al.*, 2003c). Polar auxin transport appears to play a central role in this respect, as its inhibition, either genetically or pharmacologically, results in chaotic division planes which prevents cell-to-cell adhesions at cross-walls and assembly of cells into longitudinal files (Hadfi *et al.*, 1998; Petrášek *et al.*, 2002; Geldner *et al.*, 2003). Interestingly, the putative auxin receptor ABP1 is also essential for the organisation of embryonic cells into cell files, because in the homozygous null *abp1* mutants aberrant embryos fail to assemble cells into cell files and fail to make the transition to bilateral symmetry (Chen *et al.*, 2001a,b).

So what is so special about the cross-walls? Besides high amounts of actin and myosin VIII (Ryu *et al.*, 1995; Baluška *et al.*, 2000a), they also have a unique construction. This latter feature is closely related to the high number of plasmodesmata traversing cross-walls (Ehlers & Kollmann, 2001). Generally, two cell wall polymers are synthesised by composite transmembrane synthase complexes: callose and cellulose. During cytokinesis, all cell walls are formed in the form of juvenile callosic and pectic cell plates (Samuels *et al.*, 1995; Scherp *et al.*, 2001). During maturation of cell plates into young cell walls, callose is rapidly removed and replaced with cellulose, making cell walls much more stable (Otegui & Staehelin, 2000b). Just the opposite sequence of events is accomplished in developing phloem elements during the formation of sieve pores. First, the cellulosic cell wall is replaced by a callosic cell wall which is then degraded via rapid removal of callose (Parthasarathy, 1974; Eleftheriou & Tsekos, 1982). This latter process might be accomplished via endocytosis as numerous clathrin pits are well known to be associated with callosic cell plates. Inhibition of cellulose synthesis at this stage of cell plate maturation causes failure of cytokinesis when young cell walls accumulate excessive callose deposits and fail to maintain their structural integrity (Vaughn *et al.*, 1996). Similarly, drugs inhibiting tubulin polymerisation induce excessive callose deposition at cytokinetic cell plates of the liverwort *Riella* (Scherp *et al.*, 2002). Moreover, excessive callose deposition was found also in the cytokinetic mutant *cyt1* (Nickle & Meinke, 1998). Importantly, cross-walls maintain juvenile callosic cell walls around plasmodesmata and at phloem sieve plates (Baluška *et al.*, 2000b, 2001a), both sites of active cell-to-cell transport along plant cell files. In addition, cross-walls preserve much higher amounts of callosic plasmodesmata (Juniper & Barlow, 1969; Zhu *et al.*, 1998a) while these are progressively lost from the side-walls (Zhu & Rost, 2000). In effect, plasmodesmata remain intact only in small 'islands' known as pit-fields (Juniper & Barlow, 1969; Seagull, 1983a; Zhu *et al.*, 1998b). Cell walls around plasmodesmata and in pit-fileds are also enriched with pectins (Morvan *et al.*, 1998; Orfila & Knox, 2000). In contrast to non-growing cross-walls, extending side-walls almost lack callose, and contain large amounts of cellulose as a dominant cell wall component. The only domain at side-walls lacking cellulose microfibrils are callosic pit-fields containing grouped plasmodesmata (Seagull, 1983b).

Cell walls around plasmodesmata emerge as highly specialised mini-domains which regulate the dynamic nature of these gateable 'organelles'. Plasmodesmata are firmly embedded within cell wall domains enriched in callose (Turner *et al.*, 1994; Radford *et al.*, 1998). Experimental data showed that callose participates in the gating of plasmodesmata (Wolf *et al.*, 1991; Sivaguru *et al.*, 2000). Such gating was documented by a report showing that the movement of viruses is delayed and the size exclusion limit is reduced when increasing amounts of callose are accumulated in transgenic tobacco plants deficient in β-1,3-glucanase (Iglesias & Meins, 2000). Callose assembly and breakdown was shown to participate in compartmentalisation of the shoot apical meristem into symplastic domains (Rinne & van der Schoot, 1998). Clearly, dynamic callose turnover is an important factor in gating of plasmodesmata although this is rather a long-term mechanism as callose degradation can take several hours. Several studies also reported that cell wall microdomains around plasmodesmata are characterised by a unique composition of pectin epitopes (Casero & Knox, 1995; Roy *et al.*, 1997; Orfila & Knox, 2000). Intriguingly in this respect, pectin methylesterase (PME), responsible for the *in muro* de-esterification of secreted pectins, localises also at plasmodesmata (Morvan *et al.*, 1998). Viral movement proteins of the tobacco mosaic virus interact directly with PME and this interaction is essential for the opening of plasmodesmata through the action of movement protein resulting in cell-to-cell spreading of viral particles (Dorokhov *et al.*, 1999; Chen *et al.*, 2000). On the other hand, there are several other possibilities for how PME could affect the permeability of plasmodesmata (reviewed by Jackson, 2000). One of these would be that PME rapidly changes the state of pectins at the plasmodesmata, especially at the wall-membrane interface, which would either prevent or allow dilation of plasmodesmal architecture. For instance, local de-esterification of secreted esterified pectins would allow their interactions with cell wall calcium, immediately rigidifying the whole microdomain and precluding any extensive changes in plasmodesmal architecture. In this way, direct binding of viral movement protein would inhibit PME activity and thus de-esterification of plasmodesmata pectins precluding their cross-linking with calcium. This would allow actomyosin-based dilation of plasmodesmata to be accomplished, induced perhaps via putative interactions of movement protein with actin and/or myosin.

As already mentioned above, cross-walls are typically non-growing, although there is a lot of F-actin and a vigorous vesicle-based trafficking (Baluška *et al.*, 2001b, 2003c; Geldner *et al.*, 2003). In fact, callosic cell wall domains are typically enriched with F-actin (Baluška *et al.*, 2000c) and F-actin is also required for callose deposition (Levina *et al.*, 2000; Kobayashi & Hakuno, 2003). Our testable proposal is that this abundant actin-driven trafficking is essential for maintenance of the juvenile pectic and callosic nature of the walls at the end-poles. This vesicle recycling is also closely linked to the polar transport of auxin which is accomplished across cross-walls (Geldner *et al.*, 2001, 2003; Baluška *et al.*, 2003a). We propose that the unique molecular composition of cross-walls, resembling juvenile cell plates (Baluška *et al.*, 2000d; Scherp *et al.*, 2001), is essential for auxin to cross effectively the relatively thick cell walls towards the plasma membrane of the adjacent cell within a given cell file. The reason for this might not only be the unique molecular nature of cross-walls but also an inherent tendency for vesicle recycling at these wall domains.

This feature is clearly visible not only during cytokinesis but also at the apices of tip-growing cells and at sites of attempted penetrations by fungal pathogens. Plants reacting to such infections form callosic wall depositions, known as papillae, which are assembled via vesicle-based trafficking (Stark-Urnau & Mendgen, 1995) directed by a polarised cytoskeleton (Schmelzer, 2002). Intriguingly, the composition of papillae resembles to some extent juvenile cell plates (Rodríguez-Gálvez & Mendgen, 1995) and one report even reveals that callose deposited there could be internalised via endocytosis (Xu & Mendgen, 1994). In fact, due to the local adhesion between intruder and host cells (Mellersh & Heath, 2001), one can view these highly specialised cell periphery domains, with juvenile cell wall composition, as some kind of 'plant defence synapses' analogous to 'immunological synapses' which develop between T-cells and target cells in animal immune system (Dustin & Cooper, 2000). In contrast to 'plant auxin synapses' (see below; and also Chapters 2 and 7), 'plant defence synapses' (Rodríguez-Gálvez & Mendgen, 1995; Mellersh & Heath, 2001; Schmelzer, 2002) are not specialised for polar cell-to-cell communication between adjacent plant cells of a given cell file but for extensive bi-directional communication between the intruder and host cells.

In a broad sense, synapses are defined as actin-based adhesive cell periphery domains specialised for vesicle-mediated cell-to-cell communication. Originally, this term was used exclusively for neuronal cells, but the usage of this term is currently broadening. Diverse cell types are considered to establish synaptic types of adhesive domains specialised for cell-to-cell communication. Adhesive cross-walls, enriched with specific wall components and transporting secreted auxin, may act as a plant synaptic cleft, analagous to the neuronal synaptic cleft. The unique feature of such plant synapses would be that auxin is not retrieved back into the cell from which it is released. It is rather taken up by the adjacent cell, and this allows transcytosis-like polar cell-to-cell transport of auxin (Baluška *et al.*, 2003a). An attractive view is that auxin is taken up via endocytosis together with internalised and calcium cross-linked cell wall pectins. A further speculative assumption is that calcium cross-linked pectins regulate the release of auxin from recycled secretion vesicles. Neurotransmitter release was shown to be controlled by physical properties of the intravesicular proteoglycan gel matrix which immobilises neurotransmitting molecules within synaptic vesicles (Reigada *et al.*, 2003). This proteoglycan matrix, acting as a smart ion-exchanging gel, is assembled through interactions of proteoglycan SV2 with other glycoproteins of synaptic vesicles, including synaptotagmins (Schivell *et al.*, 1996). Calcium-induced changes in the physical nature of this matrix regulate the availability of neurotransmitters for their release into the synaptic cleft (Reigada *et al.*, 2003). We speculate that cell wall pectins, which also localise to recycling vesicles (Baluška *et al.*, 2002; Yu *et al.*, 2002b), and which interact with calcium to change their physical properties (Jarvis, 1984), might also act as a similar ion-exchanging matrix regulating release of plant 'neurotransmitter-like' auxin during its polar cell-to-cell transport (Baluška *et al.*, 2003a). Intriguingly in this respect, inhibition of pectin recycling via deprivation of boron (Yu *et al.*, 2002b) is known to inhibit the polar transport of auxin (Goldbach, 1997).

Internalised cell wall pectins cross-linked with calcium and boron accumulate within the same compartments in BFA-treated cells as recycling PIN1 auxin efflux carriers, and endocytic markers such as FM dyes (Geldner *et al.*, 2003; Grebe *et al.*, 2003), indicating that cell wall pectins and PIN1 proteins take the same endocytic/recycling pathways (Baluška *et al.*, 2002; Yu *et al.*, 2002b; Geldner *et al.*, 2003). Interestingly in this respect, cells of *gnom/emb30* mutants not only fail to establish the apical-basal polarity (Shevell *et al.*, 1994; Steinmann *et al.*, 1999; Geldner *et al.*, 2001, 2003) but also show aberrant distribution of wall pectins reactive to JIM5 antibody (Shevell *et al.*, 2000) which recycle in maize and wheat root cells (Baluška *et al.*, 2002; Yu *et al.*, 2002b). Why should plant cells invest a lot of energy and other resources into such a complicated task of endocytic internalisation of cross-linked cell wall pectins? One possibility mentioned above is that pectins perform the role of a smart gel within recycling vesicles in order to control the storage and release of auxin, so resembling synaptic vesicles (Reigada *et al.*, 2003). Another plausible scenario, not excluding the former one, would be that plant cells can control local mechanical properties of the walls via local internalisation of cross-linked pectins. In fact, as discussed above, progressive cell wall stiffening at apices of tip-growing pollen tubes (Li *et al.*, 1994) and root hairs is accomplished via *in muro* activation and action of pectin methylesterase (Morvan *et al.*, 1998). This enzyme de-esterifies extracellular pectins allowing formation of calcium bridges that make cell walls mechanically more stable (Matoh & Kobayashi, 1998; Kobayashi *et al.*, 1999). Not surprisingly in this respect, it was reported that genetic downregulation of pectin methylesterase in pea inhibited root growth and had a dramatic impact on cell shapes throughout the root apex (Wen *et al.*, 1999).

Recently, the polar auxin transport inhibitor TIBA was identified as an endocytotic inhibitor for PIN1 recycling pathway (Geldner *et al.*, 2001) and our preliminary data show that TIBA also blocks endocytotic internalisation of cross-linked cell wall pectins as well as of FM dyes (Baluška, Wan and Emans, unpublished). Thus, TIBA emerges as a general inhibitor of endocytosis in plants. In support of the importance of endocytosis for cell wall assembly, *adl1A-2/adl1E-1* double mutants of the *Arabidopsis* dynamin-like proteins show severe defects in cell wall formation and cell plate assembly (Kang *et al.*, 2003a). Moreover, *adl1C-1* mutants show aberrant cell walls in pollen grains, which collapse during desiccation (Kang *et al.*, 2003b). For future research, it would be interesting to analyse changes in wall pectins and of polar transport of auxin in those mutants.

The plant plasma membrane is located at the critical interface between two opposite forces: the turgor pressure of the cell is counterbalanced by the mechanically robust cell walls. As the turgor pressure is a more-or-less stable cellular parameter, subtle changes in mechanical properties of the wall must have an immediate impact on the plasma membrane. For example, local cell wall loosening is predicted to be associated with increased stretch-stress of the plasma membrane which can be relieved by two processes: insertion of new membrane material via exocytosis (Kell & Glaser, 1993; Sheetz & Dai, 1996; Fricke *et al.*, 2000) and mechanical support of the plasma membrane via enhanced polymerisation and/or expression of the cytoskeletal components. For the latter scenario, there is a good preliminary evidence from boron-deprived plant seedlings (Yu *et al.*, 2001, 2003; Goldbach *et al.*, 2001). It might be expected that the plasma membrane of plant cells is composed of domains subjected to different stretch-stresses. For instance, recently laid-down cross (transverse) walls contain abundant plasmodesmata, which mechanically stabilise the cell periphery complex, and do not expand further while parent side walls expand dramatically when cells enter the elongation region. We suggest that actin and myosin VIII, interlinking cortical ER elements with the plasma membrane, mechanically stabilise these cell periphery domains. It can be anticipated that very low tension of the plasma membrane at these domains favours endocytosis (Kell & Glaser, 1993; Sheetz

& Dai, 1996; Fricke *et al.*, 2000). On the other hand, the plasma membrane underlying the side (longitudinal) walls can be expected to be under high tensile stress not only because of low density of plasmodesmata but also because of more loosened cell walls due to the abundance of expansins (Zhang & Hasenstein, 2000; our unpublished data). This high tensile stress of the plasma membrane then necessitates abundant exocytosis resulting in rapid growth along the longitudinal cell walls. We propose that stretch-induced stress of the plasma membrane is the basis for polarity also in plant cells. Dynamic balances of the cytoskeleton-mediated intracellular forces, buffering turgor-cell wall mechanics at the cell periphery, seem to lie at the 'heart' of plant polarity.

3.5 Cell walls and polarity at the organismal level

As the position of a cell is fixed in relation to its neighbours (Scheres, 2001), regulation of a cell's size and shape, resulting in the overall morphogenesis of plants and their organs, would be inherently related to the coordinated action of two processes: cell division and cell expansion (Martin *et al.*, 2001). In both of them, cell walls would play a crucial role. First, by fixing the mode (symmetric versus asymmetric) of cellular partitioning, which will affect the fate of individual cells and their descendants, such as during embryogenesis (Scheres & Benfey, 1999) or formation of stomata (Geisler *et al.*, 2003). Second, walls are important in determining the rate, extent and direction of cellular expansion through anisotropic differentiation of mechanical properties of wall domains around the cells. As cell walls *in planta* form a continuous network – the apoplast – this structure and its properties would affect the overall shape and functioning of the whole plant body (Wojtaszek, 2000).

Cell division follows separation of replicated nuclear genomes and involves the partitioning of cytoplasm and formation of a division machinery. It is a common feature of all organisms, among them plants (Nanninga, 2001; Guertin *et al.*, 2002). Location of division plane is crucial for further developmental fate of daughter cells, and it was realised quite early that in plants this process is dependent on the shape and positioning of the cells within the tissue. Although the cytoskeleton seems to be a crucial player there (Brown & Lemmon, 2001), both processes are also strongly related to the presence of cell walls (Smith, 2001). It should be noted, however, that in some cells, like vacuolated shoot cells (Cutler & Ehrhardt, 2002) or tissues characterised by secondary partitioning of existing multinucleated structures, like syncytial gametophytes (Brown *et al.*, 2002) or endosperm (Otegui & Staehelin, 2000a,b; Olsen, 2001), cytokinesis might follow different patterns of events. A relatively large number of genetic mutants impaired in cell division has been identified (Table 3.3; Nacry *et al.*, 2000; Assaad, 2001). It

Table 3.3 Selected cytokinesis-defective mutants

Mutant	Brief characteristics	References
cyt1	Embryo-lethal mutation; growth arrested at the early heart stage; ectopic deposition of callose, incomplete cell walls, decreased cellulose content. *CYT1* encodes mannose-1-phosphate guanylyltransferase. Weak mutation in the same gene (*vtc1*) reduces the content of vitamin C and slows down the growth, but without affecting the viability.	Nickle & Meinke (1998); Lukowitz *et al.* (2001) *vtc1* mutant: Conklin *et al.* (1999); Veljovic-Jovanovic *et al.* (2001)
discordia	Maize mutant. Disrupted guidance of phragmoplast during cytokinesis in asymmetrically dividing cells; no effect during symmetric divisions.	Gallagher & Smith (1999)
fass/tonneau	Mutants with altered organization of seedling body; thick and dwarfed seedlings. Two complementation groups: *ton1* and *ton2*, the latter allelic to *fass*. Required for control of cortical cytoskeleton. TON2 is protein phosphatase 2A subunit	Torres-Ruiz & Jürgens (1994); Camilleri *et al.* (2002)
hinkel	Enlarged cells, incomplete cell walls. HINKEL is a kinesin-related protein required for reorganization of phragmoplast microtubules.	Strompen *et al.* (2002)
keule	Enlarged cells, incomplete cell walls. KEULE is similar to Sec1 regulating vesicle trafficking.	Assaad *et al.* (1996, 2001)
knolle	Enlarged cells, incomplete cell walls. Inside the cells unfused vesicles are accumulated. KNOLLE is cytokinesis-specific syntaxin.	Lukowitz *et al.* (1996); Lauber *et al.* (1997)
root-shoot-hypocotyl-defective (rsh)	Irregular cell shape and size; effects traced back to first asymmetrical division of the zygote; RSH is a HRGP-type wall protein	Hall & Cannon (2002)
tangled1	Maize mutant. Altered orientation of cell divisions without affecting leaf shape. TAN1 is a microtubule-binding protein.	Smith *et al.* (1996); Cleary & Smith (1998) Smith *et al.* (2001)

was proposed that genes required for cytokinesis in somatic cells fall generally into two classes. The first one groups the genes required for proper orientation of division plane, while the second class contains 'executors' of cytokinesis (Söllner *et al.*, 2002). An interesting point is that the mechanism by which the division plane is determined seems to be distantly related to that explored by rod-shaped bacteria (compare, for example Fig. 3.1 in Guertin *et al.*, 2002 with Fig. 6.2 in Harold, 2001a). This observation would be in accordance with our proposal that the mechanisms determining polar cell growth in bacteria and in plants are similar and somehow related (Baluška *et al.*, 2003c). Another observation worth mentioning is the proposal that plant cytokinesis represents a specialised and polarised form of targeted secretion (Assaad, 2001), similar to polarised secretory events leading to wall remodelling or extension of tip-growing cells (Belanger & Quatrano, 2000; Hepler *et al.*, 2001; see also above).

When determining the overall morphogenesis of a plant, cell divisions are tightly coupled to differential cell expansion. It is thought that the latter process of polarised anisotropic cellular growth is largely driven by internal turgor pressure and results from local yielding of cell walls (Harold, 2002). Cell walls can be considered as a material with a structural hierarchy (Lakes, 1993) – a composite made of interweaving networks interacting with each other (Carpita & Gibeaut, 1993; Wojtaszek, 2000). From this point of view, the composition and properties of individual components are less important. In the living world, these are the physicochemical, mainly mechanical, properties of the resulting matrix, that are crucial for a given organism. It is thus not a surprise that both plants and animals construct their extracellular matrices according to the same set of rules, albeit from different components. In both cases, however, the resulting structure is resistant at the same time to tension due to fibrillar components (cellulose versus collagen) and to compression due to an amorphous matrix, made mainly from pectins in plants (Wojtaszek, 2000; Kerstens *et al.*, 2001; Chanliaud *et al.*, 2002). It is thus obvious that remodelling of the cell walls or wall domains in plants comprises at least three tightly coupled processes: directed secretion/insertion/internalisation of wall components, restructuring of interactions between existing and/or new wall components, and sensing of the mechanical environment of the cells. It should be noted here, that although most of these processes involves biochemical machinery of the protoplast, some of them, especially when the participation of ROS is considered (Wojtaszek, 1997; Fry, 1998; Chen & Schopfer, 1999; Bolwell *et al.*, 2001; Schopfer *et al.*, 2002), could be limited only to events occurring extracellularly. Telling examples come from the analysis of plant cells adapted to salt or osmotic stress or grown in the presence of 2,6-dichlorobenzonitrile (DCB – an inhibitor of cellulose biosynthesis). Transgenic plants developing high turgor pressure show an increased thickness of the walls (Hoffmann-Benning *et al.*, 1997). In NaCl-adapted tobacco cells, the cellulosic-extensin framework was identified as a primary determinant of absolute wall tensile strength (Iraki *et al.*, 1989). However, tomato cells adapted to DCB virtually lack cellulose microfibrils and accompanying xyloglucans, but are still able to divide and grow (Shedletzky *et al.*, 1990). In those cells, mechanical stability of the walls comes from an altered pattern of wall polymer cross-linking, mainly via phenolic components (Shedletzky *et al.*, 1992). This points to an extreme flexibility of the walls with respect to building components as long as the threshold mechanical properties are maintained. However, experiments with *trans*-differentiating *Zinnia* cells highlighted the importance of temporal regulation in wall synthesis and assembly. Disturbance of cellulose synthesis with DCB resulted in an aberrant deposition of lignin, which suggests that at least, the latter process is dependent on the proper construction of the cellulosic network (Suzuki *et al.*, 1992; Taylor *et al.*, 1992). On the other hand, when lignin biosynthesis is repressed, wall mechanical properties are maintained through an increased accumulation of cellulose (Hu *et al.*, 1999).

The local modulation of wall mechanical properties and thus polarised orientation of growth could be achieved either via restructuring of existing components or through directional/local secretion and endocytosis of wall building blocks. In most cases both phenomena are closely related, and the differentiation of wall properties results from either the extent or spatiotemporal regulation of individual processes. Identified wall-related mutants (Turner *et al.*, 2001; Carpita & McCann, 2002), particularly those with altered growth patterns (Table 3.2), and available biochemical data seem to confirm such conclusions (for xyloglucans see e.g. Shinohara *et al.*, 2000; Pauly *et al.*, 2001; Takeda *et al.*, 2002; Vissenberg *et al.*, 2003; for pectins see e.g. Dolan *et al.*, 1997; Jauneau *et al.*, 1997; Pilling *et al.*, 2000; Ridley *et al.*, 2001). As the polarity of growth is concerned, it is curious that in all known wall mutants, it is the elongation growth that is affected. This implies disturbed functioning of longitudinal walls. Interestingly, to our best knowledge, no mutant is as yet available showing defects in cross-walls which again implies the importance of those wall domains fulfilling the role of putative plant synapses (Baluška *et al.*, 2003c).

The body of higher plants is oriented along major polar axes (Tasaka, 2001). In roots and shoots, cells are arranged along an apical-basal axis which, in most land plants, is oriented primarily with respect to gravity with roots embedded in the soil and facing downwards, and shoots and flowers growing against a gravity vector (Cove, 2000). Moreover, there is a secondary central-peripheral axis orthogonal to the apical-basal one allowing the cells to orient their activities and to maintain radial symmetry (Tasaka, 2001). Auxin and its polar transport along the apical-basal axis is usually considered as the primary internal integrator of polar cellular alignment (Friml & Palme, 2002; Friml, 2003; Baluška *et al.*, 2003a; Niklas, 2003; Chapter 2) allowing general coordination of morphogenesis (Sachs, 2000; Berleth & Sachs, 2001; Chapters 7 and 9). As cell walls form a continuous network – the apoplast, permeating the whole plant body – the properties of such a structure, although modulated by the action of individual cells, would have an immediate bearing of the functioning of the plant as a whole (Niklas, 1992). Here, we would consider the mechanical properties of 'tubular' organs from the cell wall perspective. The wall surrounding the fertilised zygote retains its external character, and this is usually presented as evidence for the organismal nature of external walls of epidermal layers (Kaplan and Hagemann, 1991; Wojtaszek, 2001). Mechanically, the epidermis of developing plants is often regarded as fulfilling the role of the determinant of external shape, affecting also the growth polarity and mechanical properties of plant organs as a whole (Kutschera, 2001; Verbelen *et al.*, 2001). Various tissues within an organ grow at different rates due to variable mechanical properties of their walls. This could be confirmed when the plasmolytic behaviour of the cells in various tissues is observed following enzymatic digestions, affecting mechanical stability of the walls (Fig. 3.1). In addition, the construction of the walls around epidermal cells is differentiated with external walls, facing the outside world, much thicker than parallel walls facing the inner tissues (see Fig. 1 in Wojtaszek, 2000). As a result, under normal conditions, an epidermis is subjected to permanent mechanical stress (Hejnowicz *et al.*, 2000). Importantly, this stress is not only transmitted by longitudinal walls, but also by epidermal cross-walls implying that the composition and mechanical properties of the latter must be different from the analogous cross-walls of inner tissues (Niklas, 1992). It has been suggested that these stresses are sensed and transmitted back into respective tissues through membrane-bound cortical microtubules, allowing the cells to respond to changing mechanical conditions and, eventually, to modify the direction of growth (Zandomeni & Schopfer, 1994; Fischer & Schopfer, 1998). Indeed, such coordinated reorientations and establishment of a new growth polarity have been observed for example during the wound response (Hush *et al.*, 1990). Interestingly, the net orientation of cellulose microfibrils in the outer walls of epidermal layer of the root elongation zone seems to be specific for and dependent on the type of the walls building the plant. In commelinoid plants with type II walls, cellulose microfibrils are aligned parallel to the root growth axis, while in all other angiosperm plants with type I walls, the net transverse orientation is observed (Verbelen & Kerstens, 2000; Kerstens & Verbelen, 2002). Interestingly, when plants are characterised by the true lack of root elongation zone, like in *Arecaceae*, cellulose microfibrils are distributed randomly and no net orientation could be discriminated (Kerstens & Verbelen, 2002).

3.6 Cell walls as a linker between different levels of polar organisational hierarchy

Cell walls are present in the life of the plant from the beginning to the end. Formation of a cell wall around the zygote is usually the first event following fertilisation, and the degradation of the walls marks the end in the life of the plant. Cell walls are also present in all types of hierarchical organisation throughout the whole plant kingdom from unicellular algae to higher plants. They participate in the formation and maintenance of various body plans (Niklas, 1999) fulfilling their main role of mechanical support and structural integrator of plant form and function. An interesting observation by Niklas (1992) points to the unique status of the walls in the living world: 'The cell wall is functionally analogous to the skeleton in animals, but unlike metazoan evolution, in which the acquisition of a skeleton followed or closely paralleled the specialization of cells, the evolution of plant cell wall predates the appearance of multicellularity and cellular specialization.' It is worth remembering here that plants are not truly multicellular organisms, but rather supracellular (Lucas *et al.*, 1993) with partially separated protoplasts connected through plasmodesmata. These gateways allow efficient cell-to-cell exchange of signalling molecules and thus coordination of developmental pathways throughout the cells of symplasmic domains (Zambryski & Crawford, 2000; Haywood *et al.*, 2002; Heinlein, 2002).

Modern biology offers a unique opportunity to study the multiplicity of organismal organisation types and its associations with polarity with the use of related experimental model systems. At one end, there will be suspension-cultured cells whose status could be placed somewhere between individual organisms and subpopulations within organisms (Wojtaszek, 2000). The unorganised calluses and cell aggregates would present the next level, and tissues, organs and the whole plants are at the other extreme. Comparisons of data coming from those different systems provide several interesting observations and conclusions. First, at the biochemical level reactions of cell suspensions and intact plants to the imposed stresses are different (Adams *et al.*, 1992; Thomas *et al.*, 1992) indicating that many responses require the hierarchy of organised tissue to be expressed (Aon *et al.*, 1999). Second, as the cells are the products of highly ordered molecular cell engineering resulting from 'solid-state biochemistry' (Ingber, 1993b; Mathews, 1993), 'macromolecular crowding' (Luby-Phelps, 2000), and cytoskeletal organisation (Aon & Cortassa, 2002), these observations suggest that the mechanical environment of the cells could be one of the primary determinants of their function and developmental fate. For example, random cell proliferation of unorganised callus could be made highly-stereotyped in a cambium-like manner following the application of external pressure (Lintilhac & Vesecky, 1984). When individual cells or protoplasts are considered, it could also be demonstrated that they are able to detect direction and extent of mechanical stresses and align accordingly their division planes (Lynch & Lintilhac, 1997). Moreover, for protoplasts the quality and composition of the surrounding medium, either liquid or solidified becomes the developmental signal determining their fate (Roger *et al.*, 1996; Caumont *et al.*, 1997), as in dividing *Fucus* zygotes (Kropf *et al.*, 1988; Berger *et al.*, 1994; see Chapter 5). This leads to the third conclusion, corroborating earlier data summarised by Drubin & Nelson (1996). In plant cells, tissues and organs, the walls and the apoplast are spatial determinants of intracellular processes. The cues provided through vectorial mechanical stresses of the walls are recognised by and used for the cytoskeletal organisation, allowing polarised cellular reactions (Wymer *et al.*, 1996). When stabilised in any way, they are transformed into polarised growth and changes of cellular shapes (Martin *et al.*, 2001). These processes might be self-propagatory as the shape again would affect intracellular organisation (Fisher & Cyr, 1998; Roberts & Uhnak, 1998). With proper feedback mechanisms, this would lead to morphogenetic responses of the cells and the whole plants. When looked at in this way, one could say that there are at least two complementary mechanisms of intercellular communication ensuring the coordinated developmental pathways within tissues or groups of cells. The first will use the soluble signalling molecules transported either through plasmodesmata (Haywood *et al.*, 2002) or in the neuron-like mode (Baluška *et al.*, 2003a). The second will utilise mechanical (and structural) spatial cues to coordinately align cyto-skeletal elements (Overall *et al.*, 2001) and, through them, secretory events leading to remodelling of the surrounding walls. Cross-talk between both pathways could also be envisaged (Shibaoka, 1994). In all cases, however, the walls and the cytoskeleton have an indispensable role to play, although the exact relationship between both is still not clear (Lloyd *et al.*, 1980; Sugimoto *et al.*, 2003). An excellent example of such a regulatory mechanism, which is related to the formation of secondary cell wall thickenings in model systems of suspension-cultured cells, has been reviewed recently (Wojtaszek, 2000).

The transition from soluble, one-dimensional information contained in genes into three- dimensional functional structures able to persist in time still seems to be one of the most interesting, and mostly unsolved problems of modern biology (Harold, 1990, 1995). There are at least two levels at which this problem could be considered: the cellular one and the 'architectural' one dealing with the whole organs and plants (Reinhardt & Kuhlemeier, 2002). In our view, the plant cell wall, forming the apoplast of the plant body, fulfils an important role at both levels and provides a mechanistic transitory device and a linker between both levels. As mentioned above, cell walls are at the centre of both crucial processes governing plant morphogenesis: cell division and cell expansion. At the plant level, there are also mechanical properties of the walls, especially outer epidermal walls, that determine the physical properties of tissues and organs. Recalling the renewed discussion between cell and organismal theories (Kaplan & Hagemann, 1991; Wojtaszek, 2001), one could say that plant cell walls are cellular devices with organismal features. The wall proteins, expansins, can be used here to illustrate this point. Expansins are encoded by multigene families (Arabidopsis Genome Initiative, 2000; Li *et al.*, 2002) and found in most land plants, effecting the extensibility of the walls and thus participating in the control of cell growth (for reviews see Cosgrove, 1999; McQueen-Mason & Rochange, 1999; Cosgrove *et al.*, 2002). The current model proposes that they are acting in a pH-dependent mode via decreasing the extent of hydrogen bonding between cellulose microfibrils and matrix polysaccharides, especially xyloglucans. At the cellular level, they seem to play a role in root hair initiation through local weakening of the walls enabling formation of the initial bulge (Baluška *et al.*, 2000c; Cho & Cosgrove, 2002). Similarly, expansin mRNAs were found to be polarly localised either at the apical or basal ends of differentiating xylem cells suggesting the role for proteins in intrusive growth of those cells (Im *et al.*, 2000). On the other hand, permanent overexpression of expansin genes at the whole plant level led to the generation of shorter transgenics and it seems that such increased weakening of the walls by expansins invoked the activation of strong compensatory mechanisms (Rochange *et al.*, 2001). However, when placed under inducible promoters, expressed expansins have been positively correlated with enhanced plant growth (Choi *et al.*, 2003). The most interesting data come from the experiments implicating expansins in the regulation of morphogenesis. As proposed by Green (1994, 1996, 1999), generation of pattern through the formation of new primordia could be achieved via controlled extensibility of the outer epidermal walls initiating folding and subsequent tissue expansion. Indeed, local exogenous application of expansins on shoot apical meristem leads to the induction of leaf primordia and the reversal of phyllotaxis (Fleming *et al.*, 1997), although the structures forming are not phenotypically normal leaves (Fleming *et al.*, 1999). It was hypothesised that this was due to the lifting of the biophysical wall limitations only at the outermost cell layers. On the other hand, in normal plants local induction of genes coding for expansins, spanning several cell layers, predicts the site of future leaf formation (Reinhardt *et al.*, 1998), and experimental manipulation of expansin gene levels also enables induction of normal leaf formation (Pien *et al.*, 2001). When combined with the micromanipulation of expression levels for genes affecting cell division (cyclin A and yeast cdc25 – Wyrzykowska *et al.*, 2002; phragmoplastin – Wyrzykowska & Fleming, 2003), these experiments enabled for the first time a direct experimental evaluation of the relative importance of cell division and cell expansion for leaf morphogenesis (Fleming, 2002). A conclusion which could be drawn here would be that local induction of expansin, i.e. modification of wall extensibility, is a sufficient trigger for organogenesis (Pien *et al.*, 2001), while modulation of cell divisions have no effect, at least at the shoot meristem (Wyrzykowska *et al.*, 2002). Using the words of Reinhardt & Kuhlemeier (2002): 'it is likely that growth is regulated at a supracellular level, possibly through differential expansion of the apoplastic cell wall "exoskeleton" at the tissue/organ level.' A rather unexpected end to this story is provided by recent observations that particular nucleotide sequences present in the central conserved region of viroid RNAs are able to interfere with plant growth and development. Infected plants become dwarfed and this is due to restriction of cellular expansion and not cell division. Interestingly, this phenotype is also positively correlated with the downregulation of expansin gene expression (Qi & Ding, 2003).

Acknowledgements

Przemysław Wojtaszek and František Baluška would like to express their gratitude to the Alexander von Humboldt Foundation (Bonn, Germany) for support. Przemysław Wojtaszek thanks the Polish Committee for Scientific Research for funding.

References

Adams, P., Thomas, J.C., Vernon, D.M., Bohnert, H.J. & Jensen, R.G. (1992) Distinct cellular and organismic responses to salt stress. *Plant Cell Physiol.*, **33**, 1215–1223.

Akashi, T. & Shibaoka, H. (1991) Involvement of transmembrane proteins in the association of cortical microtubules with the plasma membrane in tobacco BY-2 cells. *J. Cell Sci.*, **98**, 169–174.

Akashi, T., Kawasaki, S. & Shibaoka, H. (1990) Stabilization of cortical microtubules by the cell wall in cultured tobacco cells. Effects of extensin on the cold-stability of cortical microtubules. *Protoplasma*, **182**, 363–369.

Andème-Onzighi, C., Sivaguru, M., Judy-March, J., Baskin, T.I. & Driouich, A. (2002) The *reb1-1* mutation of Arabidopsis alters the morphology of trichoblasts, the expression of arabinogalactan-proteins and the organization of cortical microtubules. *Planta*, **215**, 949–958.

Anderson, C.M., Wagner, T.A., Perret, M., He, Z.-H., He, D. & Kohorn, B.D. (2001) WAKs: cell wall-associated kinases linking the cytoplasm to the extracellular matrix. *Plant Molec. Biol.*, **47**, 197–206.

Aon, M.A. & Cortassa, S. (2002) Coherent and robust modulation of a metabolic network by cytoskeletal organization and dynamics. *Biophys. Chem.*, **97**, 213–231.

Aon, M.A., Cortassa, S., Gomez Casati, D.F. & Iglesias, A.A. (1999) Effects of stress on cellular infrastructure and metabolic organization in plant cells. *Int. Rev. Cytol.*, **194**, 239–273.

Arabidopsis Genome Initiative (2000) Analysis of the genome sequence of the flowering plant *Arabidopsis thaliana*. *Nature*, **408**, 796–815.

Arioli, T., Peng, L., Betzner, A.S., Burn, J., Wittke, W., Herth, W., Camilleri, C., Höfte, H., Plazinski, J., Birch, R., Cork, A., Glover, J., Redmond, J. & Williamson, R.E. (1998) Molecular analysis of cellulose biosynthesis in *Arabidopsis*. *Science*, **279**, 717–720.

Assaad, F.F. (2001) Plant cytokinesis. Exploring the links. *Plant Physiol.*, **126**, 509–516.

Assaad, F.F., Huet, Y., Mayer, U. & Jürgens, G. (2001) The cytokinesis gene *KEULE* encodes a

Sec1 protein which binds the syntaxin KNOLLE. *J. Cell Biol.*, **152**, 531–543.
Assaad, F.F., Mayer, U., Wanner, G. & Jürgens, G. (1996) The *KEULE* gene is involved in cytokinesis in *Arabidopsis*. *Mol. Gen. Genet.*, **253**, 267–277.
Baluška, F. & Volkmann, D. (2002) Actin-driven polar growth of plant cells. *Trends Cell Biol.* 12, 14.
Baluska, F., Barlow, P.W. & Volkmann, D. (2000a) Actin and myosin VIII in developing root cells, in Actin: a dynamic framework for multiple plant cell functions, (eds. C.J. Staiger, F. Baluška, D. Volkmann and P.W. Barlow), Kluwer Academic Publishers, Dordrecht, The Netherlands, pp 457–476.
Baluška, F., Ovečka, M. & Hirt, H. (2000b) Salt stres- and cell cycle phase-dependent changes in expression and subcellular localisation of the alfalfa mitogen-activated protein kinase SIMK. *Protoplasma*, **212**, 262–267.
Baluška, F., Šalaj, J., Mathur, J., Braun, M., Jasper, F., Šamaj, J., Chua, N.-H., Barlow, P.W. & Volkmann, D. (2000c) Root hair formation: F-actin-dependent tip growth is initiated by local assembly of profilin-supported F-actin meshworks accumulated within expansin-enriched bulges. *Devel. Biol.*, **227**, 618–632.
Baluška, F., Volkmann, D. & Barlow, P.W. (2000d) Actin-based domains of the 'cell periphery complex' and their associations with polarized 'cell bodies' in higher plants. *Plant Biol.*, **2**, 253–267.
Baluška, F., Cvrčková, F., Kendrick-Jones, J. & Volkmann, D. (2001a). Sink plasmodesmata as gateways for phloem unloading. Myosin VIII and calreticulin as molecular determinants of sink strength? *Plant Physiol.*, **126**, 39–46.
Baluška, F., Jasik, J., Edelmann, H.G., Šalajova, T. & Volkmann, D. (2001b) Latrunculin B-induced plant dwarfism: plant cell elongation is F-actin-dependent. *Devel. Biol.*, **231**, 113–124.
Baluška, F., Volkmann, D. & Barlow, P.W. (2001c) A polarity crossroad in the transition growth zone of maize root apices: cytoskeletal and developmental implications. *J. Plant Growth Reg.*, **20**, 170–181.
Baluška, F., Hlavačka, A., Šamaj, J., Palme, K., Robinson, D.G., Matoh, T., McCurdy, D.W., Menzel, D. & Volkmann, D. (2002) F-actin-dependent endocytosis of cell wall pectins in meristematic root cells. Insight from Brefeldin A-induced compartments. *Plant Physiol.*, **130**, 422–431.
Baluška, F., Šamaj, J. & Menzel, D. (2003a) Polar transport of auxin: carrier-mediated flux across the plasma membrane or neurotransmitter-like secretion? *Trends Cell Biol.*, **13**, 282–285.
Baluška, F., Šamaj, J., Wojtaszek, P., Volkmann, D. & Menzel, D. (2003b) Cytoskeleton-plasma membrane-cell wall continuum in plants: emerging links revisited. *Plant Physiol.*, **133**, 482–491.
Baluška, F., Wojtaszek, P., Volkmann, D. & Barlow, P. (2003c) The architecture of polarized cell growth: the unique status of elongating plant cells. *BioEssays*, **25**, 569–576.
Barthou, H., Petitprez, M., Brière, C., Souvré, A. & Alibert, G. (1999) RGD-mediated membrane-matrix adhesion triggers agarose-induced embryoid formation in sunflower protoplasts. *Protoplasma*, **206**, 143–151.
Baskin, T.I., Cork, A., Williamson, R.E. & Gorst, J.R. (1995) *STUNTED PLANT 1*, a gene required for expansion in rapidly elongating but not in dividing cells and mediating root growth responses to applied cytokinin. *Plant Physiol.*, **107**, 233–243.
Baumberger, N., Ringli, C. & Keller, B. (2001) The chimeric leucine-rich repeat/extensin cell wall protein LRX1 is required for root hair morphogenesis in *Arabidopsis thaliana*. *Genes Devel.*, **15**, 1128–1139.
Baumberger, N., Steiner, M., Ryser, U., Keller, B. & Ringli, C. (2003) Synergistic interaction of the two paralogous *Arabidopsis* genes *LRX1* and *LRX2* in cell wall formation during root hair development. *Plant J.*, **35**, 71–81.
Bechinger, C., Giebel, K.-F., Schnell, M., Leiderer, P., Deising, H.B. & Bastmeyer, M. (1999) Optical measurements of invasive forces exerted by appressoria of a plant pathogenic fungus. *Science*, **285**, 1896–1899.
Belanger, K.D. & Quatrano, R.S. (2000) Polarity: the role of localized secretion. *Curr. Op. Plant Biol.*, **3**, 67–72.

Berger, D. & Altmann, T. (2000) A subtilisin-like serine protease involved in the regulation of stomatal density and distribution in *Arabidopsis thaliana*. *Genes Devel.*, **14**, 1119–1131.

Berger, F., Taylor, A. & Brownlee, C. (1994) Cell fate determination by the cell wall in early *Fucus* development. *Science*, **263**, 1421–1423.

Berleth, T. & Sachs, T. (2001) Plant morphogenesis: long-distance coordination and local patterning. *Curr. Op. Plant Biol.*, **4**, 57–62.

Bernhardt, C. & Tierney, M.L. (2000) Expression of AtPRP3, a proline-rich structural cell wall protein from *Arabidopsis*, is regulated by cell-type-specific developmental pathways involved in root hair formation. *Plant Physiol.*, **122**, 705–714.

Bibikova, T.N., Zhigilei, A. & Gilroy, S. (1997) Root hair growth in *Arabidopsis thaliana* is directed by calcium and an endogenous polarity. *Planta*, **203**, 495–505.

Bibikova, T.N., Jacob, T., Dahse, I. & Gilroy, S. (1998) Localized changes in apoplastic and symplastic pH are associated with root hair development in *Arabidopsis thaliana*. *Development*, **125**, 2925–2934.

Blackman, L.M. & Overall, R.L. (2001) Structure and function of plasmodesmata. *Austr. J. Plant Physiol.*, **28**, 709–727.

Blanpied, T.A., Scott, D.B. & Ehlers, M.D. (2002) Dynamics and regulation of clathrin coats at specialized endocytic zones of dendrites and spines. *Neuron*, **36**, 435–449.

Blee, K.A., Wheatley, E.R., Bonham, V.A., Mitchell, G.P., Robertson, D., Slabas, A.R., Burell, M.M., Wojtaszek, P. & Bolwell, G.P. (2001) Proteomic analysis reveals a novel set of cell wall proteins in a transformed tobacco cell culture which synthesises secondary walls as determined by biochemical and morphological parameters. *Planta*, **212**, 404–415.

Boal, D. (2002) *Mechanics of the Cell*. Cambridge University Press, Cambridge.

Bolwell, G.P., Page, A., Piślewska, M. & Wojtaszek, P. (2001) Pathogenic infection and the oxidative defences in plant apoplast. *Protoplasma*, **217**, 20–32.

Bonin, C.P., Potter, I., Vanzin, G.F. & Reiter, W.D. (1997) The *MUR1* gene of *Arabidopsis thaliana* encodes an isoform of GDP-D-mannose-4,6-dehydratase, catalyzing the first step in the de novo synthesis of GDP-L-fucose. *Proc. Natl. Acad. Sci. USA*, **94**, 2085–2090.

Borisy, G.G. & Svitkina, T.M. (2000) Actin machinery. Pushing the envelope. *Curr. Op. Cell Biol.*, **12**, 104–112.

Bouton, S., Leboeuf, E., Mouille, G., Leydecker, M.-T., Talbotec, J., Granier, F., Lahaye, M., Höfte, H. & Truong, H.-N. (2002) *QUASIMODO1* encodes a putative membrane-bound glycosyltransferase required for normal pectin synthesis and cell adhesion in Arabidopsis. *Plant Cell*, **14**, 2577–2590.

Brett, C. & Waldron, K. (1996) *Physiology and Biochemistry of Plant Cell Walls*. 2nd Ed. Chapmann & Hall, London.

Brock, A., Chang, E., Ho, C.-C., LeDuc, P., Jiang, X., Whitesides, G.M. & Ingber, D.E. (2003) Geometric determinants of directional cell motility revealed using microcontact printing. *Langmuir*, **19**, 1611–1617.

Brown, R.C. & Lemmon, B.E. (2001) The cytoskeleton and spatial control of cytokinesis in the plant life cycle. *Protoplasma*, **215**, 35–49.

Brown, R.C., Lemmon, B. & Nguyen, H. (2002) The microtubule cycle during successive mitotic waves in the syncytial female gametophyte of ginkgo. *J. Plant Res.*, **115**, 491–494.

Brownlee, C., Goddard, H., Hetherington, A.M. & Peake, L.-A. (1999) Specificity and integration of responses: Ca^{2+} as a signal in polarity and osmotic regulation. *J. Exp. Bot.*, **50** Sp. Iss., 1001–1011.

Bucher, M., Brunner, S., Zimmermann, P., Zardi, G.I., Amrhein, N., Willmitzer, L. & Riesmeier, J.W. (2002) The expression of an extensin-like protein correlates with cellular tip growth in tomato. *Plant Physiol.*, **128**, 911–923.

Buer, C.S., Weathers, P.J. & Swartzlander Jr., G.A. (2000) Changes in Hechtian strands in cold-hardened cells measured by optical microsurgery. *Plant Physiol.*, **122**, 1365–1377.

Burk, D.H. & Ye, Z.-H. (2002) Alteration of oriented deposition of cellulose microfibrils by mutation of a katanin-like microtubule-severing protein. *Plant Cell*, **14**, 2145–2160.

Burk, D.H., Liu, B., Zhong, R., Morrison, W.H. & Ye, Z.-H. (2001) A katanin-like protein regulates normal cell wall biosynthesis and cell elongation. *Plant Cell*, **13**, 807–827.

Camacho, L. & Malhó, R. (2003) Exo/endocytosis in the pollen tube apex is differentially regulated by Ca^{2+} and GTPases. *J. Exp. Bot.*, **54**, 83–92.

Camilleri, C., Azimzadeh, J., Pastuglia, M., Bellini, C., Grandjean, O. & Bouchez, D. (2002) The Arabidopsis *TONNEAU2* gene encodes a putative novel protein phosphatase 2A regulatory subunit essential for the control of the cortical cytoskeleton. *Plant Cell*, **14**, 833–845.
Caño-Delgado, A., Metzlaff, K. & Bevan, M. (2000) The *eli1* mutation reveals a link between cell expansion and secondary cell wall formation in *Arabidopsis thaliana*. *Development*, **127**, 3395–3405.
Canut, H., Carrasco, A., Galaud, J.-P., Cassan, C., Bouyssou, H., Vita, N., Ferrara, P. & Pont-Lezica, R. (1998) High affinity RGD-binding sites at the plasma membrane of *Arabidopsis thaliana* links the cell wall. *Plant J.*, **16**, 63–71.
Carpita, N.C. & Gibeaut, D.M. (1993) Structural models of primary cell walls in flowering plants: consistency of molecular structure with the physical properties of the walls during growth. *Plant J.*, **3**, 1–30.
Carpita, N.C. & McCann, M.C. (2002) The functions of cell wall polysaccharides in composition and architecture revealed through mutations. *Plant Soil*, **247**, 71–80.
Casero, P.J. & Knox, J. P. (1995) The monoclonal antibody JIM5 indicates patterns of pectin deposition in relation to pit fields at the plasma-membrane-face of tomato pericarp cell walls. *Protoplasma*, **188**, 133–137.
Catterou, M., Dubois, F., Schaller, H., Aubanelle, L., Vilcot, B., Sangwan-Norreel, B.S. & Sangwan, R.S. (2001) Brassinosteroids, microtubules and cell elongation in *Arabidopsis thaliana*. I. Molecular, cellular and physiological characterization of the *Arabidopsis bul1* mutant, defective in the Δ^7-sterol-C5–desaturation step leading to brassinosteroid biosynthesis. *Planta*, **212**, 659–672.
Caumont, C., Petitprez, M., Woynaroski, S., Barthou, H., Brière, C., Kallerhoff, J., Borin, C., Souvré, A. & Alibert, G. (1997) Agarose embedding affects cell wall regeneration and microtubule organization in sunflower hypocotyl protoplasts. *Physiol. Plant.*, **99**, 129–134.
Chanliaud, E., Burrows, K.M., Jeronimidis, G. & Gidley, M.J. (2002) Mechanical properties of primary plant cell wall analogues. *Planta*, **215**, 989–996.
Chapman, E.R. (2002) Synaptotagmin: a Ca^{2+} sensor that triggers exocytosis? *Nature Rev. Molec. Cell Biol.*, **3**, 1–11.
Chen, J.-G., Shimomura, S., Sitbon, E., Sandberg, G. & Jones, A.M. (2001a) The role of auxin-binding protein 1 in the expansion of tobacco cell. *Plant J.*, **28**, 607–617.
Chen, J.-G., Ullah, H., Young, J.C., Sussman, M.R. & Jones, A.M. (2001b) ABP1 is required for organized cell elongation and division in *Arabidopsis* embryogenesis. *Genes Devel.*, **15**, 902–911.
Chen, M.H., Sheng, J., Hind, G., Handa, A.K. & Citovsky, V. (2000) Interaction between the tobacco mosaic virus movement protein and host cell pectin methylesterases is required for viral cell-to-cell movement. *EMBO J.*, **19**, 913–920.
Chen, S.-X. & Schopfer, P. (1999) Hydroxyl-radical production in physiological reactions. A novel function of peroxidases. *Eur. J. Biochem.*, **260**, 726–735.
Cheng, J.-C., Lertpiriyapong, K., Wang, S. & Sung, Z.R. (2000) The role of the Arabidopsis *ELD1* gene in cell development and photomorphogenesis in darkness. *Plant Physiol.*, **123**, 509–520.
Chivasa, S., Ndimba, B.K., Simon, W.J., Robertson, D., Yu, X.-L., Knox, J.P., Bolwell, G.P. & Slabas, A.R. (2002) Proteomic analysis of the *Arabidopsis thaliana* cell wall. *Electrophoresis*, **23**, 1754–1765.
Cho, H.-T. & Cosgrove, D.J. (2002) Regulation of root hair initiation and expansin gene expression in Arabidopsis. *Plant Cell*, **14**, 3237–3253.
Choi, D., Lee, Y., Cho, H.-T. & Kende, H. (2003) Regulation of expansin gene expression affects growth and development in transgenic rice plants. *Plant Cell*, **15**, 1386–1398.
Ciamporová, M., Dekánková, K., Hanáčková, Z., Peters, P., Ovečka, M. & Baluška, F. (2003) Structural aspects of bulge formation during root hair initiation. *Plant Soil*, **155**, 1–7.
Cleary, A.L. (2001) Plasma membrane-cell wall connections: roles in mitosis and cytokinesis revealed by plasmolysis of *Tradescantia virginiana* leaf epidermal cells. *Protoplasma*, **215**, 21–34.
Cleary, A.L. & Smith, L.G. (1998) The *Tangled1* gene is required for spatial control of cytoskeletal arrays associated with cell division during maize leaf development. *Plant Cell*, **10**, 1875–1888.

Conklin, P.L., Norris, S.R., Wheeler, G.L., Williams, E.H., Smirnoff, N. & Last, R.L. (1999) Genetic evidence for the role of GDP-mannose in plant ascorbic acid (vitamin C) synthesis. *Proc. Natl. Acad. Sc. USA*, **96**, 4198–4203.

Cosgrove, D.J. (1999) Enzymes and other agents that enhance cell wall extensibility. *Ann. Rev. Plant Physiol. Plant Molec. Biol.*, **50**, 391–417.

Cosgrove, D.J. (2000) Expansive growth of plant cell walls. *Plant Physiol. Biochem.*, **38**, 109–124.

Cosgrove, D.J., Li, L.C., Cho, H.-T., Hoffmann-Benning, S., Moore, R.C. & Blecker, D. (2002) The growing world of expansins. *Plant Cell Physiol.*, **43**, 1436–1444.

Cove, D.J. (2000) The generation and modification of cell polarity. *J. Exp. Bot.*, **51**, 831–838.

Craxton, M. (2001) Genomic analysis of synaptotagmin genes. *Genomics*, **77**, 43–49.

Cutler, S.R. & Ehrhardt, D.W. (2002) Polarized cytokinesis in vacuolate cells of *Arabidopsis*. *Proc. Natl. Acad. Sci. USA*, **99**, 2812–2817.

de Jong, J.C., McCormack, B.J., Smirnoff, N. & Talbot, N.J. (1997) Glycerol generates turgor in rice blast. *Nature*, **389**, 244–245.

Derksen, J., Rutten, T., Lichtscheidl, I.K., Dewin, A.H.N., Pierson, E.S. & Rongen, G. (1995) Quantitative analysis of the distribution of organelles in tobacco pollen tubes: implications for exocytosis and endocytosis. *Protoplasma*, **188**, 267–276.

Derksen, J., van Wenzel, R., Knuiman, B., Ylstra, B. & van Tunen, A.J. (1999) Pollen tubes of flavonol-deficient *Petunia*, show striking alterations in wall structure leading to tube disruption. *Planta*, **207**, 575–581.

Ding, L. & Zhu, J.-K. (1997) A role for arabinogalactan-proteins in root epidermal cell expansion. *Planta*, **203**, 289–294.

Dolan, L., Linstead, P. & Roberts, K. (1997) Developmental regulation of pectic polysaccharides in the root meristem of *Arabidopsis*. *J. Exp. Bot.*, **48**, 713–720.

Dong, C.-H., Xia, G.-X., Hong, Y., Ramachandran, S., Kost, B. & Chua, N.-H. (2001) ADF proteins are involved in the control of flowering and regulate F-actin organization, cell expansion, and organ growth in Arabidopsis. *Plant Cell*, **13**, 1333–1346.

Dorokhov, Y.L., Mäkinen, K., Frolova, O.Y., Merits, A., Saarinen, J., Kalkkinen, N., Atabekov, J.G. & Saarma, M. (1999) A novel function for a ubiquitous plant enzyme pectin methylesterase – the host-cell receptor for the tobacco mosaic virus movement protein. *FEBS Lett.*, **461**, 223–228.

Drubin, D.G. & Nelson, W.J. (1996) Origins of cell polarity. *Cell*, **84**, 335–344.

Dustin, M.L. & Cooper, J.A. (2000) The immunological synapse and the actin cytoskeleton: molecular hardware for T cell signaling. *Nature Immunol.*, **1**, 23–29.

Ehlers, K. & Kollmann, R. (2001) Primary and secondary plasmodesmata: structure, origin, and functioning. *Protoplasma*, **216**, 1–30.

Ehlers, K., Binding, H. & Kollmann, R. (1999) The formation of symplasmic domains by plugging of plasmodesmata: a general event in plant morphogenesis? *Protoplasma*, **209**, 181–192.

Eleftheriou, E.P. & Tsekos, I. (1982) Development of protophloem in roots of *Aegilops comosa*, var. *thessalica*. II. Sieve-element differetiation. *Protoplasma*, **113**, 221–233.

Fagard, M., Desnos, T., Desprez, T., Goubet, F., Refregier, G., Mouille, G., McCann, M., Rayon, C., Vernhettes, S. & Höfte, H. (2000) *PROCUSTE1* encodes a cellulose synthase required for normal cell elongation specifically in roots and dark-grown hypocotyls of Arabidopsis. *Plant Cell*, **12**, 2409–2423.

Faik, A., Labouré, A.M., Gulino, D., Mandaron, P. & Falconet, D. (1998) A plant surface protein sharing structural properties with animal integrins. *Eur. J. Biochem.*, **253**, 552–559.

Favery, B., Ryan, E., Foreman, J., Linstead, P., Boundonck, K., Steer, M., Shaw, P. & Dolan, L. (2001) KOJAK encodes a cellulose-synthase-like protein required for root hair cell morphogenesis in Arabidopsis. *Genes Devel.*, **15**, 79–89.

Feijó, J.A., Sainhas, J., Holdaway-Clarke, T., Cordeiro, M.S., Kunkel, J.G. & Hepler, P.K. (2001) Cellular oscillations and the regulation of growth: the pollen tube paradigm. *BioEssays*, **23**, 86–94.

Fischer, K. & Schopfer, P. (1998) Physical strain-mediated microtubule reorientation in the epidermis of gravitropically or phototropically stimulated maize coleoptiles. *Plant J.*, **15**, 119–123.

Fisher, D.D. & Cyr, R.J. (1998) Extending the microtubule/microfibril paradigm. Cellulose synthesis is required for normal cortical microtubule alignment in elongating cells. *Plant Physiol.*, **116**, 1043–1051.

Fleischaker, G.R. (1988) Autopoiesis: the status of its system logic. *BioSystems*, **22**, 37–49.

Fleming, A.J. (2002) The mechanism of leaf morphogenesis. *Planta*, **216**, 17–22.

Fleming, A.J., Caderas, D., Wehrli, E., McQueen-Mason, S. & Kuhlemeier, C. (1999) Analysis of expansin-induced morphogenesis on the apical meristem of tomato. *Planta*, **208**, 166–174.

Fleming, A.J., McQueen-Mason, S.J., Mantel, T. & Kuhlemeier, C. (1997) Induction of leaf primordia by the cell wall protein expansin. *Science*, **276**, 1415–1418.

Foissner, I., Grolig, F. & Obermeyer, G. (2002) Reversible protein phosphorylation regulates the dynamic organization of the pollen tube cytoskeleton: effects of calyculin A and okadaic acid. *Protoplasma*, **220**, 1–15.

Foreman, J., Demidchik, V., Bothwell, J.H.F., Mylona, P., Miedema, H., Torres, M.A., Linstead, P., Costa, S., Brownlee, C., Jones, J.D.G., Davies, J.M. & Dolan, L. (2003) Reactive oxygen species produced by NADPH oxidase regulate plant cell growth. *Nature*, **422**, 442–446.

Fowler, J.E. & Quatrano, R.S. (1997) Plant cell morphogenesis: plasma membrane interactions with the cytoskeleton and cell wall. *Ann. Rev. Cell Devel. Biol.*, **13**, 697–743.

Franklin-Tong, V.E. (1999) Signaling and the modulation of pollen tube growth. *Plant Cell*, **11**, 727–738.

Fricke, W., Jarvis, M.C. & Brett, C.T. (2000) Turgor pressure, membrane tension and the control of exocytosis in higher plants. *Plant Cell Environ.*, **23**, 999–1003.

Friml, J. (2003) Auxin transport – shaping the plant. *Curr. Opin. Plant Biol.*, **6**, 7–12.

Friml, J. & Palme, K. (2002) Polar auxin transport – old questions and new concepts? *Plant Molec. Biol.*, **49**, 273–284.

Friml, J., Benková, E., Blilou, I., Wisniewska, J., Hamann, T., Ljung, K., Woody, S., Sandberg, G., Scheres, B., Jürgens, G. & Palme, K. (2002) AtPIN4 mediates sink-driven auxin gradients and root patterning. *Cell*, **108**, 661–673.

Fry, S.C. (1998) Oxidative scission of plant cell wall polysaccharides by ascorbate-induced hydroxyl radicals. *Biochemical J.*, **332**, 507–515.

Fukuda, M. (2003) Molecular cloning, expression, and characterization of a novel class of synaptotagmin (Syt XIV) conserved from *Drosophila* to humans. *J. Biochem.*, **133**, 641–649.

Gallagher, K. & Smith, L.G. (1999) *discordia* mutations specifically misorient asymmetric cell divisions during development of the maize leaf epidermis. *Development*, **126**, 4623–4633.

Galway, M.E., Lane, D.C. & Schiefelbein, J.W. (1999) Defective control of growth rate and cell diameter in tip-growing root hairs of the *rhd4* mutant of *Arabidopsis thaliana*. *Can. J. Bot.*, **77**, 494–507.

Garcia-Gómez, B.I., Campos, F., Hernández, M. & Covarrubias, A.A. (2000) Two bean cell wall proteins more abundant during water deficit are high in proline and interact with a plasma membrane protein. *Plant J.*, **22**, 277–288.

Geiger, B., Bershadsky, A., Pankov, R. & Yamada, K.M. (2001) Transmembrane extracellular matrix-cytoskeleton crosstalk. *Nature Rev. Molec. Cell Biol.*, **2**, 793–805.

Geisler, M.J., Deppong, D.O., Nadeau, J.A. & Sack, F.D. (2003) Stomatal neighbor cell polarity and division in *Arabidopsis*. *Planta*, **216**, 571–579.

Geitmann, A. (1997) Growth and formation of the cell wall in pollen tubes of *Nicotiana tabacum*, and *Petunia hybrida*. Hänsel-Hohenhausen Verlag, Egelsbach, Frankfurt, Washington.

Geitmann, A., Wojciechowicz, K. & Cresti, M. (1996) Inhibition of intracellular pectin transport in pollen tubes by monensin, brefeldin A and cytochalasin D. *Bot. Acta*, **109**, 373–381.

Geldner, N., Anders, N., Wolters, H., Keicher, J., Kornberger, W., Muller, P., Delbarre, A., Ueda, T., Nakano, A. & Jürgens, G. (2003) The *Arabidopsis* GNOM ARF-GEF mediates endosomal reycling, auxin transport, and auxin-dependent plant growth. *Cell*, **112**, 219–230.

Geldner, N., Friml, J., Stierhof, Y.D., Jürgens, G. & Palme, K. (2001) Auxin-transport inhibitors block PIN1 cycling and vesicle trafficking. *Nature*, **413**, 425–428.

Gens, J.S., Fujiki, M. & Pickard, B.G. (2000) Arabinogalactan protein and wall-associated kinase in a plasmalemmal reticulum with specialized vertices. *Protoplasma*, **212**, 115–134.

Gibbon, B.C., Kovar, D.R. & Staiger, C.J. (1999) Latrunculin B has different effects on pollen germination and tube growth. *Plant Cell*, **11**, 2349–2363.

Gilliland, L.U., Kandasamy, M.K., Pawloski, L.C. & Meagher, R.B. (2002) Both vegetative and reproductive actin isovariants complement the stunted root hair phenotype of the Arabidopsis *act2-1* mutation. *Plant Physiol.*, **130**, 2199–2209.
Gilliland, L.U., Pawloski, L., Kandasamy, M.K. & Meagher, R.B. (2003) *Arabidopsis* actin gene ACT7, plays an essential role in germination and root growth. *Plant J.*, **33**, 319–328.
Glüsenkamp, K.-H., Kosegarten, H., Mengel, K., Grolig, F., Esch, A. & Goldbach, H.E. (1997) A fluorescein boronic acid conjugate as a marker for borate binding sites in the apoplast of growing roots of *Zea mays* L. and *Helianthus annuus* L, in *Boron in Soil and Plants* (eds. R.W. Bell & Rerkasem, B), Kluwer Academic Publishers: Dordrecht, pp 229–235.
Goff, S.A. *et al.* (2002) A draft sequence of the rice genome (*Oryza sativa* L. ssp. *japonica*). *Science*, **296**, 92–100.
Goldbach, H.E. (1997) A critical review on current hypotheses concerning the role of boron in higher plants: suggestions for further research and methodological requirements. *J. Trace Microscopy Techn.*, **15**, 51–91.
Goldbach, H.E., Yu, Q., Wingende,r R., Schulz, M., Wimmer, M., Finderklee, P. & Baluška, F. (2001) Rapid response reactions of roots to boron deprivation. *J. Plant Nutr. Soil Sci.*, **161**, 173–181.
Goubet, F., Misrahi, A., Park, S.K., Zhang, Z., Twell, D. & Dupree, P. (2003) AtCSLA7, a cellulose synthase-like putative glycosyltransferase, is important for pollen tube growth and embryogenesis in Arabidopsis. *Plant Physiol.*, **131**, 547–557.
Grebe, M., Xu, J., Möbius, W., Ueda, T., Nakano, A., Geuze, H.J., Rook, M.B. & Scheres, B. (2003) *Arabidopsis* sterol endocytosis involves actin-mediated trafficking via ARA6–positive early endosomes. *Current Biol.*, **13**, 1378–1387.
Green, P.B. (1994) Connecting gene and hormone action to form, pattern and organogenesis: biophysical transductions. *J. Exp. Bot.*, **45** Sp Iss, 1775–1788.
Green, P.B. (1996) Expression of form and pattern in plants – a role for biophysical fields. *Semin. Cell Devel. Biol.*, **7**, 903–911.
Green, P.B. (1999) Expression of pattern in plants: combining molecular and calculus-based biophysical paradigms. *Am. J. Bot.*, **86**, 1059–1076.
Grolig, F. (1998) Nuclear centering in *Spirogyra*: force integration by microfilaments along microtubules. *Planta*, **204**, 54–63.
Guertin, D.A., Trautmann, S. & McCollum, D. (2002) Cytokinesis in eukaryotes. *Microbiol. Molec. Biol. Rev.*, **66**, 155–178.
Hadfi, K., Speth, V. & Neuhaus, G. (1998) Auxin-induced developmental patterns in *Brassica juncea* embryos. *Development*, **125**, 879–887.
Hall, Q. & Cannon, M.C. (2002) The cell wall hydroxyproline-rich glycoprotein RSH is essential for normal embryo development in Arabidopsis. *Plant Cell*, **14**, 1161–1172.
Harold, F.M. (1990) To shape a cell: an inquiry into the causes of morphogenesis of microorganisms. *Microbiol. Rev.*, **54**, 381–431.
Harold, F.M. (1995) From morphogenes to morphogenesis. *Microbiology*, **141**, 2765–2778.
Harold, F.M. (2001a) *The Way of the Cell. Molecules, Organisms and the Order of Life*. Oxford University Press, Oxford and New York.
Harold, F.M. (2001b) Postscript to Schrödinger: So what is Life? *ASM News*, **67**, 611–616.
Harold, F.M. (2002) Force and compliance: rethinking morphogenesis in walled cells. *Fungal Genet. Biol.*, **37**, 271–282.
Haywood, V., Kragler, F. & Lucas, W.J. (2002) Plasmodesmata: pathways for protein and ribonucleoprotein signaling. *Plant Cell*, **Suppl.** S303–S325.
He, Z.-H., Cheeseman, I., He, D. & Kohorn, B.D. (1999) A cluster of five cell wall-associated receptor kinase genes, *Wak1-5*, are expressed in specific organs of *Arabidopsis*. *Plant Molec. Biol.*, **39**, 1189–1196.
He, Z.-H., Fujiki, M. & Kohorn, B.D. (1996) A cell wall-associated, receptor-like protein kinase. *J. Biol. Chem.*, **271**, 19789–19793.
Heath, I.B. (2001) Bridging the divide: cytoskeleton-plasma membrane-cell wall interactions in growth and development, in *The Mycota VIII. Biology of the Fungal Cell* (eds. R.J. Howard & N.A.R. Gow), Springer Verlag, Berlin, Heidelberg, pp. 201–223.
Heinlein, M. (2002) Plasmodesmata: dynamic regulation and role in macromolecular cell-to-cell signaling. *Curr. Opin. Plant Biol.*, **5**, 543–552.

Hejnowicz, Z., Rusin, A. & Rusin, T. (2000) Tensile tissue stress affects the orientation of cortical microtubules in the epidermis of sunflower hypocotyl. *J. Plant Growth Reg.*, **19**, 31–44.
Hepler, P.K., Vidali, L. & Cheung, A.Y. (2001) Polarized cell growth in higher plants. *Ann. Rev. Cell Devel. Biol.*, **17**, 159–187.
Herve, C., Dabos, P., Galaud, J.P., Rouge, P. & Lescure, B. (1996) Characterization of an *Arabidopsis thaliana* gene that defines a new class of putative plant receptor kinases with an extracellular lectin-like domain. *J. Molec. Biol.*, **258**, 778–788.
Hoffmann-Benning, S., Willmitzer, L. & Fisahn, J. (1997) Analysis of growth, composition and thickness of the cell walls of transgenic plants expressing a yeast-derived invertase. *Protoplasma*, **200**, 146–153.
Hohmann, S. (2002) Osmotic stress signaling and osmoadaptation in yeasts. *Microbiol. Molec. Biol. Rev.*, **66**, 300–372.
Howard, R.J., Ferrari, M.A., Roach, D.H. & Money, N.P. (1991) Penetration of hard substrates by a fungus employing enormous turgor pressures. *Proc. Natl. Acad. Sci. USA*, **88**, 11281–11284.
Hu, W.-J., Harding, S.A., Lung, J., Popko, J.L., Ralph, J., Stokke, D.D., Tsai, C.-J. & Chiang, V.L. (1999) Repression of lignin biosynthesis promotes cellulose accumulation and growth in transgenic trees. *Nature Biotechnol.*, **17**, 808–812.
Huang, S. & Ingber, D.E. (1999) The structural and mechanical complexity of cell-growth control. *Nature Cell Biol.*, **1**, E131–E138.
Hush, J.A., Hawes, C.R. & Overall, R.L. (1990) Interphase microtubule re-orientation predicts a new cell polarity in wounded pea roots. *J. Cell Sci.*, **96**, 47–61.
Hussey, P.J., Allwood, E.G. & Smertenko, A.P. (2002) Actin-binding proteins in the *Arabidopsis* genome database: properties of functionally distinct plant actin-depolymerizing factors/cofilins. *Phil. Trans. Royal Soc. London B*, **357**, 791–798.
Iglesias, V.A. & Meins, F. Jr. (2000) Movement of plant viruses is delayed in a β-1,3-glucanase-deficient mutant showing a reduced plasmodesmatal size exclusion limit and enhanced callose deposition. *Plant J.*, **21**, 157–166.
Im, K.-H., Cosgrove, D.J. & Jones, A.M. (2000) Subcellular localization of expansin mRNA in xylem cells. *Plant Physiol.*, **123**, 463–470.
Ingber, D.E. (1993a) Cellular tensegrity: defining new rules of biological design that govern the cytoskeleton. *J. Cell Sci.*, **104**, 613–627.
Ingber, D.E. (1993b) The riddle of morphogenesis: a question of solution chemistry or molecular cell engineering? *Cell*, **75**, 1249–1252.
Ingber, D.E. (2003a) Tensegrity I. Cell structure and hierarchical systems biology. *J. Cell Sci.*, **116**, 1157–1173.
Ingber, D.E. (2003b) Tensegrity II. How structural networks influence cellular information processing networks. *J. Cell Sci.*, **116**, 1397–1408.
Iraki, N.M., Bressan, R.A., Hasegawa, P.M. & Carpita, N.C. (1989) Alteration of the physical and chemical structure of the primary cell wall of growth-limited plant cells adapted to osmotic stress. *Plant Physiol.*, **91**, 39–47.
Jackson, D. (2000) Opening up the communication channels. Recent insights into plasmodesmal function. *Cuur. Opin. Cell Biol.*, **3**, 394–399.
Jarvis, M.C. (1984) Structure and properties of pectin gels in plant cell walls. *Plant Cell Environ.*, **7**, 153–164.
Jauneau, A., Quentin, M. & Driouich, A. (1997) Micro-heterogeneity of pectins and calcium distribution in the epidermal and cortical parenchyma cell walls of flax hypocotyl. *Protoplasma*, **198**, 9–19.
Juniper, B.E. & Barlow, P.W. (1969) The distribution of plasmodesmata in the root tip of maize. *Planta*, **89**, 352–360.
Kang, B.-H., Busse, J.S. & Bednarek, S.Y. (2003a) Members of the Arabidopsis dynamin-like gene family, ADL1, are essential for plant cytokinesis and polarized cell growth. *Plant Cell*, **15**, 899–913.
Kang, B.-H., Busse, J.S. & Bednarek, S.Y. (2003b) The dynamin-like protein ADL1C is essential for plasma membrane maintenance during pollen maturation. *Plant J.*, **35**, 1–15.
Kaplan, D.R. & Hagemann, W. (1991) The relationship of cell and organism in vascular plants. *BioScience*, **41**, 693–703.

Katembe, W.J., Swatzell, L.J., Makaroff, C.A. & Kiss, J.Z. (1997) Immunolocalization of integrin-like proteins in *Arabidopsis*, and *Chara*. *Physiol. Plant.*, **99**, 7–14.

Katsuta, J. & Shibaoka, H. (1988) The roles of the cytoskeleton and the cell wall in nuclear positioning in tobacco BY-2 cells. *Plant Cell Physiol.*, **29**, 403–413.

Katz, B.-Z., Zamir, E., Bershadsky, A., Kam, Z., Yamada, K.M. & Geiger, B. (2000) Physical state of the extracellular matrix regulates the structure and molecular composition of cell-matrix adhesions. *Molec. Biol. Cell*, **11**, 1047–1060.

Kell, A. & Glaser, R.W. (1993) On the mechanical and dynamic properties of plant cell membranes: their role in growth, direct gene transfer and protoplast fusion. *J. Theor. Biol.*, **160**, 41–62.

Kerstens, S. & Verbelen, J.-P. (2002) Cellulose orientation in the outer epidermal wall of angiosperm roots: implications for biosystematics. *Ann. Bot.*, **90**, 669–676.

Kerstens, S., Decraemer, W.F. & Verbelen, J.-P. (2001) Cell walls at the plant surface behave mechanically like fiber-reinforced composite materials. *Plant Physiol.*, **127**, 381–385.

Ketelaar, T., de Ruijter, N.C.A. & Emons, A.M.C. (2003) Unstable F-actin specifies the area and microtubule direction of cell expansion in Arabidopsis root hairs. *Plant Cell*, **15**, 285–292.

Kiba, A., Sugimoto, M., Toyoda, K., Ichinose, Y., Yamada, T. & Shiraishi, T. (1998) Interaction between cell wall and plasma membrane via RGD motif is implicated in plant defense responses. *Plant Cell Physiol.*, **39**, 1245–1249.

Kobayashi, I. & Hakuno, H. (2003) Actin-related defense mechanism to reject penetration attempt by a non-pathogen is maintained in tobacco BY-2 cells. *Planta*, **217**, 340–345.

Kobayashi, M., Nakagawa, H., Asaka, T. & Matoh, T. (1999) Borate-rhamnogalacturonan II bonding reinforced by Ca^{2+} retains pectic polysaccharides in higher-plant cell walls. *Plant Physiol.*, **119**, 199–203.

Kohorn, B.D. (2000) Plasma membrane-cell wall contacts. *Plant Physiol.*, **124**, 31–38.

Koi, S. & Kato, M. (2003) Comparative developmental anatomy of the root in three species of *Cladopus* (Podostemaceae). *Ann. Bot.*, **91**, 927–937.

Komis, G., Apostolakos, P. & Galatis, B. (2003) Actomyosin is involved in the plasmolytic cycle: gliding movement of the deplasmolyzing protoplast. *Protoplasma*, **221**, 245–256.

Kropf, D.L., Kloareg, B. & Quatrano, R.S. (1988) Cell wall is required for fixation of the embryonic axis in *Fucus*. *Science*, **239**, 187–190.

Kutschera, U. (2001) Stem elongation and cell wall proteins in flowering plants. *Plant Biol.*, **3**, 466–480.

Labouré, A.-M., Faik, A., Mandaron, P. & Falconet, D. (1999) RGD-dependent growth of maize calluses and immunodetection of an integrin-like protein. *FEBS Lett.*, **442**, 123–128.

Lakes, R. (1993) Materials with structural hierarchy. *Nature*, **361**, 511–515.

Lally, D., Ingmire, P., Tong, H.-Y. & He, Z.-H. (2001) Antisense expression of a cell wall-associated protein kinase, WAK4, inhibits cell elongation and alters morphology. *Plant Cell*, **13**, 1317–1331.

Lang-Pauluzzi, I. (2000) The behaviour of the plasma membrane during plasmolysis: a study by UV-microscopy. *J. Microscopy*, **198**, 188–198.

Lang-Pauluzzi, I. & Gunning, B.E.S. (2000) A plasmolytic cycle: the fate of cytoskeletal elements. *Protoplasma*, **212**, 174–185.

Lauber, M.H., Waizenegger, I., Steinmann, T., Schwarz, H., Mayer, U., Hwang, I., Lukowitz, W. & Jürgens, G. (1997) The *Arabidopsis* KNOLLE protein is a cytokinesis-specific syntaxin. *J. Cell Biol.*, **139**, 1485–1493.

Lazzaro, M.D., Donohue, J.M. & Soodavar, F.M. (2003) Disruption of cellulose synthesis by isoxaben causes tip swelling and disorganizes cortical microtubules in elongating conifer pollen tubes. *Protoplasma*, **220**, 201–207

Levina, N.N., Heath, I.B. & Lew, R.R. (2000) Rapid wound responses of *Saprolegnia ferax* hyphae depend upon actin and Ca^{2+}-involving deposition of callose plugs. *Protoplasma*, **214**, 199–209.

Lev-Yadun, S. (2001) Intrusive growth – the plant analog of dendrite and axon growth in animals. *New Phytol.*, **150**, 508–512.

Li, Y.Q., Chen, F., Linskens, H.F. & Cresti, M. (1994) Distribution of unesterified and esterified pectins in cell walls of pollen tube. *Sex. Plant Repr.*, **7**, 145–152.

Li, Y., Darley, C.P., Ongaro, V., Fleming, A., Schipper, O., Baldauf, S.L. & McQueen-Mason,

S.J. (2002) Plant expansins are a complex multigene family with an ancient evolutionary origin. *Plant Physiol.*, **128**, 854–864.

Li, Y.Q., Moscatelli, A., Cai, G. & Cresti, M. (1997) Functional interactions among cytoskeleton, membranes, and cell wall in the pollen tube of flowering plants. *Int. Rev. Cytol.*, **176**, 133–199.

Lintilhac, P.M. (1999) Toward a theory of cellularity Speculations on the nature of the living cell. *BioScience*, **49**, 59–68.

Lintilhac, P.M. & Vesecky, T.B. (1984) Stress-induced alignment of division plane in plant tissues grown *in vitro*. *Nature*, **307**, 363–364.

Lloyd, C.W., Slabas, A.R., Powell, A.J. & Lowe, S.B. (1980) Microtubules, protoplasts and plant cell shape. An inmmunofluorescent study. *Planta*, **147**, 500–506.

Luby-Phelps, K. (2000) Cytoarchitecture and physical properties of cytoplasm: volume, viscosity, diffusion, intracellular surface area. *Int. Rev. Cytol.*, **192**, 189–221.

Lucas, W.J., Ding, B. & Van der Schoot, C. (1993) Plasmodesmata and the supracellular nature of plants. *New Phytol.*, **125**, 435–476.

Lukowitz, W., Mayer, U. & Jürgens, G. (1996) Cytokinesis in the *Arabidopsis* embryo involves the syntaxin-related KNOLLE product. *Cell*, **84**, 61–71.

Lukowitz, W., Nickle, T.C., Meinke, D.W., Last, R.L., Conklin, P.L. & Somerville, C.R. (2001) *Arabidopsis cyt1* mutants are deficient in a mannose-1–phosphate guanylyltransferase and point to a requirement of N-linked glycosylation for cellulose biosynthesis. *Proc. Natl. Acad. Sci. USA*, **98**, 2262–2267.

Lynch, T.M. & Lintilhac, P.M. (1997) Mechanical signals in plant development: a new method for single cell studies. *Devel. Biol.*, **181**, 246–256.

McCabe, P.F., Valentine, T.A., Forsberg, L.S. & Pennell, R.I. (1997) Soluble signals from cells identified at the cell wall establish a developmental pathway in carrot. *Plant Cell*, **9**, 2225–2241.

McCann, M.C., Shi, J., Roberts, K. & Carpita, N.C. (1994) Changes in pectin structure and localization during the growth of unadapted and NaCl-adapted tobacco cells. *Plant J.*, **5**, 773–785.

McQueen-Mason, S.J. & Rochange, F. (1999) Expansins in plant growth and development: an update on an emerging topic. *Plant Biol.*, **1**, 19–25.

Majewska-Sawka, A. & Nothnagel, E.A. (2000) The multiple roles of arabinogalactan proteins in plant development. *Plant Physiol.*, **122**, 3–9.

Martin, C., Bhatt, K. & Baumann, K. (2001) Shaping in plant cells. *Curr. Opin. Plant Biol.*, **4**, 540–549.

Masuda, Y., Takagi, S. & Nagai, R. (1991) Protease-sensitive anchoring of microfilament bundles provides tracks for cytoplasmic streaming in *Vallisneria*. *Protoplasma*, **162**, 151–159.

Mathews, C.K. (1993) The cell-bag of enzymes or network of channels? *J. Bacteriol.*, **175**, 6377–6381.

Mathur, J., Mathur, N., Kernebeck, B. & Hülskamp, M. (2003a) Mutations in actin-related proteins 2 and 3 affect cell shape development in Arabidopsis. *Plant Cell*, **15**, 1632–1645.

Mathur, J., Mathur, N., Kirik, V., Kernebeck, B., Srinivas, B.P. & Hülskamp, M. (2003b) *Arabidopsis CROOKED* encodes for the smallest subunit of the ARP2/3 complex and controls cell shape by region specific fine F-actin formation. *Development*, **130**, 3137–3146.

Matoh, T. & Kobayashi, M. (1998) Boron and calcium, essential inorganic constituents of pectic polysaccharides in higher plant cell walls. *J. Plant Res.*, **111**, 179–190.

Mayer, U., Buttner, G. & Jürgens, G. (1993) Apical-basal pattern formation in the *Arabidopsis* embryo: studies on the role of the *GNOM* gene. *Development*, **117**, 149–162.

Meagher, R.B. & Fechheimer, M. (2003) The Arabidopsis cytoskeletal genome. The Arabidopsis Book (http://www.aspb.org/publications/arabidopsis/), 1–26.

Mellersh, D.G. & Heath, M.C. (2001) Plasma membrane-cell wall adhesion is required for expression of plant defense responses during fungal penetration. *Plant Cell*, **13**, 413–424.

Meyer, Y. & Abel, W.O. (1975) Importance of the wall for cell division and in the activity of the cytoplasm in cultured tobacco protoplasts. *Planta*, **123**, 33–40.

Meyer, Y. & Herth, W. (1978) Chemical inhibition of cell wall formation and cytokinesis, but not of nuclear division, in protoplasts of *Nicotiana tabacum* L. cultivated in vitro. *Planta*, **142**, 253–262.

Meyerowitz, E.M. (2002) Plants compared to animals: the broadest comparative study of development. *Science*, **295**, 1482–1485.

Miller, D.D., de Ruijter, N.C.A. & Emons A.M.C. (1997) From signal to form: aspects of the cytoskeleton-plasma membrane-cell wall continuum in root hair tips. *J. Exp. Bot.*, **48**, 1881–1896.

Mollet, J.-C., Kim, S., Jauh, G.-Y. & Lord, E.M. (2002) Arabinogalactan proteins, pollen tube growth, and the reversible effects of Yariv phenylglycoside. *Protoplasma*, **219**, 89–98.

Money, N.P. (1997) Wishful thinking of turgor revisited: the mechanics of fungal growth. *Fungal Genet. Biol.*, **21**, 173–187.

Morvan, O., Quentin, M., Jauneau, A., Mareck, A. & Morvan, C. (1998) Immunogold localization of pectin methylesterases in the cortical tissues of flax hypocotyl. *Protoplasma*, **202**, 175–184.

Nacry, P., Mayer, U. & Jürgens, G. (2000) Genetic dissection of cytokinesis. *Plant Molec. Biol.*, **43**, 719–733.

Nakagawa, N. & Sakurai, N. (2001) Cell wall integrity controls expression of endoxyloglucan transferase in tobacco BY2 cells. *Plant Cell Physiol.*, **42**, 240–244.

Nanjo, T., Kobayashi, M., Yoshiba, Y., Sanada, Y., Wada, K., Tsukaya, H., Kakubari, Y., Yamaguchi-Shinozaki, K. & Shinozaki, K. (1999) Biological functions of proline in morphogenesis and osmotolerance revealed in antisense transgenic *Arabidopsis thaliana*. *Plant J.*, **18**, 185–193.

Nanninga, N. (2001) Cytokinesis in prokaryotes and eukaryotes: common principles and different solutions. *Microbiol. Molec. Biol. Rev.*, **65**, 319–333.

Natesh, S. & Rau, M.A. (1984) The embryo, in *Embryology of Angiosperms* (ed. B.M. Johri), Berlin: Springer, pp. 377–443.

Nelson, W.J. (2003) Adaptation of core mechanisms to generate cell polarity. *Nature*, **422**, 766–774.

Nickle, T.C. & Meinke, D.W. (1998) A cytokinesis-defective mutant of Arabidopsis (*cyt1*) characterized by embryonic lethality, incomplete cell walls, and excessive callose accumulation. *Plant J.*, **15**, 321–332.

Nicol, F., His, I., Jauneau, A., Vernhettes, S., Canut, H. & Höfte, H. (1998) A plasma membrane-bound putative endo-1,4-β-glucanase is required for normal wall assembly and cell elongation in *Arabidopsis*. *EMBO J.*, **17**, 5563–5576.

Niklas, K.J. (1992) *Plant Biomechanics. An Engineering Approach to Plant Form and Function.* The University of Chicago Press, Chicago and London.

Niklas, K.J. (2000) The evolution of plant body plans – a biomechanical perspective. *Ann. Bot.*, **85**, 411–438.

Niklas, K.J. (2003) The *bio*-logic and machinery of plant morphogenesis. *Am. J. Bot.*, **90**, 515–525.

Nimnual, A.S., Taylor, L.J. & Bar-Sagi, D. (2003) Redox-dependent downregulation of Rho by Rac. *Nature Cell Biol.*, **5**, 236–241.

Okita, T.W. & Choi, S.-B. (2002) mRNA localization in plants: targeting to the cell's cortical region and beyond. *Curr. Opin. Plant Biol.*, **5**, 553–559.

Olsen, O.-A. (2001) Endosperm development: cellularization and cell fate specification. *Annu. Rev. Plant Physiol. Plant Molec. Biol.*, **52**, 233–267.

Oparka, K.J. (1994) Plasmolysis: new insights into an old process. *New Phytol.*, **126**, 571–591.

Orfila, C. & Knox, J.P. (2000) Spatial regulation of pectic polysaccharides in relation to pit fields in cell walls of tomato fruit pericarp. *Plant Physiol.*, **122**, 775–781.

Otegui, M. & Staehelin, L.A. (2000a) Syncytial-type cell plates: a novel kind of cell plate involved in endosperm cellularization of Arabidopsis. *Plant Cell*, **12**, 933–947.

Otegui, M. & Staehelin, L.A. (2000b) Cytokinesis in flowering plants: more than one way to divide a cell. *Curr. Opin. Plant Biol.*, **3**, 493–502.

Ovečka, M., Baluška, F., Nadubinská, M. & Volkmann, D. (2000) Actomyosin and exocytosis inhibitors alter root hair morphology in *Poa annua*, L. *Biológia*, **55**, 105–114.

Overall, R.L., Dibbayawan, T.P. & Blackman, L.M. (2001) Intercellular alignments of the plant cytoskeleton. *J. Plant Growth Reg.*, **20**, 162–169.

Pagant, S., Bichet, A., Sugimoto, K., Lerouxel, O., Desprez, T., McCann, M., Lerouge, P., Vernhettes, S. & Höfte, H. (2002) *KOBITO1* encodes a novel plasma membrane protein necessary for normal synthesis of cellulose during cell expansion in Arabidopsis. *Plant Cell*, **14**, 2001–2013.

Park, A.R., Cho, S.K., Yun, U.J., Jin, M.Y., Lee, S.H., Sachetto-Martins, G. & Park, O.K. (2001) Interaction of the *Arabidopsis* receptor protein kinase Wak1 with a glycine-rich protein, AtGRP-3. *J. Biol. Chem.*, **276**, 26688–26693.

Parton, R.M., Fischer-Parton, S., Trewavas, A.J. & Watahiki, M.K. (2003) Pollen tubes exhibit regular periodic membrane trafficking events in the absence of apical extension. *J. Cell Sci.*, **116**, 2707–2719.

Parton, R.M., Fischer-Parton, S., Watahiki, M.K. & Trewavas, A.J. (2001) Dynamics of the apical vesicle accumulation and the rate of growth are related in individual pollen tubes. *J. Cell Sci.*, **114**, 2685–2695.

Parthasarathy, M.V. (1974) Ultrastructure of phloem in palms. II. Structural changes and fate of the organelles in differentiating sieve elements. *Protoplasma*, **79**, 93–125.

Pauly, M., Qin, Q., Greene, H., Albersheim, P., Darvill, A. & York, W.S. (2001) Changes in the structure of xyloglucan during cell elongation. *Planta*, **212**, 842–850.

Pennell, R.I., Janniche, L., Kjellbom, P., Scofield, G.N., Peart, J.M. & Roberts, K. (1991) Developmental regulation of a plasma membrane arabinogalactan protein epitope in oilseed rape flowers. *Plant Cell*, **3**, 1317–1326.

Pennell, R.I., Janniche, L., Scofield, G.N., Booij, H., de Vries, S.C. & Roberts, K. (1992) Identification of a transitional cell state in the developmental pathway to carrot somatic embryogenesis. *J. Cell Biol.*, **119**, 1371–1380.

Perez-Moreno, M., Jamora, C. & Fuchs, E. (2003) Sticky business: orchestrating cellular signals at adherens junctions. *Cell*, **112**, 535–548.

Peters, W.S., Hagemann, W. & Tomos, D.A. (2000) What makes plants different? Principles of extracellular matrix function in 'soft' plant tissues. *Comp. Biochem. Physiol. Part A*, **125**, 151–167.

Petrášek, J., Elckner, M., Morris, D.A. & Zažímalová, E. (2002) Auxin efflux carrier activity and auxin accumulation regulate cell division and polarity in tobacco cells. *Planta*, **216**, 302–308.

Pickett-Heaps, J.D., Gunning, B.E.S., Brown, R.C., Lemmon, B.E. & Cleary, A.L. (1999) The cytoplast concept in dividing plant cells: cytoplasmic domains and the evolution of spatially organized cell division. *Am. J. Bot.*, **86**, 153–172.

Picton, J.M. & Steer, M.W. (1983) Membrane recycling and the control of secretory activity in pollen tubes. *J. Cell Sci.*, **63**, 303–310.

Pien, S., Wyrzykowska, J., McQueen-Mason, S., Smart, C. & Fleming, A. (2001) Local expression of expansin induces the entire process of leaf development and modifies leaf shape. *Proc. Natl. Acad. Sci. USA*, **98**, 11812–11817.

Pilling, J., Willmitzer, L. & Fisahn, J. (2000) Expression of a *Petunia inflata*, pectin methyl esterase in *Solanum tuberosum* L. enhances stem elongation and modifies cation distribution. *Planta*, **210**, 391–399.

Pollard, T.D. (2003) The cytoskeleton, cellular motility and the reductionist agenda. *Nature*, **422**, 741–745.

Pont-Lezica, R.F., McNally, J.G. & Pickard, B.G. (1993) Wall-to-membrane linkers in onion epidermis: some hypotheses. *Plant Cell Environ.*, **16**, 111–123.

Qi, Y. & Ding, B. (2003) Inhibition of cell growth and shoot development by a specific nucleotide sequence in a noncoding viroid RNA. *Plant Cell*, **15**, 1360–1374.

Qiu, J.-L., Jilk, R., Marks, M.D. & Szymanski, D.B. (2002) The Arabidopsis *SPIKE1* gene is required for normal cell shape control and tissue development. *Plant Cell*, **14**, 101–118.

Radford, J.E., Vesk, M. & Overall, R.L. (1998) Callose deposition at plasmodesmata. *Protoplasma*, **201**, 30–37.

Ramachandran, S., Christensen, H.E.M., Ishimaru, Y., Dong, C.-H., Chao-Ming, W., Cleary, A.L. & Chua, N.-H. (2000) Profilin plays a role in cell elongation, cell shape maintenance, and flowering in Arabidopsis. *Plant Physiol.*, **124**, 1637–1647.

Reigada, D., Diez-Perez, I., Gorostiza, P., Verdaguer, A., Gomez de Aranda, I., Pineda, O., Vilarrasa, J., Marsal, J., Blasi, J., Aleu, J. & Solsona, C. (2003) Control of neurotransmitter release by an internal gel matrix in synaptic vesicles. *Proc. Natl. Acad. Sci. USA*, **100**, 3485–3490.

Reinhardt, D. & Kuhlemeier, C. (2002) Plant architecture. *EMBO Rep.*, **3**, 846–851.

Reinhardt, D., Wittwer, F., Mandel, T. & Kuhlemeier, C. (1998) Localized upregulation of a new expansin gene predicts the site of leaf formation in the tomato meristem. *Plant Cell*, **10**, 1427–1437.

Reuzeau, C. & Pont-Lezica, R.F. (1995) Comparing plant and animal extracellular matrix-cytoskeleton connections are they alike? *Protoplasma*, **186**, 113–121.

Reuzeau, C., Doolittle, K.W., McNally, J.G. & Pickard, B.G. (1997) Covisualization in living onion cells of putative integrin, putative spectrin, actin, putative intermediate filaments, and other proteins at the cell membrane and in an endomembrane sheath. *Protoplasma*, **199**, 173–197.

Ridley, B.L., O'Neill, M.A. & Mohnen, D. (2001) Pectins: structure, biosynthesis, and oligogalacturonide-related signaling. *Phytochem.*, **57**, 929–967.

Ringli, C., Baumberger, N., Diet, A., Frey, B. & Keller, B. (2002) ACTIN2 is essential for bulge site selection and tip growth during root hair development of Arabidopsis. *Plant Physiol.*, **129**, 1464–1472.

Ringli, C., Keller, B. & Ryser, U. (2001) Glycine-rich proteins as structural components of plant cell walls. *Cell. Molec. Life Sci.*, **58**, 1430–1441.

Rinne, P.L.H. & van der Schoot, C. (1998) Symplastic fields in the tunica of the shoot apical meristem coordinate morphogenetic events. *Development*, **125**, 1477–1485.

Roberts, A.W. & Uhnak, K.S. (1998) Tip growth in xylogenic suspension cultures of *Zinnia elegans* L.: implications for the relationship between cell shape and secondary-cell-wall pattern in tracheary elements. *Protoplasma*, **204**, 103–113.

Robertson, D., Mitchell, G.P., Gilroy, J.S., Gerrish, C., Bolwell, G.P. & Slabas, A.R. (1997) Differential extraction and protein sequencing reveals major differences in patterns of primary cell wall proteins from plants. *J. Biol. Chem.*, **272**, 15841–15848.

Rochange, S.F., Wenzel, C.L. & McQueen-Mason, S.J. (2001) Impaired growth in transgenic plants over-expressing an expansin isoform. *Plant Molec. Biol.*, **46**, 581–589.

Rodríguez-Gálvez, E. & Mendgen, K. (1995) Cell wall synthesis in cotton roots after infection with *Fusarium oxysporum*. The deposition of callose, arabinogalactans, xyloglucans, and pectic components into walls, wall apositions, cell plates and plasmodesmata. *Planta*, **197**, 535–545.

Roger, D., David, A. & David, H. (1996) Immobilization of flax protoplasts in agarose and alginate beads. Correlation between ionically bound cell-wall proteins and morphogenetic response. *Plant Physiol.*, **112**, 1191–1199.

Rojo, E., Sharma, V.K., Kovaleva, V., Raikhel, N.V. & Fletcher, J.C. (2002) CLV3 is localized to the extracellular space, where it activates the Arabidopsis CLAVATA stem cell signaling pathway. *Plant Cell*, **14**, 969–977.

Roudier, F., Schindelman, G., DeSalle, R. & Benfey, P.N. (2002) The COBRA family of putative GPI-anchored proteins in Arabidopsis. A new fellowship in expansion. *Plant Physiol.*, **130**, 538–548.

Roy, S.J., Holdaway-Clarke, T.L., Hackett, G.R., Kunkel, J.G., Lord, E.M. & Hepler, P.K. (1999) Uncoupling secretion and tip growth in lily pollen tubes: evidence for the role of calcium in exocytosis. *Plant J.*, **19**, 379–386.

Roy, S., Jauh, G.Y., Hepler, P.K. & Lord, E.M. (1998) Effects of Yariv phenylglycoside on cell wall assembly in the lily pollen tube. *Planta*, **204**, 450–458.

Roy, S., Watada, A.E. & Wergin, W.P. (1997) Characterization of the cell wall microdomains surrounding plasmodesmata in apple fruit. *Plant Physiol.*, **114**, 539–547.

Ruan, Y.L., Llewellyn, D.J. & Furbank, R.T. (2001) The control of single-celled cotton fibre elongation by developmentally reversible gating of plasmodesmata and co-ordinated expression of sucrose and K^+ transporters and expansin. *Plant Cell*, **13**, 47–60.

Ryu, J.-H., Takagi, S. & Nagai, R. (1995) Stationary organization of the actin cytoskeleton in *Vallisneria*: the role of stable microfilaments at the end walls. *J. Cell Sci.*, **108**, 1531–1539.

Sachs, T. (2000) Integrating cellular and organismic aspects of vascular differentiation. *Plant Cell Physiol.*, **41**, 649–656.

Šamaj, J., Baluška, F. & Hirt, H. (2004) From signal to cell polarity: mitogen-activated protein kinases as sensors and effectors of dynamic cytoskeleton and signalling endosoems. *J. Exp. Bot.*, **55**, 189–198.

Šamaj, J., Braun, M., Baluška, F., Ensikat, H.-J., Tsumuraya, Y. & Volkmann, D. (1999)

Specific localization of arabinogalactan-protein epitopes at the surface of maize root hairs. *Plant Cell Physiol.*, **40**, 874–883.

Šamaj, J., Ovečka, M., Hlavačka, A., Lecourieux, F., Meskiene, I., Lichtscheidl, I., Lenart, P., Şalaj, J., Volkmann, D., Bögre, L., Baluška, F. & Hirt, H. (2002) Involvement of the mitogen-activated protein kinase SIMK in regulation of root hair tip growth. *EMBO J.*, **21**, 3296–3306.

Šamajová, O., Šamaj, J., Volkmann, D. & Edelmann, H. (1998) Occurrence of osmiophilic particles is correlated to elongation growth of higher plants. *Protoplasma*, **202**, 185–191.

Samuels, A.L., Giddings, T.H. & Staehelin, L.A. (1995) Cytokinesis in tobacco BY-2 and root tip cells: a new model of cell plate formation in higher plants. *J. Cell Biol.*, **130**, 1–13.

Sanders, L.C., Wang, C.-S., Walling, L.L. & Lord, E.M. (1991) A homolog of the substrate adhesion molecule vitronectin occurs in four species of flowering plants. *Plant Cell*, **3**, 629–635.

Sato, S., Kato, T., Kakegawa, K., Ishii, T., Liu, Y.-G., Awano, T., Takabe, K., Nishiyama, Y., Kuga, S., Sato, S., Nakamura, Y., Tabata, S. & Shibata, D. (2001) Role of the putative membrane-bound endo-1,4-β-glucanase KORRIGAN in cell elongation and cellulose synthesis in *Arabidopsis thaliana*. *Plant Cell Physiol.*, **42**, 251–263.

Scheres, B. (2001) Plant cell identity. The role of position and lineage. *Plant Physiol.*, **125**, 112–114.

Scheres, B. & Benfey, P.N. (1999) Asymmetric cell division in plants. *Annu. Rev. Plant Physiol. Plant Molec. Biol.*, **50**, 505–537.

Scherp, P., Grotha, R. & Kutschera, U. (2001) Occurrence and phylogenetic significance of cytokinesis-related callose in green algae, bryophytes, ferns and seed plants. *Plant Cell Rep.*, **20**, 143–149.

Scherp, P., Grotha, R. & Kutschera, U. (2002) Interaction between cytokinesis-related callose and cortical microtubules in dividing cells of the liverwort *Riella helicophylla*. *Plant Biol.*, **4**, 619–624.

Schindelman, G., Morikami, A., Jung, J., Baskin, T.I., Carpita, N.C., Derbyshire, P., McCann, M.C. & Benfey, P.N. (2001) COBRA encodes a putative GPI-anchored protein, which is polarly localized and necessary for oriented cell expansion in *Arabidopsis*. *Genes Devel.*, **15**, 1115–1127.

Schindler, M., Meiners, S. & Cheresh, D.A. (1989) RGD-dependent linkage between plant cell wall and plasma membrane: consequences for growth. *J. Cell Biol.*, **108**, 1955–1965.

Schivell, A.E., Batchelor, R.H. & Bajjalieh, S.M. (1996) Isoform-specific, calcium-regulated interaction of the synaptic vesicle proteins CV2 and synaptotagmin. *J. Biol. Chem.*, **271**, 27770–27775.

Schmelzer, E. (2002) Cell polarization, a crucial process in fungal defence. *Trends Plant Sci.*, **7**, 411–415.

Schopfer, P., Liszkay, A., Bechtold, M., Frahry, G. & Wagner, A. (2002) Evidence that hydroxyl radicals mediate auxin-induced extension growth. *Planta*, **214**, 821–828.

Schröter, K. & Sievers, A. (1971) Wirkung der Turgorreduktion auf den Golgi-Apparat und die Bildung der Zellwand bei Wurzelhaaren. *Protoplasma*, **72**, 203–211.

Seagull, R.W. (1983a) Differences in the frequency and disposition of plasmodesmata resulting from root cell elongation. *Planta*, **159**, 497–504.

Seagull, R.W. (1983b) The role of the cytoskeleton during oriented microfibril deposition. Elucidation of the possible interaction between microtubules and cellulose synthetic complexes. *J. Ultrastructure Res.*, **83**, 168–175.

Sedbrook, J.C., Carroll, K.L., Hung, K.F., Masson, P.H. & Somerville, C.R. (2002) The Arabidopsis *SKU5* gene encodes an extracellular glycosyl phosphatidylinositol-anchored glycoprotein involved in directional root growth. *Plant Cell*, **14**, 1635–1648.

Sharma, V.K., Ramirez, J. & Fletcher, J.C. (2003) The *Arabidopsis*, *CLV3*-like (*CLE*) genes are expressed in diverse tissues and encode secreted proteins. *Plant Molec. Biol.*, **51**, 415–425.

Shaw, S.L., Dumais, J. & Long, S.R. (2000) Cell surface expansion in polarly growing root hairs of *Medicago truncatula*. *Plant Physiol.*, **124**, 959–969.

Shedletzky, E., Shmuel, M., Delmer, D.P. & Lamport, D.T.A. (1990) Adaptation and growth of tomato cells on the herbicide 2,6-dichlorobenzonitrile leads to production of unique cell walls virtually lacking a cellulose-xyloglucan network. *Plant Physiol.*, **94**, 980–987.

Shedletzky, E., Shmuel, M., Trainin, T., Kalman, S. & Delmer, D. (1992) Cell wall structure in cells adapted to growth on the cellulose-synthesis inhibitor 2,6-dichlorobenzonitrile. A comparison between two dicotyledonous plants and a graminaceous monocot. *Plant Physiol.*, **100**, 120–130.

Sheetz, M.P. & Dai, J. (1996) Modulation of membrane dynamics and cell motility by membrane tension. *Trends Cell Biol.*, **6**, 85–89.

Shevell, D.E., Kunkel, T. & Chua, N.-H. (2000) Cell wall alterations in the Arabidopsis *emb30* mutant. *Plant Cell*, **12**, 2047–2059.

Shevell, D.E., Leu, W.-M., Gillmor, C.S., Xia, G., Feldman, K.A. & Chua, N.-H. (1994) *EMB30* is essential for normal cell division, cell expansion, and cell adhesion in Arabidopsis and encodes a protein that has similarity to Sec7. *Cell*, **77**, 1051–1062.

Shi, H., Kim, Y.S., Guo, Y., Stevenson, B. & Zhu, J.-K. (2002) The Arabidopsis *SOS5* locus encodes a putative cell surface adhesion protein and is required for normal cell expansion. *Plant Cell*, **15**, 19–32.

Shibaoka, H. (1994) Plant hormone-induced changes in the orientation of cortical microtubules: alterations in the cross-linking between microtubules and the plasma membrane. *Annu. Rev. Plant Physiol. Plant Molec. Biol.*, **45**, 527–544.

Shinohara, N., Demura, T. & Fukuda, H. (2000) Isolation of a vascular cell wall-specific monoclonal antibody recognizing a cell polarity by using a phage display subtraction method. *Proc. Natl Acad. Sci. USA*, **97**, 2585–2590.

Showalter, A.M. (2001) Arabinogalactan-proteins: structure, expression and function. *Cell. Molec. Life Sci.*, **58**, 1399–1417.

Sivaguru, M., Fujiwara, T., Yang, Z., Osawa, H., Šamaj, J., Baluška, F., Mori, T., Volkmann, D., Maeda, T. & Matsumoto, H. (2000) Aluminum-induced 1-3-β-glucan inhibits cell-to-cell trafficking of molecules through plasmodesmata: a new mechanism of Al toxicity in plants. *Plant Physiol.*, **124**, 991–1018.

Sivaguru, M., Ezaki, B., He, Z.H., Tong, H.Y., Osawa, H., Baluška, F., Volkmann, D. & Matsumoto, H. (2003) Aluminium-induced gene expression and protein localization of a cell wall-associated receptor kinase in *Arabidopsis*. *Plant Physiol.*, **132**, 2256–2266.

Smith, L.G. (2001) Plant cell division: building walls in the right places. *Nature Rev. Molec. Cell Biol.*, **2**, 33–39.

Smith, L.G., Gerttula, S.M., Han, S. & Levy, J. (2001) TANGLED1: a microtubule binding protein required for the spatial control of cytokinesis in maize. *J. Cell Biol.*, **152**, 231–236.

Smith, L.G., Hake, S. & Sylvester, A.W. (1996) The *tangled1* mutation alters cell division orientations throughout maize leaf development without altering leaf shape. *Development*, **122**, 481–489.

Söllner, R., Glässer, G., Wanner, G., Somerville, C.R., Jürgens, G. & Assaad, F.F. (2002) Cytokinesis-defective mutants of Arabidopsis. *Plant Physiol.*, **129**, 678–690.

Stacey, N., Roberts, K. & Knox, J.P. (1990) Patterns of expression of the JIM4 arabinogalactan-rich protein epitope in cell cultures and during somatic embryogenesis in *Daucus carota*, L. *Planta*, **180**, 285–292.

Stark-Urnau, M. & Mendgen, K. (1995) Sequential deposition of plant glycoproteins and polysaccharides into the host-parasite interface of *Uromyces vignae*, and *Vigna sinencis*. Evidence for endocytosis and secretion. *Protoplasma*, **186**, 1–11.

Stein, W.D. (1995) The sodium pump in the evolution of animal cells. *Phil. Trans. Royal Soc. Lond. B*, **349**, 263–269.

Steinmann, T., Geldner, N., Grebe, M., Mangold, S., Jackson, C.L., Paris, S., Galweiler, L., Palme, K. & Jürgens, G. (1999) Coordinated polar localization of auxin efflux carrier PIN1 by GNOM ARF GEF. *Science*, **286**, 316–318.

Strompen, G., El Kasmi, F., Richter, S., Lukowitz, W., Assaad, F.F., Jürgens, G. & Mayer, U. (2002) The *Arabidopsis*, *HINKEL* gene encodes a kinesin-related protein involved in cytokinesis and is expressed in a cell-cycle dependent manner. *Curr. Biol.*, **12**, 153–158.

Sugimoto, K., Himmelspach, R., Williamson, R.E. & Wasteneys, G.O. (2003) Mutation or drug-dependent microtubule disruption causes radial swelling without altering parallel cellulose microfibril deposition in Arabidopsis root cells. *Plant Cell*, **15**, 1414–1429.

Suzuki, K., Amino, S.-I., Takeuchi, Y. & Komamine, A. (1990) Differences in the composition of the cell walls of two morphologically different lines of suspension-cultured *Catharanthus roseus* cells. *Plant Cell Physiol.*, **31**, 7–14.

Suzuki, K., Ingold, E., Sugiyama, M., Fukuda, H. & Komamine, A. (1992) Effects of 2,6-dichlorobenzonitrile on differentiation to tracheary elements of isolated mesophyll cells of *Zinnia elegans* and formation of secondary cell walls. *Physiol. Plant.*, **86**, 43–48.

Swartz, M.A., Rschumperlin, D.J., Kamm, R.D. & Drazen, J.M. (2001) Mechanical stress is communicated between different cell types to elicit matrix remodeling. *Proc. Natl. Acad. Sci. USA*, **98**, 6180–6185.

Taiz, L. & Zeiger, E. (1998) *Plant Physiology*. 2nd Ed. Sinauer Associates, Inc., Publishers, Sunderland.

Takeda, T., Furuta, Y., Awano, T., Mizuno, K., Matsuishi, Y. & Hayashi, T. (2002) Suppression and acceleration of cell elongation by integration of xyloglucans in pea stem segments. *Proc. Natl. Acad. Sci. USA*, **99**, 9055–9060.

Tasaka, M. (2001) From central-peripheral to adaxial-abaxial. *Trends Plant Sci.*, **6**, 548–550.

Taylor, J.G., Owen Jr., T.P., Koonce, L.T. & Haigler, C.H. (1992) Dispersed lignin in tracheary elements treated with cellulose synthesis inhibitors provide evidence that molecules of the secondary cell wall mediate wall patterning. *Plant J.*, **2**, 959–970.

Thomas, J.C., De Armond, R.A. & Bohnert, H.J. (1992) Influence of NaCl on growth, proline, and phosphoenolpyruvate carboxylase levels in *Mesembryanthemum crystallinum* suspension culture. *Plant Physiol.*, **98**, 626–631.

Thompson, H.J.M. & Knox, J.P. (1998) Stage-specific responses of embryogenic carrot cell suspension cultures to arabinogalactan protein-binding β-glucosyl Yariv reagent. *Planta*, **205**, 32–38.

Torres-Ruiz, R.A. & Jürgens, G. (1994) Mutations in the *Fass* gene uncouple pattern formation and morphogenesis in *Arabidopsis* development. *Development*, **120**, 2967–2978.

Turner, A., Wells, B. & Roberts, K. (1994) Plasmodesmata of maize root tips: structure and composition. *J. Cell Sci.*, **107**, 3351–3361.

Turner, S.R., Taylor, N. & Jones, L. (2001) Mutations of the secondary wall. *Plant Molec. Biol.*, **47**, 209–219.

Ueda, T., Tsukaya, H., Anai, T., Hirata, A., Hasegawa, K. & Uchimiya, H. (1996) Brefeldin A induces the accumulation of unusual membrane structures in elongating pollen tubes of *Nicotiana tabacum*, L. *J. Plant Physiol.*, **149**, 683–689.

van Engelen, F.A. & de Vries, S.C. (1992) Extracellular proteins in plant embryogenesis. *Trends Genet.*, **8**, 66–70.

van Hengel, A.J. & Roberts, K. (2002) Fucosylated arabinogalactan-proteins are required for full root cell elongation in arabidopsis. *Plant J.*, **32**, 105–113.

van Hengel, A.J., Tadesse, Z., Immerzeel, P., Schols, H., van Kammen, A. & de Vries, S.C. (2001) N-acetylglucosamine and glucosamine-containing arabinogalactan proteins control somatic embryogenesis. *Plant Physiol.*, **125**, 1880–1890.

Vantard, M. & Blanchoin, L. (2002) Actin polymerization processes in plant cells. *Curr. Opin. Plant Biol.*, **5**, 502–506.

Varela, F.G., Maturana, H.R. & Uribe, R. (1974) Autopoiesis: the organization of living systems, its characterization and a model. *BioSystems*, **5**, 187–196.

Vaughn, K.C., Hoffman, J.C., Hahn, M.G. & Staehelin, L.A. (1996) The herbicide dichlobenil disrupts cell plate formation: immunogold characterization. *Protoplasma*, **194**, 117–132.

Veljovic-Jovanovic, S.D., Pignocchi, C., Noctor, G. & Foyer, C.H. (2001) Low ascorbic acid in the *vtc-1* mutant of Arabidopsis is associated with decreased growth and intracellular redistribution of the antioxidant system. *Plant Physiol.*, **127**, 426–435.

Verbelen, J.-P. & Kerstens, S. (2000) Polarization confocal microscopy and Congo Red fluorescence: a simple and rapid method to determine the mean cellulose fibril orientation in plants. *J. Microscopy*, **198**, 101–107.

Verbelen, J.-P., Vissenberg, K., Kerstens, S. & Le, J. (2001) Cell expansion in the epidermis: microtubules, cellulose orientation and wall loosening enzymes. *J. Plant Physiol.*, **158**, 537–543.

Vidali, L., McKenna, S.T. & Hepler, P.K. (2001) Actin polymerization is essential for pollen tube growth. *Molec. Biol. Cell*, **12**, 2534–2545.

Vissenberg, K., Fry, S.C. & Verbelen, J.-P. (2001) Root hair initiation is coupled to a highly localized increase of xyloglucan endotransglucosylase action in *Arabidopsis* roots. *Plant Physiol.*, **127**, 1125–1135.

Vissenberg, K., Van Sandt, V. & Verbelen, J.-P. (2003) Xyloglucan endotransglucosylase action is high in the root elongation zone and in the trichoblasts of all vascular plants from *Selaginella* to *Zea mays*. *J. Exp. Bot.*, **54**, 335–344.
von Groll, U., Berger, D. & Altmann, T. (2002) The subtilisin-like serine protease SDD1 mediates cell-to-cell signaling during Arabidopsis stomatal development. *Plant Cell*, **14**, 1527–1539.
Wagner, T.A. & Kohorn, B.D. (2001) Wall-associated kinases are expressed throughout plant development and are required for cell expansion. *Plant Cell*, **13**, 303–318.
Wang, X., Cnops, G., Vanderhaeghen, R., De Block, S., Van Montagu, M. & Van Lijsebettens, M. (2001) AtCLSD3, a cellulose synthase-like gene important for root hair tip growth in *Arabidopsis*. *Plant Physiol.*, **126**, 575–586.
Wayne, R., Staves, M.P. & Leopold, A.C. (1992) The contribution of the extracellular matrix to gravisensing in characean cells. *J. Cell Sci.*, **101**, 611–623.
Wehrle-Haller, B. & Imhof, B.A. (2002) The inner lives of focal adhesions. *Trends Cell Biol.*, **12**, 382–389.
Wen, F., Zhu, Y. & Hawes, M.C. (1999) Effect of pectin methylesterase gene expression on pea root development. *Plant Cell*, **11**, 1129–1140.
Whittington, A.T., Vugrek, O., Wei, K.J., Hasenbein, N.G., Sugimoto, K., Rashbrooke, M.C. & Wasteneys, G.O. (2001) MOR1 is essential for organizing cortical microtubules in plants. *Nature*, **411**, 610–613.
Wiedermeier, A.M.D., Judy-March, J.E., Hocart, C.H., Wasteneys, G.O., Williamson, R.E. & Baskin, T.I. (2002) Mutant alleles of *Arabidopsis RADIALLY SWOLLEN4* and *7* reduce growth anisotropy without altering the transverse orientation of cortical microtubules or cellulose microfibrils. *Development*, **129**, 4821–4830.
Willats, W.G.T. & Knox, J.P. (1996) A role for arabinogalactan-proteins in plant cell expansion: evidence from studies on the interaction of β-glucosyl Yariv reagent with seedlings of *Arabidopsis thaliana*. *Plant J.*, **9**, 919–925.
Williams, E.G., Knox, R.B. & Kaul, V. (1984) Post-pollination callose development in ovules of *Rhododendron* and *Ledum*, (Ericaceae): zygote special wall. *J. Cell Sci.*, **69**, 127–135.
Wojtaszek, P. (1997) Oxidative burst: an early plant response to pathogen infection. *Biochem. J.*, **322**, 681–692.
Wojtaszek, P. (2000) Genes and plant cell walls: a difficult relationship. *Biological Rev.*, **75**, 437–475.
Wojtaszek, P. (2001) Organismal view of a plant and a plant cell. *Acta Biochim. Polon.*, **48**, 443–451.
Wolf, S., Deom, C.M., Beachy, R. & Lucas, W. J. (1991) Plasmodesmatal function is probed using transgenic tobacco plants that express a virus movement protein. *Plant Cell*, **3**, 593–604.
Wyatt, S.E. & Carpita, N.C. (1993) The plant cytoskeleton-cell wall continuum. *Trends Cell Biol.*, **3**, 413–417.
Wymer, C., Bibikova, T. & Gilroy, S. (1997) Cytoplasmic free calcium distributions during the development of root hairs of *Arabidopsis thaliana*. *Plant J.*, **12**, 427–439.
Wymer, C.L., Wymer, S.A., Cosgrove, D.J. & Cyr, R.J. (1996) Plant cell growth responds to external forces and the response requires intact microtubules. *Plant Physiol.*, **110**, 425–430.
Wyrzykowska, J. & Fleming, A. (2003) Cell division pattern influences gene expression in the shoot apical meristem. *Proc. Natl. Acad. of Sci. USA*, **100**, 5561–5566.
Wyrzykowska, J., Pien, S., Shen, W.H. & Fleming, A.J. (2002) Manipulation of leaf shape by modulation of cell division. *Development*, **129**, 957–964.
Xu, H. & Mendgen, K. (1994) Endocytosis of 1,3-β-glucans by broad bean cells at the penetration site of cowpea rust fungus (haploid stage). *Planta*, **195**, 282–290.
Yan, B. & Smith, J.W. (2000) A redox site involved in integrin activation. *J. Biol. Chem.*, **275**, 39964–39972.
Yang, X.D., Sun, S.Q. & Li, Y.Q. (1999) Boron deficiency causes changes in the distribution of major polysaccharides of pollen tube wall. *Acta Botanica Sinica*, **41**, 1169–1176.
Yu, J. *et al.* (2002a) A draft sequence of the rice genome (*Oryza sativa* L. ssp. *indica*). *Science*, **296**, 79–92.
Yu, Q., Hlavačka, A., Matoh, T., Volkmann, D., Menzel, D., Goldbach, H. & Baluška, F.

(2002b) Short-term boron deprivation inhibits endocytosis of cell wall pectins in meristematic cells of maize and wheat root apices. *Plant Physiol.*, **130**, 415–421.
Yu, Q., Wingender, R., Schulze, M., Baluška, F. & Goldbach, H. (2001) Short-term boron deprivation induces increased levels of cytoskeletal proteins in *Arabidopsis* roots. *Plant Biol.*, **6**, 1–6.
Yu, Q., Baluška, F., Jasper, F., Menzel, D., Volkmann, D. & Goldbach, H.E. (2003) Short-term boron deprivation enhances levels of cytoskeletal proteins in maize, but not zucchini, root apices. *Physiol. Plant.*, **117**, 270–278.
Yuen, C.Y.L., Pearlman, R.S., Silo-suh, L., Hilson, P., Carroll, K.L. & Masson, P.H. (2003) WVD2 and WDL1 modulate helical organ growth and anisotropic cell expansion in Arabidopsis. *Plant Physiol.*, **131**, 493–506.
Zambryski, P. & Crawford, K. (2000) Plasmodesmata: gatekeepers for cell-to-cell transport of developmental signals in plants. *Annu. Rev. Cell Devel. Biol.*, **16**, 393–421.
Zamir, E. & Geiger, B. (2001) Molecular complexity and dynamics of cell-matrix adhesions. *J. Cell Sci.*, **114**, 3583–3590.
Zandomeni, K. & Schopfer, P. (1994) Mechanosensory microtubule reorientation in the epidermis of maize coleoptiles subjected to bending stress. *Protoplasma*, **182**, 96–101.
Zhang, N. & Hasenstein, K.H. (2000) Distribution of expansins in graviresponding maize roots. *Plant Cell Physiol.*, **41**, 1305–1312.
Zhong, R., Burk, D.H., Morrison III, W.H. & Ye, Z.-H. (2002a) A kinesin-like protein is essential for oriented deposition of cellulose microfibrils and cell wall strength. *Plant Cell*, **14**, 3101–3117.
Zhong, R., Kays, S.J., Schroeder, B.P. & Ye, Z.-H. (2002b) Mutation of a chitinase-like gene causes ectopic deposition of lignin, aberrant cell shapes, and overproduction of ethylene. *Plant Cell*, **14**, 165–79.
Zhong, R., Ripperger, A. & Ye, Z.-H. (2000) Ectopic deposition of lignin in the pith of stems of two *Arabidopsis* mutants. *Plant Physiol.*, **123**, 59–69.
Zhu, C., Bao, G. & Wang, N. (2000) Cell mechanics: mechanical response, cell adhesion, and molecular deformation. *Annu. Rev. Biomed. Eng.*, **2**, 189–226.
Zhu, J.-K., Bressan, R.A. & Hasegawa, P.M. (1993a) Loss of arabinogalactan-proteins from the plasma membrane of NaCl-adapted tobacco cells. *Planta*, **190**, 221–226.
Zhu, J.-K., Shi, J., Singh, U., Wyatt, S.E., Bressan, R.A., Hasegawa, P.M. & Carpita, N.C. (1993b) Enrichment of vitronectin- and fibronectin-like proteins in NaCl-adapted plant cells and evidence for their involvement in plasma membrane-cell wall adhesion. *Plant J.*, **3**, 637–646.
Zhu, J.-K., Damsz, B., Kononowicz, A.K., Bressan, R.A. & Hasegawa, P.M. (1994) A higher plant extracellular vitronectin-like adhesion protein is related to the translational elongation factor-1α. *Plant Cell*, **6**, 393–404.
Zhu, T. & Rost, T.L. (2000) Directional cell-to-cell communication in the *Arabidopsis* root apical meristem. III. Plasmodesmata turnover and apoptosis in meristem and root cap cells during four weeks after germination. *Protoplasma*, **213**, 99–107.
Zhu, T., Lucas, W.J. & Rost, T.L. (1998a) Directional cell-to-cell communication in the *Arabidopsis* root apical meristem. I. An ultrastructural and functional analysis. *Protoplasma*, **203**, 35–47.
Zhu, T., O'Quinn, R.L., Lucas, W.J. & Rost, T.L. (1998b) Directional cell-to-cell communication in the *Arabidopsis* root apical meristem. II. Dynamics of plasmodesmatal formation. *Protoplasma*, **204**, 84–93.
Zuo, J., Niu, Q.-W., Nishizawa, N., Wu, Y., Kost, B. & Chua, N.-H. (2000) KORRIGAN, an Arabidopsis endo-1,4-β-glucanase, localizes to the cell plate by polarized targeting and is essential for cytokinesis. *Plant Cell*, **12**, 1137–1152.

4 Polarity in single cells: root hairs, epidermal pavement cells and trichomes

Stefanie Falk, Jaideep Mathur and Martin Hülskamp

4.1 Three steps to single cell polarity

Any polarity seen at the level of organisms or tissues is based on the polarity of single cells. Therefore, in order to understand the origin of polarity, it is necessary to explore it at the level of single cells. How do single cells use extracellular signals to establish polarity or how do they establish polarity *de novo*? What mechanisms are involved in translating cellular polarity into asymmetric growth or asymmetric cell divisions? These aspects are well studied in animals and yeast and three major steps can be distinguished (Hülskamp *et al.*, 1998):

1. the establishment of cellular polarity by extracellular signals or by intracellular symmetry-breaking mechanisms;
2. the resulting asymmetry of the first step is fixed by an asymmetric distribution of cellular components that in turn leads to a polarization of the cell – for example, by a reorganization of the cytoskeleton;
3. polarity-dependent cellular responses such as cell growth or cell divisions are initiated in a directional manner.

In plants, surprisingly little is known about the molecular mechanisms underlying the establishment of single cell polarity. Most of our current understanding comes from the genetic and molecular analysis of a few model cell types. Root hairs and pollen tubes represent cell types that are characterized by growth that takes place exclusively at the very tip and are therefore named tip-growing cells. In these two cell types, cell polarity defines the length axis and where the tip of the cell is. Most other cell types grow by diffuse growth that occurs along the entire cell surface. Diffuse growing cells may have a single or several growth axes. For example, trichomes are single cells with a stereotyped-branching pattern and are being used to study how multiple polarity axes are established. An intermediate growth form is represented by epidermal pavement cells that have a jigsaw puzzle-like shape. In these cells the formation of lobes reflects the initiation of a new polarity and cell growth axis that are coordinated with the neighbouring epidermal cells. In this review we attempt to summarize our current knowledge on cell polarity establishment and its expression in these cell types.

4.2 Root hairs and pollen tubes

Root hair growth and pollen tube growth are characterized by a unidirectional polarized growth that is focused to one tip of the cell. In these systems the polarity of the cells is basically established by defining the growth focus. Directionality of root hair growth appears to be governed by an intrinsic mechanism whereas the growth focus in pollen tubes appears to be guided by external signals (Hülskamp *et al.*, 1995; Ray *et al.*, 1997). The mechanisms underlying the establishment of the growth point and the cells axis are not known. In light of the current data, however, it is conceivable that by analogy to other systems the induction of cell polarity begins with the local activity of small GTPases which in turn leads to local actin accumulation and the polarized organization of microtubules. This is discussed in detail in Chapter 1, but a schematic presentation of the localization of these components is shown as a summary in Fig. 4.1. In plants, the canonical Rac/Rho/Cdc42 GTPases are absent, however, a related but distinct family has been described as ROPs (*R*ho *of p*lants) (Yang, 2002). In *Arabidopsis* pollen tubes ROP1 and ROP5 were shown to be enriched at the growing tip and to be essential for proper polarized growth (Lin *et al.*, 1996; Kost *et al.*, 1999). Similarly, ROP2, ROP4 and ROP6 are involved in root hair growth and ROP2:GFP was shown to localize to the growing tip (Jones *et al.*, 2002). In both cell types ROPs are involved in the local reorganization of fine actin at the growth point.

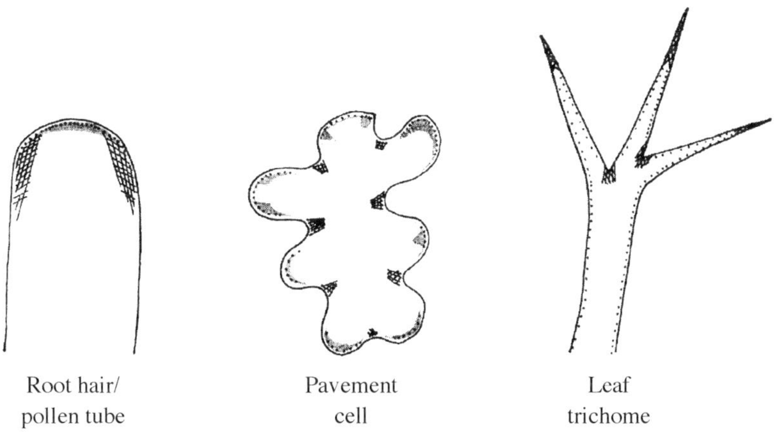

Figure 4.1 Schematic overview of ROPs and actin in different cell types. Localization data and mutant analysis suggest that localized activity of small GTPases (ROPs) leads to local accumulation of fine actin which is required for local growth of the cell. Regions with bundled actin do not grow. The intracellular localization of ROPs, fine and bundled actin is schematically shown in a tip growing cell, an epidermal pavement cell and an *Arabidopsis* leaf trichome. Light grey marks the localization of ROPs, small points the localization of fine actin, and crossing lines the localization of bundled actin.

Overexpression of an *Arabidopsis* ROP2 leads to an increase of fine actin microfilaments and conversely, the expression of the dominant negative version of ROP1 in pollen tubes and the dominant negative version of ROP2 in root hairs prevents the accumulation of fine actin filaments (Fu *et al.*, 2001; Jones *et al.*, 2002). The formation of these fine actin filaments at the tip is essential for growth to continue. This fact has become even more evident from the recent molecular characterization of three defective mutants in different components of the Arp2/3 complex (see below). This complex comprises seven protein subunits and plays a role in the region-specific nucleation of new actin filaments (Machesky and Gould, 1999). In Arabidopsis *arp2/3* mutants, in a subset of growing root hairs F-actin becomes bundled in the entire root hair cell including the very tip where normally only fine actin is found (Mathur *et al.*, 2003a,b,). As a result, mutant root hair growth is irregular, leading to wavy root hairs (Fig. 4.2). In other organisms the ARP2/3 complex is a well-established downstream target of Rho-like GTPases (Ridley, 2001). In plants, this signaling cascade is still not resolved because ROPs exhibit many differences to Rho-like proteins from animals and yeast and some downstream targets, e.g. WASP proteins, appear to be absent (Mathur & Hülskamp, 2002a).

The actual growth is likely to be targeted to the tip by the fine actin that delivers vesicles containing cell wall components to the plasma membrane. Again the characterization of several *Arabidopsis* mutants has helped to elucidate the actual growth mechanisms. These mechanisms are however

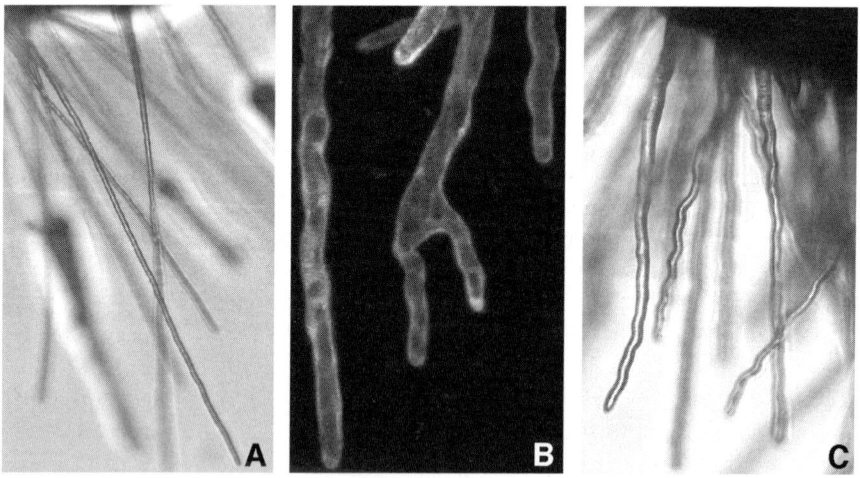

Figure 4.2 Root hair phenotypes. (A) Wild-type root hairs in *Arabidopsis*. Normally, root hairs are long and straight cellular outgrowths. (B) Root hair growth after oryzalin treatment. Note, root hairs form branches indicating that new tips have been initiated. (C) Root hair growth in the *arp2/3* mutant *crooked*. Tip growth is irregular leading to a wavy root hair.

beyond the scope of this review and are summarized elsewhere (Mathur & Hulskamp, 2001; Carol & Dolan, 2002; see also Chapter 1).

The formation of a growing tip is concommitant with a fixation of the cell axis by polarizing the remainder cell body. This can easily be seen by the longitudinally arranged actin and microtubule cytoskeleton. Two experiments indicate that microtubules are important for maintaining this polarity and for preventing the initiation of new growth sites. Treatments of growing root hairs in *Arabidopsis* with microtubule-interacting drugs results in wavy and multiple-branched root hairs (Fig. 4.2; Bibikova *et al.*, 1999). A similar phenotype is found in plants with reduced expression of α-tubulin (Bao *et al.*, 2001). How tip definition leads to the polarization of the microtubule cytoskeleton and how microtubules prevent the formation of additional tips is unknown.

4.3 Epidermal pavement cells

During development most epidermal pavement cells form large interconnecting lobes which reflect the initiation of multiple growth axes that are coordinated with respect to those of the neighbouring epidermal cells. The current data suggest that the formation of lobes also involves a ROP-Actin signalling cascade similar to that described for tip growing cells (Fig. 4.1). It was shown that GFP:ROP2 is localized in the extending lobes and that overexpression of a constitutive or dominant negative form interferes with lobe formation. In both cases, the accumulation of fine actin in the lobes was aberrant (Fu *et al.*, 2001). Also mutations in the Arp2/3 complex cause an aberrant actin distribution (Fig. 4.3) and to a failure of proper lobe formation (Fig. 4.4), indicating that localized fine F-actin is important for epidermal cell lobing (Mathur *et al.*, 2003a,b). The genetic analysis of lobe formation in epidermal pavement cells in maize has identified three *BRICK* genes, mutants which share a common phenotype of non-lobed epidermal cells (Frank *et al.*, 2003). The *BRICK1* gene has been cloned and was shown to encode a protein with sequence similarity to the human HSPC300 gene. HSPC300 is part of a large complex that also contains WAVE, a regulator of Arp2/3 complex activity. Rac binds to the HSPC300 complex, thereby causing its dissociation and resulting in a HSPC300/WAVE dimer that in turn activates Arp2/3 (Eden *et al.*, 2002). Although a WAVE homologue does not appear to be present in plants, it is tempting to speculate that the plant *BRICK1* gene is involved in a similar pathway as emerging in the human system.

A good candidate for a component that controls the spatial activity of the ROP-actin pathway is the *SPIKE* gene (Qiu *et al.*, 2002). In *spike* mutants, a variety of cellular phenotypes is seen, including the lack of lobe formation and loss of trichome branching (see below). At the cellular level both the

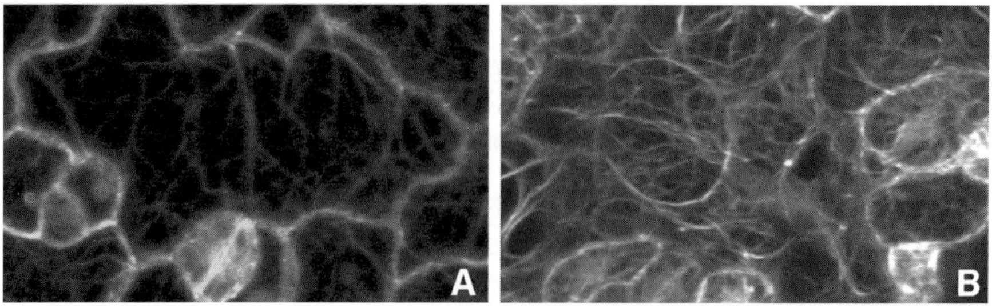

Figure 4.3 Actin distribution in wild-type and mutant epidermal pavement cells. (A) Actin distribution in a wild-type epidermal pavement cell in *Arabidopsis*. F-actin is visualized by talin:GFP. Note that actin is regularly distributed and seen as fine filaments. (B) Actin distribution of the *arp2/3* mutant *wurm*. Note that actin forms irregular and thicker bundles.

microtubule organization and the actin distribution are affected. The *SPIKE* gene encodes a CDM-family (CED5, DOCK180, MYOBLAST CITY family) protein members of which are implicated in cytoskeletal organization in response to external signals (Nolan *et al.*, 1998). As the CDM-domain is known to interact with a Rac-GTPase it is possible that the *Arabidopsis* homologue does interact with a ROP, thereby locally activating the corresponding pathways.

The establishment of multidirectional growth axes in pavement cells appears to involve additionally, a microtubule-based mechanism. This system defines non-growth regions by organizing microtubule bands that are correlated with the formation of regional cell wall thickenings (Pantres *et al.*, 1994; Jung & Wernicke, 1990). Lobes grow in between these thickenings, suggesting that growth occurs where the cell wall remains more extensible. That microtubules are important for lobe formation is supported by several lines of evidence. Mutations in the *FRAGILE FIBER2,* gene which encodes a putative microtubule-severing katanin, result in non-lobed epidermal pavement cells (Burk *et al.*, 2001). Also the *Arabidopsis angustifolia* mutant,

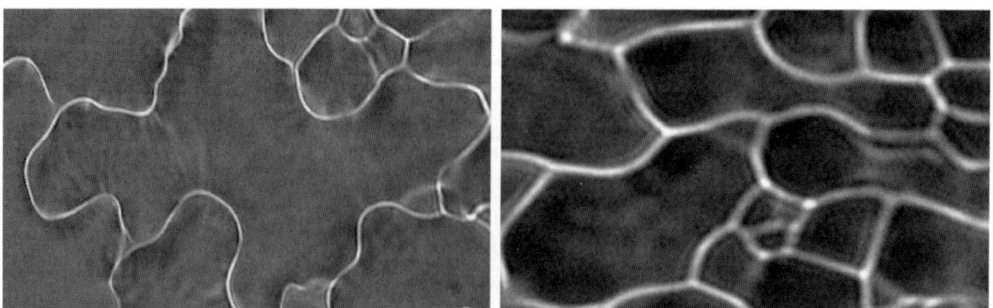

Figure 4.4 Cell shapes of wild-type and mutant epidermal pavement cells. (A) Wild type epidermal pavement cell. Note that the cell forms several irregular lobes. (B) Mutations in the *arp2/3* mutant *wurm* cause a failure of epidermal pavement cells to form lobes.

which is defective in the control of the microtubule cytoskeleton in epidermal pavement cells in *Arabidopsis*, shows reduced lobing (Tsuge *et al.*, 1996b). Though it is conceivable that the microtubule-based pathway and the ROP-Actin pathway are coordinated nothing is known about a possible connection.

4.4 Trichomes

Trichome polarity is best studied for leaf trichomes in *Arabidopsis* (Bouyer *et al.*, 2001). *Arabidopsis* leaf trichomes are single, branched cells (Marks *et al.*, 1991; Hülskamp *et al.*, 1994). Their regular branching pattern results from several polarity changes during ontogenesis (Fig. 4.5). After a protodermal cell becomes selected to develop into a trichome cell, the cell stops dividing. However, the cell continues DNA replication in the absence of cell division (endoreduplication) and thereby becomes polyploid (Hülskamp *et al.*, 1994). Concomitant with increasing ploidy levels the trichome cell increases in size. The incipient trichome cell changes its polarity and grows out of the leaf surface. The first branching event occurs when the cell has undergone two endoreduplication cycles which corresponds to a DNA content of 8C. The two branches point to the tip and the base of the leaf indicating that the initiation of the branches is aligned with the leaf axis (Folkers *et al.*, 1997). Thus branch initiation reflects the establishment of a new polarity axis within the trichome

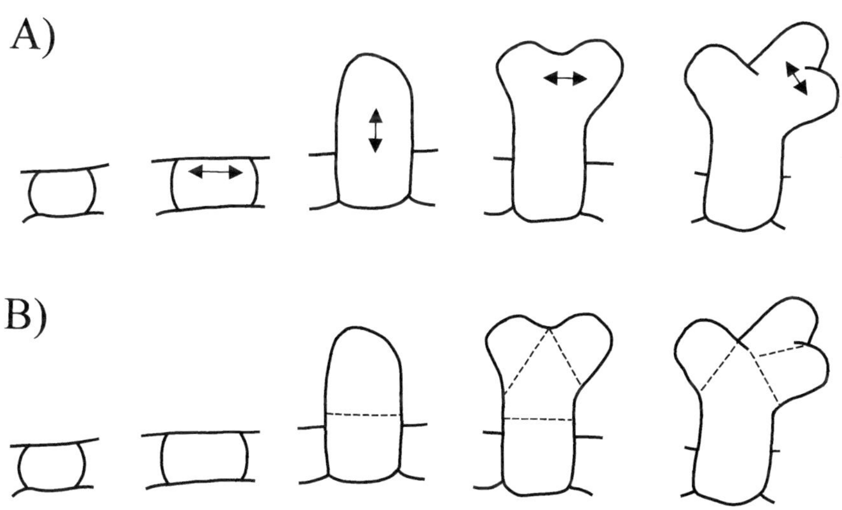

Figure 4.5 Trichome ontogenesis and its possible evolution. (A) Schematic drawings of different developmental stages of trichome development. The changes in cell polarity are indicated by arrows. (B) It is postulated that trichome cell polarity is evolutionary, derived from an underlying division pattern, which is still operating, however, in the absence of cell divisions. Hypothesized cell divisions are indicated as dotted lines.

cell. A second branching event is initiated on the branch that points towards the leaf tip. The plane of this branch is twisted by 90 degrees with respect to the first. This predictable spatial arrangement reflects an asymmetry between the initial basal and distal branch and at the distal branch another polarity axis is established. After branch initiation, the trichome cell undergoes rapid cell expansion. The mature trichome cell has on average a 16 fold DNA content of a diploid cell (Hülskamp *et al.*, 1994).

4.4.1 Does trichome cell polarity originate from the polarity of cell divisions?

How trichome cell polarity is established is still unknown. However, several lines of enquiry suggest that the underlying mechanism may be based on the machinery that controls the orientation of cell divisions (Schnittger & Hülskamp, 2002). According to this model, *Arabidopsis* trichomes are evolutionarily derived from multicellular trichomes in which branching was caused by a defined cell division pattern. Such a mechanism has been described for multicellular trichomes in other species, for example, *Verbascum* (Uphof, 1962). The branching pattern thus reflects how the polar axis changes during cell divisions. In *Arabidopsis*, where the trichome is unicellular it is assumed that the mechanism generating the polarity axis is still operating in the absence of the actual nuclear and cellular divisions (Fig. 4.5). Evidence supporting this view comes from the finding that a mutation in a single gene is sufficient to make *Arabidopsis* trichomes multicellular. Mutations in the *SIAMESE* gene result in trichomes exhibiting up to fifteen cells (Walker *et al.*, 2000). Because the phenotype is not fully penetrant it is still unclear as to what extent branching and the axis of cell divisions are linked. Also trichome-specific overexpression of two mitotic cyclins, CYCB1.2 and CYCD3.2 trigger the initiation of cell divisions (Schnittger *et al.*, 2002a,b).

One possibility to test the above model is the molecular characterization of mutants affecting trichome branching. Following the rationale of the model, one would expect that mutations in this process should result in a branching phenotype.

4.4.2 Trichome branching mutants and cell polarity

In various mutagenesis screens more than 15 branching genes were isolated in which mutations result in more or fewer trichome branches (Hülskamp, 2000; Oppenheimer, 1998; Hülskamp *et al.*, 1999). Generally, these mutants fall into two classes.

The changes in branch number of the first class of mutants are correlated with a decrease or increase in the nuclear DNA content and cell size. In mutants having a reduced DNA content e.g. *glabra3*, trichomes are smaller and show fewer branches. Trichomes with an increased DNA content, e.g. in *triptychon* mutants, are larger and show more branches (Hülskamp *et al.*, 1994; Perazza *et al.*, 1999; Traas *et al.*, 1998). This correlation holds true also for wild-type when DNA content is reduced by treatments with aphidicolin, a DNA replication inhibitor (Schwab & Hülskamp, unpublished observations) and in tetraploid plants that exhibit extra branch formation (Perazza *et al.*, 1999). These observations can easily be explained by the above mentioned model where extra rounds of endoreduplication would be considered as extra rounds of mitotic divisions in a multicellular context which in turn would generate more branches. As the primary function of this class of branching mutants lays in the regulation of cell cycle progression and not in its spatial regulation they are not considered further here.

The second class of branching mutants comprises mutants with two or no branches and one mutant which displays up to eight branches. Only three of the genes act in the pathway in which endoreduplication or cell size leads to

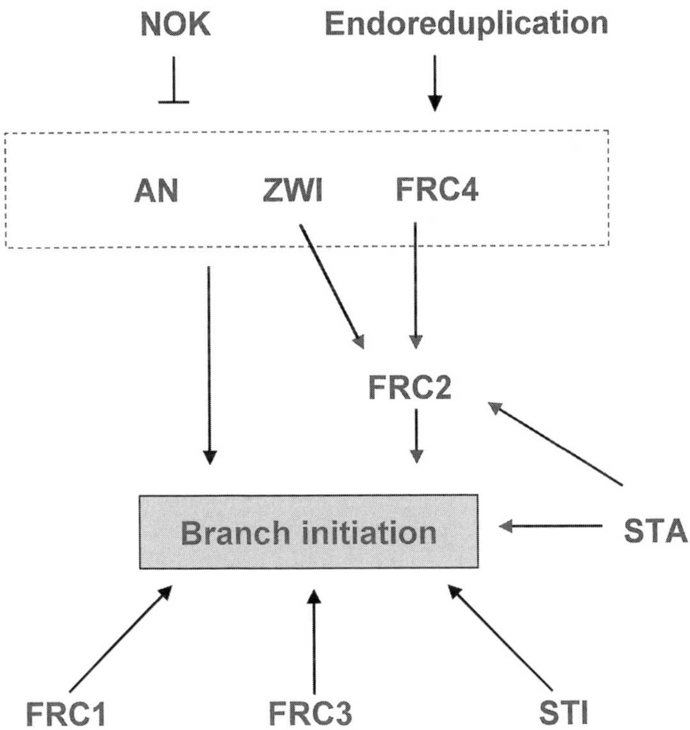

Figure 4.6 Genetic interaction of the branching genes. The genetic analysis has determined the genetic interactions between most of the known branching genes. Arrows indicate positive regulation events and blunted bars show negative interactions.

changes in branch number (Fig. 4.6; Luo & Oppenheimer, 1999; Folkers *et al.*, 1997). The genes *ANGUSTIFOLIA* (*AN*), *ZWICHEL* (*ZWI*) and *FURCA4* (*FRC4*) are required to mediate extra branch formation induced in the endoreduplication mutants and are therefore the best candidates for genes involved in cell division-related spatial organization of the trichome cell. All other branching genes that have been studied genetically in more detail appear to act in separate pathways (Luo & Oppenheimer, 1999).

4.5 Molecular analysis of the branching genes likely to be involved in polarity establishment

The genetic data that place *AN*, *ZWI* and *FRC4* in the same pathway as the endoreduplication mutants and thereby in the context of the cell division machinery are supported by the molecular analysis of *ANGUSTIFOLIA* and *ZWICHEL*.

The *ZWI* gene encodes a member of the kinesin superfamily of motor proteins and contains a calmodulin-binding site (Oppenheimer *et al.*, 1997). It turned out to be the same gene as the biochemically well characterized calmodulin-binding protein KCBP (Reddy *et al.*, 1996a,b; Song *et al.*, 1997; Deavours *et al.*, 1998). Kinesins are microtubule-based motors that play a role in intracellular transport processes and microtubule organization. Whether the *zwi* phenotype is caused by the failure of specific transport processes or due to defects in the reorientation of microtubules is unclear. The cell biological analysis revealed that ZWI function is linked to the cell cycle. It was shown in tissue culture cells that ZWI protein is localized to the preprophase band and the phragmoplast and thus specifically labels those structures in dividing cells that mark the division plane (Bowser & Reddy, 1997). This localization is also functionally relevant as the injection of ZWI antibodies in the multicellular stamen hair cells of *Tradescantia* results in metaphase arrest and abnormal cell plate formation (Vos *et al.*, 2000). While these results strongly suggest a role of ZWI in cell divisions, its role in *Arabidopsis* remains unclear as cell division defects were not described in *zwi* mutants.

The *AN* gene encodes a protein with sequence similarities to the transcription cofactor CtBP (C-terminal binding proteins) and BARS (Brefeldin A ribosylated proteins) (Folkers *et al.*, 2002; Kim *et al.*, 2002). CtBPs were identified in *Drosophila* as co-repressors of the zinc-finger transcriptional repressors KRÜPPEL, KNIRPS and SNAIL (Nibu *et al.*, 1998) where they bind to a seven amino acid long consensus sequence. BARS were identified in rat as proteins that become ADP-ribosylated after treatments with brefeldinA, a fungal toxin that interferes with Golgi dynamics (Lippincott-Schwartz *et al.*, 1989; Matteis *et al.*, 1994). More recent results showed that CtBPs and BARS can be considered homologous proteins, as BARS can bind to the same transcription factors as CtBPs and because CtBPs are ribosylated upon brefeldinA treatment (Spanfo *et al.*, 1999). The biochemical function of AN in plants has not been determined yet, although its cellular function appears to be connected with the cell division machinery. This is suggested by several lines of evidence. First, *an* mutants show not only defects in trichome development but also in all other organs including the cotyledons and rosette leaves which are more narrow than wild type. It has been shown that the decrease in leaf width is partly due to a polarity shift of the individual leaf cells and by a decrease in cell number (Tsukaya *et al.*, 1994; Tsuge *et al.*, 1996a). Second, the AN protein contains a PEST motive in the context of a consensus sequence where mitotic cyclins bind suggesting that AN protein stability is controlled in a cell cycle dependent manner (Folkers *et al.*, 2002; Kim *et al.*, 2002). Third, AN directly interacts with ZWI in a yeast two hybrid assay. It is therefore likely that AN, similar to ZWI, also has a function during cell divisions.

In addition to *AN*, *ZWI* and *FRC4*, the *STICHEL* (*STI*) gene is likely to have a regulatory role in the establishment of polarity. This is suggested by the finding that STI seems to regulate the number of trichome branches in a dosage dependent manner (Ilgenfritz *et al.*, 2003). *sti*-alleles of different strength were isolated, with strong alleles being unbranched and weaker alleles being underbranched. Conversely, overexpression of STI triggers extra branch formation (Ilgenfritz *et al.*, 2003). The *STI* gene encodes a large protein containing a domain with sequence similarities to eubacterial DNA-polymerase III subunits (Ilgenfritz *et al.*, 2003). It is, however, unlikely that STI plays a role in DNA replication as DNA-polymerases are present in *Arabidopsis* and because the STI protein contains several other protein domains other than the canonical DNA-polymerase domain. In addition, no cell division or endoreduplication phenotype is observed in *sti* mutants. It is therefore likely that STI has adopted a new role in cell morphogenesis.

4.6 Molecular analysis of branching genes likely to be involved in polarity fixation

As in other plant cell types, polarity fixation in trichomes involves the microtubule cytoskeleton (Mathur & Hülskamp, 2002b). This is evident from experiments in which the microtubule organization is disturbed by microtubule specific drugs (Mathur & Chua, 2000). Such trichomes show reduced branching and isotropic growth leading to a ball-like shape. A more specific characterization of the microtubule role in trichome branching became possible with the isolation of the two weak tubulin folding cofactor mutants TFC-A and TFC-C (Kirik *et al.*, 2002a,b). TFCA and TFCC belong to a class of proteins that associate with α and β tubulin monomers after

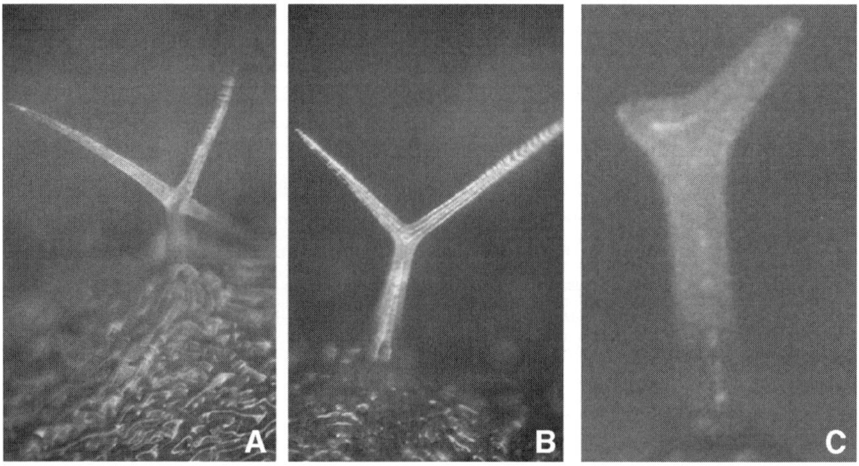

Figure 4.7 *Arabidopsis* trichome branching mutants. (A) *Arabidopsis* wild-type leaf trichome with three branches. (B) *an*-mutant trichome with two branches. (C) *tfcc*-mutant trichome with two branches. Note that the trichome has a more swollen appearance.

translation, and facilitate the formation of assembly competent α-tubulin dimers (Lewis *et al.*, 1997). In *Arabidopsis*, tubulin folding cofactors are absolutely essential as strong mutations lead to embryo lethality (Steinborn *et al.*, 2002). The weak *tfca* and *tfcc* mutants display a pleiotropic phenotype that includes a swollen and underbranched trichome phenotype (Fig. 4.7). Surprisingly, microtubules appeared to be intact. Their relative orientation was also normal except for those branches that had been initiated but failed to grow. Here the reorientation from a transverse to a longitudinal orientation did not occur. These data indicate that the formation of new branches requires the *de novo* formation of microtubules rather than the reorganisation of already existing microtubules (Kirik *et al.*, 2002a,b).

This conclusion is supported by the analysis of the katanin-p60 mutant (*ectopic root hairs 3*, *botero1*, *fat root*, *fragile fiber2*) in *Arabidopsis* (Bichet *et al.*, 2001; Burk *et al.*, 2001; Webb *et al.*, 2002). Katanins are heterodimeric proteins that cut existing microtubules into smaller fragments (Quarmby, 2000). The above mentioned mutants do not properly reorient in elongating cells and as a consequence show strongly reduced cell elongation. Trichomes appear reduced and are two-branched.

As judged by their mutant phenotype and their molecular nature, two genes could act to spatially regulate the microtubule cytoskeleton, *FASS/TONNEAU2* and *SPIKE1*. *FASS/TONNEAU2* encodes a novel protein phosphatase 2A regulatory subunit (Camilleri *et al.*, 2002). Mutations in *FASS/TONNEAU2* result – among other pleiotropic phenotypes – in reduced branching of trichomes concomitant with a randomized microtubule organization (Traas *et al.*, 1995; Torres-Ruiz & Jürgens, 1994). It is therefore likely that *FASS/TONNEAU2* controls microtubule dynamics by protein phosphorylation. The *SPIKE1* gene encodes a 207 kD protein with sequence similarity to CDM family adapter proteins (see above) (Qiu *et al.*, 2002) and is therefore a good candidate to mediate outer signals to the microtubule cytoskeleton (Nolan *et al.*, 1998).

4.7 Molecular analysis of the genes involved in growth directionality

The last step of polarity establishment is the actual growth along the polarity axis. In *Arabidopsis* trichomes, this step has been recognized as a distinct regulatory event as a class of eight mutants was discovered in which the directionality of trichome growth is disturbed. The corresponding genes have been termed *DISTORTED* genes based on the phenotype of mature trichomes which display a twisted and distorted cell form (Feenstra, 1978; Hülskamp *et al.*, 1994). Their phenotypes are expressed after branches are initiated and when trichome cells begin rapid cell expansion (Schwab *et al.*, 2003). At the cellular level this phenotype is caused by defects in the actin cytoskeleton. This is evident from the findings that on the one hand treatments with drugs specifically interfering with the actin organization result in a distorted trichome phenotype and on the other hand distorted mutants exhibit an abberant actin organization (Szymanski *et al.*, 1999, Schwab *et al.*, 2003; Mathur *et al.*, 1999). Beyond the fact that actin is required for directional growth along the established growth axis, it remained unclear how actin organization is regulated and how actin controls the actual growth. The molecular cloning of the first three *DISTORTED* genes has facilitated the analysis of these questions (Mathur *et al.*, 2003a,b). The *DISTORTED* genes *CROOKED*, *DISTORTED1* and *WURM* encode components of the Arp2/3 complex. In the corresponding mutants two surprising observations were made. First, actin defects caused by mutations in the Arp2/3 complex compromise the fusion of vacuoles that normally form the large central vacuole in trichome cells (Mathur *et al.*, 2003a). Whether this reduced membrane fusion capability affects other membrane compartments or vesicles is not known. The second observation is that massive actin accumulation is seen in non growth regions where different sub-cellular components dependent upon their transport by actin are found entrapped (Fig. 4.8). These actin-rich areas correspond to non-growing regions of a cell. By contrast, actively growing regions of the cell are characterized by the presence of very fine actin. The data suggest that dense F-actin configurations define non-growth regions, whereas fine F-actin mediates local growth (Mathur *et al.*, 2003b). How the spatial activity of the Arp2/3 complex is controlled is still unclear.

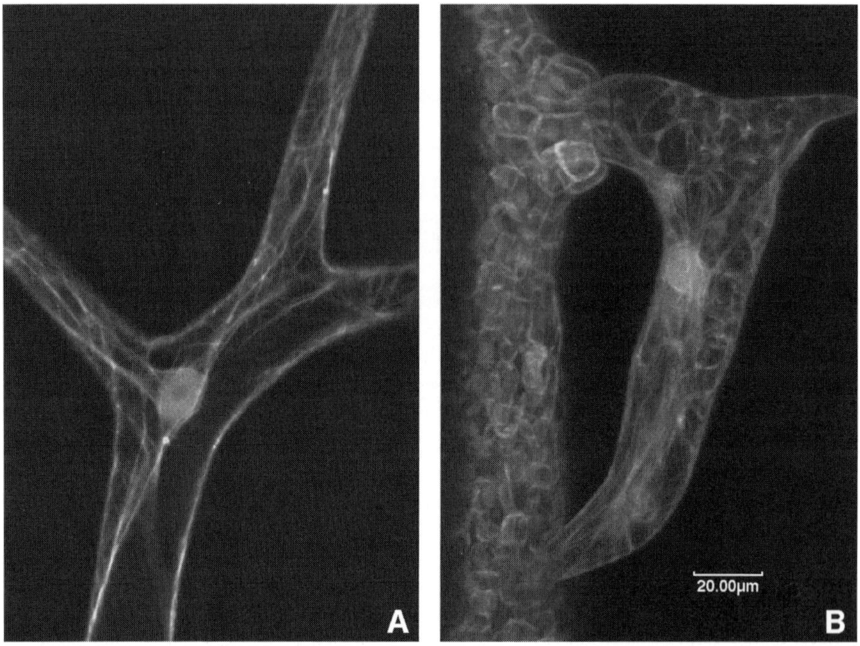

Figure 4.8 Actin organization in *Arabidopsis* and *distorted* mutant trichomes. (A) Actin distribution in an Arabidopsis wild-type trichome. The organization of F-actin is visualized by talin:GFP. (B) Actin distribution in the *arp2/3* mutant *wurm* mutant trichome. Note, actin is less organized and more bundled.

References

Bao, Y., Kost, B. & Chua, N.-H. (2001) Reduced expression of α-tubulin genes in Arabidopsis thaliana specifically affects root hair development and root gravitropism. *Plant J.*, **28**, 145–157.

Bibikova, T.N., Blancaflor, E.B. & Gilroy, S. (1999) Microtubules regulate tip growth and orientation in root hairs of *Arabidopsis thaliana*. *Plant J.*, **17**, 657–665.

Bichet, A., Desnos, T., Turner, S., Grandjean, O. & Höfte, H. (2001) BOTERO1 is required for normal orientation of cortical microtubules and anisotropic cell expansion in Arabidopsis. *Plant J.*, **25**, 137–148.

Bouyer, D., Kirik, V. & Hulskamp, M. (2001) Cell polarity in Arabidopsis trichomes. *Semin. Cell Dev. Biol.*, **12**, 353–356.

Bowser, J. & Reddy, A.S.N. (1997) Localization of a kinesin-like calmodulin-binding protein in dividing cells of *Arabidopsis* and tobacco. *Plant J.*, **12**, 1429–1437.

Burk, D. H., Liu, B., Zhong, R., Morrison, W.H. & Ye, Z.H. (2001) A katanin-like protein regulates normal cell wall biosynthesis and cell elongation. *Plant Cell*, **13**, 807–27.

Camilleri, C., Azimzadeh, J., Pastuglia, M., Bellini, C., Grandjean, O. & Bouchez, D. (2002) The Arabidopsis TONNEAU2 Gene Encodes a Putative Novel Protein Phosphatase 2A Regulatory Subunit Essential for the Control of the Cortical Cytoskeleton. *Plant Cell*, **14**, 833–845.

Carol, R.J. & Dolan, L. (2002) Building a hair: tip growth in Arabidopsis thaliana root hairs. *Phil. Trans. R. Soc. Lond. B*, **357**, 815–821.

Deavours, B.E., Reddy, A.S. & Walker, R.A. (1998) Ca^{2+}/calmodulin regulation of the

Arabidopsis kinesin-like calmodulin-binding protein. *Cell Motil. Cytoskeleton*, **40**, 408–416.

Eden, S., Rohtagi, R., Podtelejnikov, A.V., Mann, M. & Kirschner, M.W. (2002) Mechanism of regulation of WAVE1–induced actin nucleation by Rac1 and Nck. *Nature*, **418**, 790–793.

Feenstra, W. J. (1978) Contiguity of linkage groups 1 and 4, as revealed by linkage relationships of two newly isolated markers *dis-1* and *dis-2*. *Arabidopsis Info. Serv.*, **15**, 35–38.

Folkers, U., Berger, J. & Hulskamp, M. (1997) Cell morphogenesis of trichomes in Arabidopsis: differential control of primary and secondary branching by branch initiation regulators and cell growth. *Development*, **124**, 3779–3786.

Folkers, U., Kirik, V., Schobinger, U., Falk, S., Krishnakumar, S., Pollock, M.A., Oppenheimer, D. G., Day, I., Reddy, A. R., Jurgens, G. & Hulskamp, M. (2002) The cell morphogenesis gene *ANGUSTIFOLIA* encodes a CtBP/BARS-like protein and is involved in the control of the microtubule cytoskeleton. *EMBO J.*, **21**, 1280–1288.

Frank, M.J., Cartwright, H.N. & Smith, L.G. (2003) Three Brick genes have distinct functions in a common pathway promoting polarized cell division and cell morphogenesis in the maize leaf epidermis. *Development*, **130**, 753–762.

Fu, Y., Wu, G. & Yang, Z. (2001) Rop GTPase-dependent dynamics of tip-localized F-actin controls tip growth in pollen tubes. *J. Cell Biol.*, **152**, 1019–1032.

Hulskamp, M. (2000) How plants split hairs. *Curr Biol*, **10**, R308–R310.

Hülskamp, M., Folkers, U. & Grini, P. (1998) Cell morphogenesis in *Arabidopsis*. *BioEssays*, **20**, 20–29.

Hülskamp, M., Folkers, U. & Schnittger, A. (1999) Trichome development in *Arabidopsis thaliana*. *International Review of Cytology*, **186**, 147–178.

Hülskamp, M., Misera, S. & Jürgens, G. (1994) Genetic dissection of trichome cell development in *Arabidopsis*. *Cell*, **76**, 555–566.

Hülskamp, M., Schneitz, K. & Pruitt, R.E. (1995) Genetic evidence for a long range activity that directs pollen tube guidance in *Arabidopsis*. *Plant Cell*, **7**, 57.

Ilgenfritz, H., Bouyer, D., Schnittger, A., Mathur, J., Kirik, V., Schwab, B., Chua, N.-H., Jürgens, G. & Hülskamp, M. (2003) The Arabidopsis *STICHEL* Gene is a Regulator of Trichome Branch Number and Encodes a Novel Protein. Plant Physiology. *Plant Physiol.*, **131**, 643–655.

Jones, M.A., J-J, S., Fu, Y., Yang, Z. & Grierson, C.S. (2002) The Arabidopsis Rop2 GTPase is a positive regulator of both root hair initiation and tip growth. *Plant Cell*, **14**, 763–776.

Jung, G. & Wernicke, W. (1990) Cell shaping and microtubules in developing mesophyll of wheat (Triticum aestivum L.). *Protoplasma*, **153**, 141–148.

Kim, G.T., Shoda, K., Tsuge, T., Cho, K.-H., Uchimiya, H., Yokoyama, R., Nishitani, K. & Tsukaya, H. (2002) The *ANGUSTIFOLIA* gene of Arabidopsis, a plant CtBP gene, regulates leaf-cell expansion, the arrangement of cortical microtubules in leaf cells and expression of a gene involved in cell-wall formation. *EMBO J.*, **26**, 1267–1279.

Kirik, V., Grini, P.E., Mathur, J., Klinkhammer, I., Adler, K., Bechtold, N., Herzog, M., Bonneville, J. M. & Hulskamp, M. (2002a) The Arabidopsis *TUBULIN-FOLDING COFACTOR A* gene is involved in the control of the alpha/beta-tubulin monomer balance. *Plant Cell*, **14**, 2265–2276.

Kirik, V., Mathur, J., Grini, P.E., Klinkhammer, I., Adler, K., Bechtold, N., Herzog, M., Bonneville, J.M. & Hulskamp, M. (2002b) Functional analysis of the tubulin-folding cofactor C in Arabidopsis thaliana. *Curr. Biol.*, **12**, 1519–1523.

Kost, B., Lemichez, E., Spielhofer, P., Hong, Y., Tolias, K., Carpenter, C. & Chua, N.-H. (1999) Rac homologues and compartmentalized phosphatidlinositol 4,5-bisphosphate act in a common pathway to regulate polar pollen tube growth. *J. Cell Biol.*, **145**, 317–330.

Lewis, S.A., Tian, G. & Cowan, N.J. (1997) The α- and β-tubulin folding pathways. *Trends Cell Biol.*, **7**, 479–484.

Lin, Y., Wang, Y., Zhu, J. & Yang, Z. (1996) Localization of a rho GTPase implies a role in tip growth and movement of the generative cell in pollen tubes. *Plant Cell*, 8, 293–303.

Lippincott-Schwartz, J., Yuan, L.C., Bonifacino, J.S. & Klausner, R.D. (1989) Rapid redistribution of Golgi proteins into the ER in cells treated with brefeldin A: evidence for membrane cycling from Golgi to ER. *Cell*, **56**, 801–813.

Luo, D. & Oppenheimer, D.G. (1999) Genetic control of trichome branch number in Arabidopsis: the roles of the *FURCA* loci. *Development*, **126**, 5547–5557.

Machesky, L.M. & Gould, K.L. (1999) The ARP2/3 complex: a multifunctional actin organizer. *Curr. Opin. Cell Biol.*, **11**, 117–121.

Marks, M. D., Esch, J., Herman, P., Sivakumaran, S. & Oppenheimer, D. (1991) In *Molecular Biology of Plant Development* (eds. G. Jenkins & W. Schuch) Company of Biologists Ltd., Cambridge, pp. 77–87.

Mathur, J. & Chua, N.H. (2000) Microtubule stabilization leads to growth reorientation in Arabidopsis thaliana trichomes. *Plant Cell*, **12**, 465–477.

Mathur, J. & Hülskamp, M. (2001) Cell growth: how to grow and where to grow. *Curr. Biol*, **11**, R402–R404.

Mathur, J. & Hülskamp, M. (2002a) Signal transduction: rho-like proteins in plants. *Curr. Biol.*, **12**, R526–R528.

Mathur, J. & Hülskamp, M. (2002b) Microtubules and microfilaments in cell morphogenesis in higher plants. *Curr. Biol.*, **12**, R669.

Mathur, J., Mathur, N., Kernebeck, B. & Hülskamp, M. (2003a) Mutations in actin related proteins 2 and 3 affect cell shape development in Arabidopsis thaliana. *Plant Cell*, **15**, 1632–1645.

Mathur, J., Mathur, N., Kirik, V., Kernebeck, B., Srinivas, B.P. & Hülskamp, M. (2003b) Arabidopsis *CROOKED* encodes for the smallest subunit of the ARP2/3 complex and controls cell shape by region specific fine F-actin formation. *Development*, **130**, 3137–3146.

Mathur, J., Spielhofer, P., Kost, B. & Chua, N.-H. (1999) The actin cytoskeleton is required to elaborate and maintain spatial patterning during trichome cell morphogenesis in *Arabidopsis thaliana*. *Development*, **126**, 5559–5568.

Matteis, M.D., Girolamo, M.D., Colanzi, A., Pallas, M., Tullio, G.D., McDonald, L., Moss, J., Santini, G., Bannykh, S., Corda, D. & Luini, A. (1994) Stimulation of endogenous ADP-rebosylation by brefeldin A. *Proc. Natl. Acad. Sci. USA*, **91**, 1114–1118.

Nibu, Y., Zhang, H. & Levine, M. (1998) Interaction of a short-range repressors with Drosophila CtBP in the embryo. *Science*, **280**, 101–104.

Nolan, K. M., Barrett, K., Lu, Y., Hu, K.Q., Vincent, S. & Settleman, J. (1998) Myoblast city, the Drosophila homolog of DOCK180/CED-5, is required in a Rac signaling pathway utilized for multiple developmental processes. *Genes Dev.*, **12**, 3337–3342.

Oppenheimer, D. (1998) Genetics of plant cell shape. *Curr. Opin. Plant Biol.*, **1**, 520–524.

Oppenheimer, D.G., Pollock, M.A., Vacik, J., Szymanski, D.B., Ericson, B., Feldmann, K. & Marks, M.D. (1997) Essential role of a kinesin-like protein in *Arabidopsis* trichome morphogenesis. *Proc. Natl. Acad. Sci. USA*, **94**, 6261–6266.

Pantres, E., Apostolakos, P. & Galatis, B. (1994) Sinuous ordinary epidermal cells: behind several patterns of waviness, a common morphogenetic mechanism. *New Phytol.*, **127**, 771–780.

Perazza, D., Herzog, M., Hülskamp, M., Brown, S., Dorne, A. & Bonneville, J. (1999) Trichome cell growth in Arabidopsis thaliana can be depressed by mutations in at least five genes. *Genetics*, **152**, 461–476.

Qiu, J. L., Jilk, R., Marks, M. D. & Szymanski, D.B. (2002) The Arabidopsis SPIKE1 gene is required for normal cell shape control and tissue development. *Plant Cell*, **14**, 101–118.

Quarmby, L. (2000) Cellular Samurai: katanin and the severing of microtubules. *J Cell Sci*, **113** (Pt 16), 2821–2827.

Ray, S., Park, S. & Ray, A. (1997) Pollen tube guidance by the female gametophyte. *Development*, **124**, 2489–2498.

Reddy, A.S., Narasimhulu, S.B., Safadi, F. & Golovkin, M. (1996a) A plant kinesin heavy chain-like protein is a calmodulin-binding protein. *Plant J.*, **10**, 9–21.

Reddy, A.S.N., Safadi, F., Narasimhulu, S.B., Golovkin, M. & Hu, X. (1996b) A novel plant calmodulin-binding protein with a kinesin heavy chain motor domain. *J. Biol. Chem.*, **271**, 7052–7060.

Ridley, A.J. (2001) Rho-proteins:linking signaling with membrane trafficking. *Traffic*, **2**, 303–310.

Schnittger, A. & Hulskamp, M. (2002) Trichome morphogenesis: a cell-cycle perspective. *Phil. Trans. R. Soc. Lond. B* **357**, 823–826.

Schnittger, A., Schöbinger, U., Bouyer, D., Weinl, C., Stierhof, Y.-D. & Hülskamp, M. (2002a) Ectopic D-type cyclin expression induced not only DNA replication but also cell division in

Arabidopsis trichomes. *Proc. Natl. Acad. Sci. USA*, **99**, 6410–5415.

Schnittger, A., Schobinger, U., Stierhof, Y. D. & Hulskamp, M. (2002b) Ectopic B-type cyclin expression induces mitotic cycles in endoreduplicating Arabidopsis trichomes. *Curr. Biol.*, **12**, 415–420.

Schwab, B., Mathur, J., Saedler, R., Schwarz, H., Frey, B., Scheidegger, C. & Hülskamp, M. (2003) Regulation of cell expansion by the *DISTORTED* genes in *Arabidopsis thaliana*: actin controls the spatial organization of microtubules. *Molec. Genet. Genomics*, **269**, 350–360.

Song, H., Golovkin, M., Reddy, A.S. & Endow, S.A. (1997) In vitro motility of AtKCBP, a calmodulin-binding kinesin protein of *Arabidopsis*. *Proc. Natl. Acad. Sci. USA*, **94**, 322–237.

Spanfo, S., Silletta, M. G., Colanzi, A., Alberti, S., Fiucci, G., Valente, C., Fusella, A., Salmona, M., Mironov, A., Luini, A., Corda, D. & Spanfo, S. (1999) Molecular cloning and functional characterization of brefeldin A-ADP-ribosylated substrate. A novel protein involved in the maintenance of the Golgi structure. *J Biol. Chem.*, **274**, 17705–17710.

Steinborn, K., Maulbetsch, C., Priester, B., Trautmann, S., Pacher, T., Geiges, B., Kuttner, F., Lepiniec, L., Stierhof, Y.-D., Schwarz, H., Jurgens, G. & Mayer, U. (2002) The Arabidopsis PILZ group genes encode tubulin-folding cofactor orthologs required for cell division but not cell growth. *Genes Dev.*, **16**, 959–971.

Szymanski, D.B., Marks, M.D. & Wick, S.M. (1999) Organized F-actin is essential for normal trichome morphogenesis in Arabidopsis. *Plant Cell*, **11**, 2331–2348.

Torres-Ruiz, R.A. & Jürgens, G. (1994) Mutations in the FASS gene uncouple pattern formation and morphgenesis in *Arabidopsis* development. *Development*, **120**, 2967–2978.

Traas, J., Bellini, C., Nacry, P., Kronenberger, J., Bouchez, D. & Caboche, M. (1995) Normal differentiation patterns in plants lacking microtubular preprophase bands. *Nature*, **375**, 676–677.

Traas, J., Hülskamp, M., Gendreau, E. & Höfte, H. (1998) Endoreduplication and development: rule without dividing? *Curr. Opin. Plant Biol.*, **1**, 498–503.

Tsuge, T., Tsukaya, H. & Uchimiya, H. (1996a) Two independent and polarized processes of cell elongation regulate leaf blade expansion in *Arabidopsis thaliana*. *Development*, **122**, 1589–1600.

Tsuge, T., Tsukaya, H. & Uchimiya, H. (1996b) Two independent and polarized processes of cell elongation regulate leaf blade expansion in Arabidopsis thaliana (L.) Heynh. *Development*, **122**, 1589–1600.

Tsukaya, H., Tsuge, T. & Uchimiya, H. (1994) The cotyledon: a superior system for studies of leaf development. *Planta*, **195**, 309–312.

Uphof, J.C.T. (1962) *Plant Hairs,* Gebr. Bornträger, Berlin.

Vos, J.W., Safadi, F., Reddy, A.S. & Hepler, P.K. (2000) The kinesin-like calmodulin binding protein is differentially involved in cell division. *Plant Cell*, **12**, 979–990.

Walker, J.D., Oppenheimer, D.G., Concienne, J. & Larkin, J.C. (2000) SIAMESE, a gene controlling the endoreduplication cell cycle in Arabidopsis thaliana trichomes. *Development*, **127**, 3931–3940.

Webb, M., Jouannic, S., Foreman, J., Linstead, P. & Dolan, L. (2002) Cell specification in the Arabidopsis root epidermis requires the activity of ECTOPIC ROOT HAIR 3-a katanin-p60 protein. *Development*, **129**, 123–131.

Yang, Z. (2002) Small GTPases: Versatile signaling switches in plants. *Plant Cell*, **Supplement**, S375–S388.

5 From polarity to pattern: early development in fucoid algae

Colin Brownlee

5.1 Introduction

In developing multicellular organisms, cell fate decisions are made each time a cell divides. Elucidating the intra- and extracellular factors that control whether or not daughter cells adopt fates different from their parents is critical to our understanding of developmental patterning. In many developing multicellular organisms, daughter cells with different fates arise following asymmetric division. This can be seen during both somatic and zygotic development. Asymmetric divisions in the epidermis of plant leaves (Larkin *et al.*, 1997) precede differentiation of the stomatal guard cell pair from their protodermal precursors and asymmetric division is a common feature in zygotic development (Jan & Jan, 2000). Understanding the events that underlie such asymmetric divisions can provide insights into the mechanisms underlying cell fate determination (Brownlee *et al.*, 2001; Jan & Jan, 2000).

The marine intertidal brown algae (Phaeophyceae) contain several intertidal species such as *Fucus* and *Pelvetia* (now named *Silvetia*) which have for many years provided model systems for the study of polarity acquisition and maintenance in plants and algae (Brownlee & Bouget, 1998; Kropf, 1997; Kropf *et al.*, 1999; Brownlee & Berger, 1995). Adult plants of *Fucus* or *Silvetia* spp. can routinely be induced to release gametes (eggs or motile sperm) in the laboratory and it is possible to generate large populations of synchronously developing zygotes (Fig. 5.1). The relatively large size of the zygote and cells of the early embryo allow cellular approaches such as microinjection, electrophysiology and imaging. *Fucoid* zygotes and embryos have therefore provided much novel information on both the mechanisms of polarity acquisition during early development and the functional organization of polarized cells and their responses to external environmental changes. In this chapter I will describe some of the more recent work that is advancing our understanding of the acquisition of polarity and the mechanisms by which this gives rise to embryonic pattern. Some recent findings on the perception of external signals by polarized cells of the embryo will also be discussed in relation to polarized growth and the maintenance of polarity.

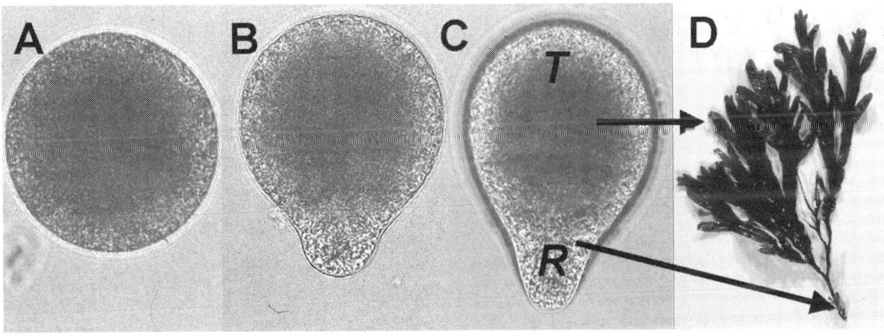

Figure 5.1 Early development in *Fucus*. (A) the initially non-polar zygote becomes polarized into rhizoid and thallus poles. (B) an apically-growing rhizoid germinates from the rhizoid pole and the first asymmetric division of the zygote occurs normal to the polar axis (C). The thallus cell goes on to produce the photosynthetic and reproductive thallus tissue while the rhizoid cell develops into the holdfast structure (D).

Polarity is frequently pre-determined in the unfertilized ovum in animals and plants. Subsequent polarized development of the zygote and embryo is often strongly influenced by the surrounding tissues. The *Arabidopsis* oocyte, for example presents a highly polarized immobile structure that makes precise contacts with surrounding cells that also dictate the site of entry of the fertilizing sperm cells delivered by the pollen tube (Brownlee, 1994; Russell, 1993; see Chapter 6). Higher plant embryos and zygotes are thus not generally amenable to direct manipulation and biochemical and physiological studies that require large numbers of synchronously developing zygotes are difficult. Polarization *de novo* can be observed in certain plant cell types. Protoplasts of mosses such as *Physcomitrella* and *Ceratodon* (Cove *et al*., 1997) and those of fucoid algae (Berger & Brownlee, 1995) can be polarized by external vectors such as the direction of unilateral light. More rarely, polarity can be acquired *de novo* following fertilization. In this respect zygotes of the fucoid algae are somewhat unique in that they can be easily polarized by external factors, leading to an asymmetric zygotic division that produces two very different embryonic cells that go on to develop with different fates.

5.2 Photopolarization: translating a unidirectional light signal into spatial information

5.2.1 Developmental changes following fertilization

The unfertilized fucoid egg is surrounded by an essentially naked plasma membrane with no cell wall. Fertilization sets off a sequence of events beginning with a transient depolarization of the egg membrane potential (the fertilization potential) and the onset of secretion of a cellulose-based cell wall

(Brawley, 1991; Roberts *et al.*, 1994). In many fertilizing animal systems, one of the first dramatic events following egg-sperm interaction is the generation of a global elevation of cytosolic Ca^{2+}. This acts as the trigger for the onset of development, including the release of the egg from cell-cycle arrest (Whitaker & Swann, 1993). So far, a fertilization-associated calcium wave has not been recorded during fertilization in *Fucus* eggs (Roberts *et al.*, 1994). However, elevations of Ca^{2+} in the cortical regions of the egg have been observed to occur within seconds of fertilization and an essential role of Ca^{2+} elevation in triggering egg development has been shown by inhibiting fertilization-associated Ca^{2+} signals (Roberts *et al.*, 1994). Following fertilization, the zygote settles on the substratum and secretes adhesive that fixes its position. It then enters a period during which it is able to perceive a range of external cues such as the direction of incident light. This is a critical period in the development of the zygote since it is at this stage that a polar axis is established. A rhizoid germinates from the side of the zygote facing away from the incident light and goes on to develop into the holdfast structure of the adult plant following an asymmetric division that divides the zygote into a rhizoid and thallus cell (Fig. 5.1).

5.2.2 Sperm entry establishes an initial axis of polarity

In the absence of a polarizing light vector, *Silvetia* zygotes will germinate a rhizoid from the site of sperm entry. This was shown in a series of elegant experiments (Hable & Kropf, 2000) that monitored the site of entry of the sperm pronucleus into the egg in relation to the site of subsequent rhizoid outgrowth and polar secretion of adhesive which occurs at the rhizoid pole prior to germination. Hable & Kropf (2000) were also able to manipulate the site of rhizoid outgrowth by selective application of sperm to one side of the eggs. This work further showed that a patch of polymerized actin could be visualized in the cell cortex at the site of sperm entry within minutes of fertilization. These observations were made possible by staining living cells that had been briefly permeabilized, with the fluorescent actin probe rhodamine phalloidin. Polarization from the site of sperm entry can be regarded as a default pathway that ensures that polarization occurs if no other external vectors are perceived.

5.2.3 Photopolarization involves re-alignment of the sperm-induced polar axis

The sperm-induced polar axis can be overridden in response to a range of external vectors or gradients (Fig. 5.2). This re-alignment of the polar axis progresses through a distinct sequence of events (Kropf, 1997; Kropf *et al.*, 1999) that begins with a selection phase, during which external polarizing

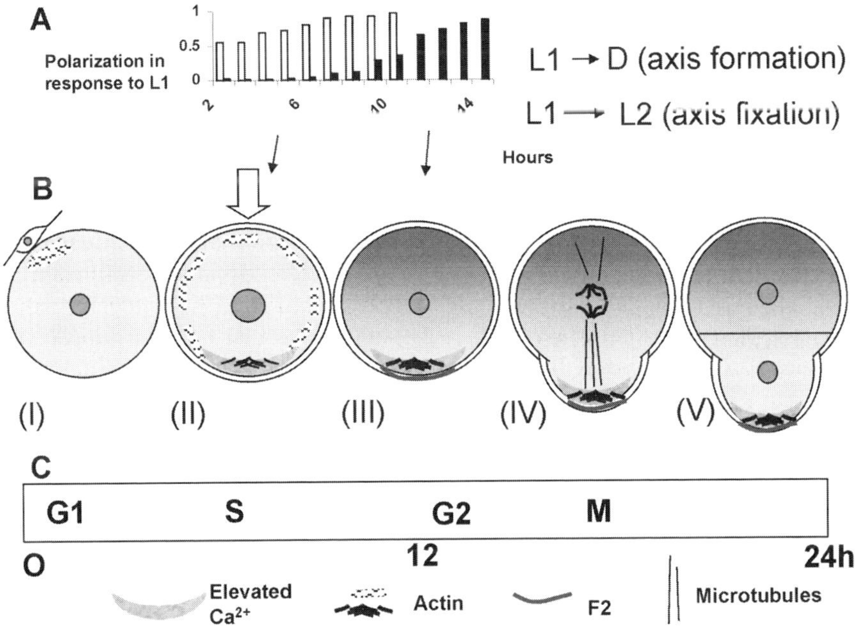

Figure 5.2 Timing of polarization events in response to unilateral light. (A) Axis formation can be determined by transferring zygotes to darkness (D) after various times in unilateral light (L1). Zygotes that germinate according to L1 are photopolarized (white bars). Fixation of the polar axis can be determined by transferring to light from the opposite direction (L2) following various times in L1. Zygotes that germinate according to L1 while in L2 are said to have fixed their polar axes while in L1 (black bars). (B) The fertilizing sperm sets up an initial axis of polarity, marked by a cortical actin patch (I). Subsequent polarization in unilateral light (II, white arrow) re-aligns the polar axis, brings about actin accumulation of in the cell cortex and localized Ca^{2+} and actin at the rhizoid pole. During axis fixation (III), localized secretion of F2 sulphated polysaccharide occurs at the rhizoid pole and gradients of mRNA can be detected (darker shading at thallus pole). The mitotic spindle aligns with the polar axis (IV) leading to the first asymmetric division (V). (C) Polarization events during normal development co-ordinate closely with cell cycle progression.

vectors are detected and translated into intracellular asymmetries. Polarizing vectors include light (Jaffe, 1958), ionic gradients (Robinson & Cone, 1980), and the proximity of other germinating zygotes (the 'group effect', Jaffe & Neuscheler, 1969). The ability to establish a polar axis in response to a wide range of polarizing stimuli is an example of convergence of signalling pathways whereby a single common downstream response results from the activation a diverse array of upstream detection pathways. The ability to polarize in response to such a diverse range of stimuli indicates an extreme sensitivity of the mechanism of polar axis fixation and the amplification of very small initial asymmetries into clear structural and physiological domains. Most studies have been carried out on photopolarization since this probably represents the most relevant physiological stimulus and it is easy to apply in the laboratory.

Fucoid zygotes are normally sensitive to polarizing light between four and ten hours following fertilization (Fig. 5.2). This is followed by an amplification phase where the initial labile asymmetries (axis formation) become increasingly irreversible (axis fixation) leading to polarized structural changes, culminating in germination of a rhizoid several hours later. The phases of polarization can be determined easily by simple re-orientation experiments (Fig. 5.2) in which zygotes that have been exposed to unidirectional light are either placed in darkness or in light from the opposite direction and the extent to which they respond to the first stimulus represents the proportion of zygotes that have respectively formed or fixed their polar axes. The asymmetric first division of the zygote occurs transverse to the polar axis (Fig. 5.2) producing the rhizoid and thallus cells.

The mechanism by which unilateral light is translated into cellular asymmetries is still unclear. It has long been accepted that photopolarization occurs in response to photoreceptor activation at the level of the plasma membrane (Jaffe, 1958). It therefore seems likely that membrane-associated events are involved in transducing the photopolarization stimulus. Indeed, photopolarization has been shown to be associated with enhanced plasma membrane redox activity, reflecting an electron transport process, at the cell surface on the side of the zygote facing away from the light source (Berger & Brownlee, 1994). Plasma membrane-associated redox activity was also shown to be enhanced by blue light in *Fucus* zygotes and uncoupling of plasma membrane electron transport by externally applied electron acceptors such as hexacyanoferrate (HCF) rendered zygotes unable to sense the direction of the incident light, giving rise to zygotes that germinated in random directions with respect to the incident light source. These results are consistent with a role of plasma membrane redox transport in photopolarization. The inhibitory effect of HCF on photopolarization was specific to the photosensitive phase of polarization. The same level of inhibitor applied at later stages of polarization did not disrupt a previously fixed polar axis and did not prevent rhizoid germination or growth. This suggests that a light-sensitive redox transport process is an early component of the signal transduction chain leading to photopolarization. Whether localized redox transport is directly linked to the perception of the light stimulus or whether it occurs in response to some other localization event (e.g. the localized elevation of Ca^{2+} or actin at the rhizoid pole, see below) remains to be determined. Thus, it would be interesting to determine whether the other early photopolarization events, such as the development of elevated Ca^{2+} or actin localization at the rhizoid pole (see below) can be established in the presence of HCF or whether inhibiting the formation of the Ca^{2+} gradient (see below) also prevents polarized redox transport.

More recently, an alternative, though not necessarily exclusive, model for fucoid photopolarization has been proposed. Robinson *et al.* (1998) demonstrated the existence of retinal in *Silvetia* zygotes. Retinal is known to be exclusively associated with opsins with which it forms rhodopsin-like light-sensitive complexes in animal cells. In animals, rhodopsin operates by interaction with a G-protein to activate a cyclic GMP phopshodiesterase that causes cyclic GMP (cGMP) levels to fall. This raised the possibility that spatial variation in the level of cyclic GMP in response to a light gradient within the *Silvetia* zygote could be one of the first asymmetries established during photopolarization. This was supported by evidence that blue light exclusively elevated cGMP levels by approximately two-fold in *Silvetia* zygotes (Robinson *et al.*, 1998). Pharmacological inhibition of guanylyl cyclase also resulted in inhibition of photopolarization, as did treatment with a cell-permeant cGMP analogue. cGMP is known to promote actin depolymerization in several systems. Robinson and co-workers proposed that rhodopsin at the membrane of the zygote facing the incident light is activated more, leading to an increase in cGMP levels in that region. While this appears to be the opposite of what occurs in the animal photoreceptor system, a mechanism for polarization involving rhodopsin has been given further support following the identification, using antibodies, of a rhodopsin-like protein in *Silvetia* eggs (Gualtieri & Robinson, 2002). Robinson & Miller (1997) proposed that elevated cGMP could result in depolymerization of actin preferentially at the 'illuminated' region of the zygote with actin becoming increasingly polymerized towards the rhizoid pole facing away from the light direction. While this can potentially explain the lack of actin localization at the thallus pole, it does not explain what promotes actin localization at the rhizoid pole (see below). Further support for these models of photopolarization will require accurate measurements of the light profile around the zygote. Photopolarization of fucoid zygotes occurs over a wide range of irradiances (Brownlee, 1990) and it is difficult to conceive how a single photoreceptor system can account for photopolarization at high photon flux densities where conceivably all photoreceptor molecules are likely to be very rapidly activated, potentially leading to uniform second messenger levels throughout the zygote, and it is possible that a more complex system operates to allow discrimination of light gradients at a wide range of irradiances.

5.3 Downstream of light perception: actin and Ca^{2+}

5.3.1 The role of actin

The role of actin in polarization is well established in yeast and animals (Goode *et al.*, 2000). In budding yeast, regulation of actin polymerization at specific sites either during cell division or during budding in response to mating pheromone is essential for polarized cell shape changes (O'Shea & Herskowitz, 2000). Organization of the actin cytoskeleton during budding is regulated by the activity of Rho family GTPases such as Cdc42, Rho and Rac. The guanidine nucleotide exchange factor (GEF) cdc24 is normally sequestered in the nucleus by complexation with a nuclear localization signal Far1. Activity of the cell cycle controller CDC28/cln in G1 phase triggers Far1 degradation and subsequently the release of cdc24 that can translocate to the plasma membrane where it associates with and activates a rho-GTPase (cdc42) involved in actin polymerization at specific sites marked by *BUD* gene products. Such detailed molecular interactions have yet to be characterized in the fucoid algal system. However, there is a considerable amount of information on the requirement of actin and the dynamics of the cytoskeleton during fucoid polarization. Moreover, the fucoid algae are revealing important information on the interactions between actin and other intracellular messengers such as Ca^{2+} (see below) that occurs during the period of axis fixation. Localization of actin at the rhizoid pole of fucoid zygotes has been well studied over recent years and is one of the earliest processes detectable during polarization. However, there is still some controversy about the exact timing of actin localization during early polarization.

Detailed observations of actin localization in response to a polarizing light vector (Alessa & Kropf, 1999) showed the presence of a cortical actin patch at the future rhizoid pole within three hours of fertilization. This coincides with the onset of the photopolarization period and is well before the period during which the axis becomes fixed. This work also showed that the light vector could override the more labile sperm-induced actin localization. After four hours in unilateral light, the position of the cortical actin patch could be reversed approximately 40 to 50 minutes after reversing the light direction. This relocation of the actin patch was shown to arise by *de novo* synthesis of a new patch at the new rhizoid pole. Hable *et al.* (2003) confirmed this by using the actin stabilizing agent jasplakinolide which stabilized existing actin arrays, promoted the formation of new arrays but prevented the dynamic rearrangement of actin. Jasplakinolide had inhibitory effects on photopolarization and re-location of the polar axis in response to a new light vector. This suggested that dynamic actin, rather than stable actin arrays, were required for polar axis formation. Jasplakinolide did not disrupt a previously formed axis, but did inhibit rhizoid germination and growth. This disrupted the alignment of the mitotic apparatus without affecting progression of the cell cycle. Together, these results indicate an essential role for dynamic actin as part of the early polarization mechanism and in the processes that lead to mitotic apparatus alignment in relation to the polar axis (see below).

A contrasting result was obtained by Pu *et al.* (2000) who imaged fluorescent phalloidin that had been microinjected into live cells prior to photopolarization of *Silvetia* zygotes. Within one hour of the onset of a polarizing light vector, cellular actin levels increased in the cell cortex. However, no cortical actin patch was observed in this study and only a very slight increase in actin levels was observed at the future rhizoid pole after two hours of unilateral light. Localized actin at the rhizoid pole was only observed after germination with the appearance of a sub-apical actin ring near the rhizoid apex. It was suggested that methodological differences could explain the appearance of cortical actin patches in the study by Alessa & Kropf (1999). Despite these differences, this work (Pu *et al.*, 2000), together with inhibitor experiments showing an absolute requirement for actin in photopolarization firmly indicates the absolute necessity for actin in photopolarization.

In addition to determining the orientation and structural asymmetries of the first asymmetric division, actin may also be critical for establishing the asymmetries of mRNA and actin RNA that have been observed during polar axis fixation (Bouget *et al.*, 1996). *Fucus* zygotes show an initially symmetrical distribution of total and actin mRNA which subsequently accumulates at the thallus pole during axis fixation. While this may be related to increased protein synthesis and metabolic activity of the thallus cell, it is also likely that this leads to asymmetric distribution of fate-determining factors between the rhizoid and thallus cells. Significantly, this asymmetric distribution of mRNA was shown to be inhibited by treatment with cytochalasin B which disrupts actin microfilament organization (Bouget *et al.*, 1996).

Further insights into the complexity of actin dynamics during polarization have come from studies of tyrosine kinase activity during polarization (Corellou *et al.*, 2000b). Tyrosine kinase-like activity appears to be required for polarity establishment, as well as for polar secretion of sulphated polysaccharide. Several studies have demonstrated an essential role for localized secretion and membrane recycling in fucoid zygote polarization (Belanger & Quatrano, 2000a,b; Quatrano & Shaw, 1997; Shaw & Quatrano, 1996). Localized secretion of jelly occurs at the presumptive rhizoid pole during the photopolarization period (Kropf, 1997). After photopolarization and prior to rhizoid germination, secretion of a sulphated fucan (F2) occurs preferentially at the rhizoid pole (Shaw & Quatrano, 1996). This continues during rhizoid growth. The presence of the cell wall is absolutely required for fixation of the polar axis (Kropf *et al.*, 1988), and the requirement for localized secretion of cell wall components in zygote polarization has led to models for polar axis determination based on the formation of axis stabilizing complexes comprising trans-plasma membrane links between a sulphated polysaccharide in the cell wall and the cytoskeleton (Brownlee & Berger, 1994; Fowler & Quatrano, 1997). Direct disruption of the actin cytoskeleton leads to inhibition of polarized F2 secretion (Brawley & Quatrano, 1979).

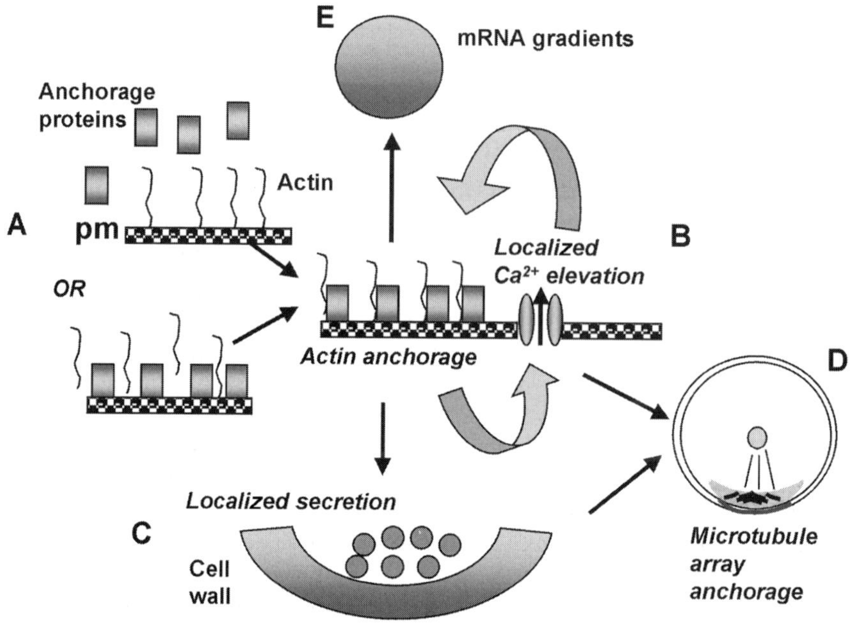

Figure 5.3 The central role of actin in polarization. (A) actin becomes localized at the rhioid pole plasma membrane (pm) by association with putative anchorage protein, leading to actin stabilization and amplification of initial asymmetries, involving the generation of an apical Ca^{2+} gradient (B). Downstream processes include localized secretion at the sites of actin anchorage (C), localized microtubule arrays between the cortical actin accumulation and nucleus and the generation of mRNA gradients (E).

Interestingly, inhibition of tyrosine kinase-like activity with a range of inhibitors did not affect actin localization during photopolarization, but did prevent the stable anchorage of actin to the rhizoid pole, so that actin could become re-localized to the opposite pole by changing the light direction up to 24 h later (Corellou *et al.*, 2000). When applied continuously to developing embryos from the photopolarization stage onwards, tyrosine kinase inhibitors produced non-polar multicellular embryos. However, inhibition of tyrosine phosphorylation did not affect the growth of pre-germinated rhizoids. These results point to separate actin localization and anchorage processes that are differentially sensitive to tyrosine phosphorylation (Fig. 5.3).

5.3.2 Ca^{2+} and zygote polarization

Ca^{2+} is a ubiquitous and essential signal in a very wide range of physiological and developmental processes (Sanders *et al.*, 2002). Elevations of cytoplasmic Ca^{2+} $[Ca^{2+}]_{cyt}$ above the resting level, which is normally very low (in the range of 0.1 μM or lower) may serve to encode specific information for signal transduction. $[Ca^{2+}]_{cyt}$ is clearly elevated at the apex of several polarized growing plant cells, including pollen tubes, root hairs and rhizoids of *Fucus* and *Silvetia*. These Ca^{2+} gradients are known to be intimately associated with apical growth (Berger & Brownlee, 1993; Bibikova & Gilroy, 1997; Holdaway-Clarke *et al.*, 1998; Messerli & Robinson, 1997) and probably serve roles in regulating secretion and ion channel activity. However, there has been some disagreement over the absolute requirements for Ca^{2+} in the photopolarization of fucoid zygotes (Brownlee & Bouget, 1998). A systematic study of the effects of Ca^{2+} buffer injection into polarizing *Silvetia* zygotes by Speksnijder *et al.* (1989) clearly demonstrated that appropriate cytoplasmic concentrations of Ca^{2+} buffers with appropriate dissociation constants could effectively prevent polarization. These buffers are thought to work as mobile chelators which rapidly dissipate locally elevated regions of $[Ca^{2+}]_{cyt}$. Speksnijder *et al.* (1989) calculated that $[Ca^{2+}]_{cyt}$ elevations up to 10 μM were required for zygote polarization. These calculations have been borne out by quantitative imaging studies of Ca^{2+} gradients in germinated rhizoid cells (e.g. Taylor *et al.*, 1996; Goddard *et al.*, 2000; Coelho *et al.*, 2002). Moreover, $[Ca^{2+}]_{cyt}$ has been shown to be elevated at the future rhizoid pole in *Fucus* zygotes prior to germination (Berger & Brownlee, 1993) and more recent confocal imaging studies (Pu & Robinson, 1998; Pu *et al.*, 2000) provided evidence for a significant Ca^{2+} gradient after only one hour of exposure to polarizing light, which was significantly earlier than any detectable asymmetries in actin distribution in that study. However, these studies also showed that the actin depolymerizing agent latrunculin B could prevent the appearance of the Ca^{2+} gradient, indicating an absolute requirement for actin. These apparently contradictory observations have led Robinson and co-workers to speculate that an actin framework is necessary for the putative asymmetric distribution of Ca^{2+} channels in the rhizoid apex but that insertion of channels into the plasma membrane at the rhizoid apex could be effected by another polarized molecule such as myosin. While localized binding of fluorescent dihydropyridines, which are known to associate with Ca^{2+} channels in animal cells, has been observed at the rhizoid pole of polarizing fucoid zygotes (Shaw & Quatrano, 1996), it is not clear whether this really represents the asymmetric distribution of Ca^{2+} channels since the specificity of dihydropyridines for Ca^{2+} channels in plants or algae has not been convincingly demonstrated. Direct electrophysiological patch clamp studies of ion channel distribution in polarized *Fucus* zygotes have so far not revealed any accumulation of Ca^{2+}-permeable channels in the rhizoid apex (Taylor *et al.*, 1997). An alternative explanation to that forwarded by Pu *et al.* (2000) is that the initial stages of Ca^{2+} gradient formation involve activation of pre-existing Ca^{2+}-permeable channels at the rhizoid pole. Actin may simply be required for the correct functioning of these channels. Localized actin occurring later during polarization may be part of the amplification process that leads to further increase in Ca^{2+} channel activity in the rhizoid region during axis stabilization.

Experiments with inhibitors have implicated calmodulin as a downstream mediator of the Ca^{2+} gradient in the rhizoid apex (Robinson, 1996; Love *et al.*, 1997). Further evidence for a role for calmodulin in mediating the Ca^{2+} gradient has come from experiments using microinjected calmodulin (Love *et al.*, 1997). This was shown to amplify the photopolarization response and to overcome the inhibition of photopolarization following injection of heterologous calmodulin antibodies. However, while these experiments indicated a role in photopolarization, localization of microinjected fluorescent calmodulin to the rhizoid pole was apparent only during the later stages of rhizoid germination, though earlier detection may have been limited by the sensitivity of the assay. Thus, while calmodulin appears to play a fundamental role in polarization, the precise stage at which it operates, or whether different calmodulin isoforms are involved in polarization and other processes, remains unclear.

5.3.3 Ca^{2+} signalling in the growing rhizoid

$[Ca^{2+}]_{cyt}$ levels in the rhizoid cell of the two-cell *Fucus* embryo have been shown to be dynamic. Thus, rather than a standing gradient of $[Ca^{2+}]_{cyt}$ at the growing rhizoid apex, $[Ca^{2+}]_{cyt}$ levels fluctuate according to changes in external osmotic conditions. Both hypo-osmotic (Taylor *et al.*, 1996; Goddard *et al.*, 2000) and hyper-osmotic (Coelho *et al.*, 2002) treatments produce $[Ca^{2+}]_{cyt}$ elevations in the rhizoid apex. Interestingly, the patterns of these elevations have been shown to be dependent on the strength of the applied stimulus. For example, weak hypo-osmotic treatments produce a sustained amplification of the apical Ca^{2+} gradient while more severe hypo-osmotic shocks produce more complex patterns of $[Ca^{2+}]_{cyt}$ elevation that can propagate through the rhizoid cell via release of Ca^{2+} from intracellular stores (Taylor *et al.*, 1996; Goddard *et al.*, 2002). These results from the polarized rhizoid cell raise the possibility that release of Ca^{2+} from intracellular store in the rhizoid apex may well play a significant role in amplifying localized plasma membrane channel-mediated elevation of Ca^{2+} levels in the polarizing zygote. A simple Ca^{2+} gradient model based on the distribution of Ca^{2+} channels in the plasma membrane may only partly explain the development of Ca^{2+} signals that may amplify initial asymmetries.

The complex spatio-temporal regulation of Ca^{2+}-dependent processes has been further indicated by studies of reactive oxygen species (ROS) production during rhizoid growth and responses to osmotic stress. Coelho *et al.* (2002) showed that extracellular production of ROS occurs very rapidly (within seconds) at the rhizoid apex following the onset of hyperosmotic stress. This is followed by Ca^{2+} elevation at the rhizoid apex which subsequently propagates throughout the entire rhizoid cell as a wave of Ca^{2+} release from intracellular stores. Experiments in which the production of ROS was inhibited by pharmacological inhibitors or the presence of catalase in the external medium showed that ROS production was necessary for the Ca^{2+} elevation. Patch clamp electrophysiology studies (Coelho *et al.*, 2002) also showed that Ca^{2+}-permeable channels in the rhizoid apex could be activated by H_2O_2. These experiments indicate a clear interaction between signalling pathways involving ROS and Ca^{2+}. These data are consistent with the notion that ROS plays a key role in Ca^{2+} signalling via the activity of NADPH oxidase as has been shown recently by the study of *Arabidopsis* mutants defective in NADPH oxidase subunits. Key physiological and development processes that involve cross-talk between Ca^{2+} and ROS signalling pathways include regulation of stomatal guard cell responses (Pei *et al.*, 2000; Kwak *et al.*, 2003) and polarized growth of root hairs (Foreman *et al.*, 2003). It remains to be seen whether the redox transport process that is required for *Fucus* zygote photopolarization (Berger & Brownlee, 1994) is linked to the activity of an electron translocating NADPH oxidase at the rhizoid pole.

5.4 Polarization and the cell cycle

5.4.1 Cell cycle regulation of polarization

Polarization and the cell cycle are tightly co-ordinated during normal fucoid development. The first asymmetric cell division occurs only after the polar axis has been established and the rhizoid has germinated (Fig. 5.2). However, cell cycle progression is not dependent on zygote polarization. Various treatments which prevent polarization and rhizoid germination do not prevent cell division. For example, inhibitors of Golgi vesicle transport (Shaw & Quatrano, 1996) tyrosine kinase activity (Corellou *et al.*, 2000b) and phospholipase C (Brownlee *et al.*, 1999) can all produce spherical embryos with no rhizoids but which have divided several times, albeit often with disrupted pattern (see Shaw & Quatrano, 1996). Taken together, these studies show that cell cycle progression does not depend absolutely on polarization but is regulated by other endogenous developmental cues. The *Fucus* cell cycle progresses through a well defined sequence of G1, S, G2 and M phases. In many animal embryos, the G1 and G2 phases are often reduced or absent. This is associated with rapid embryo cell division. The *Fucus* zygote, however shows distinct G1 and G2 phases and clear checkpoints for S and M phase progression (Corellou *et al.*, 2000a, 2001a). These checkpoints involve regulation of cyclin-dependent kinases (CDKs). The spindle assembly checkpoint in *Fucus* involves the maintenance of high levels of CDK activity and it appears that inactivation of CDKs is required for chromatin decondensation and exit from mitosis.

In contrast to the apparent independence of cell cycle progression on zygote polarization, the acquisition of polarity appears to be critically dependent on CDK activity during S-phase, though not on DNA replication (Corellou *et al.*, 2001b). Inhibition of S-phase CDK activity with pharmacological agents such as olomoucine or by injection of antibodies to the conserved CDK PSTAIRE peptide sequence, results in the block of both cell division and rhizoid polarization. This does not occur if the cell cycle is blocked in S-phase through inhibition of DNA polymerase or when CDKs are inhibited after the G1/S transition (Corellou *et al.*, 2000a, 2001b). Whether CDKs positively control polarization directly or are part of a polarity checkpoint that operates to prevent polarization when the G1/S transition is blocked is not known. However, it is clear that, a tight regulation between polarization and cell division operates during early development to ensure correct patterning of the embryo. This appears to be via cell cycle regulation of polarity rather than a control of polarity on the cell cycle.

5.4.2 Polarization and division plane alignment

While the polarization process may not directly influence the progression of the cell cycle, establishment of the polar growth axis is critical for the alignment of the division plane. How this is accomplished is becoming clear from several recent studies of division plane alignment in *Silvetia compressa* (Bisgrove *et al.*, 2003) and the related brown alga *Scytosiphon lomentaria*. In *S. compressa* centrosomes that originate from the fertilizing sperm constitute microtubule organizing centres with centrioles (Bisgrove & Kropf, 1998). These are deposited on the nuclear envelope during karyogamy and subsequently migrate to opposite poles of the nucleus (Fig. 5.4). During the early cell cycle, microtubules radiate symmetrically from the nucleus, but later form a bipolar array from the centrosomes. Following rhizoid germination and prior to, or during early M-phase, the centrosomal axis rotates such that it aligns parallel to the polar growth axis with microtubules emanating to the rhizoid apex. Precise alignment of the centrosomal axis with the growth axis is achieved in late mitosis (anaphase and telophase). The cell division plane subsequently bisects the centrosomal axis. These observations indicate that the growth axis and centrosomal axis are formed separately and only later is the centrosomal axis brought into aligment with the growth axis. More recently (Bisgrove *et al.*, 2003) used pharmacological uncoupling of the nuclear and growth axes to show that cytokinesis always occurs transverse to the telophase mitotic array and not necessarily the growth axis. The division plane was specified at cytoplast boundaries marked by interdigitating microtubules. In contrast, Nagasato & Motomura (2002) showed that determination of the cytokinetic plane in polyspermic zygotes of another brown alga, *Scytosiphon lomentaria,* depends on the position of the centrosomes after

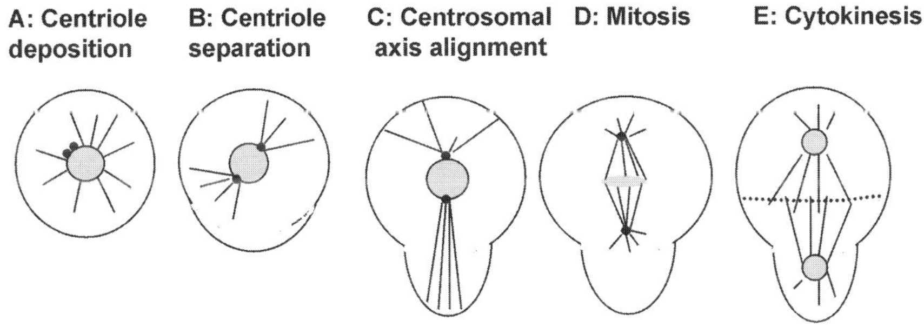

Figure 5.4 Separate formation of the polar growth axis and centrosomal axis. (A) Following fertilization the sperm-derived centrioles are deposited together on the nuclear envelope. (B) Polar axis formation and fixation occur while the centrioles move to opposite locations on the nuclear envelop that are not aligned with the polar growth axis. (C) The centrosomal axis rotates and aligns with the growth axis during rhizoid germination. (D) ensuring that the mitotic spindle is parallel with the growth axis. (E) Cell plate formation occurs in a zone of interdigitating microtubules perpendicular to the growth axis.

mitosis since the number of daughter cells of polyspermic zygotes always coincided with the number of centrosomes, rather than the number of cells with nuclei. Moreover, cytokinetic planes were always positioned between each centrosome even when one centrosome was located away from the spindle.

5.5 Polarized signalling and cell fate in the multicellular embryo: positional information and cell-cell communication

Following cytokinesis, polarity is maintained in the adult, irrespective of external cues. While the rhizoid cell remains negatively phototrophic, the polarity and pattern of the embryo remains rigidly determined during subsequent development (Fig. 5.1). In animal cells, the inheritance of fate determinants in the cytosol is an important factor in determining pattern, along with information provided by position in the embryo (Wolpert, 1998). The accessibility of the *Fucus* embryo to experimental manipulation and micro-dissection has allowed us to address the roles of cell-cell communication, cell lineage and positional information in cell fate decisions in the two-cell and multicellular embryo. Earlier work involving ablation of rhizoid or thallus cells in the two-cell embryo showed that protoplasts isolated from rhizoid or thallus cells would spontaneously lose their polarity, re-differentiate and re-polarize into rhizoid and thallus cells (i.e. thallus into rhizoid and vice-versa; Berger *et al.*, 1994). Re-differentiation of the progeny of an isolated rhizoid or thallus cell would occur only after a very significant time delay of several days. However, the presence of cell wall remaining from an ablated rhizoid

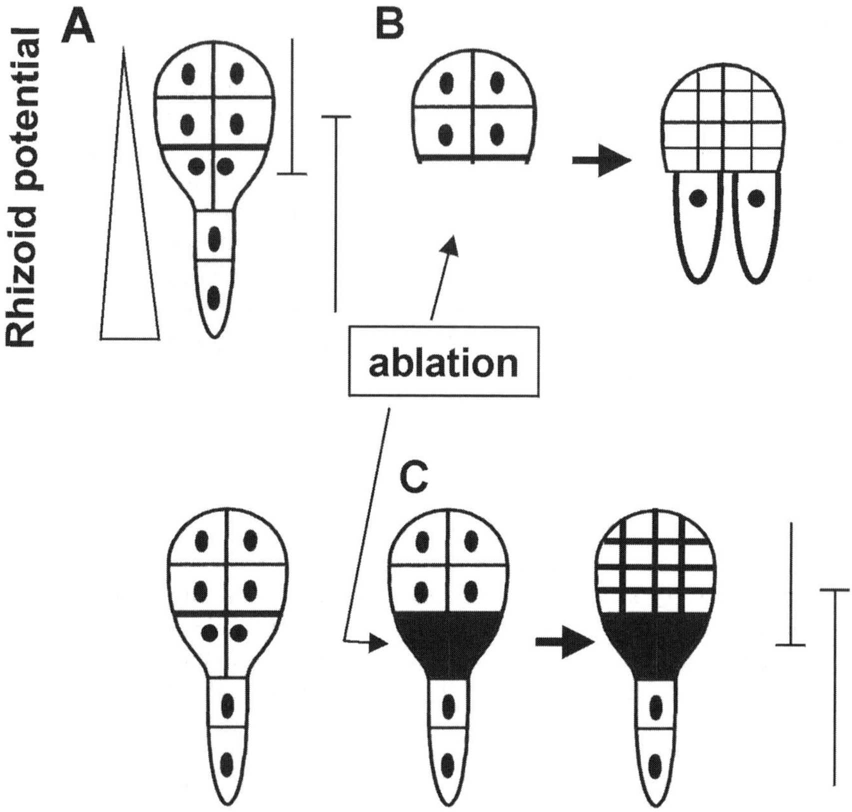

Figure 5.5 Intercellular communication via diffusible signals during multicellular development in the *Fucus* embryo. (A) During normal development, cells closer to the rhizoid pole have an increased tendency to regenerate rhizoid cells following laser ablation of their neighbours. The maintenance of rhizoid or thallus cell fate can be explained in terms of diffusible inhibitory signals produced by the thallus and rhizoid tissue which prevent rhizoid-thallus and thallus-rhizoid re-differentiation during normal development. (B) Complete ablation of rhizoid cells allows thallus cells eventually to re-differentiate and replace the missing rhizoid cells. (C) Ablation of rhizoid cells in a gap between rhizoid and thallus while leaving basal rhizoid cells intact significantly delays the appearance of new rhizoid cells from the thallus tissue, indicating that rhizoid-derived inhibitory factors can diffuse extracellularly to prevent re-differentiation of thallus cells.

cell could greatly accelerate the re-differentiation of thallus cell progeny into rhizoid cells once they made contact with the residual rhizoid cell wall. These experiments provided strong evidence that, at least in the early embryo, cell wall factors may play an important role in determining or maintaining cell fate.

Cell ablation experiments in the multicellular embryo have shown that the position of a cell in the embryo is critical for deciding its fate (Bouget *et al.*, 1998). Ablated thallus cells at the apical embryo pole are invariably replaced by new thallus cells derived from neighbouring cells. However, the tendency of thallus cells to regenerate rhizoid cells following ablation of their neighbours increases as we move down the embryo to the rhizoid pole (Fig. 5.5). These ablation experiments also showed that signals were passed from rhizoid cells to prevent their differentiation into thallus cells and vice-versa. Creating a gap of ablated cells between thallus and rhizoid tissue did not prevent this inhibitory effect, suggesting that an extracellular diffusible signal is involved in this apical-basal patterning process. The nature of this signal is unknown though it has been shown recently (Basu *et al.*, 2002) that the auxin transport inhibitor, naphthylphthalamic aid (NPA) and addition of auxin to the external medium can cause disruption of apical-basal patterning and the production of multiple rhizoids or branched rhizoids in early *Fucus disticus* embryos.

5.6 Conclusions

The fucoid zygote and embryo continue to provide new insights into polarization mechanisms from a single-celled zygote to a multicellular plant. The single most important feature that makes this possible is the ability to manipulate polarization in the zygote and its unique accessibility for cellular studies and manipulations. There are still uncertainties surrounding the precise timing of events associated with polarization and many of their interactions are still unclear. While adhesions between the plasma membrane and cell wall clearly play a critical role in polar axis specification and rhizoid germination, the unequivocal identification of the components of these complexes remains a challenge for future studies.

The fucoid system offers unique accessibility and an exceptionally detailed background of physiology and ultrastructure in relation to early zygotic and embryonic development. However, molecular genetic approaches have been lacking. While the long generation time is unlikely ever to make this system attractive for classical genetics, the application of alternative molecular and reverse genetics approaches is certainly feasible. Conventional molecular techniques are now routine in fucoid algae and potential developments in transient expression and genetic transformation should make targeted knock out and overexpression studies possible in the near future. It will then be possible to dissect out the precise roles of identified genes in the polarization process.

Acknowledgements

This work was supported by the BBSRC and NERC, UK.

References

Alessa, L.A. & Kropf, D.L. (1999) F-actin marks the rhizoid pole in living *Pelvetia compressa* zygotes. *Development*, **126**, 201–209.

Basu, S., Sun, H., Brian, L., Quatrano, R.S. & Muday, G.K. (2002) Early embryo development in *Fucus distichus* is auxin sensitive. *Plant Physiol.*, **130**, 292–302.

Belanger, K.D. & Quatrano, R.S. (2000a) Polarity: the role of localized secretion. *Curr. Opin. Plant Biol.*, **3**, 67–72.

Belanger KD. & Quatrano RS (2000b) Membrane recycling occurs during asymmetric tip growth and cell plate formation in *Fucus distichus* zygotes. *Protoplasma*, **212**, 24–37.

Berger, F. & Brownlee, C. (1993) Ratio confocal imaging of free cytoplasmic calcium in polarizing and polarized *Fucus* zygotes. *Zygote*, **1**, 9–15.

Berger, F. & Brownlee, C. (1994) Photopolarization of the *Fucus* sp. zygote by blue light involves a plasma membrane redox chain. *Plant Physiol.*, **105**, 519–527.

Berger, F. & Brownlee, C. (1995) Physiology and development of protoplasts obtained from *Fucus* embryos using laser microsurgery. *Protoplasma*, **186**, 63–71.

Berger, F., Taylor A.R. & Brownlee, C. (1994) Cell fate determination by the cell wall in early *Fucus* development. *Science*, **263**, 1421–1423.

Bibikova, T.N. & Gilroy, S. (1997) Root hair growth in *Arabidopsis thaliana* is directed by calcium and an endogenous polarity. *Planta*, **203**, 495–505.

Bisgrove, S.R. & Kropf, D.L. (1998) Alignment of centrosomal and growth axes is a late event during polarization of *Pelvetia compressa* zygotes. *Devel. Biol.*, **194**, 246–256.

Bisgrove, S.R., Henderson, D.C. & Kropf, D.L. (2003) Asymmetric division in fucoid zygotes is positioned by telophase nuclei. *Plant Cell*, **15**, 854–862.

Bouget, F.-Y., Berger, F. & Brownlee, C. (1998) Position-dependent control of cell fate in the *Fucus* embryo: Role of intercellular communication. *Development*, **125**, 1999–2008.

Bouget, F.-Y., Gerttula, S., Shaw, S.L. & Quatrano, R.S. (1996) Localization of actin mRNA during the establishment of cell polarity and early cell division in *Fucus* embryos. *Plant Cell*, **8**, 189–201.

Brawley, S.H. (1991) The fast block against polyspermy in fucoid algae is an electrical block. *Devel. Biol.*, **144**, 94–106.

Brawley, S.H. & Quatrano, R.S. (1979) Sulphation of fucoidin in *Fucus distichus* embryos. IV. Autoradiographic investigations of fucoidin sulphation and secretion during differentiation and the effect of cytochalasin treatment. *Devel. Biol.*, **73**, 193–205.

Brownlee, C. (1990) Light and development: cellular and molecular aspects of photomorphogenesis in brown algae. In P.J. Herring, A.K. Campbell, M. Whitfield & L. Maddock (eds), *Light and Life in the Sea*, pp. 115–126. Cambridge University Press: Cambridge.

Brownlee, C. (1994) Signal transduction during fertilization in algae and vascular plants. *New Phytol.*, **127**, 399–423.

Brownlee, C. & Berger, F. (1994) Extracellular matrix and pattern in plant embryos: on the look-out for developmental information. *Trends Genet.*, **11**, 345–348.

Brownlee, C. & Berger, F. (1995) Developmental signals and their transduction during *Fucus* embryogenesis. *Prog. Phycol. Res.*, **11**, 103–144.

Brownlee, C., Bouget, F-Y. & Corellou, F. (2001) Choosing sides: establishment of polarity in zyotes of fucoid algae. *Semin. Cell Devel. Biol.*, **12**, 345–351.

Brownlee, C. & Bouget F-Y (1998) Polarity determination in *Fucus*: from zygote to multicellular embryo. *Semin. Cell Devel. Biol.*, **9**, 179–185.

Brownlee, C., Goddard, H., Hetherington, A.M. & Peake, L-A. (1999) Signal specificity and integration of developmental and physiological responses. *J. Exp. Bot.*, **50**, 1001–1011.

Coelho, S., Taylor, A.R., Ryan, K.P., Sousa-Pinto I, Brown M.T. & Brownlee, C. (2002) Spatiotemporal patterning of reactive oxygen production and Ca^{2+} wave propagation in *Fucus* rhizoid cells. *Plant Cell*, **14**, 1–15.

Corellou, F., Bisgrove, S.R., Meijer, L., Kloareg, B. & Bouget, F-Y. (2000a) A S/M DNA replication checkpoint prevents nuclear and cytoplasmic events of cell division including centrosomal axis alignment and inhibits activation of cyclin-dependent kinase-like proteins in fucoid zygotes. *Development*, **127**, 1651–1660.

Corellou, F., Brownlee, C., Detivaud, L., Kloareg, B. & Bouget, F-Y. (2001a) Cell cycle in the *Fucus* zygote parallels a somatic cell cycle but displays a unique translational regulation of cyclin-dependent kinases. *Plant Cell*, **13**, 585–598.

Corellou, F., Brownlee, C., Kloareg, B. & Bouget, F-Y. (2001b) Cell cycle-dependent control of polarized development by a cyclin-dependent protein kinase in the *Fucus* zygote. *Development*, **128**, 4383–4392.

Corellou, F., Potin, P., Brownlee, C., Kloareg, B. & Bouget F-Y. (2000b) Inhibition of the establishement of zygotic polarity by protein tyrosine kinase inhibitors leads to an alteration of embryo pattern in *Fucus*. *Devel. Biol.*, **219**, 165–182.

Cove, D.J., Knight, C.D. & Lamparter, T. (1997) Mosses as model systems. *Trends Plant Sci.* **2**, 99–105.

Foreman, J., Demidchik, V., Bothwell, J.H., Mylona, P., Miedema, H., Torres, M.A., Linstead, P., Costa, S., Brownlee, C., Jones, J.D.G., Davies, J.M. & Dolan, L. (2003) reactive oxygen species produced by NADPH oxidase regulate plant cell growth. *Nature*, **422**, 442–446.

Fowler, J.E. & Quatrano, R.S. (1997) Plant cell morphogenesis: Plasma membrane interactions with the cytoskeleton and cell wall. *Ann. Rev. Cell Devel. Biol.*, **13**, 697–743.

Goddard, H., Manison, N.F.H. Tomos, D. & Brownlee, C. (2000) Elemental propagation of calcium signals in response-specific patterns determined by environmental stimulus strength. *Proc. Natl. Acad. Sci. USA*, **97**, 1932–1937.

Goode, B.L., Drubin, D.G. & Barnes, G. (2000) Functional cooperation between the microtubule and actin cytoskeletons. *Curr. Opin. Cell Biol.*, **12**, 63–71.

Gualtieri, P. & Robinson, K.R. (2002) A rhodopsin-like protein in the plasma membrane of *Silvetia compressa* eggs. *Photochem. Photobiol.*, **75**, 76–78.

Hable, W.E. & Kropf, D.L. (2000) Sperm entry induces polarity in fucoid zygotes. *Development*, **127**, 493–501.

Hable, W.E., Miller, N.R. & Kropf, D.L. (2003) Polarity establishment requires dynamic actin in fucoid zygotes. *Protoplasma*, **221**, 193–204.

Holdaway-Clarke, T., Feijo, J.A., Hackett, G.R., Kunkel, J.G. & Hepler, P.K. (1998) Pollen tube growth and the intracellular cytosolic calcium gradient oscillates in phase while extracellular calcium influx is delayed. *Plant Cell*, **9**, 1999–2010.

Jaffe, L.F. (1958) Tropistic responses of of zygotes of the fucaceae to polarized light. *Exp. Cell Res.*, **15**, 282–299.

Jaffe, L.F. & Neuschler, W. (1969) On the mutual polarization of nearby pairs of fucaceous eggs. *Devel. Biol.*, **19**, 549–565.

Jan, Y.-N. & Jan, L.Y. (2000) Polarity in cell division: What frames thy fearful asymmetry? *Cell* **100**, 599–602.

King, R.W., Jackson, P.K. & Kirschener, M.W. (1994) Mitosis in transition. *Cell*, **79**, 563–571.

Kropf, D.L. (1997) Induction of polarity in fucoid zygotes. *Plant Cell*, **9**, 1011–1020.

Kropf, D.L., Bisgrove S.R. & Hable, W.E. (1999) Establishing a growth axis in fucoid algae. *Trends Plant Sci.* **4**, 490–494.

Kropf, D.L., Kloareg, B. & Quatrano, R.S. (1988) Cell wall is required for fixation of the embryonic axis in *Fucus* zygotes. *Science*, **239**, 187–190.

Kwak, J.M., Mori, I.C., Pei, Z.M., Leonhardt, N., Torres, M.A., Dangl, J.L. Bloom, R.E., Bodde, S., Jones, J.D.G. & Schroeder, J.I. (2003) NADPH oxidase AtrbohD and AtrbohF genes function in ROS-dependent ABA signaling in Arabidopsis. *EMBO J.*, **22**, 2623–2633.

Larkin, J.C., Marks, M.D., Nadeau, J. & Sack, F. (1997) Epidemal cell fate and patternining in leaves. *Plant Cell*, **9**, 1109–1120.

Love, J., Brownlee, C. & Trewavas, A.J. (1997) Ca^{2+} and calmodulin dynamics during photopolarisation in *Fucus serratus* zygotes. *Plant Physiol.* **115**, 249–261.

Messerli, M. & Robinson, K.R. (1997) Tip localized Ca^{2+} pulses are coincident with peak pulsatile growth rates in pollen tubes of *Lilium longiflorum*. *J. Cell Sci.* **110**, 1269–1278.

Nagasato, C. & Motomura, T. (2002) Influence of the centrosome in cytokinesis of brown algae: polyspermic zygotes of *Scytosiphon lomentaria* (Scytosiphonales, Phaeophyceac). *J. Cell Sci.*, **115**, 2541–2548.

O'Shea, E.K. & Herskowitz, I. (2000) The ins and outs of cell polarity decisions. *Nature Cell Biol.*, **2**, E39–E41.

Pei, Z-M., Murata, Y., Benning, G., Thomine, W., Klusener, B., Allen, G., Grill, E. & Schroeder,

J.I. (2000) Calcium channels activated by hydrogen peroxide mediate abscisic acid signaling in guard cells. *Nature*, **406**, 731–734.
Pu, R.S. & Robinson, K.R. (1998) Cytoplasmic calcium gradients and calmodulin in the early development of the fucoid alga *Pelvetia compressa*. *J. Cell Sci.*, **111**, 3197–3207.
Pu, R.S., Wozniak, M. & Robinson, K.R. (2000) Cortical actin filaments form rapidly during photopolarization and are required for the development of calcium gradients in *Pelvetia compressa* zygotes. *Devel. Biol.*, **222**, 440–449.
Quatrano, R.S. & Shaw, S.L. (1997) Role of the cell wall in the determination of cell polarity and the plane of cell division in *Fucus* embryos. *Trends Plant Sci.*, **2**, 15–21.
Roberts, S.K., Gillot, I. & Brownlee, C. (1994) Cytoplasmic calcium and *Fucus* egg activation. *Development*, **120**, 155–163.
Robinson, K.R. (1996) Calcium and the photopolarization of *Pelvetia* zygotes. *Planta*, **198**, 378–384.
Robinson, K.R. & Cone, R. (1980) Polarization of fucoid eggs by a calcium ionophore gradient. *Science*, **207**, 77–78.
Robinson, K.R. & Miller, B.J. (1997) The coupling of cyclic GMP and photopolarization of *Pelvetia* zygotes. *Devel. Biol.*, **187**, 125–130.
Robinson, K.R., Lorenzi, R., Ceccarelli, N. & Gualtieri, P. (1998) Retinal identification in *Pelvetia fastigiata*. *Biochem. Biophys. Res. Comm.*, **243**, 776–778.
Russell, S.D. (1993) The egg cell: Development and role in fertilization and early embryogenesis. *Plant Cell*, **5**, 1349–1359.
Sanders, D., Pelloux, J., Brownlee, C. & Harper, J. (2002) Calcium at the crossroads of signaling. *Plant Cell*, **14**, S401–S417.
Shaw, S .L. & Quatrano, R.S. (1996) The role of targeted secretion in the establishment of cell polarity and the orientation of the division plane in *Fucus* zygotes. *Development*, **122**, 2623–2630.
Speksnijder, J.F., Miller, A.L., Weisenseel, M.H., Chen, T-H. & Jaffe, L.F. (1989) Calcium buffer injections block fucoid egg development by facilitating calcium diffusion. *Proc. Natl. Acad. Sci. USA*, **86**, 6607–6611.
Taylor, A.R., Manison, N.F.H. & Brownlee, C. (1997) Regulation of channel activity underlying cell volume and polarity signals in *Fucus*. *J. Exp. Bot.*, **48**, 579–588.
Taylor, A.R., Manison, N.F.H., Fernandez, C., Wood, J.W. & Brownlee, C. (1996) Spatial organization of calcium signaling involved in cell volume control in the *Fucus* rhizoid. *Plant Cell* 8, 2015–2031.
Whitaker, M. & Swann, K. (1993) Lighting the fuse at fertilization. *Development*, **117**, 1–12.
Wolpert, L. (1998) *Principles of Development*. Oxford University Press: Oxford.

6 Polarity in *Arabidopsis* embryogenesis

Ramón A. Torres Ruiz

6.1 Introduction

The generation of cellular polarity was at an early stage recognised as the prerequisite for elaborating complex organisms (e.g. Bünning, 1952, 1957). Intensive investigations in animals, fungi and plants using modern genetic and molecular techniques are revealing astonishing insights into developmental processes. The three-dimensional structures of organisms reflect a complex superimposition of asymmetries throughout the body plan. Since these ultimately originate from polarities at the cellular level, biologists have made considerable efforts to understand development from its earliest life stages, i.e. single cells.

Embryology has a long tradition in botany and has provided a plethora of observations and descriptions of plant embryos from many different genera (Wardlaw, 1955; Johri, 1984; Raghavan, 1986; Johri *et al.*, 1992; Raghavan, 2000). Due to the genetic and molecular peculiarities of *Arabidopsis thaliana*, its embryos have been the focus of intensive analyses (see earlier reviews, Mansfield & Briarty, 1991; Jürgens *et al.*, 1991; Meinke, 1991; Lindsey & Topping, 1993; West & Harada, 1993; Goldberg *et al.*, 1994; Jürgens *et al.*, 1994; Laux & Jürgens, 1997; Torres Ruiz, 1998; Souter & Lindsey, 2000; Jürgens, 2001). Though we are far from understanding plant embryo development, this work has considerably advanced our knowledge of *Arabidopsis* embryo development in the last decade. Analyses on embryos from other plants indicate that fundamental steps are shared between mono- and dicotyledonous plants (Jürgens *et al.*, 1994; Jürgens, 2001). Investigations of key genes controlling embryo development reveal that they play essential roles throughout the plant's life. The following overview is restricted to those aspects of these genes that are relevant for the *Arabidopsis* embryo. For further detailed and comprehensive discussion of special aspects and postembryonic development see the corresponding chapters of this book.

Crucial steps of the *Arabidopsis* embryo ontogeny are (intimately) linked to the generation of polarities within cells, across tissues and organs and along body axes. The following highlights recent advances in understanding establishment and maintenance of polarity with focus on:

- polar organisation of the zygote; the preceding polarity in the egg, the embryo sac and sporophytic tissues
- cellular polarity from zygotic division onwards and the role of vesicle trafficking
- polarity along the radial axis of the embryo and its implications
- polar organisation of the embryonic shoot apical meristem (SAM) region
- adaxial/abaxial polarity in the cotyledons and relation to the SAM
- polar organisation of basal region harbouring the root apical meristem (RAM).

This review does not intend to cover aspects of plant embryogenesis in general. Aspects which, at present, are not known to be linked to polarity and cell fate specification of the *Arabidopsis* embryo are excluded. This includes maturation processes, endosperm development and in part, somatic embryogenesis (for reviews see citations above and Mansfield & Briarty, 1990a,b; Berger, 1999; Brown *et al.*, 1999; Schmidt *et al.*, 1994; Mordhorst *et al.*, 1997).

6.2 Polarity from an early stage – maternal and zygotic polarities

Both the *Arabidopsis* egg cell and the zygote reveal a clear polar organisation. Since they are embedded in a highly polar structure, the embryo sac, which itself resides in a polarised ovule (Fig. 6.1), embryologists have speculated whether maternal polarity impinges on the polar organisation of the egg cell and in turn of the zygote. Such a sporophytic maternal influence would be reminiscent to the action of maternal genes found in *Drosophila melanogaster* (St. Johnston & Nüsslein-Vollhard, 1992; St. Johnston, 1995) with the implication that only a homozygous mutant plant would deliver mutant progeny even when crossed with homozygous wild-type pollen. However, in contrast to *Drosophila*, *Arabidopsis* (as with all higher plants) passes through a complete, though reduced, haploid generation, the gametophyte. It is difficult to imagine how such a control could 'survive' the gametophytic phase. In addition, one would assume that the gametophyte as the immediate precursor exerts a direct gametophytic maternal control. The latter is indeed the case. One of the first cloned examples of gametophytic maternal control has been the *MEDEA* (*MEA*) gene (also *FIS1*), required for proper embryo development (Chaudhury *et al.*, 1997; Grossniklaus *et al.*, 1998). As expected for a gametophytic maternal gene, the survival of resulting embryos depends on the presence of a wild-type *MEA* allele in the genome of the gametophyte. *MEDEA/FIS1* and several other genes seem to be preferentially transcribed from their maternal copies in the zygote though this does not apply to other cases (Vielle-Calzada *et al.*, 2000, 2001; Weijers *et al.*, 2001). Furthermore,

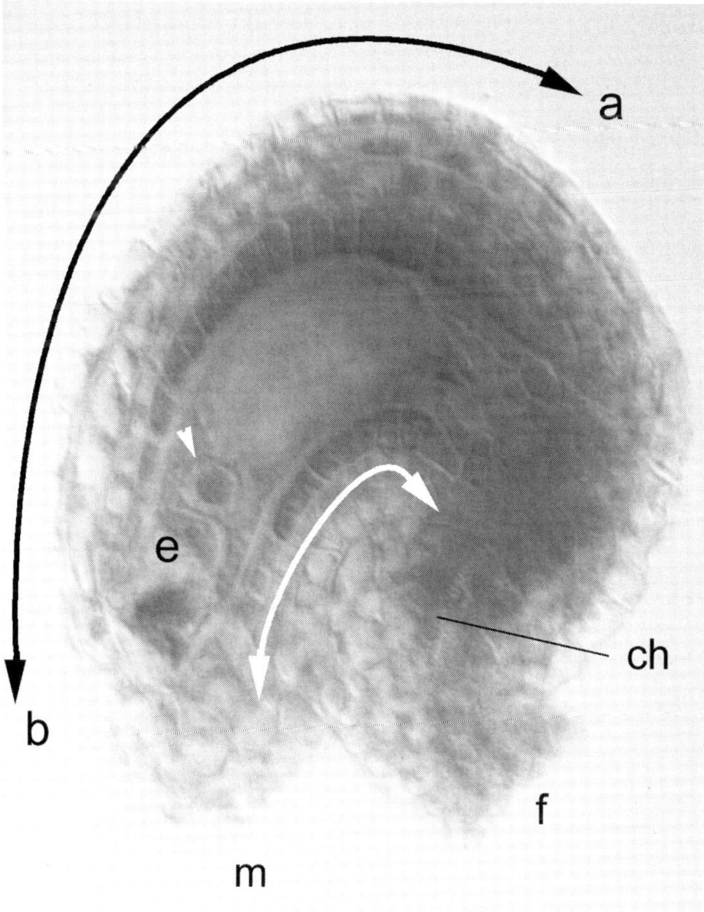

Figure 6.1 Ovule with embryo sac and egg cell. The figure illustrates the polar organisation of the ovule oriented apical (a) to basal (b) in reference to the egg cell (e) as indicated by black and white double headed arrows. Note that the nucleus localises at the uppermost (apical) end of the egg cell. White arrowhead: central cell nucleus; m: micropylar end; (ch): chalazal region; (f): funiculus. Courtesy of K. Schneitz.

maternal gametophytic control does not only apply to the zygote since *MEDEA/FIS1* and others have been found to be imprinted, such that only maternally inherited copies are expressed during endosperm development (Kinoshita *et al.*, 1999; Kiyosue *et al.*, 1999; Luo *et al.*, 1999; Vielle-Calzada *et al.*, 1999).

Though sporophytic maternal control seemed unlikely, two cases have been well documented recently. The first case has been reported for one allele (*sin1-2*) of the *SHORT INTEGUMENT1* (*SIN1*) gene (Ray *et al.*, 1996). This gene is also known as *SUSPENSOR1* or *CARPEL FACTORY* (Schwartz *et al.*, 1994; Jacobsen *et al.*, 1999). When the maternal sporophyte is homozygous *sin1-2/sin1-2*, the germination frequency of seeds is low. Most of the seedlings that germinate show serious pattern defects mainly concerning number and form of cotyledons, irrespective of whether the mother plant has been pollinated with wild-type pollen or not. In extreme cases an unstructured mass of tissue without any visible sign of polarity or pattern develops (Ray *et al.*, 1996). Homozygous mutated endosperm is not responsible for these effects (Ray *et al.*, 1996).

The *serrate* (*se*) mutant, which has been known for some time (Redei & Hirono, 1964), causes similar low penetrating effect on cotyledons when the maternal sporophyte is *se/se* and pollinated with wild-type pollen (Prigge & Wagner, 2001). Cloning of *SIN1* reveals that this gene codes for a DICER-like protein in *Arabidopsis* (Golden *et al.*, 2002). *SERRATE* encodes a protein with a single C2H2–type zinc-finger and multiple bipartite nuclear localisation motifs (Prigge & Wagner, 2001). Both their structure and additional observations raise the possibility that these genes are involved, at least partly, in epigenetic control mechanisms (Prigge & Wagner, 2001). *SIN1* seems to participate in dsRNA processing. *SE* might regulate transcription by altering chromatin structure (Prigge & Wagner, 2001) since it not only exhibits similarity to known zinc-finger proteins linked to chromatin remodelling, *FIS2* (Luo *et al.*, 1999) and GAGA factor (Trithorax-like gene; Farkas *et al.*, 1994), but also enhances mutations in the largest subunit of chromatin assembly factor I (*FAS1*; Kaya *et al.*, 2001). These findings prompt Prigge & Wagner (2001) to discuss whether imprinting might be a possible mode of sporophytic maternal control. In this case the chromatin status of relevant genes would be established prior to, and persist through, gametogenesis and the gametophytic phase. Other alternatives are also possible but remain speculative. For instance, the sporophyte could contribute a protein or mRNA with the property to persist through the gametophyte and its early zygotic action would be essential, since the paternally inherited copy remains inactive during the first divisions as observed in some cases (see above). The third possibility involves direct signalling between maternal tissue (ovule) and the developing zygote/embryo. The baso-lateral contacts of the egg with the synergids as well as the entrance point of the generative nucleus could provide additional orienting cues. The latter would be reminiscent of the 'cortical' or 'actin patch', which in *Fucus* and *Pelvetia* serves as 'polar information' in absence of any other orienting cue from the environment (Alessa & Kropf, 1999; Hable & Kropf, 2000; see also Chapter 5). The embryo then remains tightly connected with maternal tissue at the micropylar end through the basal suspensor cell until the suspensor degenerates during the end of the torpedo stage (see below). A signal could also pass through the endosperm and reach the zygote or embryo. At present there is no experimental support for any of these possibilities.

On the other hand, while *serrate* and *sin1-2* effects call for a molecular explanation, the plasticity and regenerative capabilities of plant cells argue against a significant contribution from the sporophyte to guide polarity in the embryo. It is not yet clear to what extent both somatic and zygotic embryogenesis are comparable. However, the ability of single cells or undifferentiated cell masses to undergo induced or spontaneous polarisation and subsequent development to complete polar organised plants is a major point in this discussion (Luo & Koop, 1997; Ikeda-Iwai *et al.*, 2003; for a review see Schmidt *et al.*, 1994; Mordhorst *et al.*, 1997). Apart from somatic embryogenesis, the development of perfectly polarised but inverted embryos of *twin* mutants arising from the suspensor (Schwartz *et al.*, 1994; Zhang & Somerville, 1997) also indicates independency from maternal cues in zygotic embryogenesis. Furthermore, embryos of *Korthasella dacrydii*, *Rafflesia* and *Balanophora* species have been reported to develop from unpolarised zygotes (Rutishauser, 1969).

Examining the pros and cons of possible sporophytic maternal control of zygotic polarity, the available information indicates independent establishment by intrinsic cues in the zygote. However, this scenario does not necessarily exclude sporophytic maternal cues (in the form of inherited and/or extrinsic signals) which might support or reinforce the final specification and orientation of the embryo, for instance by forcing the embryo to develop in the usual orientation, rather than upside-down, in the ovule. The significance of sporophytic maternal contribution to embryogenesis and the mechanism of how *SIN1* and *SE* exert this effect, remains unknown, and a final discussion will only be possible after evaluation of a systematic maternal screen.

6.3 Cell division planes and cell polarity

Arabidopsis displays a predictable pattern of cell division in its (early) developmental stages. A number of cell-specific or tissue-specific markers allow us to analyse and to recognise different cell fates. However, it is clear that cell lineage is not instrumental in determining cell polarity and, ultimately, embryonic pattern. The observations supporting this view come from classical embryology (Johri, 1984, 1992; Rhagavan, 2000), *Arabidopsis* embryo culture (Luo & Koop, 1997), analysis of *fass* mutants (Torres Ruiz & Jürgens, 1994) and laser ablation experiments (van den Berg *et al.*, 1995). In addition, certain *Arabidopsis* strains sometimes produce minor deviations from the known division pattern and still result in wild-type looking seedlings (Willemsen *et al.*, 1998; unpublished observations). This shows that the cell division plane alone is not instrumental in the determination of polarity, and

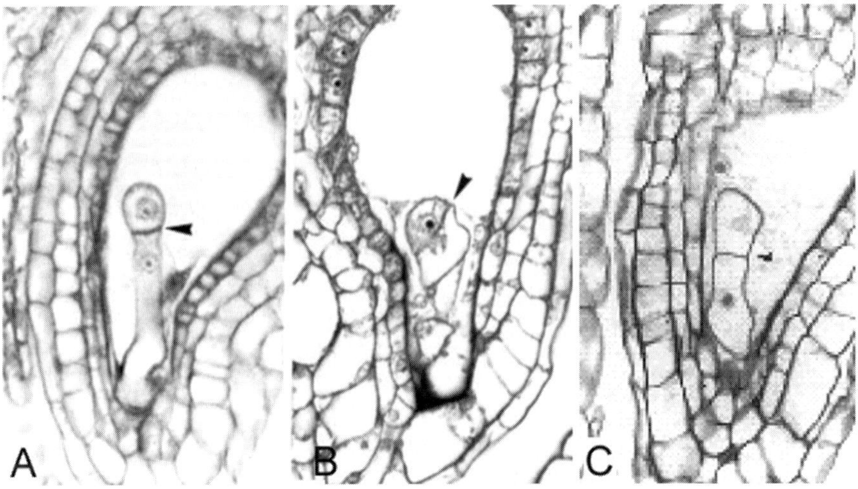

Figure 6.2 Position and orientation of cell walls after zygotic division. Compared are wild-type (A), *fass* mutant (B) and *gnom* mutant (C). Arrowheads indicate cell wall of zygotic division; cell above the arrowhead is the apical cell (ac). (A) and (B) from Torres & Jürgens (1994); (C) from Mayer *et al.* (1993); Company of Biologists Ltd. with permission.

its perturbation is not sufficient to break it down. A comparison of wild-type, *fass* (Torres Ruiz & Jürgens, 1994) and *gnom* (Mayer *et al.*, 1993) illustrates this point. The cell division plane can be nearly oblique in *fass* but nevertheless polarity remains intact, as indicated by cytoplasmic density and later by correct apical-basal pattern of *fass* mutants (Fig. 6.2). As shown below, disturbed vesicle trafficking seems responsible for a break down of polarity after the first division (Geldner *et al.*, 2003).

In the context of these observations it might be asked why it is that embryogenesis follows such a regular pattern in *Arabidopsis*. An interesting proposition is that the observed cell division pattern resulted from an optimisation process, which accommodates for the small embryo size by using a minimum number of cells (Kaplan & Cooke, 1997; Berleth & Chatfield, 2002). This attractive explanation awaits a careful comparative analysis of species following different developmental patterns and generating embryos of different sizes.

6.4 Stages in *Arabidopsis* embryogenesis

The following gives a brief overview of the different *Arabidopsis* embryonic stages (Fig. 6.3) in order to clarify subsequent discussion.

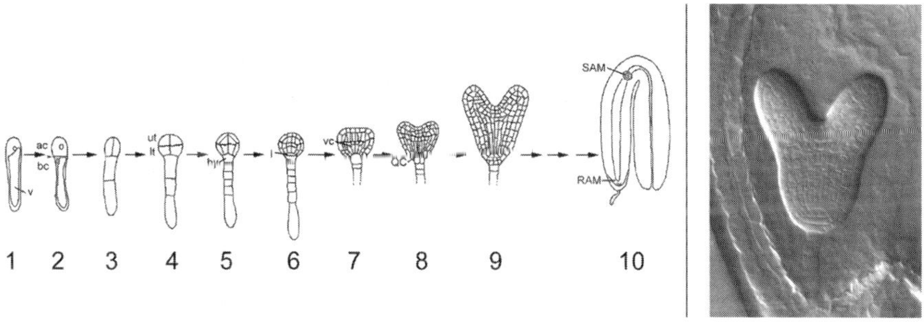

Figure 6.3 Stages of *Arabidopsis* embryogenesis. Left side shows (1): zygote, (2): two cell stage; ac: apical cell, bc: basal cell; (3): quadrant stage proembryo (only two derivatives of ac are visible; below two suspensor cells), (4): octant stage proembryo (only four cells visible; ut: upper tier; lt: lower tier), (5): 16-cell stage embryo (with epidermis precursor, hy: hypophysis cell). (6): globular stage (l: lens shape cell), (7): transition stage (vc: vascular cell), (8): early heart stage (QC: quiescent centre), (9): heart stage, (10): mature embryo last torpedo stage with shoot apical (SAM) and root apical (RAM) meristem respectively; right side shows heart stage within the ovule (Nomarski optics); from Torres Ruiz (2000), BIOspektrum with permission.

On pollination, the microspore generates the pollen tube which enters the stigma and grows towards the ovule. It leaves the placental tissue at the base of the funiculus and enters the micropyle (Mansfield *et al.*, 1991; Mascarenhas, 1993; Russel, 1993; Wilhelmi & Preuss, 1996). The pollen tube intrudes into one of the two synergids and releases the two generative cells possibly into a specialised structure called the filiform aparatus from where they actively proceed to the egg and the central cell nuclei (Mansfield & Briarty, 1990a,b; Lopes & Larkins, 1993; Russel, 1993). For several plants including *Arabidopsis*, so-called post-fertilisation shifts have been reported which reorganise the polar distribution of cytoplasm, organelles and vacuoles (Schulz & Jensen, 1968; Natesh & Rau, 1984; Mansfield & Briarty, 1991). The zygote then undergoes the first unequal division, which follows the polar organisation of the zygote (now in the two-cell stage; Fig. 6.3). *Arabidopsis* then follows the so-called Onagraceen type of development. The cytoplasmically dense apical cell (ac) is the progenitor of the larger part of the embryo. One horizontal and two vertical divisions generate a sphere of eight cells, the octant stage (Fig. 6.3). The next divisions are tangential and separate the earliest and most constant tissue of the plant, the epidermis precursor, such that a 16-cell embryo appears. This layer, which is equivalent to L1, will enlarge predominantly if not exclusively by anticlinal division throughout the plant's life. Formally the O′-boundary between the upper cells (upper tier, ut) and the lower cells (lower tier, lt) separates the cotyledons, SAM and epicotyl from the hypocotyl and root. This boundary is visible even in later stages due to somewhat thicker cell walls. However, clonal analysis showed that the cotyledons recruit cells originating from the lower cell group (Scheres *et al.*, 1994). This shows again that cell walls do not delineate borders between cell fates. Additionally it indicates that the fate of cells generating cotyledons must be traceable to very early stages.

At the same time the basal cell of the two-cell stage (bc) has performed a series of divisions leading to a filamentous organ of 7–9 cells in a row. The so-called suspensor is in contact with the maternal tissue at its base, suggesting a possible nurturing function. At the same time it can be seen that the uppermost suspensor cell intrudes into the upper globular cell group and adopts a distinct shape. This hypophysis undergoes a further division and produces an isodiametric cell adjacent to a suspensor cell beneath and an upper lens shape cell. The latter will lead to the quiescent centre (QC; see below). At the heart stage, the hypophysis derivatives have established a layered cell group with the QC as its core, around which adjacent cells are maintained as undifferentiated initials for different root tissues; together they constitute the root apical meristem (RAM).

Along the radial axis of the embryo further internal divisions discern cortical precursors and precursors of the central cylinder (Fig. 6.3). Their fate remains recognisable by their elongated shape throughout embryogenesis. This elongation in concert with further cell divisions means that the hypocotyl region comprises two thirds of the length of the embryo.

The globular embryo shows a polar organisation at the shoot apical meristem (SAM) region, which is morphologically recognisable only at late stages. However, centred expression domains of essential SAM genes show the existence of SAM anlagen (Barton & Poethig, 1993; Long *et al.*, 1996; Mayer *et al.*, 1998; see also Chapter 8). This polar distribution of SAM determinants in the upper region interface with a polar distribution of other factors at the periphery, which determine cotyledonary fate. Additional mechanisms ensure first, a precise opposed position of two primordia and second, act to suppress a fusion of cotyledonary tissue at the bases of the cotyledons such that the SAM is not enclosed (see below). Fate mapping has shown the generation of cotyledon primordia to be rather, a sequential, than a simultaneous series of events (Woodrick *et al.*, 2000). Increased cell division events at the two opposing points are followed by further overgrowing of the centre, thus, profoundly changing the organisation and generating the sole embryo-specific organs. This heart-stage embryo represents the basic unit of the plant. Additional divisions and cell enlargement, especially in the lower region, enlarge the hypocotyl considerably. The further stages, called torpedo stages, do not add qualitatively to the basic organisation but prepare the embryo for the dessication stage by accumulating storage proteins and starch. During torpedo stages the endosperm is fully consumed, the suspensor degenerates, the SAM becomes morphologically recognisable and the cotyledons bend, giving the seed its characteristic shape. Only two to three cells of mature embryonic SAM contribute to the upper stem and floral structures. The remainder of the SAM zone contributes to the rosette leaves (Irish, 1991; Irish & Sussex, 1992).

6.5 Zygotic polarity – vesicle trafficking in the cellular rush hour

Polarity of zygotes or single cells (during somatic embryogenesis) might be established under quite diverse conditions but are maintained and perpetuated throughout the plant's life. In recent years, some knowledge has been obtained of early factors essential for apical-basal polarity. In *Arabidopsis*, a systematic screen has revealed a single gene *GNOM* (also *EMB30*), which exerts profound influence on establishment of polarity at zygotic division (Mayer *et al.*, 1991, 1993). Loss-of-function mutations of this gene display striking pattern phenotypes including lack of embryonic root and variable deletion or fusion of hypocotyl and cotyledons. In extreme cases, the loss of any apical-basal polarity results in ball-shaped seedlings with differentiated and randomly oriented vascular cells in their centre (Figs 6.2 and 6.4; Mayer *et al.*, 1993; Steinmann *et al.*, 1999). In extreme cases, even strong markers of polarity such as *AtLP1* and *POLARIS* fail to indicate any sign of correct apical-basal polarity (Vroemen *et al.*, 1996; Topping & Lindsey, 1997). Similar defects can be mimicked by treating *in vitro* cultured *Brassica juncea* embryos (a species related to *Arabidopsis*) with high amounts of auxin analogues or auxin transport blockers (Liu *et al.*, 1993; Hadfi *et al.*, 1998). Note, however, that these 'phenocopies' are similar but not identical to *gnom* mutants; for instance ball-shaped embryos induced by these treatments do not seem to possess differentiated vascular cells (Hadfi *et al.*, 1998). The results of such experiments suggested an involvement of auxin metabolism, transport and/or localisation in the establishment of early embryo polarity. Much of the significance of auxin transport has been partly understood by application of substances like TIBA and NPA, which inhibit directional, i.e., polar transport in stem segments (Estelle, 1998; Palme & Gälweiler, 1999). Briefly the key to this directional streaming lies in the asymmetric distribution of the carriers (ibid.). Generally auxin is thought to be produced in young organs or near the apex and then transported to other parts of the plant by influx and efflux carriers located at the plasma membrane, thus moving the hormone through the different cell files by successive rounds of directional transport. The details of this process and the complexity of directing and canalising the appropriate auxin streams within the plant to direct development of vascular tissues are discussed elsewhere in this book (see Chapters 2 and 9).

The *GNOM* gene encodes a GDP/GTP exchange factor for small G-proteins of the ARF class (ARF-GEF; Shevell *et al.*, 1994; Busch *et al.*, 1996; Steinmann *et al.*, 1999). It is known that such molecules are necessary for vesicle trafficking by recruiting vesicle coats necessary for vesicle budding (Donaldson & Jackson, 2000). The GEF-ARF encoded by *GNOM/EMB30* has functional similarity to ARF GEFs from budding yeast and is sensitive to the fungal toxin brefeldin A (BFA) in a concentration-dependent manner (Steimann *et al.*, 1999). Application of BFA not only affects GNOM but

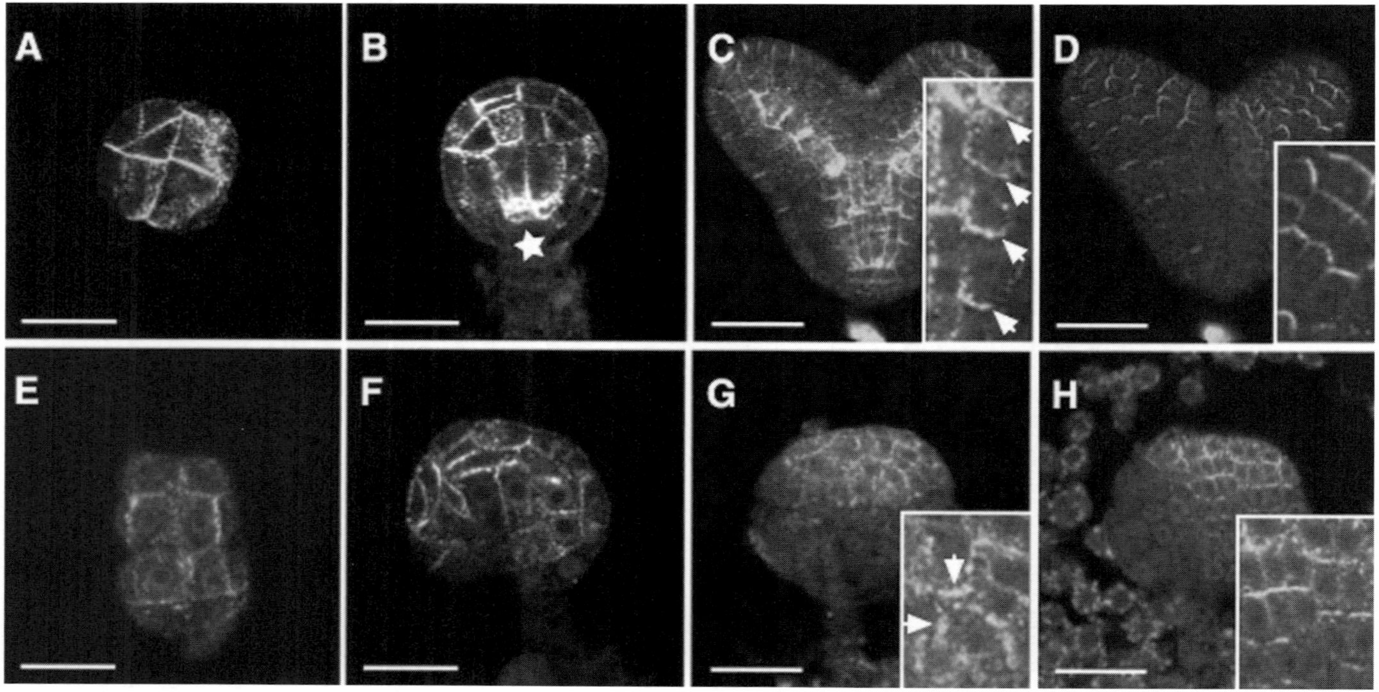

Figure 6.4 PIN1 protein localisation in *Arabidopsis* embryogenesis. Confocal microscopy of wild-type (A–D) and *gnom* (E–H) embryos (scale barrs 20m). A–C and E–G are internal optical sections; D and H are surface (epidermal) views. Insets show magnifications. Asterisk and arrows mark polar localisation of PIN1. Note disturbed internal versus correct surface localisation in *gnom*. Reprinted from Steinmann *et al.* (1999). Copyright 1999 American Association for the Advancement of Science; with permission.

also co-ordinated polar localisation of PIN1 (Geldner *et al.*, 2001). This links GNOM action to auxin transport since PIN1 is one member of putative efflux carriers and is localised at the basal or downstream side of transporting cells in the stem of adult plants (Gälweiler *et al.*, 1998). The localisation is not fixed, but instead, PIN1 is dynamically recycled and moves between the plasma membrane and an intercellular compartment not identical to the Golgi (Geldner *et al.*, 2001). This cycling is also dependent on actin filaments. A similar asymmetric localisation of PIN1 as in the adult plant has been shown in the *Arabidopsis* embryo (Fig. 6.4; Steinmann *et al.*, 1999). Thus, by analogy, auxin might have very similar roles as in post-germination stages, stressing the significance of the heart stage embryo as a basic unit, which is perpetuated in the adult.

Many components might affect cellular vesicle movement, therefore the question is whether GNOM has a specific role in the localisation of PIN1 (and other proteins) and thus in the maintenance of polarity. Specificity seemed probable since *gnom* suspension cultures grow well and localisation of compartment-specific membrane markers such as the KNOLLE syntaxin and plasma membrane H^+-ATPase are not altered in *gnom* embryos (Steinmann *et al.*, 1999). Very recently the answer has been given with the aid of an engineered BFA-resistant GNOM variant (Geldner *et al.*, 2003). This shows that BFA-resistant GNOM is sufficient for rendering PIN1 recycling to the plasma membrane insensitive to BFA, i.e. in transgenic plants auxin transport and auxin-mediated growth responses become BFA-resistant and grow (Fig. 6.5). Furthermore, immuno-localisation with different markers specific for endomembrane subsets revealed that GNOM localises to a subset of endomembranes which partially overlaps with a FM4-64-labelled compartment but not to ER membranes, intra-Golgi vesicles or the trans-Golgi network (Fig. 6.5; Geldner *et al.*, 2003). The localisation of other proteins like PIN2, a PM-ATPase and KNOLLE are partially or entirely independent from GNOM and indicates the existence of additional vesicle recycling routes in plants similar to the different routes described in animals (Mostov *et al.*, 2000; Gruenberg, 2001). So far then, it seems that GNOM directs apical-basal polarity in the embryo by specifically supporting basal PIN1 localisation starting from the 16- to 32-cells stage. In the subsequent stages this accompanies cell elongation in accordance with the theory that polar auxin transport is necessary for this process (see Estelle, 1998 for review). However, this cannot explain the breakdown of polarity up to the 16-cell stage, since PIN1 accumulation at the basal poles of the inner embryonic cells does not begin before this stage (Fig. 6.4; Steinmann *et al.*, 1999). In mammalian cells, directed vesicle transport to distinct plasma membrane sub-domains establishes cell polarity (Mostov *et al.*, 2000; Gruenberg, 2001). Thus GNOM-mediated localised secretory vesicle targeting of other factors might be necessary in early embryo stages and be compromised in *gnom* mutants.

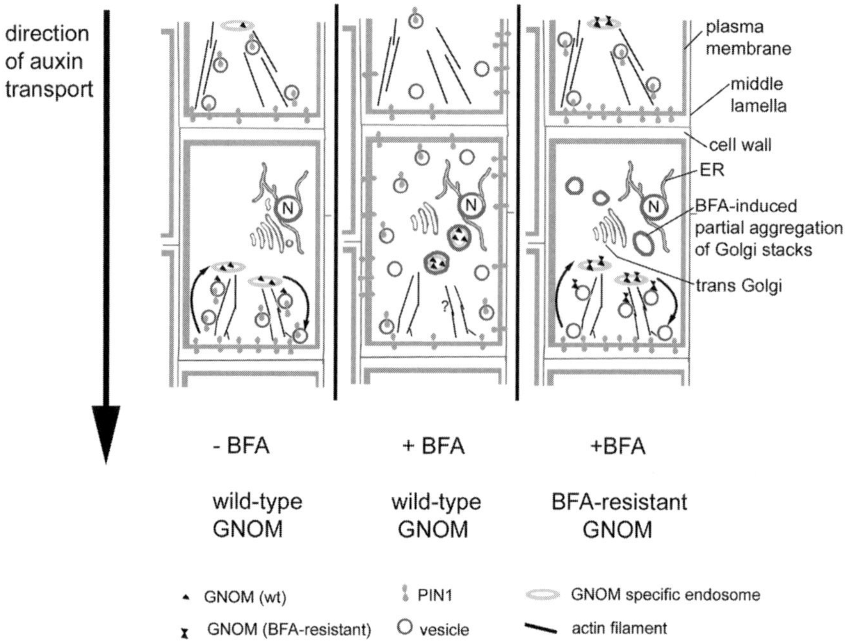

Figure 6.5 Schematic representation of sensitive (wild-type) and brefeldin A (BFA)-resistant GNOM in cells. Left: Wild-type GNOM determines cargo of vesicles with PIN1, which are transported to the basal pole of the cell. GNOM is localised in an endosome compartment not identical with the Golgi. The GNOM specific endosome is also labelled by the marker FM4-64. Directional transport of vesicles is thought to be supported by actin filaments. PIN1 is dynamically cycled and recycled between GNOM-specific endosome and plasma membrane. In wild-type GNOM is predominantly found in the cytosol and PIN1 in the plasma membrane. Middle: BFA-treatment leads to break-down of polar PIN1 localisation. ER and trans Golgi network seem unaffected as indicated by the marker AtSEC12 and TLG2a respectively. Some Golgi stacks aggregate to larger clusters as indicated by the marker gamma-COP. They surround but do not overlap with endosomal patches where GNOM is localised (fate of actin filaments was not analysed). Right: Engineered BFA-resistant GNOM reconstitutes PIN1 cycle and polar localisation. Enlarged aggregated Golgi stacks remain. For details see text and Geldner *et al.* (2003). (N = nucleus).

This poses the question whether zygotic polarity is disturbed in *gnom* starting from fertilisation. Alternatively the zygote might carry over some polarity from the egg cell and then polar breakdown arises continuously during the (disturbed) generation of the cell plate in the first division.

The significance of directed vesicle trafficking for cellular polarity in higher plants finds striking parallels in some algae (see Chapter 5). These similarities involve a crucial role of actin filaments as well as BFA-sensitive targeted secretion in the generation of polarity (Shaw & Quatrano, 1996; Belanger & Quatrano, 2000; Chapter 5). Further parallels are conspicuous, as for instance, the uncoupling of cell division plane and establishment of polarity as seen in *fass* mutants and BFA-treated *Fucus* embryos (Torres Ruiz & Jürgens, 1994; Shaw & Quatrano, 1996) or the striking similarity of the first cell division routes in *Arabidopsis* and *Fucus* (Brownlee & Berger, 1995). A further focus is directed to the cytoskeleton in directing polarity which is reminiscent of its crucial role for anterior-posterior and dorso-ventral polarity in *Drosophila* eggs (St. Johnston, 1995). The prominent role of the cytoskeleton is discussed in Chapters 1 and 4 of this book.

6.6 Epidermis and radial axis/polarity

The quadrant embryo initiates the earliest tissue delineation by undergoing tangential divisions. This process generates an inner cell group and the outer epidermis precursor, one of the most constant cell lines (also known as L1), which maintains itself by anticlinal divisions. The early marker AtML1 also indicates an early epidermis-specific fate from the early globular stage onwards (Fig. 6.6; Lu *et al.*, 1996; Sessions *et al.*, 1999). This does not exclude determination of cell fate specification by position in the epidermis, as evidenced by laser ablation (Berger *et al.*, 1998). A further radial polarity is visible in the 16-cell stage by a clear concentration of GNOM protein in the inner cell walls (see above). Does GNOM participate in epidermis

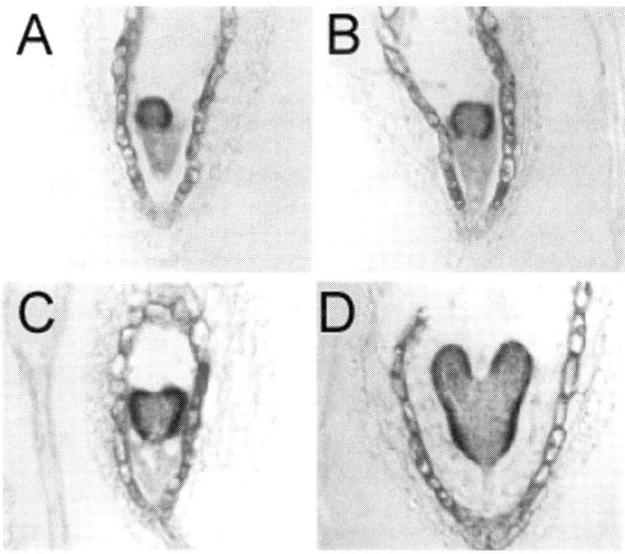

Figure 6.6 Epidermal specific gene expression. The *MERISTEM LAYER1 (ATML1)* promoter was fused to the *GUS* reporter gene. (A–D): *in situ* hybridisation (*GUS* probe) of different embryo stages of plants transgenic for *ATML1*promoter::*GUS* construct. Note absence of signal at the apical and basal poles. From Sessions *et al.* (1999), *The Plant Journal*, Blackwell Publishing, with permission.

development by its ability to direct PIN1 localisation? This is contradicted by the fact that *gnom* mutants possess an epidermis (Mayer *et al*., 1993) and PIN1 seems to be correctly localised in *gnom* epidermal cells (Fig. 6.4; Steinmann *et al*., 1999). How correct epidermal position is achieved while the inner localisation is disturbed is not known. Moreover *gnom* epidermis is capable of undergoing asymmmetric cell divisions to generate stomatal cells. The special status of the L1 is also stressed by the fact that in adult stages it is symplastically isolated (Ducket *et al*., 1994). So far there is no mutation known which interferes with this step and the question arises if such mutants would be viable. It is also conspicuous that subepidermal cells can express epidermal markers as for instance in *knolle* mutants (Lukowitz *et al*., 1996; Vroemen *et al*., 1996) or can adopt epidermal cell fates as seen upon *GLABRA1* overexpression in absence of its negative regulator *TRIPTYCHON* (Schnittger *et al*., 1998). The reverse has not been reported. So one possible model is that outside cells adopt epidermal fates because they receive a signal by contacting 'outer space', as the embryo sac interior does during embryogenesis. In this model, any outermost cell adopts an epidermal fate by default. Though this would explain the absence of specific epidermis mutations an inverse model cannot be excluded, i.e. that internal cues specify internal cell fates and epidermal cells do not sense (enough) signal to adopt internal fates. Whether the outer cell walls carry an epidermis fate determining signal at least in early development (similar to *Fucus*, see Chapter 5), which is unable to influence the fate of inner cells is not known. However, cell wall components such as arabino-galactan proteins are unequally distributed between different cells as indicated by specific antibodies (for further discussion of this aspect see Souter & Lindsey, 2000).

Radial polarity in terms of epidermal specification is clearly not well understood. A further element of this polarity, much better analysed, found a starting point in the analysis of two loss-of-function mutants, *scarecrow* (*scr*) and *short-root* (*shr*) which exhibit absence of cortex and endodermis identity in the ground tissue. Here the initial daughter cell is not capable of asymmetric division, and so a single tissue layer in place of cortex and endodermis layer results (Benfey *et al*., 1993; Di Laurenzio *et al*., 1996; Helariutta *et al*., 2000). *scr* mutants possess a ground tissue layer which adopts characteristics of cortex and endodermis cells whereas *shr* solely exhibits cortex features. The conclusions from these observations are first, that *SCR* is required for correct unequal division to give cortex and endodermis. Second, *SHR* seems to be responsible for cell division and the endodermal cell fate specification. Both genes encode putative transcription factors of the GRAS family (Di Laurenzio *et al*., 1996; Pysh *et al*., 1999; Helariutta *et al*., 2000) but do not possess the same capabilities since SHR has the potential for intercellular movement whereas SCR does not (Nakajima *et al*., 2001). The expression patterns of both genes reflect a striking polarity along the radial axis and simultaneously

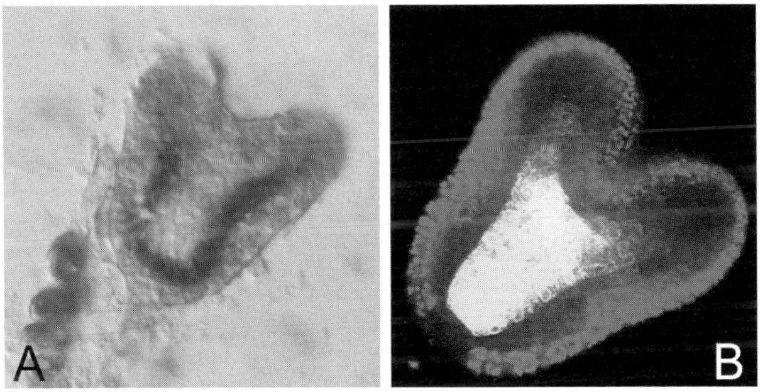

Figure 6.7 SCR and SHR expression in the *Arabidopsis* embryo. (A): *SCR* anti sense probe hybridised in situ. Signal is recognisable in the endodermis, cortex/endodermis precursor and QC. (B): SHR::GFP transcriptional fusion construct indicates presence of SHR mRNA in the central cylinder (bright fluorescence). Courtesy of Phil Benfey and Hidehiro Fukaki.

indicate non cell-autonomy of *SHR*. *SCR* expression is confined to the endodermis the cortex/endodermis initial and its daughter and the QC (Fig. 6.7). *SHR* transcript is found in the entire stele comprising pericle, xylem and phloem precursors (Fig. 6.7). Thus *SHR* exerts effects in cell groups (endodermis; cortex/endodermis initial daughter) where its transcript is not present (Figs 6.7 and 6.8). Localisation of SHR-GFP fusion protein, as well as SHR antibodies, shows that SHR protein is transported from the stele to the endodermis cortex/endodermis initial and daughter and the QC where it is found in the nuclei (Fig. 6.8; Nakajima *et al.*, 2001). This strongly suggests that SHR, as a transcription factor, controls the fate of the endodermis and its precursor cell, and that one likely target might be *SCR,* whose expression level is at least reduced in *shr* mutants. The phenotype of plants expressing *SHR* under the *SCR* promoter do also indicate the target cells of *SHR*, *pSCR::SHR* plants exhibit supernumerary layers of endodermis and QC (Nakajima *et al.*, 2001). Though mainly analysed in adult roots, the basic steps of *SCR* and *SHR* action are very likely the same in the embryo, since *SCR* mutations reveal the same effects at the heart stage of the embryo as they do in adults. *SCR* is transcribed in embryonic tissues (Fig. 6.7) and the expression pattern of *SHR* is also confined to the QC, cortex/endodermis initials and endodermis respectively. *SHR* seems to be expressed in the embryonic stele precursor (Figs 6.7 and 6.8; Scheres *et al.*, 1995; Fukaki *et al.*, 1998; Helariutta *et al.*, 2000; Nakajima *et al.*, 2001). The radial polar distribution of key transcription factors is extended by the localisation of WOODEN-LEG/CRE1, a newly identified cytokinin receptor responsible cell division, within the procambium of the embryo (and later vascular cylinder of the root; Scheres *et al.*, 1995; Mähönen *et al.*, 2000; Hwang & Sheen, 2001; Inoue *et al.*, 2001). Apparently the generation and patterning of this region requires more factors since its

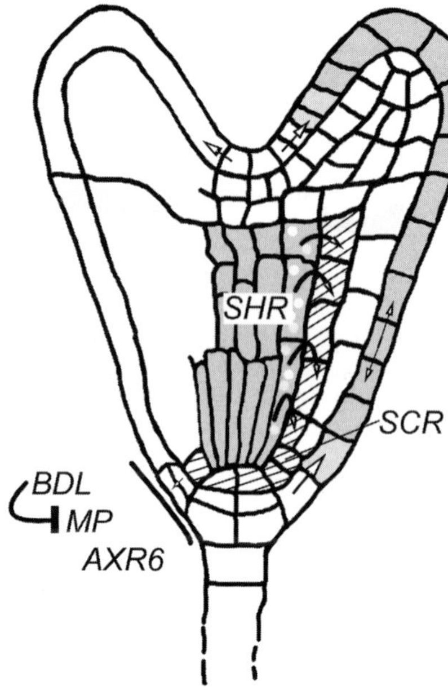

Figure 6.8 Schematic representation of radial polarity. SHR is expressed in the central cylinder (grey shaded) whose outermost cell row is the pericycle (grey with white spots). SHR protein is moved to the endodermis and its precursor as well as the QC (bended arrows). The special 'autonomy' of the epidermis is indicated by small arrows. Epidermis generation or enlargement results predominantly or exclusively from epidermis cells or epidermis initials at the meristematic regions. The genes *MP*, its antagonist *BDL* and *AXR6* are included to indicate their role in generation of the embryonic root.

structure is more complex as judged from the number and fates of cells. In this context it is interesting that the expression of a number of genes including some involved in axialisation and vascular cell development, like *MONOPTEROS*, are also confined to the central cylinder in advanced torpedo stages (see below and Chapter 9).

6.7 Embryonic shoot pole – pole position for polar activities

During adult plant development, a proliferative centre at the tip of the main stem provides new cells for a number of organs, while the centre maintains itself throughout development (details of these processes are discussed in Chapter 8).

The basis for adult growth is laid down in the form of the SAM and RAM at the opposite poles of the embryo. A number of transcription factors not only define the prospective SAM during embryogenesis, but are also active in terms of determining the size of the SAM. In transition stage embryos, the upper tier (ut) additionally elaborates two cotyledon primordia which have an essential task in sustaining and supporting seedling/and early adult plant development. At the end of its development, the upper region has passed through a number of polarities reflected by overlapping and partly transient activities of genes in different zones. Initially these domains are sometimes relatively broad and are then refined to appear in smaller regions.

A precise size restriction of the SAM is essential for the shape of the plant. Mutations of key genes lead to ablation or increase of the SAM zone, which in the mature embryo comprises about 100 cells (Irish & Sussex, 1992). Consequently positive and negative regulators of the SAM interact through activating and repressing feedback interaction loops already present during embryogenesis. At the 16-cell stage *WUSCHEL (WUS)*, a novel homeodomain transcription factor, is expressed in the inner cells of the upper tier (Mayer *et al.*, 1998). *WUS* expression is then restricted to a few cells (L3 layer cells) beneath the subepidermal cells (L2 layer cells) at the summit (Fig. 6.9). The former are believed to represent an organising centre through which *WUS* action induces stem cell identity in the cells above (Mayer *et al.*, 1998). How the cells below this centre escape *WUS* meristem induction is not clear. The polar influence towards the cells above is responded by expression of *CLAVATA3* (CLV3; Fletcher *et al.*, 1999), a protein ligand which activates a membrane receptor serine/threonine kinase *CLAVATA1* (*CLV1*) whose activity represses *WUS* (Brand *et al.*, 2000; Schoof *et al.*, 2000; Trotochaud *et al.*, 2000). Early embryonic SAM cells also express *STM* (*SHOOT MERISTEMLESS*), a *KNOTTED*-type homeodomain transcription factor and *CUP SHAPED COTYLEDON1* and *2* (*CUC1, CUC2*), redundant NAM-domain transcription factors (Fig. 6.10; Long *et al.*, 1996; Aida *et al.*, 1996, 1999; Takada *et al.*, 2001). The former is believed to be essential for undifferentiated stem cell maintenance (Barton & Poethig, 1993; Long *et al.*, 1996). Plants mutant in *stm1* and *asymmetric leaves1* (*as1*) reflect a partial rescue of the *stm* single mutant phenotype. The conclusion is that *STM* maintains undifferentiated status of the SAM by suppressing *AS1* (a MYB domain transcription factor) and is not strictly necessary for SAM formation (Byrne *et al.*, 2000). Conversely *AS1* is necessary for specifying organ founder cell identity. *AS1* as well as *AS2* (a putative transcription factor with a leucine zipper) both repress class I *KNOX* homeobox genes to maintain leaf identity (Byrne *et al.*, 2000; Semiarti *et al.*, 2001; Iwakawa *et al.*, 2002 and references therein). This is probably also the case in heart stage embryos where *AS1* is strongly expressed in cotyledon primordia (ibid.). The fusion of cotyledons at their base in *stm1* mutants are taken as a hint to the suppressing function of *STM*. During development from globular stage onwards, additional domains are delineated by genes whose activities are only partially understood. This complex pattern will be extended by adaxial and abaxial expression patterns of new players (Figs 6.11 and 6.12; see below). While *STM* is expressed as a

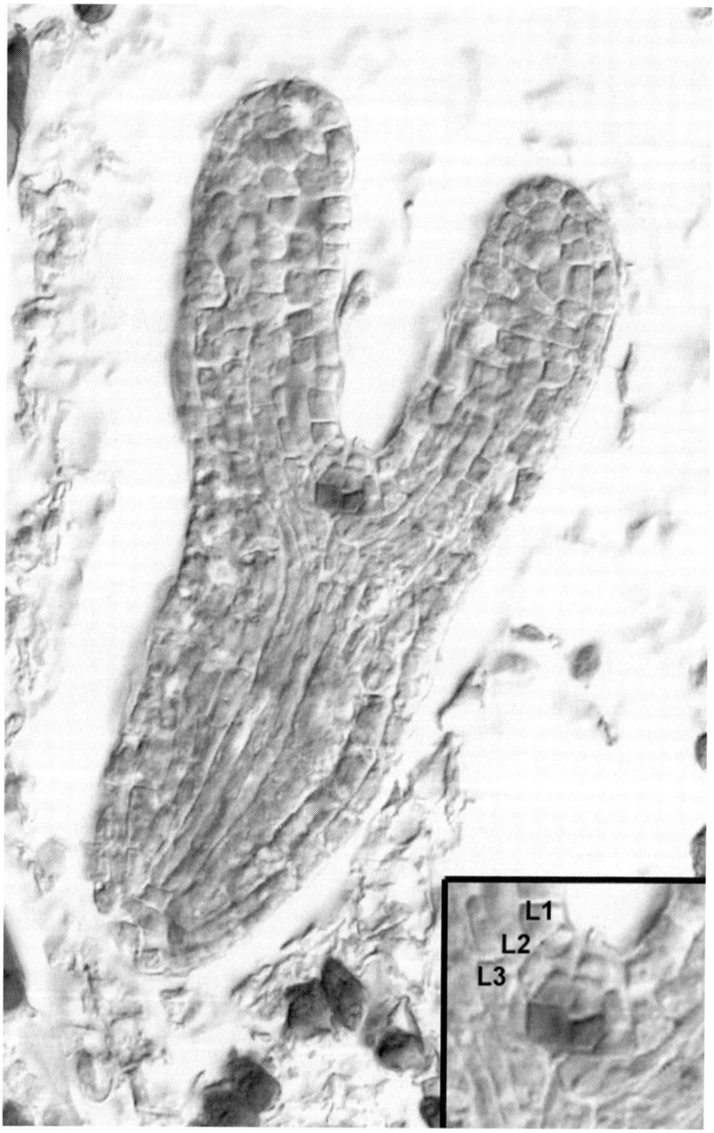

Figure 6.9 *WUS* expression in torpedo stage embryo. The *in situ* hybridisation shows strong expression in few cells in the L3. Courtesy of Holger Breuninger, Enno Rikirsch and Thomas Laux.

central stripe at the apical pole, a further gene, *ANT*, elaborates a partially overlapping ring-shaped expression domain (see below, Fig. 6.13). Though *ANT* expression concentrates at presumptive cotyledon anlagen (Elliott *et al.*, 1996), *ant* mutants have not been reported to display cotyledon defects (Klucher *et al.*, 1996; Elliott *et al.*, 1996). The genes *CUC1* and *CUC2*,

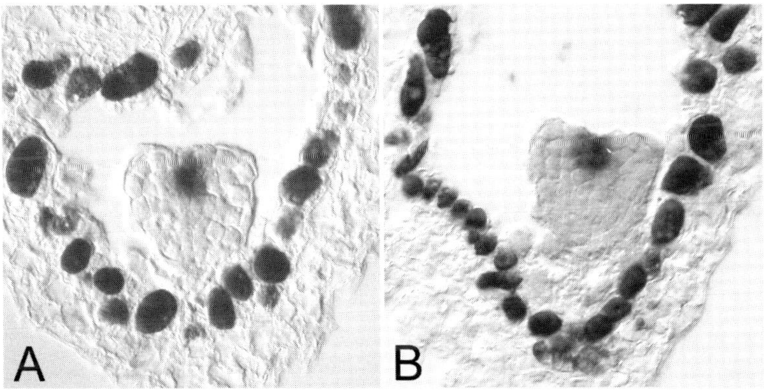

Figure 6.10 SAM region specific gene expression. (A) transition stage embryo showing localisation of *CUC2* expression. (B) Similar expression domain of *STM*. Courtesy of Masahiko Furutani and Masao Tasaka.

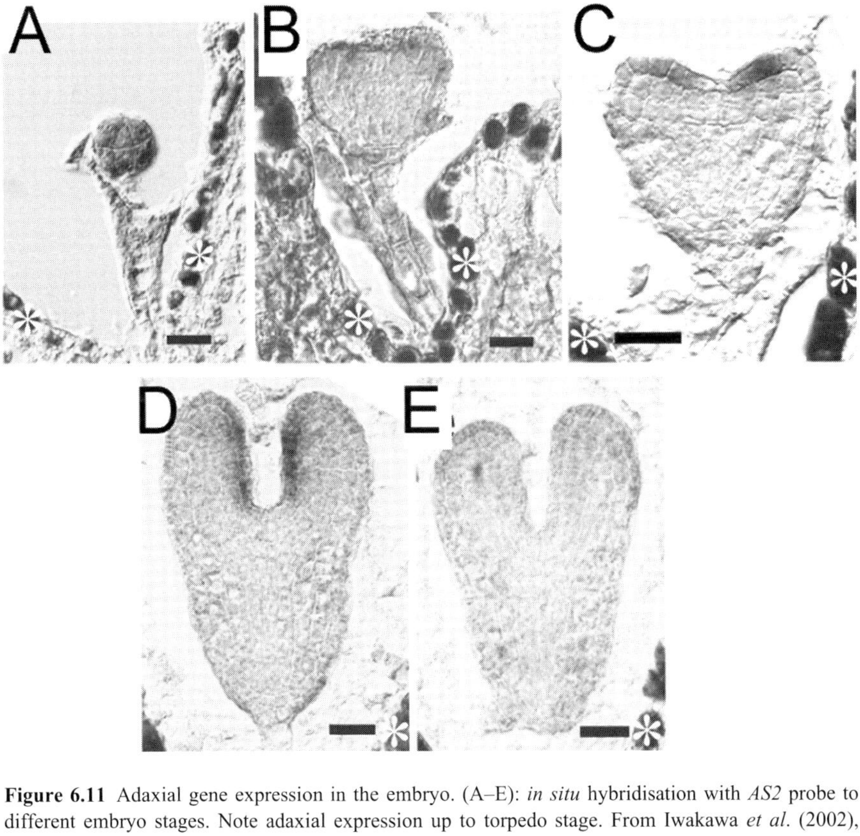

Figure 6.11 Adaxial gene expression in the embryo. (A–E): *in situ* hybridisation with *AS2* probe to different embryo stages. Note adaxial expression up to torpedo stage. From Iwakawa *et al*. (2002), *Plant Cell Physiol*., Oxford University Press; courtesy of Yasunori Machida.

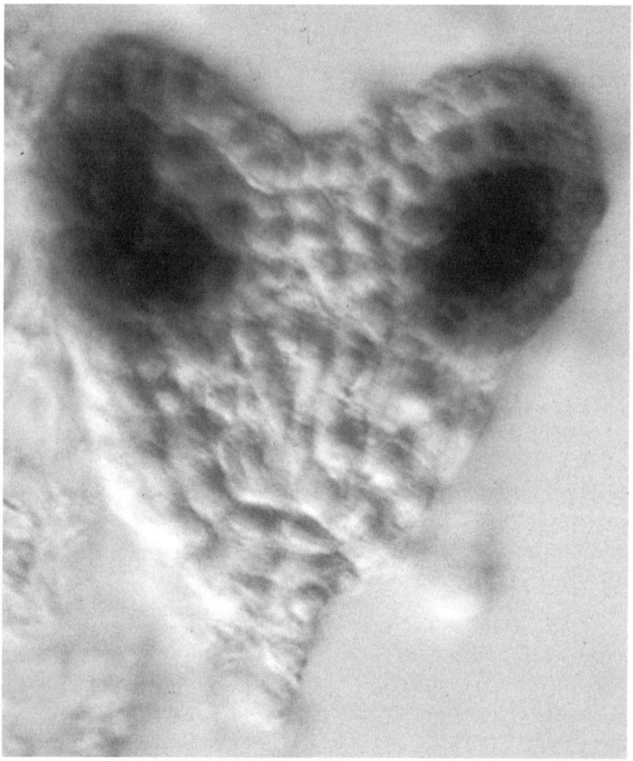

Figure 6.12 Abaxial gene expression in the embryo. The figure shows the abaxial expression domain of *FIL* at early heart stage. From Siegfried *et al.* (1999); Company of Biologists Ltd. with permission; courtesy of John Bowman.

representing NAC domain transcription factors, obviously function redundantly to suppress fusion of the cotyledons as well as induction of *STM* expression (Aida *et al.*, 1996, 1999; Takada *et al.*, 2001). *CUC2* exhibits an expression pattern similar to *STM*, but later on separates such that *STM* is centrally found while *CUC2* defines the peripheral borders between the bases of cotyledons (Aida *et al.*, 1997, 1999). *STM* expression is dependent on *CUC1* and *2* as seen in the *cuc1 cuc2* double mutant, but recently a positive influence on *CUC1* expression by *STM* has also been shown (Fig. 6.13; Aida *et al.*, 1996, 2002). Bearing in mind the expression of the other SAM promoting/inhibiting genes, the apical pole of a transition stage embryo exhibits an astonishingly complex polarity of gene activities. These are gradually organised (Long & Barton, 1998), may partially arise independently, and then become dependent on each other in later stages to give a structured shoot apical region (Fig. 6.13).

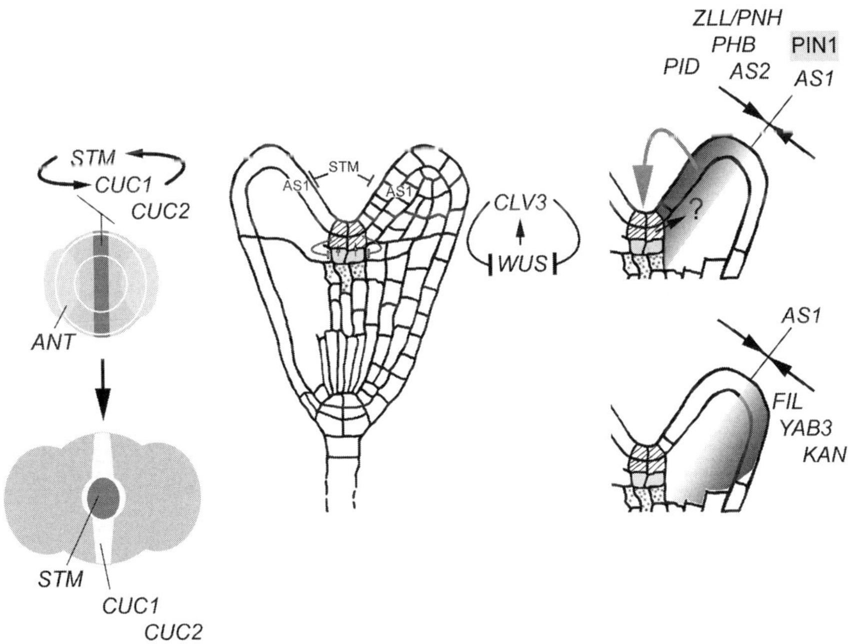

Figure 6.13 Schematic representation of polar activities at the apical pole. Left: Initial overlapping *STM* and *CUC1,2* domain develops to separated domains of *STM* (centre) and *CUC1,2* expression (between incipient cotyledon primordia). *STM* and *CUC1* are thought to support each other through a positive feedback loop. Peripheral expression domain of *ANT* is indicated. In later stages the periphery harbour further expression domains of other genes. Middle: Size of apical stem cell group is adjusted by a positive (*WUS*) and negative (*CLV3*) interaction loop between *WUS* and *CLAVATA* genes respectively. WUS expressing cells (grey) are possibly instructed in this function by vascular precursors beneath (pointed cells). *STM* expression domain may embrace grey and hatched areas and their immediate neighbours. *CLV3* domain hatched cells, *CLV1* domain subepidermal hatched cells and grey cells. *STM* also functions to repress *AS1* and thus maintains undifferentiated stage of SAM cells. Right: Adaxial and abaxial expression of genes in the growing cotyledon are indicated. Juxtaposition of these genes is thought to promote or enable lamina outgrowth. *AS1* is expressed throughout the primordium. The adaxial side of the cotyledon is thought to promote correct SAM development (grey arrow). Whether the SAM provides essential signals for cotyledon development is not clear (small arrow with question mark). Cotyledons might posses higher auxin concentrations transported by *PIN1*, which concentrates in their centres (see Figs 6.4 and 6.15). Also *PID* is initially concentrated at the tips of incipient primordia and then restricts to the adaxial side of cotyledons. For details see text.

6.8 Adaxial/abaxial polarity in the cotyledons

The characteristics of cotyledon dorsoventral polarity are repeated and refined postembryonically to give a complex network of dependencies and interactions. The details of dorsoventral leaf polarity are discussed in Chapter 10. The polar localisation of gene activities discussed above clearly indicate an early separation of domains for SAM versus regions harbouring anlagen for cotyledon primordia. In fact this separation might occur as early as the 32-cell stage when *WUS* expression becomes restricted to the inner subepidermal cells of the upper tier. At least they are not derived from cells that previously expressed *STM* (Long *et al.*, 1996) in contrast to primordia of adult leaves which are generated by cells displaced from the centre of the SAM post-embryonically (see Chapter 8).

On the one hand, cotyledons lack specific features of adult leaves and possess unique characteristics (for instance, a lack of trichomes, simple tracheary network, storage reserve metabolites, etc.). On the other hand, it is clear that leaves and cotyledons are homologous structures (Kaplan & Cooke, 1997). Mutations of genes responsible for maturation processes do partially convert cotyledons to leave-like organs (Meinke, 1992; Lotan *et al.*, 1998; Stone *et al.*, 2001). The reverse phenotypes exist as well (Conway & Poethig, 1997).

The basic organisation of cotyledons reflects similar polarities as adult leaves, as indicated by gene expression patterns. However, not all of these seem to be essential for cotyledon development *per se*. For instance mutations in *ANT, ZLL/PNH (ZWILLE/PINHEAD), REV (REVOLUTA), AS1, AS2* do not markedly affect cotyledon structure (Elliott *et al.*, 1996; Moussian *et al.*, 1998; Lynn *et al.*, 1999; Otsuga *et al.*, 2001; Byrne *et al.*, 2000; Semiarti *et al.*, 2001; Iwakawa *et al.*, 2002). Nevertheless, they adopt adaxial or abaxial polarity of expression respectively in the cotyledon (Figs 6.11 and 6.12). Some activities probably influence post-embryonic development. Some observations suggest that correct adaxial cotyledon identity promotes normal SAM development. Mutations in *ZLL/PNH,* which is adaxially expressed, lead to early termination of the SAM, producing a terminal leaf-like organ. Since *ZLL/PNH* exhibit a central expression domain, this could also be responsible for this effect (Moussian *et al.*, 1998; Lynn *et al.*, 1999). A positive effect on SAM generation is also evidenced from a double mutant in *ZLL/PNH* and the related gene *ARGONAUTE*, which fails to express *STM* (Lynn *et al.*, 1999). SAM-promoting activity of adaxial identity is also seen in the dominant *phb* mutation, which leads to adaxialisation and simultaneously induces ectopic meristems (McConnell & Barton, 1998; McConnell *et al.*, 2001). Conversely overexpression of the putative transcription factors *FIL* (*FILAMENTOUS FLOWER*) and other *YABBY* (*YAB*) genes, which are expressed abaxially (Fig. 6.12) lead to SAM termination (Sawa *et al.*, 1999; Siegfried *et al.*, 1999). Consequently, the opposite situation is found when *FIL* and *YAB3* are both mutated: ectopic meristems are produced already in cotyledons (Kumaran *et al.*, 2002; see also Chapter 10). A further gene, *KANADI*, also promotes abaxialisation and additionally influences radial polarity (Kerstetter *et al.*, 2001). The conclusion from this information is that the young SAM takes reference from cotyledon polarity and receives positive or negative signals from adaxial and abaxial sites respectively (Fig. 6.13). Furthermore, the juxtaposition of adaxial and abaxial cell fates are considered as a requirement for leaf lamina outgrowth promoted by genes whose ancestral function specified abaxial cell fate of plant lateral organs (Waites & Hudson, 1995; Bowman, 2000; see Chapter 10). The same function also seems to apply for cotyledons. An impact in the opposite direction, from the SAM to the cotyledons, might also exist (Fig. 6.13). Classical experiments indicate that leaves fail to develop adaxial/abaxial polarity when separated from meristems (Sussex, 1954). However, cotyledons remain relatively normal in *stm* mutants except in having the small region of fusion at their bases.

Analyses of adult plants suggests that polar distribution of auxin induces the generation of leaves (Reinhardt *et al.*, 2000). It is probable that the significance of this hormone extends to an involvement in generation of cotyledons. This is deduced from the application of auxin or its transport inhibitors to plant embryos. When *Brassica* embryos are subjected to these agents, deviations from normal cotyledon pattern result mostly as seedlings/embryos with collar or fused cotyledons (Liu *et al.*, 1993; Hadfi *et al.*, 1998). As shown above, the polar organisation of PIN1 is established during early steps of development, and accumulates in central regions of the emerging cotyledons (Fig. 6.4; Steinmann *et al.*, 1999). The role of auxin in this process is also stressed by the fused or collar cotyledon pattern of single or double mutation(s) in genes which are involved in sensing and/or response signalling to auxin (*pinformed1, gn, monopteros/bodenlos*; Mayer *et al.*, 1993; Hamann *et al.*, 1999). Alterations of cotyledon numbers are observed in *pin1* and *pinoid* mutants. The *PINOID* (*PID*) gene is an auxin response-related protein kinase. Its transcript is seen at incipient cotyledon primordia in globular stage embryos. Later it is found at the hypocotyl flanks and adaxial cotyledon regions (Bennet *et al.*, 1995; Okada *et al.*, 1991; Vernoux *et al.*, 2000; Christensen *et al.*, 2000). Therefore, several activities (and polarities) are revealed in the cotyledon: those which promote organ identity, those related to auxin, some inhibitory functions (*KNOTTED*-like gene suppressors) and factors which specify adaxial versus abaxial sides and promote/suppress SAM development (Fig. 6.13).

These advances provide considerable information on embryonic SAM and cotyledon formation. However, the lack of cotyledon specific mutants is conspicuous. Mutants in the *GURKE, PASTICCINO1* and *PEPINO* (also *PASTICCINO2*) genes affect cotyledon formation, but seem to control proliferation processes especially of aerial structures (Torres Ruiz *et al.*, 1996a,b; Faure *et al.*, 1998). The cloning of *PASTICCINO1* and *PEPINO* indicates that these are members of a highly conserved pathway (Vittotrioso *et al.*, 1998; Haberer *et al.*, 2002; Bellec *et al.*, 2002). Thus, at present, there is no comprehensive molecular model for the organisation of the early embryonic shoot apical pole.

6.9 Polarity of the basal embryonic region

While the polarity of the shoot pole becomes evident during cotyledon generation the opposite end elaborates a morphological polarity much earlier. These processes generate a distinctive layered group of cells constituting the RAM region whose activity produces the root post-embryonically. (See Chapter 7 for details of root meristem development.)

The course of cell divisions in the embryo show that this region is built up by two clonally different cell lines (lower tier and hypophysis; see above) whose integration is thought to be co-ordinated initially by signals emanating from the pro-embryo (discussed in Jürgens, 2001). Their influence directs the unequal division of the hypophysis to give the lens-shaped cell (the precursor of the quiescent centre cells (QC)) and initials for the central columella cells (Jürgens, 2001). This proposition is based on the fact that mutants of the auxin response factor 5 (ARF5) gene *MONOPTEROS (MP),* the MP-inhibiting IAA protein encoded by *BODENLOS* (*BDL*), and *AXR6* affect the embryonic root but not root formation *per se* (Fig. 6.8). In the *mp* mutant, postembryonic rooting can be induced by wounding (Berleth & Jürgens, 1993). Also *mp* and *bdl* exhibit abnormalities in the pro-embryo before the hypophysis shows any deviation (Berleth & Jürgens, 1993; Hamann *et al*., 1999). The core of the root region is taken up by the four quiescent centre cells, which exhibit a marked polarity in several respects. First, these cells will rarely, if at all, undergo further cell divisions in embryonic and postembryonic development. Second, these cells seem to maintain the undifferentiated state of surrounding initials for all other root tissues (Fig. 6.14). Though the corresponding experiments have been performed only on postembryonic roots (van den Berg *et al*., 1997) the division behaviour and expression of cell markers indicate that this role might function already at heart stage embryo. However, QC does not instruct the initials to produce daughter cells with distinct fates. This information seems to come from cells already differentiated (Fig. 6.15; van den Berg, 1995). A third characteristic is manifested as a strong accumulation of auxin within QC cells, as seen in embryos transgenic with the DR5::GUS construct which indicates auxin accumulation (Sabatini *et al*., 1999). The link to auxin is also stressed by the fact that phenotypes of the embryonic root result, following the mutation of genes related to auxin sensing or response, i.e. *MP* (Hardtke & Berleth, 1998), *BDL* (Hamann *et al*., 1999; Hamann *et al*., 2002) and *AXR6* (Hobbie *et al*., 2000). However, these genes seem to be further involved in global axialisation and tracheid differentiation (see Chapters 7 and 9). Surprisingly, the activities of these genes might also comprise a support of *CUC1* and *2* expression (Aida *et al*., 2002).

The application of high doses of auxin or auxin transport blockers cause dramatic morphogenetic alterations, and in extreme cases the complete reorganisation of the root axis (Sabatini *et al*., 1999). This indicates the

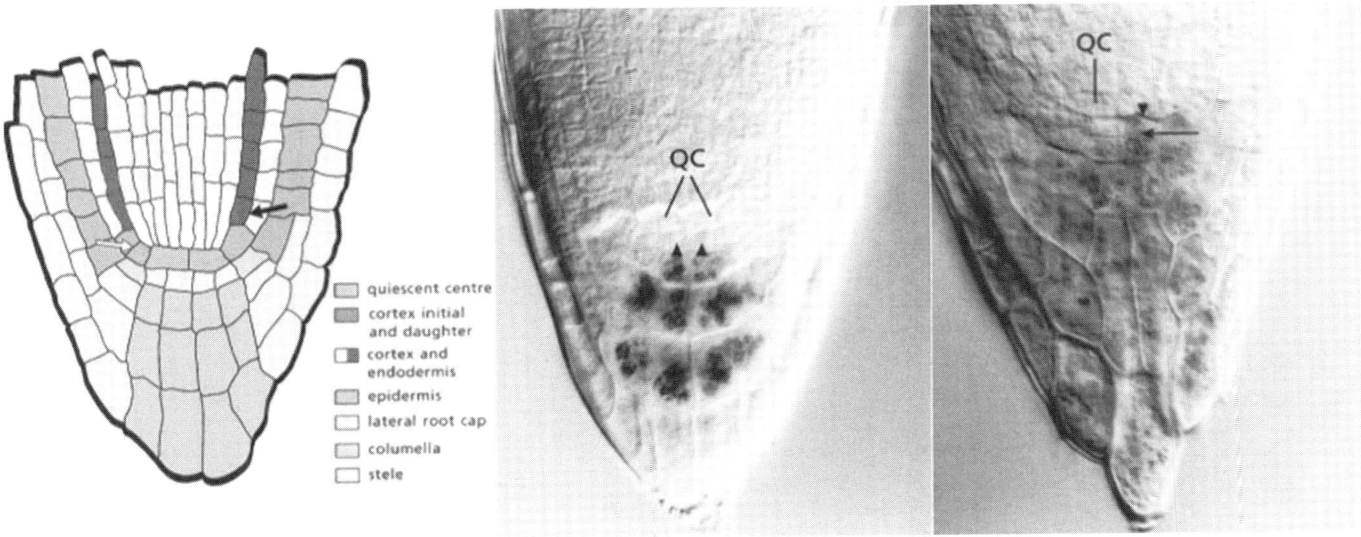

Figure 6.14 Short range control of cell (un-)differentiation. Left: Schematic representation of cell types in the *Arabidopsis* root. Middle: Starch granules (darkened cells with dark granules) indicate differentiated columella cells. Columella initials are not differentiated and lack starch granules (arrowheads). Right: Ablation eliminated one QC cell (arrowhead). Columella initial beneath former QC cell has differentiated to columella cell (arrow). From van den Berg *et al.* (1997); Nature Publishing Group, with permission.

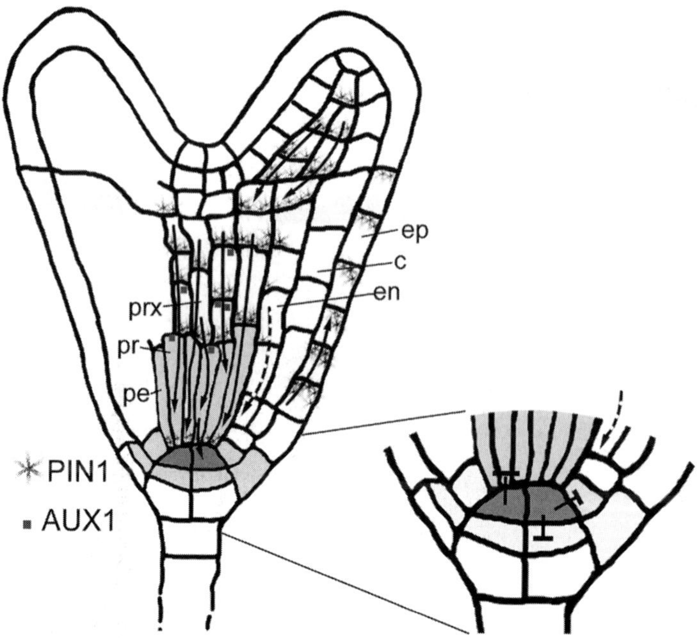

Figure 6.15 Schematic summary of polarity at the basal embryonic pole. QC cells (dark grey) accumulate high concentrations of auxin which is transported (arrows) with the aid of asymmetrically localised PIN1 (star shaped spots). PIN1 epidermal localisation, suggest auxin transport in opposite direction. Note the concentration of PIN1 in cotyledons whose tips might provide auxin. The QC cells maintain adjacent initial cells in an undifferentiated status (block arrows). Stippled arrow indicates signalling of differentiated cells, which is required for correct unequal division of cortical initial daughter cells thus enabling differentiation into endodermis and cortex. Note the included polar localisation of AUX1 (an auxin permease) only given for completion. AUX1 has not been analysed in embryos in detail.

prominent role of correct auxin flow and distribution as suggested by PIN1 localisation and function as well as by other PIN-carriers or permeases such as AUX1 (Swarup *et al.*, 2001; Friml *et al.*, 2002a,b). The latter has recently been shown to adopt asymmetric localisation with reference to PIN1 in protophloem and represents a second auxin transport pathway (Swarup *et al.*, 2001). It remains open whether auxin itself is the morphogenetic signal or simply the mediator of such signals, which now are evidently acting in different regions of the embryo. A recent report on the *topless* mutant shows that auxin is not required for all cases of root specification. *topless* mutants sometimes elaborate an apical root instead of a SAM and cotyledons without exhibiting an auxin maximum as indicated by DR5::GUS expression (Fig. 6.16; Long *et al.*, 2002; Sabatini *et al.*, 1999). This is reminiscent of the embryonic root dependence on genes such as *MP* but post-embryonic independence from the same and poses the question: which sort of polarity will show up in these secondary roots?

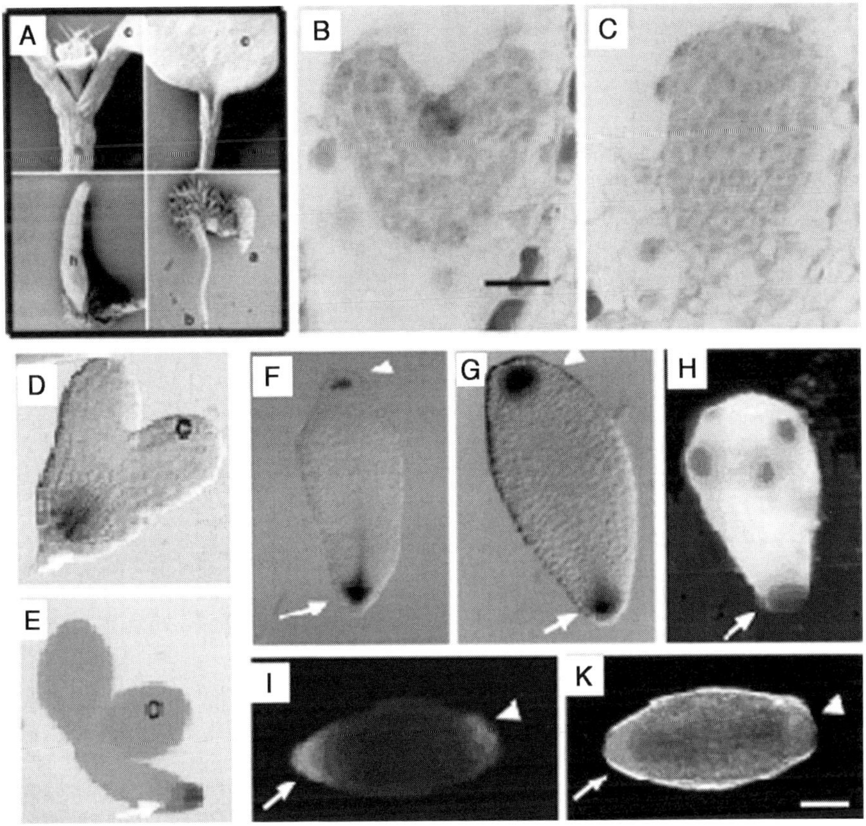

Figure 6.16 Inversion of polarity in the *topless* mutant. (A): Wild-type seedling (upper left), fused cotyledon seedling (upper right), *topless* seedling (lower left), double root seedling (lower right). (B): STM in wild-type heart stage (in situ hybridisation; scale barr 25μm). (C): *topless* embryo (same *in situ* experiment). (D–G): root specific marker *LENNY* localises to root poles in wild-type embryo (D), mature stage embryo (E), *topless* embryos (F and G). (H): rare case where the QC and columella marker DR5::GUS, which indicates auxin maxima, appears at both poles of topless embryo (one pole always shows a patchy distribution). (I and K): Localisation of the root cap marker *J1092* (K superimposition on Nomarski foto, scale bar 50μm). From Long *et al.* (2002), Company of Biologists Ltd. with permission.

6.10 Conclusions

The foregoing discussion showed that the embryo elaborates from numerous polarities, which serve as archimedic points for the generation of a patterned body, which itself is a reference for post-embryonic development. At present the focus of investigation centres on several aspects.

At the cellular level, ongoing work aims to elucidate how cell polarities are established and how they are translated into asymmetric cell fates. The role of the cytoskeleton and further details of vesicle generation and transport are of major importance in this context (see Chapters, 1, 2, 4 and 5). For instance, the alteration of sterol biosynthesis and its disturbed incorporation into membranes seems to profoundly affect morphogenesis and cell polarity by interfering with vesicle trafficking. This can be observed in the mutants *cvp-1, fackel (fk), hydra1 (hyd1)* and *sterol methyltransferase1* (*smt1*); Diener *et al.*, 2000; Schrick *et al.*, 2000; Carland *et al.*, 2002; Souter *et al.*, 2002; Willemsen *et al.*, 2003). The next question is how cellular polarities once established are maintained, transferred and perpetuated to daughter cells. How do some of them remain localised? Understanding how hormones are distributed to exert their effects is emerging (Friml *et al.*, 2002a,b), but will only provide part of the answer. Auxin (or auxin maxima) alone is not sufficient to explain root development as seen in *topless* mutants which tells us that other clues have to be unravelled until we understand inverted polarities. However, advances in these fields might significantly help us understand the similarities and differences between the SAM and RAM. Both seem to operate in a comparable fashion, whereby a fixed centre induces stem cell fate (see Chapters 7 and 8). Interestingly, the *Arabidopsis* CLV homolog CLV40 is ubiquitously expressed in root and other tissues and is able to replace CLAVATA3 (Hobe *et al.*, 2003). The similarity of SAM and RAM extends to the fact that both cores, WUS expressing cells and QC, are adjacent to vascular precursors (see above). In the embryo, this links the development of meristem precursors to vascular cells (see Chapter 9), which is the core of the radial pattern. Is radial polarity a reference for apical basal polarity? The robustness of radial pattern and polarity is intriguing in *gurke* and *gnom* mutants which completely lack either apical or basal or both structures (Mayer *et al.*, 1993; Torres Ruiz *et al.*, 1996b). If apical-basal polarity refers to radial, the next reference might be the octant stage. In the end we will arrive where we started – at the single cell.

Acknowledgements

I would like to thank the following colleagues for support and providing figures: Kay Schneitz, Gerd Jürgens, Ulrike Mayer, M. Yanofsky, Philip Benfey, Hidehiro Fukaki, Holger Breuninger, Enno Rikirsch, Thomas Laux, Masahiko Furutani, Masao Tasaka, Yasunori Machida, John Bowman, Ben Scheres, Kathryn Barton and co-workers; special thanks to Alfons Gierl for consistent support of our work and Ulrich Genschel for reading parts of the manuscript. I also thank the following publishing agencies for permissions: Company of Biologists Ltd., Nature Publishing Group, Oxford University Press, Blackwell Publishing, BIOspektrum and American Association for the Advancement of Science.

References

Aida, M., Ishida, T. & Tasaka, M. (1999) Shoot apical meristem and cotyledon formation during *Arabidopsis* embryogenesis: interaction among the *CUP-SHAPED COTYLEDON* and *SHOOT MERISTEMLESS* genes. *Development*, **126**, 1563–1570.

Aida, M., Ishida, T., Fukaki, H., Fujisawa, H. & Tasaka, M. (1997) Genes involved in organ separation in *Arabidopsis*: an analysis of the *cup-shaped cotyledon* mutant. *Plant Cell*, **9**, 841–857.

Aida, M., Vernoux, T., Furutani, M., Traas, J. & Tasaka, M. (2002) Roles *of PIN-FORMED* and *MONOPTEROS* in pattern formation of the apical region of the *Arabidopsis* embryo. *Development*, **129**, 3965–3974.

Alessa, L. & Kropf, D. (1999) F-actin marks the rhizoid pole in living *Pelvetia compressa* zygotes. *Development*, **126**, 201–209.

Barton, M.K. & Poethig, R.S. (1993) Formation of the shoot apical meristem in *Arabidopsis thaliana*: an anlysis of development in the wild-type and in the shoot meristemless mutant. *Development*, **119**, 823–831.

Belanger, K.D. & Quatrano, R.S. (2000) Polarity: the role of localized secretion. *Curr. Opin. Plant Biol.*, **3**, 67–72.

Bellec, Y., Harrar, Y., Butaeye, C., Darnet, S., Bellini, C. & Faure, J.D. (2002) *Pasticcino2* is a protein tyrosine phosphatase-like involved in cell proliferation and differentiation in *Arabidopsis*. *Plant J.*, **32**, 713–22.

Benfey, P.N., Linstead, P.J., Roberts, K., Schiefelbein, J.W., Hauser, M.-Th. & Aeschbacher, R.A. (1993) Root development in *Arabidopsis*: four mutants with dramatically altered root morphogenesis. *Development*, **119**, 57–70

Bennett, S.R.M., Alvarez, J., Bossinger, G. & Smyth, D.R. (1995) Morphogenesis in *pinoid* mutants of *Arabidopsis thaliana*. *Plant J.*, **8**, 505–520.

Berger, F. (1999) Endosperm development. *Curr. Op. Plant Dev.*, **2**, 28–32.

Berger, F., Haseloff, J., Schiefelbein, J. & Dolan, L. (1998) Positional information in root epidermis is defined during embryogenesis and acts in domains with strict boundaries. *Current Biol.*, **8**, 421–430.

Berleth, T. & Chatfield, S. (2002) Embryogenesis: pattern formation from a single cell. In: The *Arabidopsis* book. American Society of Plant Biologists.

Berleth, T. & Jürgens, G. (1993) The role of the *monopteros* gene in organising the basal body region of the *Arabidopsis* embryo. *Development*, **118**, 575–587.

Bowman, J.L. (2000) The *YABBY* gene family and abaxial cell fate. *Curr. Opin. Plant Biol.*, **2**, 17–22.

Brand, U., Fletcher, J.C., Hobe, M., Meyerowitz, E.M. & Simon, R. (2000) Dependence of the stem cell fate in *Arabidopsis* on a feedback loop regulated by CLV3 activity. *Science*, **289**, 817–819.

Brown, R.C., Lemmon, B.E., Nguyen, H. & Olsen, O.-A. (1999) Development of endosperm in *Arabidopsis thaliana*. *Sex. Plant Reprod.*, **12**, 32–34.

Brownlee, C. & Berger, F. (1995) Extracellular matrix and pattern in plant embryos: on the lookout for developmental information. *Trends Genet.*, **11**, 344–348.

Bünning, E. (1952) Morphogenesis in plants. *Surv Biol Progr.*, **2**, 105–140.

Bünning, E. (1957) Polarität und inaquale Teilung des pflanzlichen Protoplasten. *Protoplasmatologia* **VIII**, 9a.

Busch, M., Mayer, U. & Jürgens, G. (1996) Molecular analysis of the *Arabidopsis* pattern formation gene *GNOM*: gene structure and intragenic complementation. *Mol. Gen. Genet.*, **250**, 681–691.

Byrne, M.E., Barley, R., Curtis, M., Arroyo, J.M., Dunham, M., Hudson, A. & Martienssen, R.A. (2000) *Asymmetric leaves1* mediates leaf patterning and stem cell function in *Arabidopsis*. *Nature*, **408**, 967–971.

Carland, F.M., Fujioka, S., Takatsuto, S., Yoshida, S. & Nelson, T. (2002) The identification of CVP1 reveals a role for sterols in vascular patterning. *Plant Cell*, **14**, 2045–2058.

Chaudhury, A.M., Ming, L., Miller, C., Dennis, E.S. & Peacock, W.J. (1997) Fertilization-independent seed development in *Arabidopsis thaliana*. *Proc. Natl. Acad. Sci. USA*, **94**,

4223–4228.

Christensen, S.K., Dagenais, N., Chory, J. & Weigel, D. (2000) Regulation of auxin response by the protein kinase *PINOID*. *Cell*, **100**, 469–478.

Conway, L.J. & Poethig, R.S. (1997) Mutations of *Arabidopsis thaliana* that transform leaves into cotyledons. *Proc. Natl. Acad. Sci. USA*, **94**, 10209–10214.

Di Laurenzio, L., Wysocka-Diller, J., Malamy, J.E., Pysh, L., Helariutta, Y., Freshour, G., Hahn, M.G., Feldmann, K.A. & Benfey, P.N. (1996) The *SCARECROW* gene regulates an asymmetric cell division that is essential for generating the radial organization of the *Arabidopsis* root. *Cell*, **86**, 423–433.

Diener, A.C., Li, H., Zhou, W., Whoriskey, W.J., Nes, W.D. & Fink, G.R. (2000) Sterol methyltransferase 1 controls the level of cholesterol in plants. *Plant Cell*, **12**, 853–870.

Donaldson, J.G. & Jackson, C.L. (2000) Regulators and effectors of the ARF GTPases. *Curr. Opin. Cell Biol.*, **12**, 475–482.

Duckett, C.M., Oparka, K.J., Prior, D.A.M., Dolan, L. & Roberts, K. (1994) Dye-coupling in the root epidermis of *Arabidopsis* is progressively reduced during development. *Development*, **120**, 3247–3255.

Elliott, R.C., Betzner, A.S., Huttner, E., Oakes, M.P., Tucker, W.Q., Gerentes, D., Perez, P. & Smyth, D.R. (1996) *AINTEGUMENTA*, **an** *APETALA-2* like gene in *Arabidopsis* with pleiotropic roles in ovule development and floral organ growth. *Plant Cell*, **8**, 155–168.

Estelle, M. (1998) Polar auxin transport: new support for an old model. *Plant Cell*, **10**, 1775–1778

Farkas, G., Gausz, J., Galloni, M., Reuter, G., Gyurkovics, H. & Karch, F. (1994) The *Trithorax*-like gene encodes the *Drosophila* GAGA factor. *Nature*, **371**, 806–808.

Faure, J-D., Vittorioso, P., Santoni, V., Fraisier, V., Prinsen, E., Barlier, I., van Onckelen, H., Caboche, M. & Bellini, C. (1998) The *PASTICCINO* genes of *Arabidopsis thaliana* are involved in the control of cell division and differentiation. *Development*, **125**, 909–918.

Fletcher, J.C., Brand, U., Running, M.P., Simon, R. & Meyerowitz, E.M. (1999) Signalling of cell fate desicion by CLAVATA-3 in *Arabidopsis* shoot meristems. *Science*, **283**, 1911–1914.

Friml, J., Benkova, E., Blilou, I., Wisniewska, J., Hamann, T., Ljung, K., Woody, S., Sandberg, G., Jürgens, G. & Palme, K. (2002a) *AtPIN4* mediates sink-driven auxin gradients and root patterning in *Arabidopsis*. *Cell*, **108**, 661–673.

Friml, J., Wisniewska, J., Benkova, E., Mendgen, K. & Palme, K. (2002b) Lateral relocation of auxin efflux regulator PIN3 mediates tropism in *Arabidopsis*. *Nature*, **415**, 806–809.

Fukaki, H., Wysocka-Diller, J., Kato, T., Fujisawa, H., Benfey, P.N. & Tasaka, M. (1998) Genetic evidence that the endodermis is essential for root gravitropism in *Arabidopsis thaliana*. *Plant J.*, **14**, 425–430.

Gälweiler, L., Guan, C., Müller, A., Wisman, E., Mendgen, K., Yephremov, A. & Palme, K. (1998) Regulation of polar auxin transport by AtPIN1 in *Arabidopsis* vascular tissue. *Science*, **282**, 2226–2230.

Geldner, N., Anders, N., Wolters, H., Keicher, J., Kornberger, W., Muller, P., Delbarre, A., Ueda, T., Nakano, A. & Jürgens, G. (2003) The *Arabidopsis* GNOM ARF-GEF mediates endosomal recycling, auxin transport, and auxin-dependent plant growth. *Cell*, **112**, 19–30.

Geldner, N., Friml, J., Stierhof, Y.-D., Jürgens, G. & Palme, K. (2001) Auxin transport inhibitors block PIN1 cycling and vesicle trafficking. *Nature*, **413**, 425–428.

Goldberg, R.B., de Paiva, G. & Yadegari, R. (1994) Plant embryogenesis: zygote to seed. *Science*, **266**, 605–614.

Golden, T.A., Schauer, S.E., Lang, J.D., Pien, S., Mushegian, A.R., Grossniklaus, U., Meinke, D.W. & Ray, A. (2002) *SHORT INTEGUMENT1/ SUSPENSOR1/CARPEL FACTORY* a DICER homolog, is a maternal effect gene required for embryo development in *Arabidopsis*. *Plant Physiol.*, **130**, 808–822.

Grossniklaus, U., Vielle-Calzada, J.-P., Hoeppner, M. A. & Gagliano, W.B. (1998) Maternal control of embryogenesis by *MEDEA*, a *Polycomb* group gene in *Arabidopsis*. *Science* 280, 446–450.

Gruenberg, J. (2001) The endocytic pathway: a mosaic of domains. *Nat. Rev. Mol. Cell Biol.*, **2**, 721–730.

Haberer, G., Erschadi, S. & Torres Ruiz, R.A. (2002) The *Arabidopsis* gene *PEPINO/*

PASTICCINO2 is required for proliferation control of meristematic and non-meristematic cells and encodes a putative anti-phosphatase. *Dev. Genes Evol.*, **212**, 542–550.
Hable, W.E. & Kropf, D.L. (2000) Sperm entry induces polarity in fucoid zygotes. *Development*, **127**, 493–501.
Hadfi, K., Speth, V. & Neuhaus, G. (1998) Auxin-induced developmental patterns in *Brassica juncea* embryos. *Development*, **125**, 879–887
Hamann, T., Benkova, E., Baurle, I., Kientz, M. & Jürgens, G. (2002) The *Arabidopsis BODENLOS* gene encodes an auxin response protein inhibiting *MONOPTEROS*-mediated embryo patterning. *Genes Dev.*, **16**, 1610–1615.
Hamann, T., Mayer, U. & Jürgens, G. (1999) The auxin-insensitive *bodenlos* mutation affects primary root formation and apical-basal patterning in the *Arabidopsis* embryo. *Development*, **126**, 1387–1395.
Hardtke, C. & Berleth, T. (1998) The *Arabidopsis* gene *MONOPTEROS* encodes a transcription factor mediating embryo axis formation and vascular development. *EMBO J.*, **17**, 1405–1411.
Helariutta, Y., Fukaki, H., Wysocka-Diller, J., Nakajima, K., Jung, J., Sena, G., Hauser, M. & Benfey, P. (2000) The *SHORT-ROOT* gene controls radial patterning of the *Arabidopsis* root through radial signalling. *Cell*, **101**, 555–567.
Hobbie, L., McGovern, M., Hurwitz, L.R., Pierro, A., Liu, N.Y., Bandyopadhyary, A. & Estelle, M. (2000) The *axr6* mutants of *Arabidopsis thaliana* define a gene involved in auxin response and early development. *Development*, **127**, 23–32.
Hobe, M., Muller, R., Grunewald, M., Brand, U. & Simon, R. (2003) Loss of CLE40, a protein functionally equivalent to the stem cell restricting signal CLV3, enhances root waving in *Arabidopsis. Dev. Genes Evol.*, **213**, 371–381.
Hwang, I. & Sheen, J. (2001) Two-component circuity in *Arabidopsis* cytokinin signal transduction. *Nature*, **413**, 383–389.
Ikeda-Iwai, M., Umehara, M., Satoh, S. & Kamada, H. (2003) Stress-induced somatic embryogenesis in vegetative tissues of *Arabidopsis thaliana. Plant J.*, **34**, 107–14.
Inoue, T., Higuchi, M., Hashimoto, Y., Seki, M., Kobayashi, M., Kato, T., Tabata, S., Shinozaki, K. & Kakimoto T. (2001) Identification of *CRE1* as a cytokinin receptor from *Arabidopsis. Nature*, **409**, 1060–1063.
Irish, V. (1991) Cell lineage in plant development. *Current Opin. Genet. Dev.*, **1**, 169–173.
Irish, V. & Sussex, I.M. (1992) A fate map of the *Arabidopsis* embryonic shoot apical meristem. *Development*, **115**, 745–753.
Iwakawa, H., Ueno, Y., Semiarti, E., Onouchi, H., Kojima, S., Tsukaya, H., Hasebe, M., Soma, T., Ikezaki, M., Machida, C. & Machida, Y. (2002) The *ASYMMETRIC LEAVES2* gene of *Arabidopsis thaliana*, required for formation of a symmetric flat leaf lamina, encodes a member of a novel family of proteins characterized by cysteine repeats and a leucine zipper. *Plant Cell Physiol.*, **5**, 467–478.
Jacobsen, S.E., Running, M.P. & Meyerowitz, E.M. (1999) Disruption of an RNA helicase/RNAseIII gene in *Arabidopsis* causes unregulated cell division in floral meristems. *Development*, **126**, 5231–5243.
Johri, B.M. (ed.) (1984) *Embryology of Angiosperms*. Springer: Berlin, New York, Tokyo.
Johri, B.M., Ambegaokar, K.B. & Srivastava, P.S. (1992) *Comparative Embryology of Angiosperms* Vol 1 & 2, Springer: Berlin.
Jürgens, G. (2001) Apical-basal pattern formation in *Arabidopsis* embryogenesis. *EMBO J.*, **20**, 3609–3616.
Jürgens, G., Mayer, U., Torres Ruiz, R.A., Berleth, T. & Misera, S. (1991) Genetic analysis of pattern formation in the *Arabidopsis* embryo. *Development* (Supplement), **1**, 27–38.
Jürgens, G., Torres Ruiz, R.A. & Berleth, T. (1994) Embryonic pattern formation in flowering plants. *Ann. Rev. Genet.*, **28**, 351–371.
Kaplan, D.R. & Cooke, T.J. (1997) Fundamental concepts in the embryogenesis of ditcotyledons: A morphological interpretation of embryo mutants. *Plant Cell*, **9**, 1903–1919.
Kaya, H., Shibahara, K., Taoka, K., Iwabuchi, M., Stillman, B. & Araki, T. (2001) *FASCIATA* genes for chromatin assembly factor-1 in *Arabidopsis* maintain the cellular organization of apical meristems. *Cell*, **104**, 131–142.
Kerstetter, R.A., Bollman, K., Taylor, R.A., Bomblies, K. & Poethig, R.S. (2001) *KANADI*

regulates organ polarity in *Arabidopsis*. *Nature*, **411**, 706–709.
Kinoshita, T., Yadegari, R., Harada, J.J., Goldberg, R.B. & Fischer, R.L. (1999) Imprinting of the *MEDEA* polycomb gene in the *Arabidopsis* endosperm. *Plant Cell*, **11**, 1945–1952.
Kiyosue, T., Ohad, N., Yadegari, R., Hannon, M., Dinnery, J., Wells, D., Katz, A., Margossian, L., Harada, J.J., Goldberg, R.B. & Fischer, R.L. (1999) Control of fertilization-independent endosperm development by the *MEDEA polycomb* gene in *Arabidopsis*. *Proc. Natl. Acad. Sci. USA*, **96**, 4186–4191.
Klucher, K.M., Chow, H., Reiser, L. & Fischer, R.L. (1996) The *AINTEGUMENTA* gene of *Arabidopsis* required for ovule and female gametophyte development is related to the floral homeotic gene *APETALA2*. *Plant Cell*, **8**, 137–153.
Kumaran, M.K., Bowman, J.L. & Sundaresan, V. (2002) *YABBY* polarity genes mediate the repression of *KNOX* homeobox genes in *Arabidopsis*. *Plant Cell*, **14**, 2761–2770.
Laux, T. & Jürgens, G. (1997) Embryogenesis: a new start in life. *Plant Cell*, **9**, 989–1000.
Lindsey, K. & Topping, J.F. (1993) Embryogenesis: a question of pattern. *J. Exp. Bot.*, **44**, 359–374.
Liu, C.M., Xu, Z.H. & Chua, N.H. (1993) Auxin polar transport is essential for the establishment of bilateral symmetry during early plant embryogenesis. *Plant Cell*, **5**, 621–630.
Long, J.A. & Barton, M.K. (1998) The development of apical embryonic pattern in *Arabidopsis*. *Development*, **125**, 3027–3035.
Long, J.A., Moan, E.I., Medford, J.I. & Barton, M.K. (1996) A member of the *KNOTTED* class of homeodomain proteins encoded by the *STM* gene of *Arabidopsis*. *Nature*, **379**, 66–69.
Long, J.A., Woody, S., Poethig, S., Meyerowitz, E.M. & Barton, M.K. (2002) Transformation of shoots into roots in *Arabidopsis* embryos mutant at the *TOPLESS* locus. *Development*, **129**, 2797–2806.
Lopes, M.A. & Larkins, B.A. (1993) Endosperm origin, development and function. *Plant Cell*, **5**, 1383–1399.
Lotan, T., Ohto, M., Yee, K.M., West, M.A.L., Lo, R., Kwong, R., Yamagishi, K., Fischer, R.L., Goldberg, R.B. & Harada, J.J. (1998) *Arabidopsis LEAFY COTYLEDON1* is sufficient to induce embryo development in vegetative cells. *Cell*, **93**, 1195–1205.
Lu, P., Porat, R., Nadeau, J.A. & O'Neill, S.D. (1996) Identification of a meristem L1 layer-specific gene in *Arabidopsis* that is expressed during embryonic pattern formation and defines a new class of homeobox genes. *Plant Cell*, **8**, 2155–2168.
Lukowitz, W., Mayer, U. & Jürgens, G. (1996) Cytokinesis in the *Arabidopsis* embryo involves the syntaxin-related *KNOLLE* gene product. *Cell*, **84**, 61–71.
Luo, M., Bilodeau, P., Koltunow, A., Dennis, E.S., Peacock, W.J. & Chaudhury, A.M. (1999) Genes controlling fertilization-independent seed development in *Arabidopsis thaliana*. *Proc. Natl. Acad. Sci. USA*, **96**, 296–301.
Luo, Y. & Koop, H.-U. (1997) Somatic embryogenesis in cultured immature zygotic embryos and leaf protoplasts of *Arabidopsis thaliana* ecotypes. *Planta*, **202**, 387–396.
Lynn, K., Fernandez, A., Aida, M., Sedbrook, J., Tasaka, M., Masson, P. & Barton, M.K. (1999) The *PINHEAD/ZWILLE* gene acts pleiotropically in *Arabidopsis* development and has overlapping functions with the *ARGONAUTE1* gene. *Development*, **126**, 469–481.
McConnell, J.R. & Barton, M.K. (1998) Leaf polarity and meristem formation in *Arabidopsis*. *Development*, **125**, 2935–2942.
McConnell, J.R., Emery, J., Eshed, Y., Bao, N., Bowman, J. & Barton, M.K. (2001) Role of *PHABULOSA* and *PHAVOLUTA* in determining radial patterning in shoots. *Nature*, **411**, 709–713.
Mähönen, A.P., Bonke, M., Kauppinen, L., Riikonen, M., Benfey, P.N. & Helariutta, Y. (2000) A novel two-component hybrid molecule regulates vascular morphogenesis of the *Arabidopsis* root. *Genes Dev.*, **14**, 2938–2943.
Mansfield, S.G. & Briarty, L.G. (1990a) Development of free-nuclear endosperm in *Arabidopsis thaliana* L. *Arabid. Inform. Serv.*, **27**, 53–65
Mansfield, S.G. & Briarty, L.G. (1990b) Endosperm cellularization in *Arabidopsis thaliana* L. *Arabid. Inform. Serv.*, **27**, 65–72.
Mansfield, S.G. & Briarty, L.G. (1991) Early embryogenesis in *Arabidopsis thaliana*. II. The developing embryo. *Can. J Bot.*, **69**, 461–476.
Mansfield, S.G., Briarty, L.G. & Erni, S. (1991) Early embryogenesis in *Arabidopsis thaliana*. I

The mature embryo sac. *Can. J Bot.*, **69**, 447–460.
Mascarenhas, J.P. (1993) Molecular mechanisms of pollen tube growth and differentiation. *Plant Cell*, **5**, 1303–1314.
Mayer, K.F.X., Schoof, H., Haecker, A., Lenhard, M., Jürgens, G. & Laux, T. (1998) Role of *WUSCHEL* in regulating stem cell fate in the *Arabidopsis* shoot meristem. *Cell*, **95**, 805–815.
Mayer, U., Büttner, G. & Jürgens, G. (1993). Apical-basal pattern formation in the *Arabidopsis* embryo: studies on the role of the *GNOM* gene. *Development*, **117**, 149–162.
Mayer, U., Torres Ruiz, R.A., Berleth, T., Miséra, S. & Jürgens, G. (1991) Mutations affecting body organization in the *Arabidopsis* embryo. *Nature*, **353**, 402–407.
Meinke, D.W. (1991) Perspectives on genetic analysis of plant embryogenesis. *Plant Cell*, **3**, 857–866.
Meinke, D.W. (1992) A homeotic mutant of *Arabidopsis thaliana* with leafy cotyledons. *Science*, **258**, 1647–50.
Mordhorst, A.P., Toonen, M.A.J. & de Vries, S.C. (1997) Plant embryogenesis. *Crit. Rev. Plant Sci.*, **16**, 535–576.
Mostov, K.E., Verges, M. & Altschuler, Y. (2000) Membrane traffic in polarized epithelial cells. *Curr. Opin. Cell Biol.*, **12**, 483–490.
Moussian, B., Schoof, H., Haecker, A., Jürgens, G. & Laux, T. (1998) Role of the *ZWILLE* gene in the regulation of central shoot meristem cell fate during *Arabidopsis* embryogenesis. *EMBO J.*, **17**, 1799–1809.
Nakajima, K., Sena, G., Nawy, T. & Benfey, P. (2001) Intercellular movement of the putative transcription factor *SHR* in root patterning. *Nature*, **413**, 307–311.
Natesh, S. & Rau, M.A. (1984) The embryo, in *Embryology of Angiosperms* (ed. B.M. Johri), Springer, Berlin, pp 377–443.
Okada, K., Ueda, J., Komaki, M.K., Bell, C.J. & Shimura, Y. (1991) Requirement of the auxin polar transport system in early stages of *Arabidopsis* floral bud formation. *Plant Cell*, **3**, 677–684.
Otsuga, D., DeGuzman, B., Prigge, M.J., Drews, G.N. & Clark, S.E. (2001) *REVOLUTA* regulates meristem initiation at lateral positions. *Plant J.*, **25**, 223–236.
Palme, K. & Gälweiler, L. (1999) PIN-pointing the molecular basis of auxin transport. *Curr. Opin. Plant Biol.*, **2**, 375–381.
Prigge, M.J. & Wagner, D.R. (2001) The *Arabidopsis SERRATE* gene encodes a zinc-finger protein required for normal shoot development. *Plant Cell*, **13**, 1263–1279.
Pysh, L., Wysocka-Diller, J., Camilleri, C., Bouchez, D. & Benfey, P. (1999) The GRAS gene family in *Arabidopsis*: Sequence characterization and basic expression analysis of the *SCARECROW-LIKE* genes. *Plant J.*, **18**, 111–119.
Raghavan, V. (1986) *Embryogenesis in Angiosperms. A developmental and experimental study.* Cambridge University Press: Cambridge.
Raghavan, V. (2000) *Developmental Biology of Flowering Plants.* Springer: New York.
Ray, S., Golden, T. & Ray, A. (1996) Maternal effects of the *short integument* mutation on embryo development in *Arabidopsis. Dev. Biol.*, **180**, 365–369.
Redei, G.P. & Hirono, Y. (1964) Linkage studies. *Arabidopsis Inf. Serv.*, **1**, 9–10.
Reinhardt, D., Mandel, T. & Kuhlemeier, C. (2000) Auxin regulates the initiation and radial position of plant lateral organs. *Plant Cell*, **12**, 507–518.
Russell, S.D. (1993) The egg cell: Development and role in fertilization and early embryogenesis. *Plant Cell*, **5**, 1349–1359.
Rutishauser, A. (1969) *Embryologie und Fortpflanzungsbiologie der Angiospermen.* Springer: New York, Vienna.
Sabatini, S., Beis, D., Wolkenfelt, H., Murfett, J., Guilfoyle, T., Malamy, J., Benfey, P., Leyser, O., Bechtold, N., Weisbeek, P. & Scheres, B. (1999) An auxin-dependent distal organizer of pattern and polarity in the *Arabidopsis* root. *Cell*, **99**, 463–472.
Sawa, S., Watanabe, K., Goto, K., Kanaya, E., Morita, E.H. & Okada, K. (1999) *FILAMENTOUS FLOWER* a meristem and organ identity gene of *Arabidopsis* encodes a protein with a zinc finger and HMG-related domains. *Genes Dev.*, **13**, 1079–1088.
Scheres, B., Di Laurenzio, L., Willemsen, V., Hauser, M.-T., Janmaat, K., Weisbeek, P. & Benfey, P. (1995) Mutations affecting the radial organisation of the *Arabidopsis* root display specific defects throughout the embryonic axis. *Development*, **121**, 53–62.

Scheres, B., Wolkenfelt, H., Willemsen, V., Terlouw, M., Lawson, E., Dean, C. & Weisbeek, P. (1994) Embryonic origin of the *Arabidopsis* primary root meristem initials. *Development*, **120**, 2475–2487.

Schmidt, E.D., Dejong, A.J. & de Vries, S.C. (1994) Signal molecules involved in plant embryogenesis. *Plant Mol. Biol.*, **26**, 1305–1313.

Schnittger, A., Jürgens, G. & Hülskamp, M. (1998) Tissue layer and organ specificity of trichome formation are regulated by *GLABRA1* and *TRIPTYCHON* in *Arabidopsis*. *Development*, **125**, 2283–2289.

Schoof, H., Lenhard, M., Haecker, A., Mayer, K.F.X., Jürgens, G. & Laux, T. (2000) The stem cell population of *Arabidopsis* shoot meristems is maintained between a regulatory loop between the *CLAVATA* and *WUSCHEL* genes. *Cell*, **100**, 635–644.

Schrick, K., Mayer, U., Horrichs, A., Kuhnt, C., Bellini, C., Dangl, J., Schmidt, J. & Jürgens, G. (2000) FACKEL is a sterol C-14 reductase required for organized cell division and expansion in *Arabidopsis* embryogenesis. *Genes Dev.*, **14**, 1471–1484.

Schulz, R. & Jensen, W.A. (1968) *Capsella* embryogenesis: the egg, zygote and young embryo. *Am. J Bot.*, **55**, 807–819.

Schwartz, B.W., Yeung, E.C. & Meinke, D.W. (1994) Disruption of morphogenesis and transformation of the suspensor in abnormal suspensor mutants of *Arabidopsis*. *Development*, **120**, 3235–3245.

Semiarti, E., Ueno, Y., Tsukaya, H., Iwakawa, H., Machida, C. & Machida, Y. (2001) The *ASYMMETRIC LEAVES2* gene of *Arabidopsis thaliana* regulates formation of a symmetric lamina, establishment of venation and repression of meristem-related homeobox genes in leaves. *Development*, **128**, 1771–1783.

Sessions, A., Weigel, D. & Yanofsky, M.F. (1999) The *Arabidopsis thaliana MERISTEM LAYER 1* promoter specifies epidermal expression in meristems and young primordia. *Plant J.*, **20**, 259–263.

Shaw, S.L. & Quatrano, R.S. (1996) The role of targeted secretion in the establishment of cell polarity and the orientation of the division plane in *Fucus* zygotes. *Development*, **122**, 2623–2630.

Shevell, D.E., Leu, W.-M., Gillmor, C.S., Xia, G., Feldmann, K.A. & Chua, N.-H. (1994) *EMB30* is essential for normal cell division, cell expansion, and cell adhesion in *Arabidopsis* and encodes a protein that has similarity to Sec7. *Cell*, **77**, 1051–1062.

Siegfried, K.R., Eshed, Y., Baum, S.F., Otsuga, D., Drews, G.N. & Bowman, J.L. (1999) Members of the *YABBY* gene family specify abaxial cell fate in *Arabidopsis*. *Development*, **126**, 4117–4128.

Souter, M. & Lindsey, K. (2000) Polarity and signalling in plant embryogenesis. *J. Exp. Biol.*, **51**, 971–983.

Souter, M., Topping, J., Pullen, M., Friml, J., Palme, K., Hackett, R., Grierson, D. & Lindsey, K. (2002) *hydra* mutants of *Arabidopsis* are defective in sterol profiles and auxin and ethylene signaling. *Plant Cell*, **15**, 1017–1031.

St. Johnston, D. (1995) The intracellular localization of messenger RNAs. *Cell*, **81**, 161–170.

St. Johnston, D. & Nüsslein-Volhard, C. (1992) The origin of pattern and polarity in the *Drosophila* embryo. *Cell*, **68**, 201–219.

Steeves, T.A. & Sussex, I.M. (1989) *Patterns in Plant Development*. Cambridge University Press: Cambridge.

Steinmann, T., Geldner, N., Grebe, M., Mangold, S., Jackson, C.L., Paris, S., Gälweiler, L., Palme, K. & Jürgens, G. (1999) Coordinated polar localization of auxin efflux carrier PIN1 by GNOM ARF GEF. *Science*, **286**, 316–318.

Stone, S.L., Kwong, L.W., Yee, K.M., Pelletier, J., Lepiniec, L., Fischer, R.L., Goldberg, R.B. & Harada, J.J. (2001) *LEAFY COTYLEDON2* encodes a B3 domain transcription factor that induces embryo development. *Proc. Natl. Acad. Sci. USA*, **98**, 11806–11811.

Sussex, I.M. (1954) Experiments on the cause of dorsiventrality in leaves. *Nature*, **174**, 351–352.

Swarup, R., Friml, J., Marchant, A., Ljung, K., Sandberg, G., Palme, K. & Bennet, M. (2001) Localisation of the auxin permease AUX1 suggests two functionally distinct hormone transport pathways operate in the *Arabidopsis* root apex. *Genes Dev.*, **15**, 2648–2653.

Takada, S., Hibara, K.-I., Ishida, T. & Tasaka, M. (2001) The *CUP-SHAPED COTYLEDON1* gene of *Arabidopsis* regulates shoot meristem formation. *Development*, **128**, 1127–1135.

Topping, J.F. & Lindsey, K. (1997) Promoter trap markers differentiate structural and positional componants of polar development in *Arabidopsis*. *Plant Cell*, **9**, 1713–1725.
Torres Ruiz, R.A. (1998) *Embryogenesis*, in *Arabidopsis* (eds. J.A. Roberts & M. Anderson). Annual Plant Reviews Vol. 1., Sheffield Academic Press: Sheffield, pp. 223–261.
Torres Ruiz, R.A. (2000) Molekulare Entwicklungsbiologie des *Arabidopsis*-Embryos. *BIOspektrum*, **4**, 267–270.
Torres Ruiz, R.A. & Jürgens, G. (1994) Mutations in the FASS gene uncouple pattern formation and morphogenesis in *Arabidopsis* development. *Development*, **120**, 2967–2978.
Torres Ruiz, R.A., Fischer, T. & Haberer, G. (1996a) Genes involved in the elaboration of apical pattern and form in *Arabidopsis thaliana*: Genetic and molecular analysis, in *Embryogenesis: The Generation of a Plant.* (eds. E. Cummings & T. Wang). Bios Scientific Publishers: Oxford, pp. 15–34.
Torres Ruiz, R.A., Lohner, A. & Jürgens, G. (1996b) The *GURKE* gene is required for normal organization of the apical region in the *Arabidopsis* embryo. *Plant J.*, **10**, 1005–1016.
Trotochaud, A.E., Jeong, S. & Clark, S.E. (2000) CLAVATA3, a multimeric ligand for the CLAVATA1 receptor- kinase. *Science*, **289**, 613–617.
van den Berg, C., Willemsen, V., Hage, W., Weisbeek, P. & Scheres, B. (1995) Cell fate in the *Arabidopsis* root meristem determined by directional signalling. *Nature*, **378**, 62–65.
van den Berg, C., Willemsen, V., Hendriks, G., Weisbeek, P. & Scheres, B. (1997) Short-range control of cell differentiation in the *Arabidopsis* root meristem. *Nature*, **390**, 287–289.
Vernoux, T., Kronenberger, J., Grandjean, O., Laufs, P. & Traas, J. (2000) *PIN-FORMED1* regulates cell fate at the periphery of the shoot apical meristem. *Development*, **127**, 5157–5165.
Vielle-Calzada, J.P., Baskar, R. & Grossniklaus, U. (2000) Delayed activation of the paternal genome during seed development. *Nature*, **404**, 91–94.
Vielle-Calzada, J.P., Baskar, R. & Grossniklaus, U. (2001) Early paternal gene activity in *Arabidopsis* (response). *Nature*, **414**, 710.
Vielle-Calzada, J.P., Thomas, J., Spillane, C., Coluccio, A., Hoeppner, M.A. & Grossniklaus, U. (1999) Maintenance of genomic imprinting at the *Arabidopsis medea* locus requires zygotic *DDM1* activity. *Genes Dev.*, **13**, 2971–2982.
Vittorioso, P., Cowling, R., Faure, J.-D., Caboche, M. & Bellini, C. (1998) Mutation in the *Arabidopsis PASTICCINO1* gene, which encodes a new FK506-binding protein-like protein, has a dramatic effect on plant development. *Mol Cell Biol.*, **18**, 3034–3043.
Vroemen, C.W., Langeveld, S., Mayer, U., Ripper, G., Jürgens, G., Van Kammen, A. & de Vries, S. (1996) Pattern formation in the *Arabidopsis* embryo revealed by position-specific lipid transfer protein gene expression. *Plant Cell*, **8**, 783–791.
Waites, R. & Hudson, A. (1995) *phantastica*: a gene required for dorsiventrality of leaves in *Antirrhinum majus*. *Development*, **121**, 2143–2154.
Wardlaw, C.W. (1955) *Embryogenesis in Plants.* Methuen: London.
Weijers, D., Geldner, N., Offringa, R. & Jürgens, G. (2001) Early paternal gene activity in *Arabidopsis*. *Nature*, **414**, 709–710.
West, M.A.L. & Harada, J.J. (1993) Embryogenesis in higher plants: an overview. *Plant Cell*, **5**, 1361–1369.
Wilhelmi, L.K. & Preuss, D. (1996) Self-sterility in *Arabidopsis* due to defective pollen tube guidance. *Science*, **274**, 1535–1537.
Willemsen, V., Friml, J., Grebe, M., Van Den Toorn, A., Palme, K. & Scheres, B. (2003) Cell polarity and PIN protein positioning in *Arabidopsis* require *STEROL METHYL-TRANSFERASE1* function. *Plant Cell*, **15**, 612–625.
Willemsen, V., Wolkenfelt, H., de Vrieze, G. Weisbeek, P. & Scheres, B. (1998) The *HOBBIT* gene is required for formation of the root meristem in the *Arabidopsis* embryo. *Development*, **125**, 521–531.
Woodrick, R., Martin, P.R., Birman, I. & Pickett, F.B. (2000) The *Arabidopsis* embryonic shoot fate map. *Development*, **125**, 521–531.
Zhang, J.Z. & Somerville, C.R. (1997) Suspensor derived polyembryony caused by altered expression of *valyl tRNA synthetase* in the *twn2* mutant of *Arabidopsis*. *Proc. Natl. Acad. Sci. USA*, **94**, 7349–7355.

7 Polarity in roots

Peter W. Barlow, Dieter Volkmann and František Baluška

7.1 Introduction

The introduction of a growth differential upon the surface of a sphere creates the possibility of change to some other geometrical form. This breaking of spherical symmetry is the situation in many cases of early embryogenesis in plants, especially at the time when the pro-embryo and globular embryo advance to the heart- and torpedo-shaped stages. Two separate, more rapidly growing zones arise just below the surface of the spherical early embryo, marking out the sites of the future shoot and root apices. The more slowly growing portion which separates the apices becomes the hypocotyl. The presence of this organ defines not only the growth axis of the embryo but also its organographic polarity (shoot-hypocotyl-root) into which is built a root-directed polarity of auxin transport (Fry & Wangermann, 1976; see Chapter 6). It is the coupling of this physiological asymmetry to the geometry of the system that leads to a polarisation of growth and development.

When the radicle emerges at germination, it comprises a meristematic zone where cells divide. Although the extension growth associated with this division activity is not particularly rapid, it accelerates to a maximum just behind the meristem and then declines to zero in the more basal zone of the root. By the time the cells of the root cease elongating, most of the cell differentiation associated with the primary root tissues is nearing completion. Development of these tissues is associated with what here is termed the 'primary polarity' of the root, characterised by an elongating root tip and non-elongating basal portion. Not only can this primary polarity be detected in the various physico-chemical properties surrounding a root (Miller & Gow, 1988), as well as in the siting of zones of new growth (lateral roots, root nodules), but it can also be inferred at an ethological level from the feeding pattern of certain rodent species who show preference for either the root tip or the root base (Burda *et al.*, 1999).

Of great interest for studies of polarity are the sites along the root where primary meristematic growth is re-established, leading to the development and emergence of lateral roots with axes of growth perpendicular to that of their parent root. The sites of lateral root primordium initiation break an existing primary polarity and establish a new locality of primary polarity. And even within the non-elongating primary tissues of many species it is possible for yet another type of growth – secondary growth – to develop in association with a secondary meristem, or cambium. It is accompanied by the formation of secondary tissues. As a result, a 'secondary polarity' comes into existence which is orthogonal (i.e., in the radial and circumferential planes) to the primary axial polarity upon which it is superimposed.

The key to both the establishment and the resetting of polarity is the orientation and reorientation of the 'plant cell body' resident in every cell (Baluška *et al.*, 1998, 2001a). The cell body is composed of a nucleus and a complement of perinuclear microtubules inherently associated with the nuclear surface (during interphase) and mitotic chromosomes (during mitosis). The cell body interacts with, as well as organises, the actin cytoskeleton and membranous organelles, primarily those derived from the endoplasmic reticulum. This multifunctional cellular apparatus is responsible for the execution of mitosis, cytokinesis, cell wall synthesis and, as we shall argue, for intra- and intercellular transport of auxin. Its reorientation is the result of an inherent sensitivity to the immediate environment. The consequence of this becomes especially apparent when there is a switch in the plane of cell division, from anticlinal, say, to periclinal, as happens during early embryogenesis and initiation of lateral root primordia. The realisation that the cell body reorients like a compass needle in response to supracellular directional stimuli, and that it possesses its own intrinsic, growth-directed activities, which include support of polar auxin transport, all helps provide a rational framework concerning organ and cell polarity in plants.

This chapter discusses two important aspects of polarity in relation to the primary tissues of roots: the subcellular structures, particularly those of the cell body and its associated cytoskeletal system of actin filaments and endomembranes, which maintain the primary polarity of actively growing root cells and the establishment of new sites of polarised growth, focusing on the initiation of lateral root primordia and their transformation into emerging lateral roots via the reorientation of cell bodies. What is known of the secondary polarity of secondary meristems (cambia) in roots and the tissues to which they give rise, is also discussed below. Although less well studied, secondary polarity may utilise much of the apparatus responsible for primary polarity. We now present a general hypothesis for polarised root growth centred on actin filaments and cell-to-cell sites (plant synapses) specialised for the polar transmission of auxin.

7.2 Polarity within primary root tissues is an expression of the actin cytoskeleton and of polar auxin transport

The polarity of root cells and their cell bodies conforms to that same polarity which shapes the whole plant body. Evidence of polarity appears during early embryogenesis (Grebe *et al.*, 2001; Jürgens, 2001) with the division of the zygote into two daughter cells which differ in fate, aligned along the future apical-basal axis of the plant body. Even prior to fertilisation there is an inherent asymmetry of the embryo sac, aspects of which are derived from asymmetric distribution of the F-actin cytoskeleton (Pónya & Barnabás, 2001; and for a review of how embryonic asymmetry impinges on root development, see Barlow, 2002). In fact, any cell body derived by division of a parent cell body has an inherent asymmetry because, at its division, one end was associated with the phragmoplast and the other end with one pole of the mitotic spindle. Moreover, the nucleus is asymmetric on account of the telophase configuration of the chromatids. Not only does the cytoplasm bear an asymmetry but so, too, do the walls at either end of the cell since these differ in age. Some, or all, of these features may account for why *in vitro* protoplast-derived embryoids have an inherent asymmetry (Luo & Koop, 1997), protoplasts being considered the cell type least likely to be imprinted with polarising information. Successive divisions of the embryo or embryoid consequently propagate the inherent polarity of the zygote or mother cell on account of the self-generating asymmetries of the newly divided cell bodies. At the same time, however, the propagation process admits the formation of new sites of polarity at what will become branching points along the plant body. Thus, also of fundamental interest are the conditions and molecules that are instrumental in breaking a pre-existing polarity of plant cell lineages.

Older studies of plant polarity gave primacy to the polar transport of auxin. As will be discussed, there is now mounting evidence that this trans-organ phenomenon is underpinned, at the level of the individual cell, by processes localised within a dynamic actin-based cytoskeleton and a vesicle-trafficking system based within the cell body. Together with strategically sited scaffolds of signalling proteins, these three items (actin, vesicles and signalling proteins) represent the three major modules which generate and maintain polarity throughout the eukaryotic super-kingdom (Nelson, 2003). In plants, these three modules interact to move auxin (indole-3-acetic acid, IAA), a mobile plant-specific signalling molecule of ancient origin (Cooke *et al.*, 2002).

Auxin was originally considered to be a plant growth 'hormone', a concept first proposed by Fitting (1910) and thereafter pursued largely by analogy with animal hormones (Thimann, 1988). Auxin has also been regarded as a 'morphogen' (Warren Wilson & Warren Wilson, 1984), though the concept of a morphogen as an organ-determining substance, can be traced back to ideas of Julius von Sachs in the nineteenth century. Its role as morphogen rests on the theory that auxin, by its presence, serves as one of the preconditions for the activation of a particular pathway of plant development. The history of these ideas indicates a continual refinement of the observational level at which this molecule is studied: the auxin-as-hormone concept emerges from the whole plant level and concerns the integration of shoot and root growth, whereas the auxin-as-morphogen concept emerges from the organ/tissue level and involves cell differentiation and development. Recent discoveries at the sub-cellular and molecular level invoke yet another appraisal of auxin action, and the search for further correspondences, and even homologies, with other living systems. The interactions of an actin-based cytoskeleton and vesicle-mediated membrane trafficking with the processes of intra- and intercellular auxin movement illustrate this new approach since it brings the plant auxin concept close to the neurotransmitter concept encountered in animal systems (see Lodish *et al.*, 2000). It is on this basis – i.e., roots as organs which possess a quasi-neurotransmitter system – that the generation and maintenance of polarity in root tissues can be fruitfully discussed (Baluška *et al.*, 2003a,b).

Insight into the neurotransmitter-like world of auxin transport is provided by the *gnom* mutants of *Arabidopsis thaliana* (Mayer *et al.*, 1991, 1993). The *gnom* mutations are allelic to *emb30* studied by Shevell *et al.* (1994, 2000), also in *Arabidopsis*. In *gnom/emb30* mutants, the polarity of the premitotic zygote and of the early embryo, including the hypophysis from which the root arises, is severely disturbed. Not only are the embryonic cells of irregular shape, but their usual arrangement into cell files is lacking, suggesting that cellular polarity is disturbed and that cell-to-cell adhesion is affected. The strongest *gnom* line shows much reduced axial growth, resulting in globular plantlets which do, nevertheless, preserve the usual radial patterning of tissues. The *GNOM/EMB30* gene encodes for a brefeldin A-sensitive small G-protein (Steinmann *et al.*, 1999), a molecule which mediates plasma membrane–endosomal recycling of PIN1 protein (Bonifacino & Jackson, 2003; Geldner *et al.*, 2003). PIN1 is a regulator of auxin efflux from cells (see Chapter 2). It is located mainly in vascular parenchyma where it is crucial for the transport of auxin from cell to cell within the longitudinal files of cells (Gälweiler *et al.*, 1998). Thus, the *gnom/emb30* mutation undermines the basis of polar auxin transport and the consequent polarised development of embryos. Moreover, GNOM might affect the recycling of other components of the polar auxin transport machinery, besides PIN1. The PIN1–GNOM system together appear to regulate auxin efflux, and thus act like a part of a plant neurotransmitter system.

7.2.1 Recent insights into actin-auxin feedback interactions driving polarity in root cells

Transversely arranged cortical microtubules (CMTs) underlying the side walls of root cells are essential for developing a characteristic cylindrical cell shape (Baluška *et al.*, 1992, 1993a; Barlow & Baluška, 2000) and for organising the direction of root growth, here referred to as 'axiality'. Even though the CMTs have different arrangements at the end walls compared with those at the side walls, CMTs are not instrumental in generating cell polarity as they cannot define two different ends, or poles, of the cells. True polarity requires that the two poles are distinguished by different molecular assemblies which support distinct processes (Baluška *et al.*, 2003a). Just as neuronal synapses have a unique molecular composition at the presynaptic and postsynaptic termini (Dresbach *et al.*, 2001), the polar nature of plant cells is guaranteed by an auxin-influx molecular system at the one end, and an auxin efflux system at the opposite end (Jones, 1998; Muday & Murphy, 2002; Baluška *et al.*, 2003b). In a cell file, the influx and efflux systems lie on opposite sides of each cross wall. These two systems utilise different carriers: respectively, they involve products of the *AUX1* (influx) and the *PIN1* (efflux) genes (Swarup *et al.*, 2001; Gälweiler *et al.*, 1998). In addition, both influx and efflux of auxin require a dynamic actin cytoskeleton combined with active vesicular membrane trafficking (Baluška *et al.*, 2003a,b). Polar auxin transport thus resembles the cell-to-cell transport of the neurotransmitter, glutamate (Nedergaard *et al.*, 2002), and of the classical morphogen, Dpp, characteristic of animal epithelia (Entchev & González-Gaitán, 2002; Vincent & Dubois, 2002).

What is the evidence that an auxin-transport system, comprising AUX1 and PIN proteins, is an integral component of an already operating complex of actin cytoskeleton and vesicle-based membrane trafficking? First, F-actin is essential for endocytosis (Qualmann *et al.*, 2000; Schafer, 2002). Second, the surfaces of endosomes and endocytic vesicles promote the assembly of F-actin from G-actin by providing the proper micro-environment for this process (Taunton, 2001; Schafer, 2002; Orth & McNiven, 2003). Third, although studies on vesicle recycling in higher plants are still in their infancy, it is becoming clear that intact F-actin is essential for this process (Muday, 2000; Geldner *et al.*, 2001, 2003; Baluška *et al.*, 2002) since recycling is inhibited by agents which disassemble F-actin. These same agents also affect polar transport of auxin (Cande *et al.*, 1973; Butler *et al.*, 1998; Muday, 2000). By contrast, anti-microtubular drugs do not affect polar auxin transport (Hasenstein *et al.*, 1999) and do not inhibit endocytosis (Baluška *et al.*, 2002). Fourth, auxin has been shown to accumulate within vesicular cytoplasmic structures (Shi *et al.*, 1993). It is logical, therefore, to propose that where both auxin and F-actin-related endocytosis co-exist, the auxin-containing vesicles become 'caught up' in the endocytotic stream. From these points of view it is now possible to re-appraise the action of some of the classical inhibitors of polar auxin transport, such as NPA and TIBA. These turn out to inhibit endocytosis and vesicle recycling (Geldner *et al.*, 2001), evidence which strengthens the supposition that polar auxin transport requires endocytotic events.

Inhibition of vesicle recycling by means of brefeldin A (BFA) treatment likewise rapidly halts polar auxin transport (Robinson *et al.*, 1999; Morris, 2000). BFA, however, also affects the organisation of F-actin (Waller *et al.*,

2002). These two observations harmonise with other observations on root growth effects where both polar transport of auxin and the presence of intact F-actin have been shown to be essential for the accomplishment of the rapid elongation of plant cells (Thimann *et al.*, 1992; Baluška *et al.*, 2001b,c; Baluška & Volkmann, 2002). Moreover, both BFA and actin inhibitors block the gravitropic response of root apices (Geldner *et al.*, 2001), a growth event dependent upon the rearrangement of the direction of auxin transport by PIN proteins.

The link between the actin cytoskeleton and polar auxin transport may be mediated by the actin isovariant, ACT7. Of the eight actin genes expressed in *Arabidopsis* (McDowell *et al.*, 1996b), *ACT7* appears to be critical for the organisation of the auxin-transmitting 'plant synapses' (Baluška *et al.*, 2003a,b; see also below) in root apices (Gilliland *et al.*, 2003). *ACT7* is placed in its own subclass of actin genes, its defining feature being that its expression is responsive both to auxin (McDowell *et al.*, 1996a; Kandasamy *et al.*, 2001) and to several environmental signals which affect the polar transport of auxin (McDowell *et al.*, 1996a). This feature is a consequence of the presence of a core auxin-responsive sequence, TGTCTC, in the 5′ untranslated region of the gene (McDowell *et al.*, 1996b). Repeats of this sequence are sufficient for the induction of gene expression via auxin (Ulmasov *et al.*, 1999), and they also form a basis of the *DR5–GUS* construct used for tissue-specific visualisation of auxin-induced gene expression (Sabatini *et al.*, 1999). *act7* mutants show distinct root phenotypes, such as severe reduction of growth, twisting and waving roots, irregular cell files, and the chaotic placement of division walls which mature into aberrant plant synapses (Gilliland *et al.*, 2003). By contrast, above-ground organs are not affected in these *act7* mutants.

As already suggested, besides its role in polar auxin transport, F-actin appears to be essential for the proper assembly of new division walls and for the subsequent formation and maintenance of plant synapses. It is well known that inhibitors of actin polymerisation permit cytokinesis but result in misoriented division walls. The new walls are often inserted obliquely, rather than perpendicularly, upon the side walls of the mother cell (for results on maize roots, see Baluška *et al.*, 2001b; and for a review, see Barlow & Baluška, 2000). Mutants which affect the organisation of the actin cytoskeleton also show aberrant distribution and positioning of division walls (Ramachandran *et al.*, 2000; Dong *et al.*, 2001; Baluška *et al.*, 2001b; Mathur *et al.*, 2003a,b). These features are seen in the roots of three *Arabidopsis* mutants which prescribe the loss of the actin-organising complex, Arp2/3 (Machesky & Gould, 1999). The last-mentioned mutant plants also exhibit disassembly of division-walls in the epidermis, resulting in the loss of individual cell-to-cell contacts (Mathur *et al.*, 2003a,b).

7.2.2 Plant synapses as represented by the actin/myosinVIII-enriched and auxin-transporting cross walls

Asymmetric adhesion domains specialised for rapid cell-to-cell communication are defined as synapses (Dustin & Colman, 2002). This term was originally applied to neurones which inter-communicate using two types of synapse, electrical and chemical. Later, the term 'synapse' was extented to other cell-to-cell contacts, such as between neurons and muscle cells, and even between non-neuronal cells of which the 'immunological synapse' is the best understood (Dustin & Colman, 2002). Recently, the term 'infectious synapse' has been used to describe adhesion domains between monocyte-derived dendritic cells and T cells. These synapses facilitate endocytosis-based cell-to-cell transport of HIV virions (McDonald *et al.*, 2003).

Electrical synapses allow direct electrical coupling between adjacent cells, whereas chemical synapses transform an electrical message into a chemical signal via actin-driven and vesicle-mediated membrane trafficking (Morales *et al.*, 2000; Shupliakov *et al.*, 2002). After receipt of an electrical action potential, which locally increases the level of cytoplasmic calcium, chemical synapses perform a rapid, vesicle-mediated release of neurotransmitters from their presynaptic terminals into the synaptic cleft. Here, the transmitter molecules bind to receptors in the postsynaptic zones and thereby induce a signal. Chemical synapses are characterised by extensive vesicle recycling which is absolutely dependent upon dense meshworks of F-actin. Those cross walls of root apices which are active in auxin transport closely resemble chemical synapses (Baluška *et al.*, 2003a,b). They not only represent local adhesion domains specialised for cell-to-cell communication, but they are also sites where both electrical and chemical signals converge and can be transduced from one form to the other. The cross walls are also traversed by abundant plasmodesmata, channels which allow electrical coupling between cells (Spanswick, 1972) and may serve as electrical mini-synapses. Cross walls are the sites which actively drive the polar transport of auxin, thereby moving this plant neurotransmitter-like molecule from one cell to the next (Geldner *et al.*, 2001, 2003; Baluška *et al.*, 2003a,b).

Visualisation of the actin cytoskeleton within the elongating cells of intact root apices (Parthasarathy & Pesacreta, 1980; Parthasarathy *et al.*, 1985) indicates the potential abundance of such plant synapses and hence the potential transcellular pathway of polar auxin flow (Fig. 7.1). Detailed analysis of cells in maize root apices (Baluška *et al.*, 1997) revealed that a characteristic apical-basal arrangement of F-actin is established in cells as they leave the meristem and enter the transition zone prior to the phase of rapid cell elongation. In the transition zone, the former division walls of the meristematic zone complete their maturation and attain the status of an end wall (with or without their synapse feature, depending on the tissue with

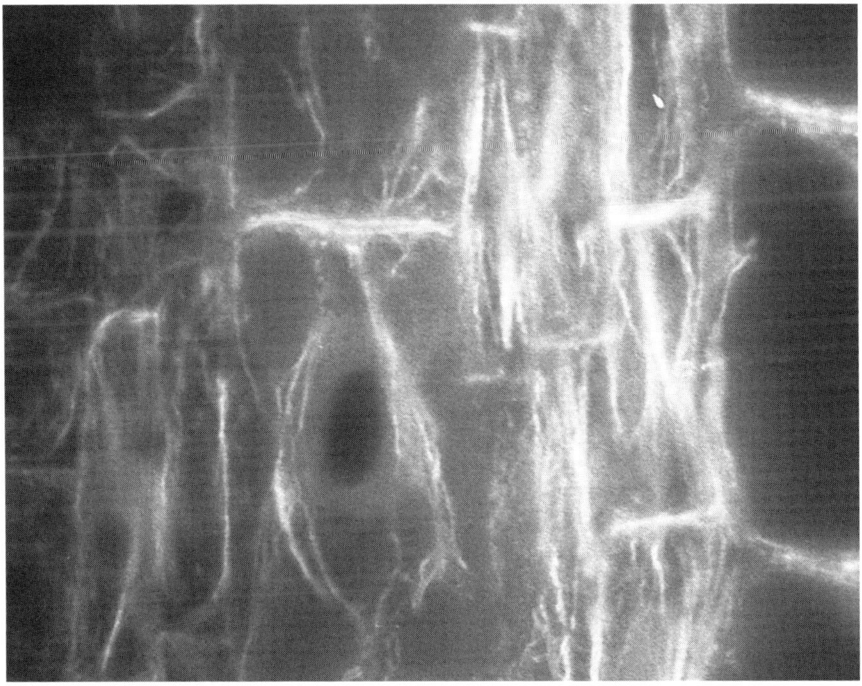

Figure 7.1 Outer stelar cells of a maize primary root, about 1 mm from the apex. Cells were stained with fluorescent anti-actin antibody. This reveals an extensive complement of F-actin at end walls (plant synapses) and longitudinal bundles of F-actin which encase the nuclei and interconnect the opposing end walls.

which they are associated). Importantly, the end walls do not increase in surface area, and axial cellular growth is accomplished solely at the side-walls with the assistance of the CMTs (Barlow & Baluška, 2000). The non-growing status of end walls is, surprisingly, associated with abundant F-actin, whereas the extending side-walls are depleted of F-actin (Fig. 2d in Baluška *et al.*, 1997). A similar pattern applies for plant-specific myosin VIII (Reichelt *et al.*, 1999; Baluška *et al.*, 2000a,d), this molecule also being a characteristic marker of plasmodesmata. The non-growing end walls are sites of extensive endocytosis/vesicle recycling (Fig. 7.2) which in turn leads to cell-to-cell communication (Baluška *et al.*, 2003a,b), and in this respect they bear distinct similarities with neuronal and immunological synapses.

In pursuing the parallels with neurotransmission, it is interesting to find that, just as classical inhibitors of the polar auxin transport have proved to be inhibitors of endocytosis (Geldner *et al.*, 2001), so auxin has been shown to accumulate in what may be endo- and exocytotic vesicles (Shi *et al.*, 1993). Thus, auxin and its transport share many of the features of animal neurotransmitter systems (Baluška *et al.*, 2003b). Several other recent data

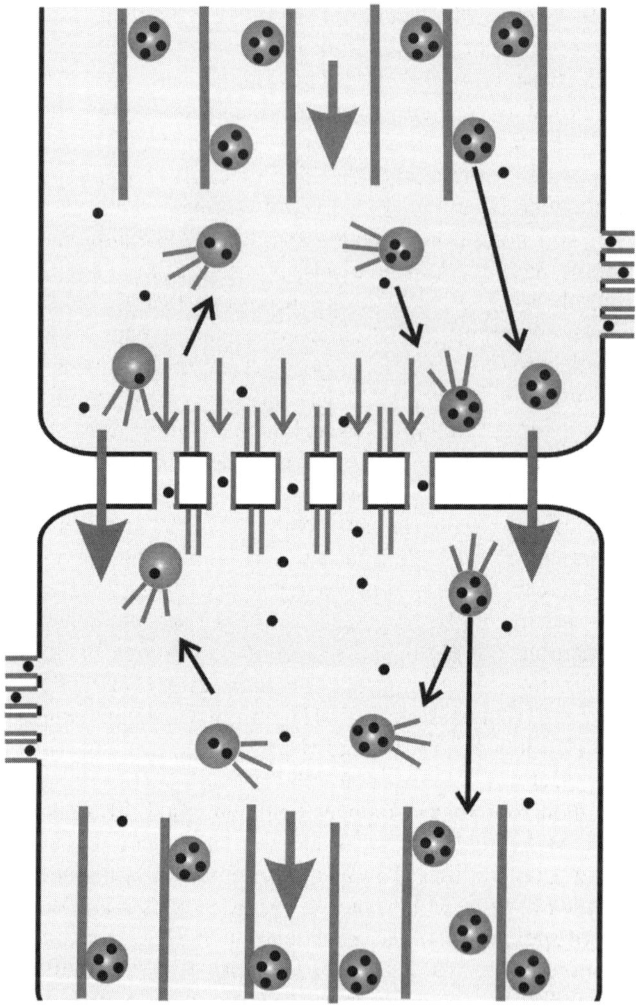

Figure 7.2 Scheme showing the interrelationships between F-actin, endocytotic vesicles, auxin carriers PIN1 and AUXI that set up a polarised auxin transport system. Transport is towards the root tip, in the direction indicated by the four thicker arrows. Cell-to-cell transport of auxin occurs at specially adapted plant synapse sites (plasmodesmata) at the end walls. In this scheme, some additional termini of actin filaments – plant mini-synapses – synapses are located upon the side walls. These represent pit fields where plasmodesmata cluster.

(reviewed in Baluška *et al.*, 2003a) also imply that the non-growing end walls are auxin-transporting plant synapses, thereby maintaining both apical-basal polarity and cell-to-cell communication. From an anatomical point of view, acropetal polar auxin transport takes place in the stele (Greenwood & Ycas,

1975; Saugy & Pilet, 1984) and this is where abundant F-actin is found organised in the form of longitudinal F-actin bundles (Fig. 7.1; see also Parthasarathy & Pesacreta, 1980; Parthasarathy *et al.*, 1985; Baluška *et al.*, 1997). So important is this auxin transporting property of the stele to the growth of the root that the cortex (which mostly lacks prominent F-actin bundles) is required to play the role of a specialised protective tissue, the vacuolated cortical cells acting like impact-absorbing 'bubble wrap' material. When compact soil conditions threaten the stele, for example, the cortex is ready to protect it, having responsive cells that will expand and force apart the resistant soil (Baluška *et al.*, 1993b; Huang *et al.*, 1998).

As is the case in neurons, the 'strength' of a plant synapse is expected to be determined by the rate of auxin transport across it. It is well-known that a polarised actin system is necessary for the regular development of longitudinally aligned cell files in the root meristem (see above). However, the auxin which such a system carries is also necessary for the development of the vascular strands within the stele. The strands are themselves adapted to carry this polarised auxin flow (reviewed by Sachs, 2000; Berleth, 2000; Berleth & Mattsson, 2000; Chapter 9). The development of the system is thus self-reinforcing, or 'canalised'. The same morphogenetic principle may play a part in the formation of all plant tissues, as well as in the shaping of plant organs. For instance, it is well known that a good deal of the patterning of root apical tissues is determined by auxin gradients dynamically maintained by acropetal polar transport of auxin (Sabatini *et al.*, 1999; Friml *et al.*, 2002a,b; Barlow, 2003), and that mutants defective in polar auxin transport not only fail to develop the proper continuous vascular strands (Berleth, 2000; Berleth & Mattsson, 2000) but also show defects in the organisation of their root apices (Jiang & Feldman, 2003; Friml *et al.*, 2002a,b). But how can the self-reinforcing property of auxin transport be accounted for? How can auxin apparently 'create' a sink for more auxin (Rayle *et al.*, 1969; Veen, 1969)? In the case of the growing root apex, an auxin sink might be related to the ability of the root apex, and the quiescent centre in particular, to catabolise auxin (Pilet & Lance-Nougarède, 1967; Kerk *et al.*, 2000; Jiang & Feldman, 2003). By so doing, a demand for auxin is created, perhaps on account of the strong gradient developed by the presence of a catabolic sink at the terminus of the organ. This problem will be discussed again below.

7.2.3 Spatially distinct acropetal and basipetal auxin transport systems and their consequences for cell body (nucleus) localisation

The auxin transport system discussed above is characteristic of the stele and is directed towards the root tip. The source of auxin is likely to be mainly from the caryopsis (as in the early growth of graminaceous plantlets such as maize) or from the developing shoot (Martin *et al.*, 1978; Bhalerao *et al.*, 2002). A few early studies, however, had indicated that during the gravireaction there was also a basipetal flow of some mobile growth regulator from the root cap into, and through, the epidermis and outer cortex (Pilet, 1977). At that time, this regulator was thought to be a growth inhibitor such as abscisic acid, but recent evidence suggests that the molecule transported basipetally is auxin (Muday, 2001; Rashotte *et al.*, 2001). It is the basipetal flow of auxin in epidermal and outer cortex that regulates the differential growth upon which root gravitropism depends (Rashotte *et al.*, 2000). An important question, therefore, is whether the polarised intracellular distributions of auxin influx and efflux carriers in these lateral portions of the roots and root cap are inverted with respect to the orientation of these carriers in the stele and, if so, how this comes about.

The data on the distinctive polarised auxin fluxes in stele and cortex of maize roots appears to have a counterpart in the polarised division sequences in young root tissue of this species. Related cells (sisters, cousins, grand-cousins, etc.) within longitudinally oriented cell packets have strong tendencies to divide in an acropetal sequence in the stele, but in a basipetal sequence in the cortex (Barlow, 1987; Lück *et al.*, 1994). This effect can persist over at least the first four cell generations following germination and has been inferred to be present in the embryonic radicle also. The various polarised division pathways are due to the inequality of sister cell size following cytokinesis. In the case of the cortex, the basal sister cell is usually larger (longer) than the apical sister. The converse is the case in the stele. The unequal length of sister cells may be determined by the position of the cell body within the mother cell prior to mitosis: thus, in the stele, the cell bodies would tend be located towards the apical end of the cells, but in the cortex they would be located at the basal end of the cell. It might be, therefore, that the auxin flux, according to whether it is acropetal or basipetal, 'drags' the nucleus towards the efflux end of the cell. This effect clearly has implications for nuclear movements in pericycle cells during lateral root primordium development (see below).

Cell division polarity leads to daughter cells of unequal size, and these can be distributed not only along cell files, as described (Lück *et al.*, 1994), but also across a given file to generate a second file of cells, as in the cortex of maize roots (Barlow, 1987). In the cortex, it is the inner daughter cell which is the larger of the two sisters. This again indicates a tendency for the cell body to be located nearer one wall than another – the outer periclinal wall in this case. Again, this would be in conformity with the nucleus being attracted to the proposed site of efflux of auxin through these cells.

7.2.4 Actin-auxin-based cell body rotations induce morphogenetic periclinal cell divisions

In connection with the setting up of the correct number of cell files at the tip of the root for subsequent tissue differentiation, the main question must concern how the polar auxin transport and the system of plant synapses bring about changing orientations of the cell division apparatus so that suitable proportions of transverse, radial and periclinal divisions occur. Inevitably, there must be a dynamic interaction between the actin cytoskeleton-enriched cell cortex (Baluška *et al.*, 2000b) and the perinuclear microtubules which position the pre- and post-mitotic plant cell body (Baluška *et al.*, 1998, 2001d). The means by which the proliferative divisions (transverse with respect to the root axis) are oriented could be quite straightforward. Our prediction would be that the mitotic cell body, comprised at first of mitotic spindle (metaphase) and then of cytokinetic apparatus (telophase), is oriented in the same direction as the polar flux of auxin. The division wall which forms following cytokinesis is enriched with actin and components of the vesicle-recycling machinery. These elements of the plant synapse attract the perinuclear microtubules prior to the next mitosis and then hold the forming mitotic spindle in this orientation. The axis of the mitotic cell body thus conforms to the direction of polar auxin transport.

It is the longitudinal (radial and periclinal) formative divisions which have the greatest developmental impact because they generate the additional cell files required for the prospective plant tissues. In the root apex, especially if the apex is of the ‘closed’constructional type (as in grasses), periclinal divisions occur only in certain localised regions (Luxová, 1975) and probably indicate where auxin levels are relatively high (Feldman, 1981; Jiang & Feldman, 2003). Similarly, an increase of cellular auxin, elicited either by the inhibition of auxin efflux or by the addition of auxin, induces a switch from anticlinal to periclinal divisions (Zhang & Hasenstein, 2000; Petrásek *et al.*, 2002). It is suggested that the local level of auxin might play an important role in the switch from anticlinal to periclinal divisions, but the mechanism of how this is accomplished remains obscure. We propose that an elevated level of auxin, as well as its actin-based vesicular transporting machinery, concentrated at the plasma membrane and/or plasmodesmata at one side of a cell (e.g., the inner periclinal face) may be sufficient to initiate a new plant synapse. This site with its abundant actin filaments may accordingly attract one pole of the mitotic cell body which, when it subsequently divides, produces a periclinal division wall. In this way, the same elements participate in the orientation of both transverse and periclinal divisions. It is the local circumstances that determine asymmetries in the cells involved in these divisions. The direction of auxin transport may be critical in this respect as it seems that it is integral to the polarity of each individual cell. Moreover, whenever wounds are made in the root apex (or elsewhere) cell divisions occur with orientations strongly suggestive of the mitotic spindle having been aligned in the same direction as a flow (of auxin?) towards the wound surface (Hardham & McCully, 1982; Warren Wilson & Warren Wilson, 1984; Barlow, 1984, Fig. 10.9).

It is conceivable that radial divisions, which accommodate the increasing girth of the root, are regulated in the same way. But there could be another way – one that involves the sensitivity of premitotic cells to the strain (stretch) imposed upon them by the expansion of some underlying tissue. Similarly, in the root cortex, where the cells do not display prominent actin bundles and where acropetal auxin flow is not supported, transverse divisions may be regulated in response to the strain sensed in the longitudinal direction. In the root cap meristem also, prominent F-actin is not evident (Baluška *et al*., 1997). Here, however, both auxin and strain may regulate the pattern of division (Barlow, 2003).

Early embryogenesis represents another excellent model system for testing the interactions between polar auxin transport and cytoskeletal elements. In *Arabidopsis*, cell divisions are tightly regulated during embryogenesis (Barlow, 2002, Fig. 3) and they are also closely linked to the direction of auxin transport (Hamann, 2001). Mutations in genes affecting either the polar transport of auxin or the auxin response pathway result in abnormal orientation of the division walls (reviewed by Geldner *et al*., 2000; Jürgens, 2001; see Chapter 6). In *monopteros* and *bodenlos* mutants, the first longitudinal cell division fails and all divisions are anticlinal. In the *gnom* mutant, this first longitudinal division is misoriented (see Fig. 1 in Geldner *et al*., 2000). Results from all three mutants suggest the importance of the early longitudinal division for embryonic root fomation. The crucial role of periclinal divisions is also obvious from the *fist* mutant which lacks these divisions after the eight-cell stage; the embryo then develops as a disorganised cluster of cells (Dunn *et al*., 1997). Such aberrant spherical embryos can be phenocopied by applying inhibitors of polar auxin transport, thereby confirming that, even at the early stage of embryogenesis, auxin transport regulates the orientation of new division walls (Hadfi *et al*., 1998).

7.2.5 Hypothetical model explaining how the polar transport of auxin confers synaptic status upon non-growing cross walls

To interpret satisfactorily the above-mentioned findings, we postulate that auxin fulfils the criteria of a hormone (long-distance transport culminating in a developmental event), a morphogen (director of differentiation and determiner of plant form), and a neurotransmitter-like signalling molecule (active movement of auxin from cell to cell via vesicular trafficking). All these properties are crucial for establishing the polarised apical-basal axis inherent to roots and most other plant cells and organs. Two central properties of auxin are:

1. Its mobility within and along specially adapted cell files of the stele.
2. Its association with the actin cytoskeleton-driven membrane trafficking across non-growing cross walls.

Further postulates can be made which integrate current knowledge of auxin into a coherent hypothetical model open for experimental evaluation.

- The non-growing end walls are considered to be 'plant synapses' which drive electrical and chemical communication between adjacent cells within longitudinal cell files. These subcellular domains, as well as the stability of the end walls, are actively maintained by endocytic, vesicular recycling events which rely on structural support from dense F-actin meshworks and unconventional myosin of class VIII.
- There are feedback interactions between the vesicular auxin transport machinery and the actomyosin-based cytoskeleton. The amount of auxin transported across a given synapse not only determines its synaptic strength but also actively attracts F-actin to the end wall. At this site, F-actin drives endocytosis and vesicle recycling. Auxin, because it is contained within vesicles, is one of the molecules moved by endocytosis – hence its active transport. Vesicular trafficking and actomyosin stabilise the synapse and exclude cortical MTs from the cross walls of meristematic cells during G2 phase and prophase, while at the same time they attract endoplasmic MTs which radiate from mitotic spindle poles. New division walls are aligned perpendicularly to the direction of polar auxin transport.
- As long as auxin is transported longitudinally within the cell files, it induces anticlinal divisions because the premitotic cell bodies are oriented in the same direction as the auxin flow. However, when the level of endogenous auxin is locally increased, and the auxin-transporting capacity of synapses is overloaded, auxin starts to be transported out of a given cell file via a side wall. The premitotic cell bodies become aligned perpendicularly to the axis of the cell file and periclinal cell divisions are induced. These new divisions follow the new direction of auxin efflux from the cell file. If a new periclinal synapse proves to be sustainable, due to the sufficient strength of the new auxin sink generated by the periclinal division wall, then a new cell file can be maintained. If the lateral efflux is sufficiently prolonged, the new files may grow perpendicularly to the pre-existing mother cell files, as is proposed to happen, for instance, during lateral root primordia initiation.

7.3 Establishment of new polarities and new root organs

Plant organs branch repeatedly to produce new organs similar to those already present. Primary root systems are made up of many generations of branched roots (lateral roots): for example, up to seven orders of lateral root branching have been recorded during the first year of growth of apple tree seedlings (Kolesnikov, 1930). Both the site of branching upon a parent root axis and the time of its inception are variable (see Appendix 1, Table 1). Lateral roots may arise upon a parent root either in acropetal succession, apparently keeping pace with the elongation of the root apex, or else new branch roots arise sporadically as adventitious or adventive roots without any apparent relationship to the growth of the main apex. In the first case, the initiation of branching may occur within or just behind the elongating zone of the root (as in angiosperms and gymnosperms), within the meristem (in ferns), or can occur at the root apex itself (in *Selaginella*), where the branching leads to an apparent dichotomy of the apex. In the second case, branching can occur anywhere along the root axis (Barlow & Palma, 1997) and is often a response to wounding or some major change to the structure of the primary tissue, such as the loss of the cortex and the simultaneous development of secondary tissues (Paolillo & Zobel, 2002). In both cases, a branch root develops from a group of cells embedded within the existing polarity of the parent axis. The influence of the parental tissue has to be broken before the primordium can establish its own axis of polarity, and thus enable it to emerge, as a new root, from the parent axis. Subsequent re-orientation of growth comes about when the new root acquires the ability to perceive and respond to gravity.

There are, therefore, two aspects to the problem of how a new branch root is established:

1. The selection, or determination, of sites where a primordium will arise.
2. The polarisation of the primordium and the acquisition of a growth axis.

Thereafter, the primary polarity of the new lateral root is maintained in the manner described above.

7.3.1 Sites for organogenesis

A number of studies have examined the spacing of lateral roots along the parent axis (Barlow, 1984; Barlow & Adam, 1988; Newson *et al.*, 1993) as well as circumferentially around that axis (Pulgarín *et al.*, 1988). In all cases, the aim has been to uncover the principles that underlie the siting of primordia not only as a set of unique, independent siting events, but also as a result of interactions of the primordia with each other and with the anatomy of the parental apex. Although reviewed on many occasions (see Charlton, 1996), none of the theories for the siting of primordia can be regarded as generally applicable. For example, looking at the range of species from angiosperms to pteridophytes, it seems that:

1. The pericycle is not an essential tissue for lateral root formation because it does not participate in primordium initiation in fern roots. In these plants, lateral roots arise from endodermis (a tissue which contributes to the cap of lateral roots in many species of higher plants). However, in ferns, root and root cap originate in a different way from that found in angiosperms. Ferns, in keeping with a monoplex organisation of their root meristem, require only a single cell from which to organise their root primordia,

whereas angiosperm root apices require many cells for this purpose (see Philipson, 1990).

2. Primordium initiation does not require the presence of an underlying differentiated vascular tissue (e.g., primary xylem and phloem). Again, in ferns (e.g., *Marsilea*), the apical cell for the new lateral root originates within the meristem of the parent root at a position where vascular tissue is itself just beginning to differentiate (Lin & Raghavan, 1991). Lateral root primordia in a number of angiosperm species also begin to develop prior to the formation of primary xylem in the supporting parent root (Mallory *et al.*, 1970).
3. Where an underlying vasculature is present, primordia can arise adjacent to protophloem (wheat), adjacent to either protophloem or protoxylem (maize) or between the two (tomato) or, in most of the reported cases, adjacent to the protoxylem poles (pea, onion) (Mallory *et al.*, 1970). The significance of the neighbouring differentiated vascular strand(s) is probably not because it has any inductive influence upon primordium initiation, but because the strand is able to resist the pressure exerted upon the stele by the enlarging primordium (see Barlow & Adam, 1989a, Figs 1d and e). This may be why protoxylem has been most frequently selected to underlie the primordia of dicot roots.

The three above-mentioned areas of variability in lateral root primordium initiation suggest that some regulatory factor common to all cases of primordium initiation has so far eluded identification. We shall now try and identify this factor and so satisfy all the conditions not only for primordium formation but also for the establishment of a new axis of polarity. The proposed solution to this problem again involves the concepts of the plant cell body (Baluška *et al.*, 1998) and of directed auxin transport. Although some of the ideas are speculative, they nevertheless provide a hypothesis concerning the conditions for primordium formation and are therefore open for experimental analysis.

7.3.2 Hypothesis for the siting of lateral root primordia

Generally, the cooperation of at least two levels of organisation (e.g., the cellular and organ levels) is required for the completion of a developmental event in plants. At the cellular level, the simplest mechanism of primordium siting would be one that has a strong deterministic component. That is, it should be the unique outcome of a prior event in the cellular construction of the root itself. The seemingly disparate areas of apical cell development in ferns and the simulation of their cell patterns by means of D0L-system algorithms (Barlow *et al.*, 2001) provide a clue as to what this event may be. Deterministic simulation of cellular arrays reveals that it is possible for new organogenetic centres to arise in response to information propagated by the cellular system itself. This principle works, in theory, for the generation of leaves and new apices on the shoots of both lower and higher plants (Lück & Lück, 1993; Barlow *et al.*, 2001). We hypothesise that this also applies to the origin of organogenetic centres in roots. The only requirement is that the cell layer (pericycle or endodermis) which initiates the primordium follows the predetermined algorithm that specifies its development. In many species of ferns, organogenetic centres correspond to tetrahedral (or apical) cells on or below the surface of the parent organ. These cells are the outcome of a series of deterministic cell divisions, the critical division (from the point of view of generating a new organogenetic centre) being an asymmetric but not necessarily unequal division. The nascent apical cells so produced are sited at specific locations which comprise a particular configuration of cell walls of different age. Cell walls are now accepted as being repositories of informational molecules, ranging from simple ions, such as Ca^{2+}, to complex oligosaccharides (Wojtaszek, 2000; and see also Chapter 3). The cell body with its associated cytoskeleton and endocytotic/recycling membrane system is the means by which these molecules are accessed and cell wall components such as calcium- and boron-cross linked pectins are internalised (Baluška *et al.*, 2002; Yu *et al.*, 2002).

A mechanism for specifying organogenetic sites, similar to the fern system, is proposed to occur in the roots of higher (angiosperm) plants. Our hypothesis is that the meeting place of the corners of four pericycle cell packets (groups of cells of common descent), each with information-bearing end walls, defines an organogenetic centre at which a new root primordium will form. Configured like a cross, +, the four walls demarcate the four quadrants of cells which will construct the future root. We propose that the end walls of the four packets are able to attract to themselves the cell bodies resident in the cells to which these walls belong (Fig. 7.3). The attraction may be mediated by actin cables which actively pull and/or push the cell bodies away from their previous resting site at the cell centre (Fig. 7.4). When the cell bodies reach the vicinity of their respective end walls, they and their nuclei then divide and cytokinesis follows (Casero *et al.*, 1993). A small cell and a larger cell are thus produced in each cell file (Figs 7.5 and 7.6). Following three further divisions of the nucleus in this larger cell, a file of four small cells is formed. Similar quartets of cells form in adjacent cell files belonging to the same cell packet. The four groups of newly divided cells in the angles of the four packet walls constitute the pericyclic ‘primordial plaque’. Their cell bodies are now ready to respond to repolarising stimuli. At least one gene, *SLR*, which codes for a protein that helps regulate auxin-induced gene expression, affects these early unequal anticlinical cell divisions (Fukaki *et al.*, 2002). Interestingly, these early divisions, including the preceding polarised nuclear movements, were first described by Kawata and Shibayama (1965) in their detailed study of lateral root primordium development in rice roots; this was nearly 30 years before their ‘re-discovery’ by Casero *et al.* (1993).

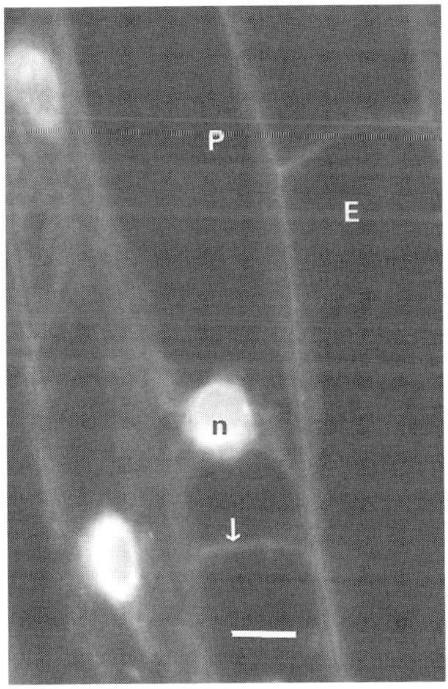

Figure 7.3 Stages of lateral root primordium development in maize (*Zea mays*) root. Longitudinal section. A nucleus at one end of a pericycle cell, near the apical end wall (arrowed). DAPI staining. Abbreviations: E – endodermis, n – nucleus, P – pericycle. Scale bar = 10 μm.

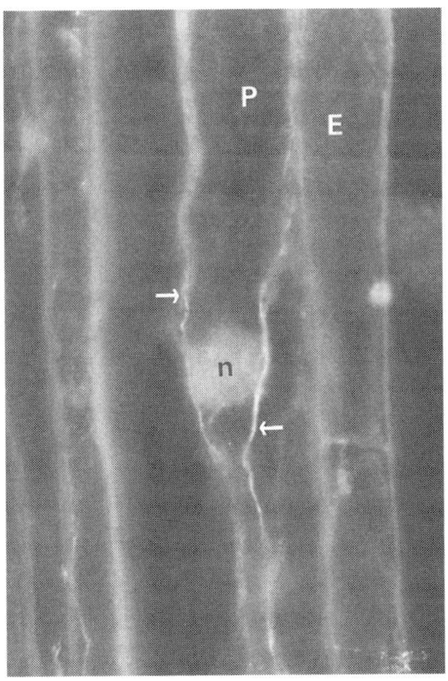

Figure 7.4 Stages of lateral root primordium development in maize (*Zea mays*) root. Longitudinal section. Two actin cables (arrows) support a nucleus in a pericycle cell. The nucleus may be being pushed/pulled to one end of the cell. Anti-actin staining. Abbreviations: E – endodermis, n – nucleus, P – pericycle.

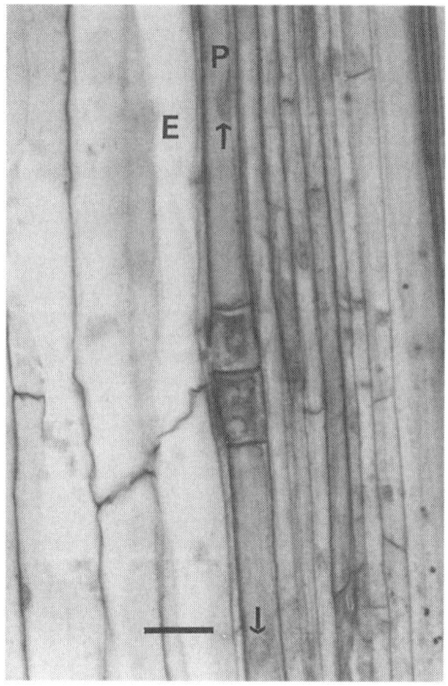

Figure 7.5 Stages of lateral root primordium development in maize (*Zea mays*) root. Longitudinal section. First unequal division of two pericycle cells. Nuclei in the longer daughter cells are marked with arrows. Abbreviations: E – endodermis, P – pericycle. Scale bar = 20 μm.

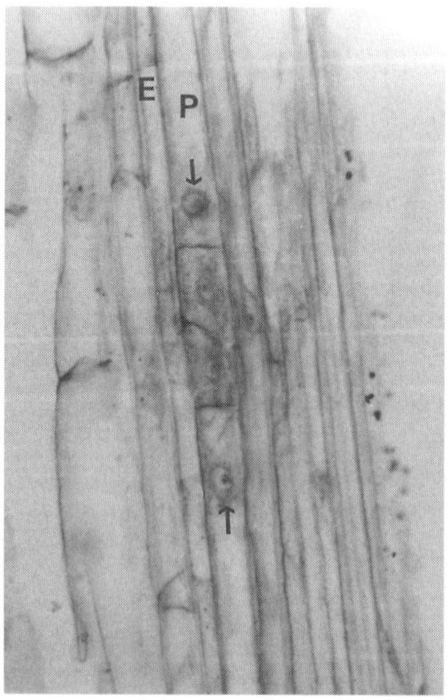

Figure 7.6 Stages of lateral root primordium development in maize (*Zea mays*) root. Longitudinal section. Preparation for the second division in each of the two longer pericycle cells. Their nuclei (arrowed) are now closer to the short cells. Abbreviations: E – endodermis, P – pericycle.

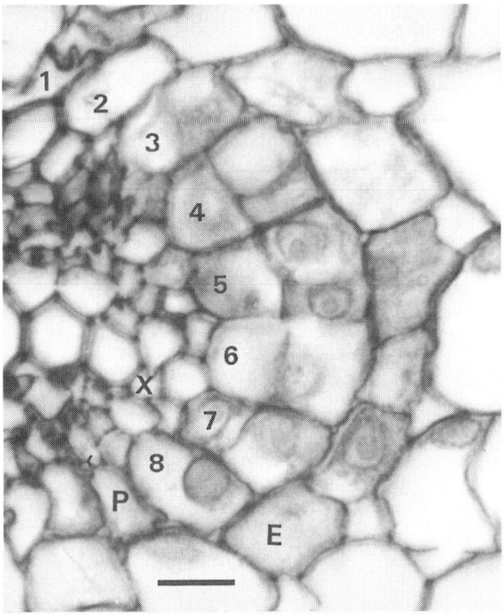

Figure 7.7 Stages of lateral root primordium development in an *in vitro*-grown root of tomato (*Lycopersicon esculentum*). Transverse section. Eight cells (numbered) are evident at the base of the primordium. Abbreviations: E – endodermis, P – pericycle, X – protoxylem pole. Scale bar = 20 μm. Same magnification applies to Figs 7.8 and 7.9.

A similar plaque (though possibly with fewer cells) develops simultaneously within the endodermal cells overlying the pericyclic plaque. In *Zea mays* and other species, these endodermal cells give rise a cap-like structure (Figs 7.7 and 7.8). The importance of this endodermal 'proto-cap' in establishing the polarity of the lateral root primordium will be discussed below.

The location of the organogenetic centres at intervals along the length of the pericycle determines both the longitudinal and circumferential spacing of primordia. The longitudinal spacing reflects the length of the pericycle cell packet (Barlow & Adam, 1988). It is not a scalar, but a cell-based, distance. In pea roots, Hinchee & Rost (1992b) found 91 to 136 pericycle cells between each successive primordium along a given rank of primordia. These numbers correspond to 6.5–7 cell doublings and probably reflect the cellular products of the final round of transverse cell division enclosed within each cell packet. Likewise, circumferential spacing reflects the circumferential arc lengths of the cell packets and the number of radial divisions that occurred within them. The activation of the organogenetic centres, which involves nuclear and cell body migration in the pericycle, may be a feature of the age-related state of the packet end walls, though there is good evidence that activation is also related to the presence of some other regulatory material, such as auxin (Bhalerao *et al.*, 2002; Marchant *et al.*, 2002). The formation of

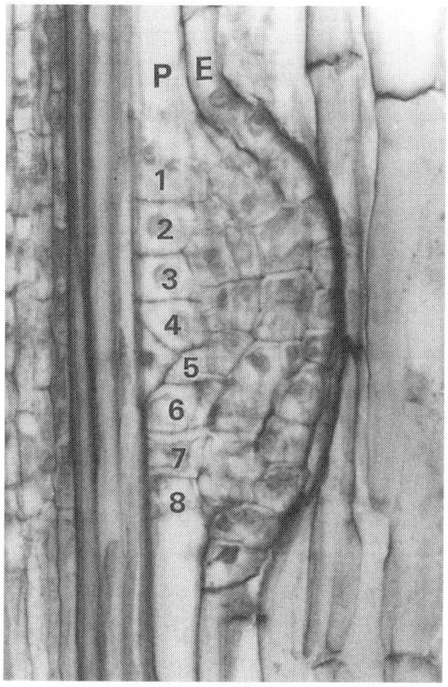

Figure 7.8 Stages of lateral root primordium development in an *in vitro*-grown root of tomato (*Lycopersicon esculentum*). Longitudinal section. The inner cell layers of the young primordium are derived from periclinal divisions of the pericycle. The outer layer is derived from endodermis. The section is median and shows that descendents of eight pericycle cells (numbered) form the base of the primordium. Abbreviations: E – endodermis, P – pericycle.

the primordial plaque involves first anticlinal and then periclinal divisions. These processes may therefore be indicators of the accumulation of auxin at the organogenetic centre, together with the establishment of new plant synapses, as described earlier. The walls at the organogenetic centre may favour localised auxin accumulation and radial auxin movement on account of the age-related occlusion of plasmodesmata, particularly on the anticlinal cross walls.

Coming now to the organ level of organisation, it is not necessary to invoke any causative (e.g., inductive) association between the siting of primordia and the vascular strands of protoxylem or primary phloem. The vascular tissue enclosed by arcs of primordium-bearing pericycle cell packets is itself part of an even larger and older cell packet. Within this tissue there are other forms of information (loosely termed 'positional information') that specify the location of phloem and xylem (Warren Wilson & Warren Wilson, 1984). One contribution of the vascular cylinder to primordium initiation is through the supply of auxin (e.g., indoleacetic acid, IAA and maybe indolebutyric acid, IBA). Auxins are well known promoters of mitosis, and IAA can be detected in pericycle cells (Mathesius *et al.*, 1998). As suggested above, auxin may also promote the putative actin-regulated movements of cell bodies in pericycle cells (see Waller *et al.*, 2002). Cell body migration (Baluška *et al.*, 2001a), is perhaps a type of intracellular chemotaxis in response to auxin, the attractive site being associated with the end walls of the cell packet.

Anatomically, the pericycle is part of the vascular cylinder, and hence it has access to auxin derived from the shoot. How could auxin be channelled towards the organogenetic centres in the pericycle from which the primordia arise. Could it be that the endodermis holds the key? Endodermis expresses SCARECROW, a transcription factor which is also found in the quiescent centre (Di Laurenzio *et al.*, 1996; Lim *et al.*, 2000; Wysocka-Diller *et al.*, 2000) and thus could maintain the stem-cell identity of the neighbouring pericycle (Sabatini *et al.*, 2003). Moreover, endodermis is usually thought of as regulating the inflow of solutes to the stele by means of its Casparian band. But it can equally well regulate outflow from the stele. Endodermal passage cells are sites where Casparian band development is delayed relative to neighbouring non-passage cells (Cutter, 1971). In regions where primordia form, their frequency may be quite high (Ma & Peterson, 2000), probably in excess of the estimated 1 per 1000 endodermal cells, stated in Kolek & Kozinka (1992) for wheat roots. The passage cells may therefore be localised ports of outflow from stele to cortex, and hence channel auxin through the pericycle underlying such sites. This suggests the siting of endodermal passage cells opposite the protoxylem – confirmed for roots of *Ranunculus*, maize and onion, for example (Cutter, 1971; Ma & Peterson, 2000) – this location being where lateral root primordia usually form in each of these species.

The role of auxin in lateral root development in ferns is unknown. The small size and dense cytoplasm of their endodermal cells means that no special stimulus is required for the induction of the rapid cell growth and mitoses necessary for primordium development, as is the case for the more vacuolate and elongated pericycle cells of angiosperms. Nor do they have need of a primordial plaque since the new apical cell of the primordium arises directly during an unbroken predetermined sequence of divisions (cf. Barlow *et al.*, 2001). Some ferns, e.g. *Azolla*, do not form lateral roots. This may be a result of a failure to form either the correct primordial precursor cell or to contain sufficient auxin within their roots.

The proximate stimulus for the onset of division and growth in the pericycle is derived from genetic control. In *Arabidopsis thaliana,* two genes, *LIN1* and *ALF-4*, affect the earliest stages, possibly by regulating the availability of auxin within the vascular cylinder (Celenza *et al.*, 1995; Malamy & Ryan, 2001). In our proposal for the siting of primordia, it is the cell bodies associated with the organogenetic centre which are sensitive to auxin, and these three items – end cell wall, cell body and auxin – provide the conditions for the initiation of primordial growth and division.

7.3.3 Repolarising the primordial plaque

The first signs of repolarisation within the pericyclic primordial plaque are periclinal divisions, though, as Dubrovsky *et al.* (2001) have shown for *Arabidopsis thaliana,* anticlinal divisions continue to occur after the commencement of periclinal divisions. Divisions of each type commence in the first new cells to form at the organogenetic centre. Divisions then occur in cells located away from the centre, in both the axial and circumferential directions. Observation of both longitudinal and transverse sections through young primordia of many species usually show that there are eight pericycle cells at their base (Figs 7.7 and 7.8). Thus, each primordium would seem to rest upon a grid of $8 \times 8 = 64$ cells which are the descendents of the primordial plaque. Anticlinal divisions establish 64 cell files in the new root. These 64 files are sufficient to block out the concentrically arranged four major tissues of the new root: epidermis, outer and inner cortex, vascular tissue including the pericycle.

The acquisition of newly synthesised cytoplasm by cells of the primordial plaque may be an important event in the restoration of meristematic growth in the formerly vacuolate pericycle cells in maize (Bell & McCully, 1970) and other roots. However, this new cytoplasmic synthesis follows upon the enlargement of the cells which develop the primordial plaque. Although the plane of division appears to be regulated by the principal direction of growth (Hofmeister's division rule – see Appendix II), according to our hypothesis the direction of growth could correspond to the direction of auxin transport. The earliest stage of primordium growth, however, tends to be isotropic, and is accompanied by the occasional random arrangement of cell division walls. Later growth is strongly unidirectional (centrifugal) in accordance with a newly established polar auxin flow within the primordium. It may be that initially, the growth of the primordial plaque can only be centrifugal, in the radial plane with respect to the parent root axis, since its growth in the circumferential and axial planes is limited by a non-expanding perimeter. The strain resulting from growth under such conditions may also favour periclinal divisions in these pericycle cells.

Experimental evidence from cultured roots of *Arabidopsis* suggested to Laskowski *et al.* (1995) that when a primordium reaches the stage of having 3–5 cells layers it then becomes independent of exogenously supplied auxin for its development. A similar inference of an early auxin-dependent initiation phase and a later auxin-independent growth phase is evident from other studies of *in vitro* auxin-treated roots of lettuce (MacIsaac *et al.*, 1989) and eucalyptus (Pelosi *et al.*, 1995). The establishment of a flow of auxin between the primordium and the stele of the parent root may be one factor that makes it appear that the primordium becomes auxin-sufficient. This early flow of auxin in concert with auxin-transporting and actin-enriched plant synapses on former periclinal division walls may, in addition, be instrumental in establishing a polarised vascular cylinder in the new root, and in effecting a vital connection between the developing primordium and the pre-existing vasculature of the parent root (Byrne *et al.*, 1982). Development of successive orders of laterals is a function of the connectedness of all plant synapses in the chain of successively formed root axes.

7.3.4 The emerging primordium: formation of the quiescent centre

The emergence of a young lateral root from the body of the parent root, with the rupture of intervening cortical and epidermal tissues, coincides with the acquisition of a quiescent centre at its apex (Clowes, 1978). The slow rate of volumetric growth in the quiescent zone need not affect the already established growth of the young primordium since the area of slow surface growth is confined to the dome of the root (Barlow, 2003). The region of maximal extension growth has, by this stage of development, been displaced to behind the apex.

A quiescent centre is, apparently, a universal feature of angiosperm roots, and its development is probably crucial to the polarisation of primordium growth. Early work on auxin catabolism suggested that the root tip was implicated in setting up a gradient of auxin within the root as a whole (Pilet, 1967). Recent work confirms this and suggests that the quiescent centre is where auxin is catabolised by the action of ascorbic acid oxidase (AA oxidase; Kerk & Feldman, 1995), thus preventing auxin accumulation therein. After its establishment, the new quiescent centre could serve as an 'attractor' for auxin supplied by the steles of both the parent root and of the young lateral root itself. According to older ideas about auxin transport, auxin would be encouraged to flow acropetally along a concentration gradient, from the base to the tip of the primordium where the quiescent centre is located. Such an acropetal flow of auxin would be substantially aided by being polarised through the cell bodies and plant synapses in the stele of the new root.

If a quiescent centre is critical for the axiality and then the polarity of the pre-emergent lateral, it is important to discover how this group of cells comes to be established. According to Clowes (1978), the quiescent centre of pre-emergent maize lateral roots develops at the same time as the endodermal cells differentiate to form a cap-like tissue which covers the primordial dome (Fig. 7.9). This 'proto-cap', however, is ephemeral: after a brief period of mitotic activity its cells stop dividing, the quiescent centre disappears, and the proto-cap is replaced by another cap derived by the division of cells of the underlying primordial hemisphere (Fig. 7.10). When these new cap cells form their set of dividing cap initials, a quiescent centre then re-appears in the primordium. The correlation between the commencement of cell divisions in

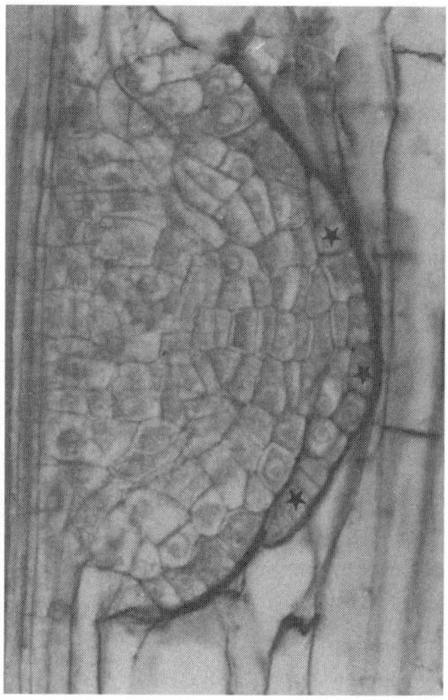

Figure 7.9 Stages of lateral root primordium development in an *in vitro*-grown root of tomato (*Lycopersicon esculentum*). Longitudinal section. Primordium showing a proto-cap (asterisks) derived from endodermis.

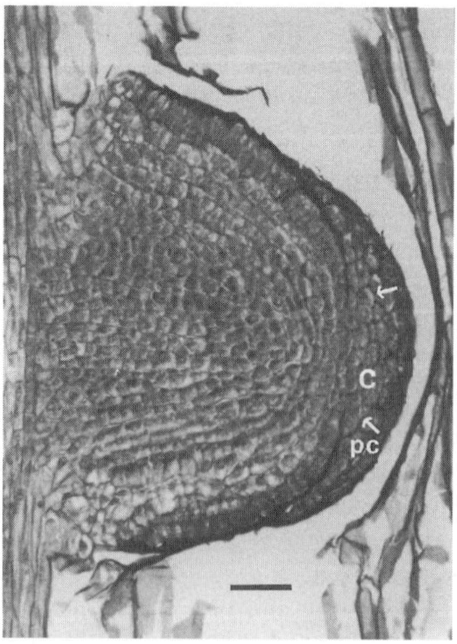

Figure 7.10 Stages of lateral root primordium development in an *in vitro*-grown root of tomato (*Lycopersicon esculentum*). Longitudinal section. Older primordium than that shown in Fig. 7.9. A new cap has formed under the proto-cap. Cap/proto-cap junction is marked with arrows. Abbreviations: C – root cap, pc – proto-cap. Scale bar = 40 μm.

the cap (or proto-cap) meristem, the onset of quiescence of the underlying primordial cells (as well as the converse situation of non-division of the cap initials and the onset of divisions within the quiescent centre leading to its disappearance) suggests either that a quiescent zone in the primordium stimulates the overlying cap cells to divide, or that divisions in the cap provoke quiescence in the underlying primordial cells.

It has been suggested that the relationship between an increasing auxin level in the quiescent centre and the induction thereby of AA oxidase, as well as the consequent decrease in the intensity of cell cycling due to a decrease of ascorbic acid, may show a natural rhythmic activity (Kerk *et al.*, 2000). The appearance, disappearance, and then reappearance of the quiescent centre during the course of primordium development may be a manifestation of this cycling of auxin level and auxin activity. The production of a second cap from a previously quiescent centre in the primordium would be a consequence of the latter's resumed mitotic activity on account of its relatively low auxin level.

How does the proposed cyclical activity of AA oxidase in the quiescent centre come about? That is, why should there be cyclical activity in contradistinction to a continuous intermediate level of AA oxidase activity that would result in an intermediate rather than a low level of mitotic activity? To achieve cyclical enzyme activity, it must be assumed that there are abrupt step-wise changes in auxin level. When auxin reaches a particular threshold level, AA oxidase then becomes active and catabolises auxin in a cataclysmic fashion. Auxin level falls precipitously and AA oxidase disappears. There is then again a steady increase in auxin level within the quiescent centre, due to its continual acropetal transport, until the threshold level is reached once more; and the cycle then repeats. Should the first of the two quiescent centres fail to form in a young primordium, as seems to be the case in *Vicia faba* (MacLeod & McLachlan, 1974), then the first of the two putative increases in auxin level may continue unregulated and result in the observed decline in mitotic index within the late-stage, pre-emergent primordium (MacLeod, 1972). The first quiescent centre also appears to be absent in young primordia of *Malva sylvestris* (Byrne, 1973). Roots of both *Vicia* and *Malva* have an 'open' type of apex, and their lateral root primordia may show different patterns of division activity compared to primordia of species with a 'closed' apex (e.g., maize, onion). For example, in *Malva*, and also in *Ipomoea purpurea* (Seago, 1973), the endodermis appears to make little contribution to a proto-cap. Hence, the absence of a proto-cap may preclude the development of the first quiescent centre and so permit the 'open' type of apex. A similar regulation of apical activity presumably occurs in the embryo, during the formation of the primary root.

7.3.5 Polarised inductive events in lateral root initiation

That the endodermis could play an even more substantial role in the morphogenesis of lateral root primordia, besides providing a proto-cap, is suggested by an anatomical study of primordia developing in *Convolvulus* root segments grown in liquid medium *in vitro* (Bonnett & Torrey, 1966). These segments have an intrinsic polarity, developing adventitious shoot primordia mainly at their proximal ends. Adventitious root primordia, however, form all the way along the segments. The two types of primordia, both of which develop opposite protoxylem poles, can be distinguished early in their development on account of the behaviour of the endodermis. Root primordia possess a proto-cap formed by periclinal divisions of endodermal cells (Fig. 7.11), whereas shoot primordia show no comparable endodermal development (Fig. 7.12) except for a few radial divisions, possibly in response to the stretching of this layer by the expanding derivatives of pericycle. An intriguing possibility is that the endodermis is able to provide the primordium with an inductive stimulus which determines root development. Should the endodermis be unable to deliver this stimulus, then development, by default, is towards a shoot. It may not be too far-fetched to suppose that the quiescent centre, which the endodermis seems normally to induce, plays some vital role of initiating a rhizogenetic developmental pathway in the primordium.

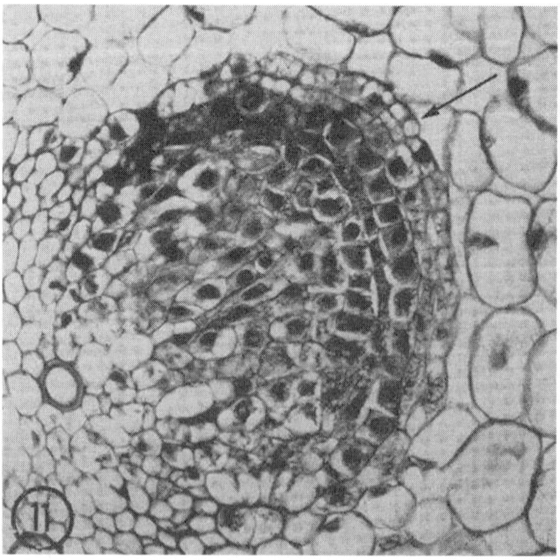

Figure 7.11 Contrasting patterns of primordial development within cultured roots of *Convolvulus arvensis*. Developing lateral root primordium. Note the active involvement of the basophilic endodermis (arrowed) which has formed a bilayer of cells. Reproduced from Bonnett and Torrey (1966) with the permission of *American Journal of Botany*. Magnification ×260.

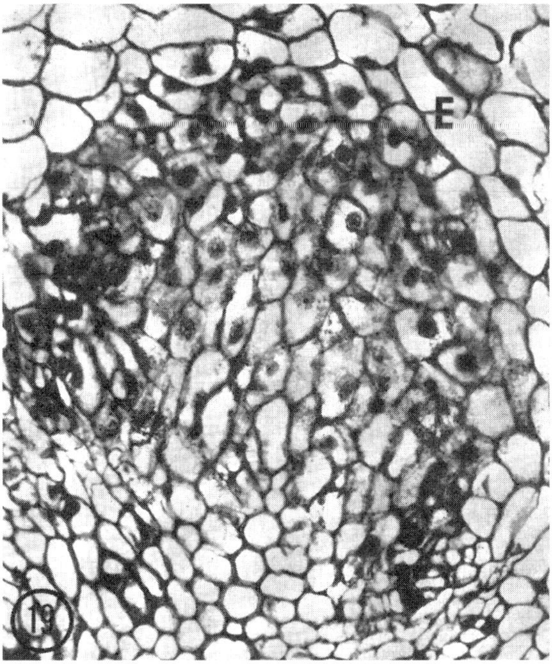

Figure 7.12 Contrasting patterns of primordial development within cultured roots of *Convolvulus arvensis*. Developing shoot primordium of approximately the same height as the root primordium in Fig. 7.11. The endodermis (E) is not basophilic and shows little involvement on primordium formation. Reproduced from Bonnett and Torrey (1966) with the permission of *American Journal of Botany*. Magnification ×270.

7.3.6 Disturbances to the usual pattern of lateral root initiation

It is well known that treatment of roots with auxin induces mitosis in cells which, in the normal course of development, would have never divided again (Levan, 1939). Likewise, auxin, as well as amputation of the root tip, induces the formation of additional lateral root primordia (Lloret *et al*., 1988; Hinchee & Rost, 1992b). Some auxins are particularly potent in this regard: IBA, for example, induces lateral roots in a variety of lettuce (*Lactuca sativa* cv Baijianye) which normally lacks them (Zhang & Hasenstein, 1999). The fact that the additional primordia are not hapahazardly arranged but lie in rows could suggest that the anatomy of the stele restricts primordia to particular files of pericycle cells. However, it may be that only certain cell files are able to support primordia. Additional, exogenously supplied auxin may simply reduce the number of divisions that occur within the files of pericycle, thus reducing the spacing, in terms of cell numbers, between successive primordia (Hinchee & Rost, 1992b).

Often, the additional primordia are discrete and do not merge with one another. However, the primordia can also be peculiarly enlarged in the axial direction. This could arise if their base comprised an asymmetric grid of cells: with 16 × 8 cells, for example, instead of the usual 8 × 8 cells. The resultant primordia would appear to be fasciated, as is the case in roots treated with inhibitors of polar auxin transport (Hinchee & Rost, 1992a). Laskowski *et al.* (1995) examined transverse sections of such auxin-stimulated roots and found that the enlarged primordia were also unusually wide and could even encompass about half of the pericycle perimeter: here, the root bases were composed of 16 × 16 = 256 cells (Laskowski *et al.*, 1995, Fig. 2B). Whether such auxin-induced primordia of *Arabidopsis* roots actually convert to emergent lateral roots, as happens in the case of the fasciated primordia of pea roots (Hinchee & Rost 1992a), has not been studied. Published photographs suggest that at least some of them may not grow (e.g. Boerjan *et al.*, 1995, Fig. 1G). If so, then the lateral roots which do emerge may develop only from those primordia which would have normally been formed in the acropetal sequence, having commenced development at one of the prescribed organogenetic centres. It is unclear whether the auxin-induced primordial plaque is formed in the normal way, with anticlinal divisions preceding the periclinal divisions. Evidence suggests that there is a preponderance of periclinal divisions in such primordia (Pulgarín *et al.*, 1996).

7.3.7 Root nodules and other types of root primordia

Uridine was identified as a cell division factor involved in rhizobium-induced nodule formation in pea root cortex (Smit *et al.*, 1995). The source of uridine is the stele. When infected with *Rhizobium leguminosarum*, the effect of uridine liberation into the cortex from sites opposite the protoxylem becomes apparent. The result is a pattern of cortical cell divisions that seems to follow the flow of this mitogen (Libbenga & Harkes, 1973). These observations on early nodulation events, as well as those already mentioned for lateral root primordium initiation, suggest that fluxes of physiologically active chemicals (such as auxin and uridine) set up polarised morphogenetic gradients and pre-patterns within the root. Indeed, van Fleet (1948) observed many different, species-specific patterns of redox gradients across the root cortex whose significance are hardly understood. In pea roots, rhizobia induce the cortical cell divisions for nodule formation in a zone located between that where lateral root primordia and secondary vascular development each separately take place. The anatomical/physiological features which prepare these two sites for cell division, each having a distinct developmental outcome, has not been systematically examined.

Enlarged, pericycle-derived root primordia are found in tap roots of *Quercus ilex* following amputation of the tip (Beissalah *et al.*, 1988). Here they are called 'regeneration roots'. Because they occur near the cut stump, these regeneration primordia are likely to mark sites where auxin has accumulated to unusually high levels, hence accounting for their enlarged size (cf. Boerjan *et al.*, 1995). A similar accumulation of auxin at the root tip may account for bifurcation of roots, such as found in *Selaginella* spp. Possibly the apices of these pteridophyte roots have no quiescent centre and hence no means of breaking down the auxin which is arriving at the tip as the result of its active transport. Rhythmic bifurcation of the tip may be one means of regulating the auxin content of the tip, by splitting the flow path; the high auxin content may also ensure that each succeeding branch order of root is no thinner than the last (A. Jackson & P. Barlow, unpublished observations), unlike the case in angiosperms where successive orders of lateral root decrease in diameter according to a power law. Another interesting pattern of division is that which accompanies adventitious root primordium formation in explants of walnut (Ermel *et al.*, 2000). A circular island of primordial cells develops to one side of the procambium. The circular pattern is due to the orientation of new division walls and may indicate an accumulating pool of auxin in much the same way that circular xylem vessels indicate a previous circular auxin flow in vessel precursor cells (Sachs & Cohen, 1982).

7.4 Polarity of secondary growth

Secondary tissue in roots displays three types of polarity. First, there is the longitudinal polarity that is characteristic of the original primary tissue. It maintains the acropetal transport of auxins (in the vascular cambium) and of sucrose (in the secondary phloem). Second, there is a radial polarity manifested in the radial production, from the vascular cambium, of new cells that differentiate as secondary xylem and phloem. Third, there is a circumferential polarity manifested upon the increasing cambial circumference as a regular spacing of tissues, such as parenchymatous rays and phloem fibre bundles. Recent evidence suggests there is an active genetic control of these circumferential spacing patterns (Parker *et al.*, 2003).

7.4.1 Origin and polarised behaviour of the vascular cambium

Circumferential and radial polarities are evident in the earliest stages of secondary development, when cambium forms within undifferentiated parenchymatous tissue. In both *in vitro*-cultured roots of *Raphanus sativus* (Torrey & Loomis, 1967) and soil-grown seedling roots of the tree *Aesculus hippocastanum* (Stain, 2002), periclinal cell divisions mark out the cambium in procambial cells to the inside of primary phloem bundles (Fig. 7.13). Periclinal divisions then proceed circumferentially until two moving fronts of

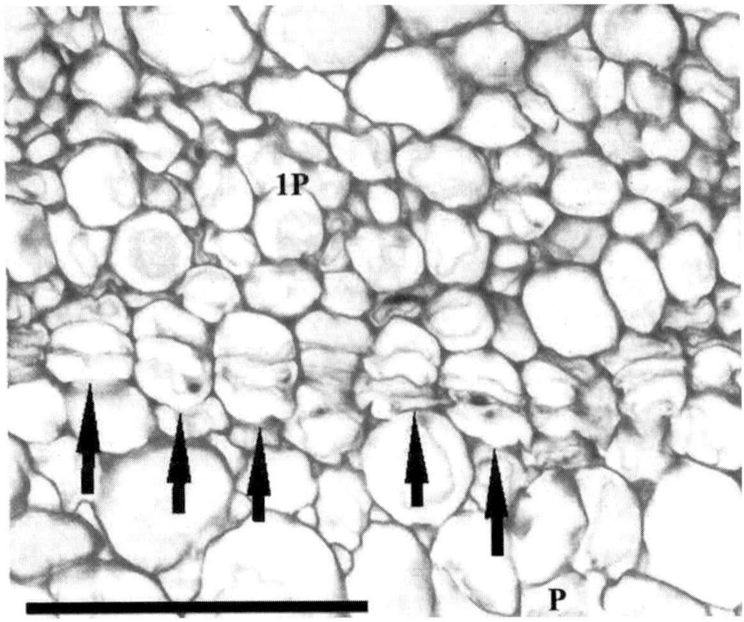

Figure 7.13 Cambial development in taproot of horse-chestnut (*Aesculus hippocastanum*). First periclinal divisions (arrowheads) in procambial cells lying between pith and primary phloem. Abbreviations: 1P – primary phloem, P – pith. Scale bar = 100 μm. Reproduced from Stain (2002), with permission.

division meet in the interfascicular zones (Fig. 7.14) or arrest at the poles of the protoxylem. In these latter locations, cells are produced centripetally from the inner face of the pericycle and establish files of radially oriented parenchymatous, multiseriate primary rays (Fig. 7.15). After its establishment, the band of cambium consists of radial files of axially elongated and mitotically active fusiform cells (Fig. 7.16). In *Aesculus* roots, when cambium is maximally active, there may be up to ten such fusiform cells in each radial file (Stain, 2002). Cell division in the cambium is probably supported by axial auxin transport, though whether the route of the auxin is through the fusiform cambial cells is not precisely known. Axial auxin flow may also encourage the elongation of the fusiform cells in this axial plane, though there is also some radial movement of auxin which influences cell enlargement in this plane (summarised by Kramer, 2001).

Whether there is a true directionality of cell production in the cambium – that is, whether new cells are produced 'towards' the xylem domain, or 'towards' the phloem – remains a theoretical possibility (cf. Barlow *et al.*, 2002), particularly since the cell divisions involved are periclinal and are not marked by any obvious asymmetry. Directionality of cell production could involve particular orientations of oblique division walls. In fact, sections in the

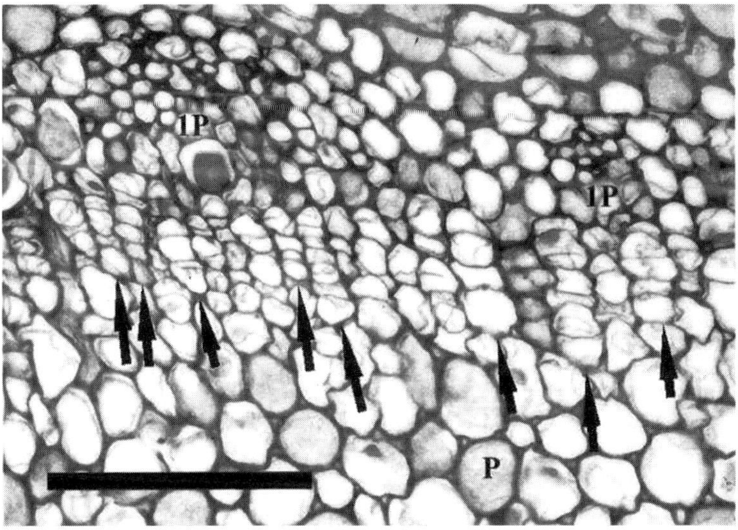

Figure 7.14 Cambial development in taproot of horse-chestnut (*Aesculus hippocastanum*). Joining of two circumferentially advancing vascular cambial fronts. A few cells between the two fronts (indicated by third arrowhead from the right) have not yet divided periclinally. Abbreviations: 1P – primary phloem, P – pith. Scale bar = 100 μm. Reproduced from Stain (2002), with permission.

tangential plane reveal S- and Z-type pseudo-transverse anticlinal divisions, with the upper tip of the division wall pointing, respectively either to the left (\) or to the right (/). S- and Z-types of anticlinal cambial division are multiplicative and contribute to circumferential cell increase. Where one or other type of such divisions are clustered in domains, as is well known in stems of trees, they introduce a chirality to the structure of the wood (spiral grain).

It is difficult to quantify accurately the relative amounts of axial, radial and circumferential growth in a non-storied cambial zone because of non-symplastic growth, the fusiform shape of its cells and their oblique divisions. An oblique anticlinal division wall between two fusiform daughter cells indicates that the cell body within the mother cell was itself oriented obliquely prior to mitosis. Observations of dividing fusiform cells in roots and shoots of conifers (Barghoorn, 1940) suggest that the cell bodies of their non-storied cambia are constantly moving up and down the interior of these cells. S and Z anticlinal cell divisions (and maybe S and Z periclinal divisions also) in the cambium could arise because, at the time of mitosis, the tip of the motile cell body, with its nucleus, was moving upon one or other of the radial (or tangential) walls. The new division wall then has alternative S and Z orientations. It is also notable that long bundles of actin run the length of the cambial cells (Chaffey *et al.*, 1997a). These bundles may promote both polarised axial flow of auxin and nuclear (cell body) movement, as proposed

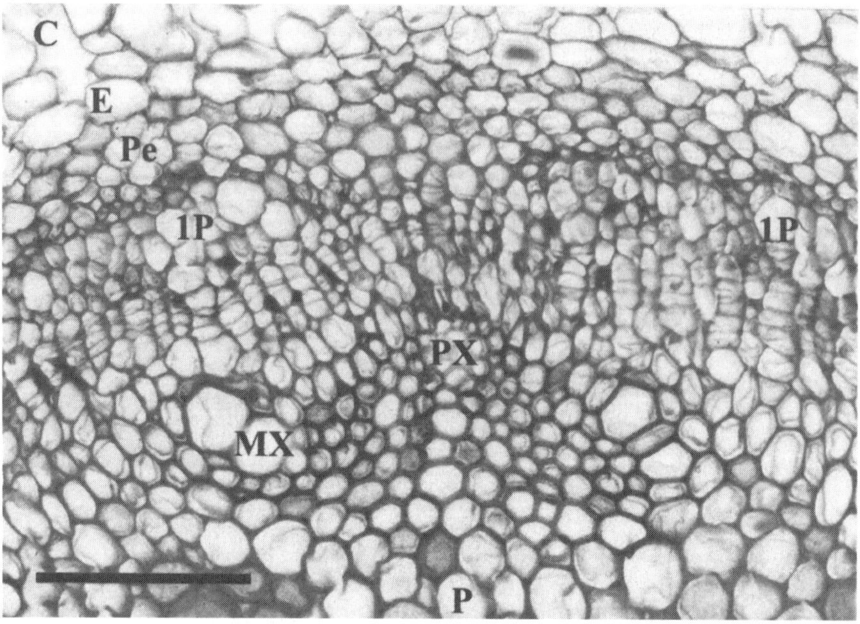

Figure 7.15 Cambial development in taproot of horse-chestnut (*Aesculus hippocastanum*). Young vascular cambium established close to each area of primary phloem, but absent at the primary xylem pole. Here, a primary multiseriate ray will be formed from derivatives of the pericycle. Abbreviations: C – cortex, E – endodermis, MX – metaxylem, 1P – primary phloem, P – pith, Pe – pericycle, PX – primary xylem pole. Scale bar = 100 μm. Reproduced from Stain (2002), with permission.

earlier for stelar cells of primary tissue. Up and down movement (cyclosis) of the cell body, from one end of the cell to the other, would also support bidirectional cell extension: that is, the two tips at opposite ends of the fusiform cell extend intrusively upwards and downwards, in parallel with the axis of the organ. Cyclosis may continue when the xylem and phloem derivatives of the cambium differentiate as fibres, as it is a feature of these cells to continue elongating during maturation. Xylem vessels, by contrast, show limited elongation of their tips, probably because these cells expand greatly in girth during their phase of differentiation.

An obvious radial polarity of cell differentiation is evident: immature cells are closest to the cambium and mature cells are further away. This sequence of differentiation within a radial file, and the continuous radial development of these files, make it appear that there is an advancement of differentiation towards the cambium – centrifugal advancement in the case of xylem, and centripetal in the case of phloem – which is balanced against the rate of cell production. In the case of a radial file of xylem fibre cells, for example, when the rate of new cell production from the cambium and the rate of onset of fibre cell differentiation are in balance, the cambium remains a constant width,

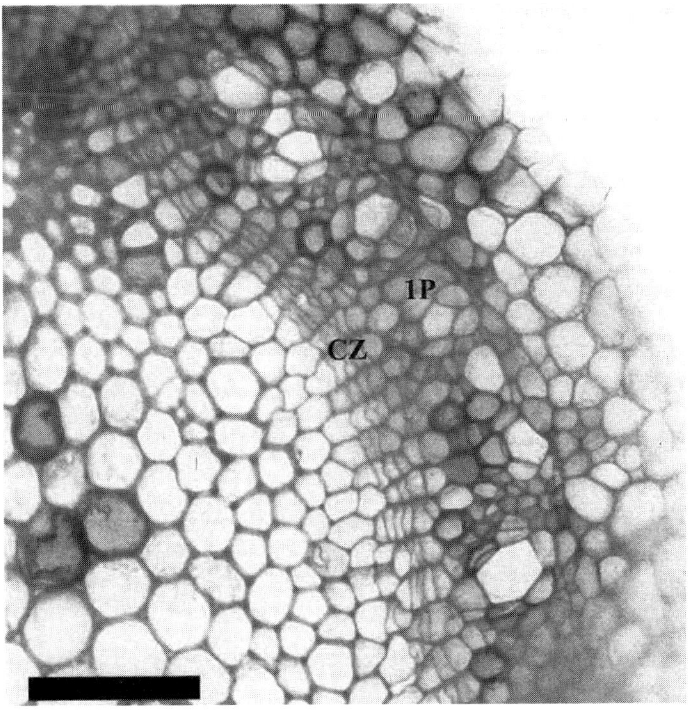

Figure 7.16 Cambial development in taproot of horse-chestnut (*Aesculus hippocastanum*). Young vascular cambium with single or pairs of cells having undergone 2–3 periclinal divisions to contribute to each radial file. Abbreviations: CZ – cambial zone, 1P – primary phloem. Scale bar = 100 μm. Reproduced from Stain (2002), with permission.

even though the xylem file radius steadily increases. But when the rate of onset of cell differentiation outpaces that of cell production, then the cambium shrinks, and can even be lost, as at the onset of winter dormancy. The rate of cell production in the cambium is, then, regulated by the number of dividing cells within its radial files.

7.4.2 Sites of mixed polarity

Within wood, rays can be composed of two types of cells, procumbent and upright (Panshin & de Zeeuw, 1980). The former are radially elongated (radial width > vertical height), the latter are vertically elongated (vertical height > radial width). In hardwood species, rays may be composed entirely of one or other type, or show mixtures of both types. The mixed type of ray, such as found in the wood of sweetgum, *Liquidambar styraciflua*, is of interest because it highlights the possibility that different axialities and polarities might coexist within ray tissue. In sweetgum, the upright cells lie upon the upper and lower margins of the ray which otherwise consists of procumbent cells. The upright cells, besides widening in the radial plane and dividing periclinally, also grow in the axial plane and presumably divide anticlinally. The procumbent cells only widen radially and divide periclinally. In species with procumbent ray cells only, it may be that axial growth and anticlinal divisions are distributed along the whole axial length of the ray and are not confined just to the tip cells (as may be the case in rays with marginal upright cells). Axial growth and anticlinal division of a ray initial cell may be a feature of a cell body that is oriented in line with the direction of auxin transport, whereas a periclinal division is a response to radial stress or strain. Alteration from an upright to a procumbent cellular form can occur during the early development of rays, due to segmentation and repolarisation of fusiform cells, in the roots of various gymnosperm species in the Pinaceae (Bannan, 1934), those of *Cedrus libanotica* being especially informative in this respect (Barghoorn, 1940). Cells in transitional stages between the two types have curious amoeboid shapes, suggesting unstable patterns of plasma membrane outgrowth. Given that the balance between rates of cell growth and cell division are quite delicately regulated in response to the concentration of auxin (Tao & Verbelen, 1996), differences in wood anatomy in roots (and shoots) may be manifestations of this, but displayed in the context of an extremely complex system of cellular interactions, auxin fluxes and the various potential pathways for cellular differentiation.

It is worth remarking that mixed polarities may also exist within the contractile roots of monocotyledonous geophytes. In the contractile zone of the cortex, cells rearrange their shape so that they become longer in the radial direction than previously, and shorter in the axial direction. How this comes about is controversial (Pütz & Froebe, 1995). The contractile phenomenon may be an example of a rupture in the primary polarity and the establishment of a new growth field. Any change in the axiality of growth is likely to be preceded by an alteration to polarised flux of a growth regulator.

7.4.3 *Comparison of primary and secondary meristems and their polarity*

Unfortunately knowledge of the physiology and cytology of cambium in roots is far from perfect (reviewed by Chaffey, 2002), especially as we wish now to determine whether the ideas about auxin transport, cell body, cell division, and so on, that were developed above for primary root tissue, also apply to their counterparts in the secondary tissues. As already indicated, the two meristems, one primary, the other secondary, differ in important respects: in the primary root meristem cell growth is symplastic with the cells growing in conformity with each other, whereas in non-storied root cambium this is not so and the tips of the fusiform cells can glide past each other. Moreover, the orientation of new division walls in primary meristems is correlated with the orientation of a preprophase band of microtubules, whereas in fusiform cambial cells this class of microtubules is absent. (Cambial ray cells, however, are similar to cells in primary meristems with respect to the above-mentioned features; their growth is radial as well as axial, and they also fabricate pre-prophase bands.) Cambial fusiform cells contain extensive F-actin microfilaments (Chaffey *et al.*, 1997a; Chaffey & Barlow, 2000), but whether these filaments are associated with axial auxin transport is not known but seems likely. There are also actin cables which extend radially through the ray system (Chaffey & Barlow, 2000, 2001), and these may transport auxin deep into the secondary xylem and phloem. Actin is densely aggregated at pit fields of both secondary xylem and phloem fibres. The presence of myosin and plasmodesmata at the pit fields (Chaffey & Barlow, 2002) fosters the idea that these structures are plant mini-synapses (Fig. 7.2) which transport auxin from one cell to the next. Apparently, pit fields within the primary walls of cambial cells are mobile (Barnett, 1987), a finding offering intriguing possibilities in terms of their possible function as plant mini-synapses. Although actin is present in the phragmoplast of dividing fusiform cells, whether it contributes to the formation of plant synapses upon their division wall has not been investigated. Abundant vesicles, some of which were presumed to be endocytotic, have also been recognised in cambium of *Aesculus* taproots (Chaffey *et al.*, 1997b). The sliding growth shown by such cells may preclude the establishment of such synaptic structures. In cambium of poplar stems, for example, the number of plasmodesmata traversing the tangential walls is only 14 per cent of the number traversing the radial walls (J. Barnett & K. Pickering, personal communication). Possibly, the difference in origin of these two types of walls (from either a periclinal or a pseudotransverse division, respectively), as well as the non-symplastic growth of the cambial tissue, lead to this spatial asymmetry of communication channels. The axial xylem parenchyma cells are derived from fusiform cells, and these divide transversely. Little is known of them; but they may also be candidates for axial auxin transport on account of their longevity (they outlive the surrounding fibre cells), their pit fields, and their transverse division walls which might be capable of supporting plant synapses.

7.4.4 Comparison of sites of renewed cell division in roots

How can the extensive development of vascular cambium in mature regions of a root be reconciled with the more localised, pointillistic development of lateral root primordia? Both sets of meristematic cells may depend upon auxin for their divisions. The possible role of endodermal cells in primordium initiation was remarked upon earlier. Pericycle cells neighbouring the endodermal passage cells may co-localise with a bottle-neck for auxin as it flows towards and then out through the passage cells. Cell division could be stimulated in those auxin-enriched pericycle cells lying on its path. Further back along the root, endodermis is more mature and the passage cells become sealed.

Endodermis is now a barrier to solute outflow, and auxin becomes confined within the vascular cylinder, having no outlet from the stele. The sequence of cambial initiation to the inside of the stele perimeter may reflect the pattern of auxin build-up. The absence of vascular cambium opposite the primary protoxylem poles (Fig. 7.15) could still have some relationship with the former presence of endodermal passage cells. Here there could still be some leakage, or transport, of auxin to induce the development of primary rays from the pericycle, particularly since these rays also extend into the secondary phloem. Barriers to auxin flow that result in auxin accumulation may also be set up by the distribution of naturally occurring substances which disable PIN protein-dependent auxin transport systems. Flavonoids are powerful inhibitors of longitudinal auxin transport (Brown *et al.*, 2001), and certain of them, located, for example, within the endodermis or cortex, may be able to impair the radial efflux of auxin from the stele of the root and, hence, by the consequent retention of auxin, encourage the formation of cambium. It is also possible that the presence of flavonoids nearer to the root tip could help regulate the positioning of lateral root primordia, as suggested by Brown *et al.* (2001).

The orientation of the first periclinal cambial divisions may be a response to the physical pressure exerted by the potentially expansive parenchymatous procambial cells upon the relatively inextensible endodermis which encircles them. Pressure has been found to orient cell division in such a way that the phragmoplast and new division walls form at right angles to the direction of its application (Brown & Sax, 1962; Lintilhac & Vesecky, 1984; Lynch & Lintilhac, 1997). Similar considerations also apply to the development of the cambial phellogen whose radial cell files develop from pericycle beneath the endodermis (Fig. 7.17). From these cells a suberised layer (periderm) is differentiated.

It should not be overlooked that cell divisions somewhat similar to cambial divisions also occur in primary root apices. As already noted, there are the periclinal divisions which widen the inner cortex (Barlow, 1987; Barlow & Adam, 1989b) and, in some cases (e.g. maize), the outer cortex as well (Seago & Wolniak, 1976; Barlow, 1987). Again, auxin amounts and fluxes across the cell wall boundary between stele and cortex, and the backing up of auxin levels at the exodermal wall, may, together with the cells' responsiveness to the pattern of physical forces inherent to root tissue, orient cell growth and division at the root apex.

7.5 Cell polarity and organ growth

The major concern of the present chapter has been to adduce contemporary information concerning cellular polarity within root tissues. Acropetal polarity of auxin flow is actively maintained and, given auxin's capacity to promote

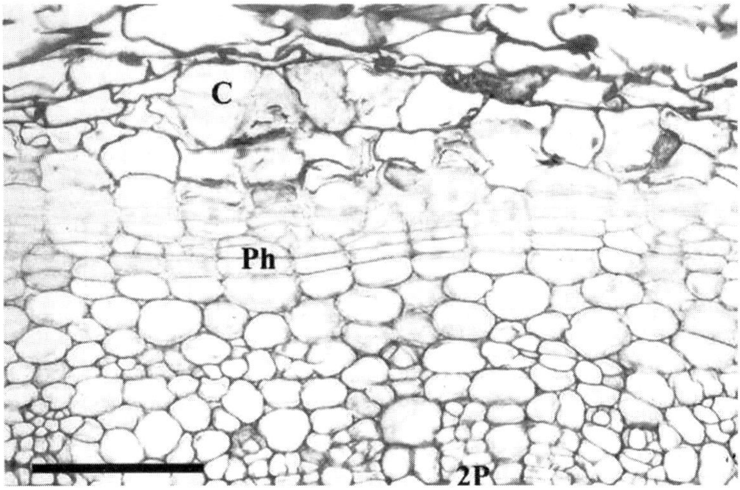

Figure 7.17 Periclinal divisions in the pericycle contribute to a phellogen in taproot of horse-chestnut (*Aesculus hippocastanum*). Abbreviations: C – cortex, 2P – secondary phloem, Ph – phellogen. Scale bar = 100 μm. Reproduced from Stain (2002), with permission.

both cell growth and cell division (for which cytokinins are also required), it seems probable that wherever auxin accumulates as a result of its transport, some sort of meristem will develop. Whether the meristem becomes organised as an apical meristem or remains disorganised may, in part, depend on whether or not the cells are able to develop collectively around themselves a physical boundary. This is evident in the experimental example of the carrot root used by Pilet (1967) for observations upon its polarity. This root with its stelar auxin flow is a familiar and organised organ. Isolated cubes of carrot root stelar tissue, by contrast, although they still retain polar auxin flow, form callus at their apical end. Here, there is no information provided by a structural boundary that can be utilised to regulate the orientation of their cell divisions.

Acropetal auxin transport occurs in the stele of the root. It was proposed previously (Barlow, 1989) that the stele is the portion of the root which drives root growth. The cortex and epidermis are skin-like tissues which are stretched protectively over the extending stelar core. The extension of this core forces the root into a physically resistant soil medium where the cone-shaped root cap and sloughed-off cap cells (border cells) then reduce the friction between root and soil. Just as the vascular tissue, the protoxylem poles in particular, counters the force of the growing lateral primordium, so it also helps to make the root an efficient penetrative tool by countering the tendency of the root to buckle. Indeed, mechanical impedance can induce alterations to the pattern of the primary xylem strands (Jaramillo *et al.*, 1992) so that the stele, thus modified, possesses an increased ability to overcome this stricture. As already suggested, orientation of cell body division in the stele and the consequent process of cytokinesis may be regulated in separate ways in both stele and cortex: by auxin polarisation in the stele, and by mechanical stretching in the cortex. As long as sufficient auxin and sucrose can pass from the stele into the surrounding tissues, then the cell cycle can be maintained throughout the root apex.

7.6 Conclusions

Excepting cases where polarised flows of solutes (auxins, xylem sap, etc.) can be experimentally determined, polarised developmental phenomena in roots are often inferred from the patterns and sequences of cell divisions, and the changing appearance of the cells as they participate in histogenesis. Many of these patterns may be traced back to probable patterns of auxin flow, and to the physical stresses and constraints within the developing tissues. As we have attempted to show, the community of plant cell bodies is collectively responsible not only for executing many root-specific developmental programmes, but also for sensing the inductive flows and stress/strain patterns which are their trigger. Polarity is the evidence, or fingerprint, of the cell bodies' ability to integrate the physico-chemical fields of the living plant system. Polarity is also a direct outcome of the orientation of actin microfilaments upon which the auxin transport system is built. Sometimes it is forgotten that axiality and polarity are distinct aspects of cell growth and that, together, they participate in growth and morphogenesis. Now that the structure of the plant cytoskeleton has become more comprehensible in recent years, it is appropriate to summarise its importance by suggesting that the actin cytoskeleton contributes to polarity whereas the cortical microtubular cytoskeleton contributes to axiality, and that these attributes apply not only to the cell but, by their collective properties, to the organ also. In the development of growing cellular systems in plants, establishment of polarity is the fundamental event that leads to the establishment of axiality.

References

Baluška, F. & Volkmann, D. (2002) Actin-driven polar growth of plant cells. *Trends Cell Biol.*, **12**, 14.

Baluška, F., Parker, J.S. & Barlow, P.W. (1992) Specific patterns of cortical and endoplasmic microtubules associated with cell growth and tissue differentiation in roots of maize (*Zea mays* L.). *J. Cell Sci.*, **103**, 191–200.

Baluška, F., Parker, J.S. & Barlow, P.W. (1993a) A role for gibberellic acid in orienting microtubules and regulating cell growth polarity in the maize root cortex. *Planta*, **191**, 149–157.

Baluška, F., Brailsford R.W., Hauskrecht M., Jackson, M.B. & Barlow, P.W. (1993b) Cellular dimorphism in the maize root cortex: involvement of microtubules, ethylene and gibberellin in the differentiation of cellular behaviour in postmitotic growth zones. *Bot. Acta*, **106**, 394–403.

Baluška, F., Vitha, S., Barlow, P.W. & Volkmann, D. (1997) Rearrangements of F-actin arrays in growing cells of intact maize root apex tissues: a major developmental switch occurs in the postmitotic transition region. *Eur. J. Cell Biol.*, **72**, 113–121.
Baluška, F., Barlow, P.W., Lichtscheidl, I.K. & Volkmann, D. (1998) The plant cell body: a cytoskeletal tool for cellular development and morphogenesis. *Protoplasma*, **202**, 1–10.
Baluška, F., Barlow, P.W. & Volkmann, D. (2000a) Actin and myosin VIII in developing root cells, in *Actin: A Dynamic Framework for Multiple Plant Cell Functions* (eds C.J. Staiger, F. Baluška, D. Volkmann & P.W. Barlow) (Developments in Plant and Soil Sciences, vol. 89), Kluwer Academic Publishers, Dordrecht, pp. 457–476.
Baluška, F., Volkmann, D. & Barlow, P.W. (2000b) Actin-based domains of the 'cell periphery complex' and their associations with polarized 'cell bodies' in higher plants. *Plant Biol.*, **2**, 253–267.
Baluška, F., Volkmann, D. & Barlow, P.W. (2001a) Motile plant cell body: a 'bug' within a 'cage'. *Trends Plant Sci.*, **6**, 104–111.
Baluška, F., Jásik, J., Edelmann, H.G., Šalajová, T. & Volkmann, D. (2001b) Latrunculin B induced plant dwarfism: plant cell elongation is F-actin dependent. *Devel. Biol.*, **231**, 113–124.
Baluška, F., Busti, E., Dolfini, S., Gavazzi, G. & Volkmann, D. (2001c) *Lilliputian* mutant of maize shows defects in organization of actin cytoskeleton. *Devel. Biol.*, **236**, 478–491.
Baluška, F., Volkmann, D. & Barlow, P.W. (2001d) A polarity crossroad in the transition growth zone of maize root apices: cytoskeletal and developmental implications. *J. Plant Growth Reg.*, **20**, 170–181.
Baluška, F., Hlavačka, A., Šamaj, J., Palme, K., Robinson, D.G., Matoh, T., McCurdy, D.W., Menzel, D. & Volkmann, D. (2002) F-actin-dependent endocytosis of cell wall pectins in meristematic root cells: insights from brefeldin A-induced compartments. *Plant Physiol.*, **130**, 422–431.
Baluška, F., Wojtaszek, P., Volkmann, D. & Barlow, P.W. (2003a) The architecture of polarized cell growth: the unique status of elongating plant cells. *BioEssays*, **25**, 569–576.
Baluška, F., Šamaj, J. & Menzel, D. (2003b) Polar transport of auxin: carrier-mediated flux across the plasma membrane or neurotransmitter-like secretion? *Trends Cell Biol.*, **13**, 282–285.
Bannan, M.W. (1934) Origin and cellular character of xylem rays in gymnosperms. *Bot. Gaz.*, **96**, 260–280.
Barghoorn, E.S., Jr. (1940) Origin and development of the uniseriate ray in the Coniferae. *Bull. Torrey Bot. Club*, **67**, 303–328.
Barlow, P.W. (1984) Positional controls in root development, in *Positional Controls in Plant Development* (eds P.W. Barlow & D.J. Carr), Cambridge University Press, Cambridge, pp. 281–318.
Barlow, P.W. (1987) Cellular packets, cell division and morphogenesis in the primary root meristem of *Zea mays* L. *New Phytol.*, **105**, 27–56.
Barlow, P.W. (1989) Anatomical controls of root growth. *Asp. Appl. Biol.*, **22**, 57–66.
Barlow, P.W. (2002) Cellular patterning in root meristems: its origin and significance, in *Plant Roots. The Hidden Half*, 3rd edn (eds Y. Waisel, A. Eshel & U. Kafkafi), Marcel Dekker, New York, pp. 49–82.
Barlow, P.W. (2003) The root cap: cell dynamics, cell differentiation and cap function. *J. Plant Growth Reg.*, **21**, 261–286.
Barlow, P.W. & Adam, J.S. (1988) The position and growth of lateral roots on cultured root axes of tomato, *Lycopersicon esculentum* (*Solanaceae*). *Plant Systemat. Evol.*, **158**, 141–158.
Barlow, P.W. & Adam, J.S. (1989a) Anatomical disturbances in primary roots of *Zea mays* following periods of cool temperature. *Environ. Exp. Bot.*, **29**, 323–336.
Barlow, P.W. & Adam, J.S. (1989b) Experimental control of cellular patterns in the cortex of tomato roots, in *Structural and Functional Aspects of Transport in Roots* (eds B.C. Loughman, O. Gašparíková & J. Kolek) (Developments in Plant and Soil Sciences, vol. 36), Kluwer Academic Publishers, Dordrecht, pp. 21–24.
Barlow, P.W. & Baluška, F. (2000) Cytoskeletal perspectives on root growth and morphogenesis. *Ann. Rev. Plant Physiol. Plant Molec. Biol.*, **51**, 289–322.
Barlow, P.W. & Palma, B. (1997) The place of roots in plant development, in *Biology of Root Formation and Development* (eds A. Altman & Y. Waisel), Plenum Press, New York, pp. 1–11.

Barlow, P.W., Lück, H.B. & Lück, J. (2001) The natural philosophy of plant form: cellular autoreproduction as a component of a structural explanation of plant form. *Ann. Bot.*, **88**, 1141–1152.

Barlow, P.W., Brain, P. & Powers, S.J. (2002) Estimation of directional division frequencies in vascular cambium and in marginal meristematic cells of plants. *Cell Proliferation*, **35**, 49–68.

Barnett, J.R. (1987) Changes in the distribution of plasmodesmata in developing fibre-tracheid pit membranes of *Sorbus aucuparia* L. *Ann. Bot.*, **59**, 269–279.

Basu, S., Sun, H., Brian, L., Quatrano, R.L. & Muday, G.K. (2002) Early embryo development in *Fucus distichus* is auxin sensitive. *Plant Physiol.*, **130**, 292–302

Beissalah, Y., Amin, T., El Hajzein, B. & Neville, P. (1988) Formation des racines de régénération chez *Quercus ilex* L. *Bull. Soc. Bot. France*, **135**, *Lettres Bot.*, 4/5, 333–342.

Bell, J.K. & McCully, M.E. (1970) A histological study of lateral root initiation and development in *Zea mays*. *Protoplasma*, **70**, 179–205.

Berleth, T. (2000) Plant development: hidden networks. *Curr. Biol.*, **10**, R658–R661.

Berleth, T. & Mattsson, J. (2000) Vascular development: tracing signals along veins. *Curr. Opin. Plant Biol.*, **3**, 406–411.

Bhalerao, R.P., Eklöf, J., Ljung, K., Marchant, A., Bennett, M. & Sandberg, G. (2002) Shoot-derived auxin is essential for early lateral root emergence in *Arabidopsis* seedlings. *Plant J.*, **29**, 325–332.

Boerjan, W., Cervera, M.-T., Delarue, M., Beeckman, T., Dewitte, W., Bellini, C., Caboche, M., Van Onkelen, H., Van Montagu, M. & Inzé, D. (1995) *superroot*, a recessive mutation in *Arabidopsis* confers auxin overproduction. *Plant Cell*, **7**, 1405–1419.

Bonifacino, J.S. & Jackson, C.L. (2003) Endosome-specific localization and function of the ARF activator GNOM. *Cell*, **112**, 141–145.

Bonnett, H.T., Jr. & Torrey, J.G. (1966) Comparative anatomy of endogenous bud and lateral root formation in *Convolvulus arvensis* roots cultured in vitro. *Am. J. Bot.*, **53**, 496–507.

Brown, C.L. & Sax, K. (1962) The influence of pressure on the differentiation of secondary tissues. *Am. J. Bot.*, **59**, 311–316.

Brown, D.E., Rashotte, A.M., Murphy, A.S., Normanly, J., Tague, B.W., Peer, W.A., Taiz, L. & Muday, G.K. (2001) Flavonoids act as negative regulators of auxin transport in vivo in *Arabidopsis*. *Plant Physiol.*, **126**, 524–535.

Burda, H., Begall, S., Grütjen, O., Scharff, A., Nevo, E., Beiles, A., Cerveny, J. & Prucha, K. (1999) How to eat a carrot? Convergence in the feeding behavior of subterranean rodents. *Naturwiss.*, **86**, 325–327.

Butler, J.H., Hu, S., Brady, S.R., Dixon, M.W. & Muday, G.K. (1998) *In vitro* and *in vivo* evidence for actin association of the naphthylphthalamic acid-binding protein from zucchini hypocotyls. *Plant J.*, **13**, 291–301.

Byrne, J.M. (1973) The root apex of *Malva sylvestris*. III. Lateral root development and the quiescent center. *Am. J. Bot.*, **60**, 657–662.

Byrne, J.M., Byrne, J.M. & Emmit, D.P. (1982) Development and structure of the vascular connection between primary and lateral root of *Lycopersicon esculentum*. *Am. J. Bot.*, **69**, 287–297.

Cande, W.Z., Goldsmith, M.H. & Ray, P.M. (1973) Polar auxin transport and auxin-induced elongation in the absence of cytoplasmic streaming. *Planta*, **111**, 279–296.

Casero, P.J., Casimero, I., Rodríguez-Gallardo, L., Martín-Partido, G. & Lloret, P.G. (1993) Lateral root initiation by asymmetrical transverse divisions of pericycle cells in adventitious roots of *Allium cepa*. *Protoplasma*, **176**, 138–144.

Celenza, J.L., Jr., Grisafi, P.L. & Fink, G.R. (1995) A pathway for lateral root formation in *Arabidopsis thaliana*. *Genes Devel.*, **9**, 2131–2142.

Chaffey, N.J. (2002) Secondary growth in roots: a cell biological perspective, in *Plant Roots. The Hidden Half*, 3rd edn (eds Y. Waisel, A. Eshel & U. Kafkafi), Marcel Dekker, New York, pp. 93–111.

Chaffey, N. & Barlow, P.W. (2000) Actin in the secondary vascular system of woody plants, in *Actin: A Dynamic Framework for Multiple Plant Cell Functions* (eds C.J. Staiger, F. Baluška, D. Volkmann & P.W. Barlow) (Developments in Plant and Soil Sciences, vol. 89), Kluwer Academic Publishers, Dordrecht, pp. 587–600.

Chaffey, N.J. & Barlow, P. (2001) The cytoskeleton facilitates a three-dimensional symplasmic continuum in the long-lived ray and axial parenchyma cells of angiosperm trees. *Planta*, **213**, 811–823.

Chaffey, N. & Barlow, P. (2002) Myosin, microtubules, and microfilaments: co-operation between cytoskeletal components during cambial cell division and secondary vascular differentiation in trees. *Planta*, **214**, 526–536.

Chaffey, N.J., Barlow, P.W. & Barnett, J.R. (1997a) Microtubules rearrange during differentiation of vascular cambial derivatives, microfilaments do not. *Trees*, **11**, 333–341.

Chaffey, N., Barnett, J. & Barlow, P.W. (1997b) Endomembranes, cytoskeleton, and cell walls: aspects of the ultrastructure of the vascular cambium of taproots of *Aesculus hippocastanum* L. (Hippocastanaceae). *Int. J. Plant Sci.*, **158**, 97–109.

Charlton, W.A. (1996) Lateral root initiation, in *Plant Roots. The Hidden Half*, 2nd edn, (eds Y. Waisel, A. Eshel & U. Kafkafi), Marcel Dekker, New York, pp. 149–173.

Clowes, F.A.L. (1978) Origin of the quiescent centre in *Zea mays*. *New Phytol.*, **80**, 409–419.

Cooke, T.J., Poli, D., Sztein, A.E. & Cohen, J.D. (2002) Evolutionary patterns in auxin action. *Plant Molec. Biol.*, **49**, 319–338.

Cutter, E.G. (1971) *Plant Anatomy. Experiment and Interpretation. Part 2. Organs*, Edward Arnold, London.

Di Laurenzio, L., Wysocka-Diller, J., Malamy, J.E., Pysh, L., Helariutta, Y., Freshour, K.A. & Benfey, P.N. (1996) The *SCARECROW* gene regulates an asymmetric cell division that is essential for generating the radial organization of the *Arabidopsis* root. *Cell*, **86**, 423–433.

Dong, C.-H., Xia, G.-X., Hong, Y., Ramachandran, S., Kost, B. & Chua, N.-H. (2001) ADF proteins are involved in the control of flowering and regulate F-actin organization, cell expansion, and organ growth in *Arabidopsis*. *Plant Cell*, **13**, 1333–1346.

Dresbach, T., Qualmann, B., Kessels, M.M., Garner, C.G. & Gundelfinger, E.D. (2001) The presynaptic cytomatrix of brain synapses. *Cell. Molec. Life Sci.*, **58**, 94–116.

Dubrovsky, J.G., Rost, T.L., Colón-Carmona, A. & Doerner, P. (2001) Early primordium morphogenesis during lateral root initiation in *Arabidopsis thaliana*. *Planta*, **214**, 30–36.

Dunn, S.M., Drews, G.N., Fischer, R.L., Harada, J.J., Goldberg, R.B. & Koltunow, A.M. (1997) *fist*: an *Arabidopsis* mutant with altered cell division planes and radial pattern disruption during embryogenesis. *Sex. Plant Repr.*, **10**, 358–367.

Dustin, M.L. & Colman, D.R. (2002) Neural and immunological synaptic relations. *Science*, **298**, 785–789.

Entchev, E.V. & González-Gaitán, M.A. (2002) Morphogen gradient formation and vesicular trafficking. *Traffic*, **3**, 98–109.

Ermel, F.F., Vizoso, S., Charpentier, J.-P., Jay-Allemand, C., Catesson, A.-M. & Couée, I. (2000) Mechanisms of primordium formation during adventitious root development from walnut cotyledon explants. *Planta*, **211**, 563–574.

Feldman, L.J. (1981) Effect of auxin on acropetal auxin transport in roots of corn. *Plant Physiol.*, **67**, 278–281.

Fitting, H. (1910) Weitere entwicklungsphysiologie Untersuchungen an Orchideenblüten. *Z. Bot.*, **2**, 225–267.

Friml J., Wisniewska, J., Benková, E., Mendgen, K. & Palme, K. (2002a) Lateral relocation of auxin efflux regulator PIN3 mediates tropism in *Arabidopsis*. *Nature*, **415**, 806–809.

Friml, J., Benková, E., Blilou, I., Wisniewska, J., Hamann, T., Ljung, K., Woody, S., Sandberg, G., Scheres, B., Jürgens, G. & Palme, K. (2002b) AtPIN4 mediates sink-driven auxin gradients and root patterning. *Cell*, **108**, 661–673.

Fry, S.C. & Wangermann, E. (1976) Polar transport of auxin through embryos. *New Phytol.*, **77**, 313–317.

Fukaki, H., Tameda, S., Masuda, H. & Tasaka, M. (2002) Lateral root formation is blocked by a gain-of-function mutation in the *SOLITARY-ROOT/IAA14* gene of *Arabidopsis*. *Plant J.*, **29**, 153–168.

Gälweiler, L., Guan, C., Müller, A., Wisman, E., Mendgen, K., Yephremov, A. & Palme, K. (1998) Regulation of polar auxin transport by AtPIN1 in *Arabidopsis* vascular tissue. *Science*, **282**, 2226–2230.

Geldner, N., Hamann, T. & Jürgens, G. (2000) Is there a role for auxin in early embryogenesis? *Plant Growth Reg.*, **32**, 187–191.

Geldner, N., Friml, J., Stierhof, Y.D., Jürgens, G. & Palme, K. (2001) Auxin transport inhibitors block PIN1 cycling and vesicle trafficking. *Nature*, **413**, 425–428.

Geldner, N., Anders, N., Wolters, H., Keicher, J., Kornberger, W., Muller, P., Delbarre, A., Ueda, T., Nakano, A. & Jürgens, G. (2003) The *Arabidopsis* GNOM ARF-GEF mediates endosomal recycling, auxin transport, and auxin-dependent plant growth. *Cell*, **112**, 219–230.
Gilliland, L.U., Pawloski, L.C., Kandasamy, M.K. & Meagher, R.B. (2003) *Arabidopsis* actin gene *ACT7* plays an essential role in germination and root growth. *Plant J.*, **33**, 319–328.
Grebe, M., Xu, J. & Scheres, B. (2001) Cell axiality and polarity in plants: adding pieces to the puzzle. *Curr. Opin. Plant Biol.*, **4**, 520–526.
Greenwood, M.S. & Ycas, J. (1975) The role of stelar auxin in the development of the primary root of *Zea mays*. *Planta*, **122**, 311–314.
Hadfi, K., Speth, V. & Neuhaus, G. (1998) Auxin-induced developmental patterns in *Brassica juncea* embryos. *Development*, **125**, 879–887.
Hamann, T. (2001) The role of auxin in apical-basal pattern formation during *Arabidopsis* embryogenesis. *J. Plant Growth Reg.*, **20**, 292–299.
Hardham, A.R. & McCully, M.E. (1982) Reprogramming of cells following wounding in pea (*Pisum sativum* L.) roots. I. Cell division and differentiation of new vascular elements. *Protoplasma*, **112**, 143–151.
Hasenstein, K.H., Blancaflor, E.B. & Lee, J.S. (1999) The microtubule cytoskeleton does not integrate auxin transport and gravitropism in maize roots. *Physiol. Plant.*, **105**, 729–738.
Hinchee, M.A.W. & Rost, T.L. (1992a) The control of lateral root development in cultured pea seedlings II. Root fasciation induced by auxin inhibitors. *Bot. Acta*, **105**, 121–126.
Hinchee, M.A.W. & Rost, T.L. (1992b) The control of lateral root development in cultured pea seedlings III. Spacing intervals. *Bot. Acta*, **105**, 127–131.
Huang, Y.F., Jordan, W.T., Wing, R.A. & Morgan, P.W. (1998) Gene expression induced by physical impedance in maize roots. *Plant Molec. Biol.*, **37**, 921–930.
Jaramillo, G., White, J.W. & de la Cruz, G. (1992) The effect of soil compaction on differentiation of late metaxylem in common bean (*Phaseolus vulgaris* L.). *Ann. Bot.*, **70**, 105–110.
Jiang, K. & Feldman, L.J. (2003) Root meristem establishment and maintenance: the role of auxin. *J. Plant Growth Reg.*, **21**, 432–440.
Jones, A.M. (1998) Auxin transport: down and out and up again. *Science*, **282**, 2201–2203.
Jürgens, G. (2001) Apical-basal pattern formation in *Arabidopsis* embryogenesis. *EMBO J.*, **20**, 3609–3616.
Kandasamy, M.K., Gilliland, L.U., McKinney, E.C. & Meagher, R.B. (2001) One plant actin isovariant, ACT7, is induced by auxin and required for normal callus formation. *Plant Cell*, **13**, 1541–1554.
Kawata, S. & Shibayama, H. (1965) On the lateral root primordia formation in the crown roots of rice plants. *Proc. Crop Sci. Soc. Japan*, **33**, 423–431.
Kerk, N.M. & Feldman, L.J. (1995) A biochemical model for the initiation and maintenance of the quiescent center: implications for the organization of root meristems. *Development*, **121**, 2825–2833.
Kerk, N.M., Jiang, K. & Feldman, L.J. (2000) Auxin metabolism in the root apical meristem. *Plant Physiol.*, **122**, 925–932.
Kolek, J. & Kozinka, V. (eds) (1992) *Physiology of the Plant Root System*. (Developments in Plant and Soil Sciences, vol. 46), Kluwer Academic Publishers, Dordrecht.
Kolesnikov, V.A. (1930) The root system of fruit tree seedlings. *J. Pomol.*, **8**, 197–203.
Kramer, E.M. (2001) A mathematical model of auxin-mediated radial growth in trees. *J. Theor. Biol.*, **208**, 387–397.
Laskowski, M.J., Williams, M.E., Nusbaum, H.C. & Sussex, I.M. (1995) Formation of lateral root meristems is a two-stage process. *Development*, **121**, 3303–3310.
Levan, A. (1939) Cytological phenomena connected with root swelling caused by growth substances. *Hereditas*, **25**, 87–96.
Libbenga, K.R. & Harkes, P.A.A. (1973) Initial proliferation of cortical cells in the formation of root nodules in *Pisum sativum* L. *Planta*, **114**, 17–28.
Lim, J., Helariutta, Y., Specht, C.D., Jung, J., Sims, L., Bruce, W.B., Diehn, S. & Benfey, P.N. (2000) Molecular analysis of the *SCARECROW* gene in maize reveals a common basis for radial patterning in diverse meristem. *Plant Cell,* **12**, 1307–1318.
Lin, B.-L. & Raghavan, V. (1991) Lateral root initiation in *Marsilea quadrifolia*. I. Origin and histogenesis of lateral roots. *Can. J. Bot.*, **69**, 123–135.

Lintilhac, P.M. & Vesecky, T.B. (1984) Stress-induced alignment of division plane in plant tissues grown *in vitro*. *Nature*, **307**, 363–364.

Lloret, P.G., Casero, P.J., Navascués & Pulgarín, A. (1988) The effects of removal of the root tip on lateral root distribution in adventitious roots of onion. *New Phytol.*, **110**, 143–149.

Lodish, H., Berk, A., Zipurski, S.L., Matsudaira, P. Baltimore, D. & Darnell, J. (2000) *Molecular Cell Biology*, 4th edn, Chapter 21, Nerve Cells, W.H. Freeman, New York.

Lück, H.B. & Lück, J. (1993) About automata theoretical models for phyllotaxis in a global morphogenetic framework, in *Some Physicochemical and Mathematical Tools for Understanding Living Systems* (eds H. Greppin, M. Bonzon & R. Degli Agosti), University of Geneva Press, Geneva, pp. 333–359.

Lück, J., Barlow, P.W. & Lück, H.B. (1994) Cell genealogies in a plant meristem deduced with the aid of a 'bootstrap' L-system. *Cell Prolif.*, **27**, 1–21.

Luo, Y. & Koop, H.U. (1997) Somatic embryogenesis in cultured immature zygotic embryos and leaf protoplasts of *Arabidopsis thaliana* ecotypes. *Planta*, **202**, 387–396.

Luxová, M. (1975) Some aspects of the differentiation of primary root tissues, in *The Development and Function of Roots* (eds J.G. Torrey & D.T. Clarkson), Academic Press, London and New York, pp. 73–90.

Lynch T.M. & Lintilhac, P.M. (1997) Mechanical signals in plant development: a new method for single cell studies. *Devel. Biol.*, **181**, 246–256.

Ma, F. & Peterson, C.A. (2000) Plasmodesmata in onion (*Allium cepa* L.) roots: a study enabled by improved fixation and embedding techniques. *Protoplasma*, **211**, 103–115.

McDonald, D., Wu, L., Bohks, S.M., KewalRamani, V.N., Unutmaz, D. & Hope, T.J. (2003) Recruitment of HIV and its receptors to dendritic cellT cell junctions. *Science*, **300**, 1295–1297.

McDowell, J.M., An, Y.-Q., McKinney, E.C., Huang, S. & Meagher, R.B. (1996a) The *Arabidopsis ACT7* gene is expressed in rapidly developing tissues and responds to several external stimuli. *Plant Physiol.*, **111**, 699–711.

McDowell, J.M., Huang, S., McKinney, E.C., An, Y.-Q. & Meagher, R.B. (1996b) Structure and evolution of the actin gene family in *Arabidopsis thaliana*. *Genetics*, **142**, 587–602.

MacIsaac, S.A., Sawhney, V.K. & Pohorecky, Y. (1989) Regulation of lateral root formation in lettuce (*Lactuca sativa*) seedling roots: interacting effects of α-naphthaleneacetic acid and kinetin. *Physiol. Plant.*, **77**, 287–293.

MacLeod, R.D. (1972) Lateral root formation in *Vicia faba* L. I. The development of large primordia. *Chromosoma*, **39**, 341–350.

MacLeod, R.D. & McLachlan, S.M. (1974) The development of the quiescent centre in lateral roots of *Vicia faba*. *Ann. Bot.*, **38**, 535–544.

Machesky, L.M. & Gould, K.L. (1999) The Arp2/3 complex: a multifunctional actin organizer. *Curr. Opin. Cell Biol.*, **11**, 117–121.

Malamy, J.E. & Ryan, K.S. (2001) Environmental regulation of lateral root initiation in *Arabidopsis*. *Plant Physiol.*, **127**, 899–909.

Mallory, T.E., Chiang, S.-H., Cutter, E.G. & Gifford, E.M., Jr. (1970) Sequence and pattern of lateral root formation in five selected species. *Am. J. Bot.*, **57**, 800–809.

Marchant, A., Bhalerao, R., Casimiro, I., Eklöf, J., Casero, P.J., Bennett, M. & Sandberg, G. (2002) AUX1 promotes lateral root formation by facilitating indole-3-acetic acid distribution between sink and source tissues in the *Arabidopsis* seedling. *Plant Cell*, **14**, 589–597.

Martin, H.V., Elliott, M.C., Wangermann, E. & Pilet, P.-E. (1978) Auxin gradient along the root of the maize seedling. *Planta*, **141**, 179–181.

Mathesius, U., Schlaman, H.R.M., Spaink, H.P., Sautter, C., Rolfe, B.G. & Djordjevic, M.A. (1998) Auxin transport inhibition precedes root nodule formation in white clover roots and is regulated by flavonoids and derivatives of chitin oligosaccharides. *Plant J.*, **14**, 23–34.

Mathur, J., Mathur, N., Kernebeck, B. & Hülskamp, M. (2003a) Mutations in actin-related proteins 2 and 3 affect cell shape development in *Arabidopsis*. *Plant Cell*, **15**, 1632–1645.

Mathur, J., Mathur, N., Kirik, V., Kernebeck, B., Srinivas, B.P. & Hülskamp, M. (2003b) *Arabidopsis CROOKED* encodes for the smallest subunit of the ARP2/3 complex and controls cell shape by region specific for fine F-actin formation. *Development*, **130**, 3137–3146.

Mayer, U., Büttner, G. & Jürgens, G. (1991) Mutations affecting body organization in the *Arabidopsis* embryo. *Nature*, **353**, 402–407.

Mayer, U., Büttner, G. & Jürgens, G. (1993) Apical-basal pattern formation in the *Arabidopsis* embryo: studies on the role of the *gnom* gene. *Development*, **117**, 149–162.

Miller, A.L. & Gow, N.A.R. (1988) Correlation between root-generated ionic currents, pH, fusicoccin, indoleacetic acid, and growth of the primary root of *Zea mays*. *Plant Physiol.*, **89**, 1198–1206.

Morales, M., Colicos, M.A. & Goda, Y. (2000) Actin-dependent regulation of neurotransmitter release at central synapses. *Neuron*, **27**, 539–550.

Morris, D. (2000) Transmembrane auxin carrier systems: dynamic regulators of polar auxin transport. *Plant Growth Reg.*, **32**, 161–172.

Muday, G.K. (2000) Maintenance of asymmetric cellular localization of an auxin transport protein through interaction with the actin cytoskeleton. *J. Plant Growth Reg.*, **19**, 385–396.

Muday, G.K. (2001) Auxins and tropisms. *J. Plant Growth Reg*, **20**, 226–243.

Muday, G.K. & Murphy, A.S. (2002) An emerging model of auxin transport regulation. *Plant Cell*, **14**, 293–299.

Nedergaard, M., Takano, T. & Hansen, A.J. (2002) Beyond the role of glutamate as a neurotransmitter. *Nature Revi. Neurosci.*, **3**, 748–755.

Nelson, W.J. (2003) Adaptation of core mechanisms to generate cell polarity. *Nature*, **422**, 766–774.

Newson, R.B., Parker, J.S. & Barlow, P.W. (1993) Are lateral roots of tomato spaced by multiples of a fundamental distance? *Ann. Bot.*, **71**, 549–557.

Orth, J.D. & McNiven, M. (2003) Dynamin at the actin-membrane interface. *Curr. Opin. Cell Biol.*, **15**, 31–39.

Panshin, A.J. & De Zeeuw, C. (1980) *Textbook of Wood Technology*, 4th edition. McGraw-Hill, Inc., New York.

Paolillo, D.J., Jr. & Zobel, R.W. (2002) The formation of adventitious roots on root axes is a widespread occurrence in field-grown dicotyledonous plants. *Am. J. Bot.*, **89**, 1361–1372.

Parker, G., Schofield, R., Sundberg, B. & Turner, S. (2003) Isolation of *COV1*, a gene involved in the regulation of vascular pattern in stems of *Arabidopsis*. *Development*, **130**, 2139–2148.

Parthasarathy, M.V. & Pesacreta, T.C. (1980) Microfilaments in plant vascular cells. *Can. J. Bot.*, **58**, 807–815.

Parthasarathy, M.V., Perdue, T.D., Witztum, A. & Alvernaz, J. (1985) Actin network as a normal component of the cytoskeleton in many vascular plant cells. *Am. J. Bot.*, **72**, 1318–1323.

Pelosi, A., Lee, M.C.S., Chandler, S.F. & Hamill, J.D. (1995) Hormonal control of root primordia differentiation and root formation in cultured explants of *Eucalyptus globulus* seedlings. *Austr. J. Plant Physiol.*, **22**, 409–415.

Petrásek, J., Elckner, M., Morris, D.A. & Zazímalová, E. (2002) Auxin efflux carrier activity and auxin accumulation regulate cell division and polarity in tobacco cells. *Planta*, **216**, 302–308.

Philipson, W.R. (1990) The significance of apical meristems in the phylogeny of land plants. *Plant System. Evol.*, **173**, 17–38.

Pilet, P.-E. (1967) Auxin content and auxin catabolism in relation to the growth polarity. *Physiol. Plant.*, **20**, 285–290.

Pilet, P.-E. (1977) Growth inhibitors in growing and geostimulated maize roots, in *Plant Growth Regulation* (ed. P.E. Pilet), Springer-Verlag, Berlin and Heidelberg, pp. 115–128.

Pilet, P.-E. & Lance-Nougarède, A. (1967) Structure histologique et catabolisme auxinique des méristèmes radiculaires de *Lens*, *Pisum* et *Zea*. *Bull. Soc. Bot. Suisse*, **77**, 156–172.

Pónya, Zs. & Barnabás, B. (2001) Microinjected fluorescent phalloidin *in vivo* reveals F-actin dynamics in isolated egg cells of wheat (*Triticum aestivum*, L.) developed *in situ* and fertilised *in vitro*. *J. Plant Physiol.*, **158**, 1527–1539.

Pulgarín, A., Navascués, J., Casero, P.J. & Lloret, P.G. (1988) Branching pattern in onion adventitious roots. *Am. J. Bot.*, **75**, 425–432.

Pulgarín, A., Casero, P.J., Casimero, I. & Lloret, P.G. (1996) Polar transverse expansion of pericycle cells in auxin-treated adventitious roots of onion (*Allium cepa*). *Can. J. Bot.*, **74**, 544–552.

Pütz, N. & Froebe, H.A. (1995) A re-evaluation of the mechanism of root contraction in Monocotyledons using the example of *Arisarum vulgare* Targ.-Tozz. (*Araceae*). *Flora*, **190**, 285–297.

Qualmann, B., Kessels, M.M. & Kelly, R.B. (2000) Molecular links between endocytosis and the actin cytoskeleton. *J. Cell Biol.*, **150**, F111–F116.

Ramachandran, S., Christensen, H.E.M., Ishimaru, Y., Dong, C.-H., Chao-Ming, W., Cleary, A.L. & Chua, N.-H. (2000) Profilin plays a role in cell elongation, cell shape maintenance, and flowering in *Arabidopsis*. *Plant Physiol.*, **124**, 1637–1647.

Rashotte, A.M., Brady, S.R., Reed, R.C., Ante, S.J. & Muday, G.K. (2000) Basipetal auxin transport is required for gravitropism in roots of *Arabidopsis*. *Plant Physiol.*, **122**, 481–490.

Rashotte, A.M., DeLong, A. & Muday, G. (2001) Genetic and chemical reductions in protein phosphatase activity alter auxin transport, gravity response and lateral root growth. *Plant Cell*, **13**, 1683–1697.

Rayle, D.L., Ouitrakul, R. & Hertel, R. (1969) Effects of auxin on the auxin transport system in coleoptiles. *Planta*, **87**, 49–53.

Reichelt, S., Knight, A.E., Hodge, T.P., Baluška, F., Šamaj, J., Volkmann, D. & Kendrick-Jones, J. (1999) Characterization of the unconventional myosin VIII in plant cells and its localization at the post-cytokinetic cell wall. *Plant J.*, **19**, 555–569.

Robinson, J.S., Albert, A.C. & Morris, D.A. (1999) Differential effects of brefeldin A and cycloheximide on the activity of auxin efflux carriers in *Cucurbita pepo* L. *J. Plant Physiol.*, **155**, 678–684.

Sabatini, S., Beis, D., Wolkenfelt, H., Murfett, J., Guilfoyle, T., Malamy, J., Benfey, P., Leyser, O., Bechtold, N., Weisbeek, P. & Scheres B. (1999) An auxin-dependent distal organizer of pattern and polarity in the *Arabidopsis* root. *Cell*, **99**, 463–472.

Sabatini, S., Heidstra, R., Wildwater, M. & Scheres, B. (2003) SCARECROW is involved in positioning the stem cell niche in the *Arabidopsis* root meristem. *Genes Devel.*, **17**, 354–358.

Sachs, T. (2000) Integrating cellular and organismic aspects of vascular differentiation. *Plant Cell Physiol.*, **41**, 649–656.

Sachs, T. & Cohen, D. (1982) Circular vessels and the control of vascular differentiation in plants. *Differentiation*, **21**, 22–26.

Saugy, M. & Pilet, P.-E. (1984) Endogenous indol-3yl-acetic acid in stele and cortex of gravistimulated maize roots. *Plant Sci. Lett.*, **37**, 93–99.

Schafer, D.A. (2002) Coupling actin dynamics and membrane dynamics during endocytosis. *Curr. Opin. Cell Biol.*, **14**, 76–81.

Seago, J.L. (1973) Developmental anatomy in roots of *Ipomoea purpurea*. II. Initiation and development of secondary roots. *Am. J. Bot.*, **60**, 607–618.

Seago, J.L. & Wolniak, S.M. (1976) Cortical ontogeny in roots. I. *Zea mays*. *Am. J. Bot.* **63**, 220–225.

Shevell, D.E., Leu, W.-M., Gillmor, C.S., Xia, G., Feldmann, K.A. & Chua, N.-H. (1994) *EMB30* is essential for normal cell division, cell expansion, and cell adhesion in *Arabidopsis* and encodes a protein that has similarity to Sec7. *Cell*, **77**, 1051–1062.

Shevell, D.E., Kunkel, T. & Chua, N.-H. (2000) Cell wall alterations in the *Arabidopsis emb30* mutant. *Plant Cell*, **12**, 2047–2059.

Shi, L., Miller, I. & Moore, R. (1993) Immunocytochemical localization of indole-3–acetic acid in primary roots of *Zea mays*. *Plant Cell Environ.*, **16**, 967–973.

Shupliakov, O., Bloom, O., Gustafsson, J.S., Kjaerulff, O., Löw, P., Tomilin, N., Pieribone, V.A., Greengard, P. & Brodin, L. (2002) Impaired recycling of synaptic vesicles after acute perturbation of the presynaptic actin cytoskeleton. *Proc. Natl. Acad. Sci. USA*, **99**, 14476–14481.

Smit, G., de Koster, C.C., Schripsema, J., Spaink, H.P., van Brussel, A.A. & Kijne, J.W. (1995) Uridine, a cell division factor in pea roots. *Plant Molec. Biol.*, **29**, 869–873.

Spanswick, R.M. (1972) Electrical coupling between cells of higher plants: a direct demonstration of intercellular communication. *Planta*, **102**, 215–227.

Stain, C.J. (2002) The Vascular Cambium: Establishment, Activity and Woody Biomass Production, Ph.D. Thesis, University of Reading, Reading, UK.

Steinmann, T., Geldner, N., Grebe, M., Mangold, S., Jackson, C.L., Paris, S., Gälweiler, L., Palme, K. & Jürgens, G. (1999) Coordinated polar localization of auxin-efflux carrier PIN1 by GNOM ARF GEF. *Science*, **286**, 316–318.

Swarup, R., Friml, J., Marchant, A., Ljung, K., Sandberg, G., Palme, K. & Bennett, M. (2001) Localization of the auxin permease AUX1 suggests two functionally distinct hormone transport pathways operate in the *Arabidopsis* root apex. *Genes Devel.*, **15**, 2648–2653.

Tao, W. & Verbelen, J.-P. (1996) Switching on and off cell division and cell expansion in

cultured mesophyll protoplasts from tobacco. *Plant Sci.*, **134**, 207–215.
Taunton, J. (2001) Actin filament nucleation by endosomes, lysosomes and secretory vesicles. *Curr. Opin. Cell Biol.*, **13**, 85–91.
Thimann, K.V. (1988) A history of the knowledge of auxin, in *Physiology and Biochemistry of Auxin in Plants* (eds. M. Kutáček, R.S. Bandurski & J. Krekule), Academia, Prague, pp. 3–20.
Thimann, K.V., Reese, K. & Nachmias, V.T. (1992) Actin and the elongation of plant cells. *Protoplasma*, **171**, 151–166.
Torrey, J.G. & Loomis, R.S. (1967) Ontogenetic studies of vascular cambium formation in excised roots of *Raphanus sativus* L. *Phytomorph.*, **17**, 401–409.
Ulmasov, T., Hagen, G. & Guilfoyle, T.J. (1999) Activation and repression of transcription by auxin-response factors. *Proc. Natl Acad. Sci. USA*, **96**, 5844–5849.
van Fleet, D.S. (1948) Cortical patterns and gradients in vascular plants. *Am. J. Bot.*, **35**, 219–227.
Veen, H. (1969) Auxin transport, auxin metabolism and aging. *Acta Bot. Neer.*, **18**, 447–454.
Vincent, J.-P. & Dubois, L. (2002) Morphogen transport along epithelia, an integrated trafficking problem. *Devel. Cell*, **3**, 615–623.
Waller, E., Riemann, M. & Nick, P. (2002) A role for actin-driven secretion in auxin-induced growth. *Protoplasma*, **219**, 72–81.
Warren Wilson, J. & Warren Wilson, P.M. (1984) Control of tissue patterns in normal development and in regeneration, in *Positional Control in Plant Development* (eds P.W. Barlow & D.J. Carr), Cambridge University Press, Cambridge, pp. 225–280.
Wysocka-Diller, J., Helariutta, Y., Fukaki, H., Malamy, J.E. & Benfey, P.N. (2000) Molecular analysis of SCARECROW function reveals a radial patterning mechanism common to root and shoot. *Development*, **127**, 595–603.
Wojtaszek, P. (2000) Genes and plant cell walls: a difficult relationship. *Biol. Rev.*, **75**, 437–475.
Yu, Q., Hlavačka, A., Matoh, T., Volkmann, D., Menzel, D., Goldbach, H.E. & Baluška, F. (2002) Short-term boron deprivation inhibits endocytosis of cell wall pectins in meristematic cells of maize and wheat root apices. *Plant Physiol.*, **130**, 415–421.
Zhang, N. & Hasenstein, K.H. (1999). Initiation and elongation of lateral roots in *Lactuca sativa*. *Int. J. Plant Sci.*, **160**, 511–519.
Zhang, N. & Hasenstein, K.H. (2000) Halogenated auxins affect microtubules and root elongation in *Lactuca sativa*. *J. Plant Growth Reg.*, **19**, 397–405.

Appendix I

Position and timing of breaks in primary polarity during the formation of new root primordia

We have described a complex, subcellular system for the transport of auxin, indoleacetic acid, within the primary tissues of roots. The auxin so transported maintains cell growth and cell division within the root apices as well as within the secondary meristems and their derivative tissues behind the apex. Auxin also regulates the polarised spatial pattern of cell divisions in these locations.

The sites where new roots can arise are an emergent property of the cellular construction of the parent root. The relative timing and position of events initiating these new roots, although complex, are summarised in Table 7A.1. Of the eight possible rooting types, new lateral root development corresponds to Type 1 in this table. This table also corrects the misprinted Table 1 in Barlow & Palma (1997).

Table 7A.1. Eight types of new lateral root development on a parent root (or shoot) axis. Each type differs in regard to the timing and location of initiation and outgrowth of the new root.

	Root primordium initiation			
	Usual timing		Unusual timing*	
Root primordium outgrowth	Usual location	Unusual location	Usual location	Unusual location
Immediate (sylleptic)	1	3	5	7
Delayed (proleptic)	2	4	6	8

* Delayed or stimulated in relation to the usually observed, species-specific temporal sequence of primordium initiation. Morphologically, the eight types of root would be: 1 – Sylleptic; 2 – Proleptic; 3 – Sylleptic adventitious; 4 – Proleptic adventitious; 5 – Sylleptic adventive; 6 – Proleptic adventive; 7 – Induced sylleptic adventitious; 8 – Induced proleptic adventitious.

The fact that different types of root formation can be discerned at various stages within the lives of plant species is indicative of the plasticity of root development which, in turn, suggests an interaction between the predictable physiological and anatomical states of the plant and the unpredictable physical environment in which it lives (Paolillo & Zobel, 2002). In the Table, emphasis is placed upon the birth of roots, though there would be an equally varied and significant pattern of root death. Together, the birth and death of roots create patterns of root turnover. The cell body, as well as the cell and organ polarities of which the cell body is both the servant and the master, are all parts of this grand developmental scheme.

Clearly, the plant invests a great deal of its own inherent technology and energy resources to construct a polarised auxin transport system within the root and shoot. Through the action of this system, roots keep on growing and new roots keep being born. The roots comprise an expanding exploratory interface between the vegetative and reproductive shoot and the nutritive soil environment. The auxin transport system constructs a network of new meristematic rooting sites. One may ask about the true significance of the pattern of these sites upon the plant. Plant meristem networks are only just now beginning to be explored, not only with respect to their anatomical location and physiological function (Vogt *et al.*, 1995; Amzallag, 1999) but also in connection with their more general configurations (Milo *et al.*, 2002) which may be of importance for information-gathering and signalling.

References

Amzallag, G.N. (1999) Regulation of growth: the meristem network approach. *Plant Cell Environ.*, **22**, 483–493.

Barlow, P.W. & Palma, B. (1997) The place of roots in plant development, in *Biology of Root Formation and Development* (eds A. Altman & Y. Waisel), Plenum Press, New York, pp. 1–11.
Milo, R., Shen-Orr, S., Itzkovitz, S., Kashtan, N., Chklovskii, D. & Alon, U. (2002) Network motifs: simple building blocks of complex networks. *Science*, **298**, 824–827.
Paolillo, D.J., Jr. & Zobel, R.W. (2002) The formation of adventitious roots on root axes is a widespread occurrence in field-grown dicotyledonous plants. *Am. J. Bot.*, **89**, 1361–1372.
Vogt, K.A., Vogt, D.J., Asbjornsen, H. & Dahlgren, R.A. (1995) Roots, nutrients and their relationship to spatial patterns. *Plant Soil*, **168–169**, 113–123.

Appendix II

Rules for cell division

During the early growth of a lateral root primordium, the number of cells increases in both the anticlinal and periclinal directions. When the new primordium acquires its own polarity, the width of the root meristem stabilises but its length continues to increase and cell divisions in the anticlinal plane predominate. These changes in cell numbers and division cycles in the anticlinal and periclinal planes have been carefully documented by López-Garrido *et al.* (1993) for lateral root primordia of onion during the first 120 hours of their development. By the time the new root meristem has become established, the planes of cell division conform to the pattern already displayed in the meristem of the supporting axis. For some reason, this pattern of growth and division is optimal for the species in question. But what are the factors that regulate the planes of cell division at this early stage of primordium development, especially prior to its polarisation?

The geometry of volumetric growth and, in particular, the distribution of growth upon the curved surface of the early hemispherical and of the later paraboloidal primordium, imposes a self-regulating growth axis upon the emerging cellular construction. The pattern of overall volumetric growth, to which the individual cellular units contribute, may result in the cells falling under the influence of Hofmeister's rule: that is, the cells and their cell bodies divide perpendicularly to the principal direction of their growth. We propose that, when the root meristem has become established, this 'rule' is an expression of polar auxin transport (in the vascular cylinder) and, perhaps, of strain (in the cortex). The link between auxin transport and division plane is provided by the cell body. Its orientation determines the direction of auxin flow and the orientation of the new division wall via the phragmoplast. Strain (and its counterpart, stress) may also regulate division plane through the medium of the cell body. But in the early stages, when primordial growth is isotropic, the cells are approximately isodiametric (except near the base of the primordium). At this stage, there is a somewhat ambiguous principal direction due to the uniform expansion of the hemispherical primordium. Division orientation may here be guided by a default principle whereby a new division wall attaches to the one last formed. A new cell wall affords some stability to the dividing system since it does not start to expand until the next cell cycle commences. This default rule may apply when neither auxin transport nor a preferential direction of strain have been developed within the primordium.

It is convenient to tabulate the various rules of plant cell division orientation (Table 7A.2) and thereby see how they connect with polarised root growth. The various rules suggest that regulation of division plane is hierarchically organised, there being different degrees of complexity which are reflected in the increasing levels of organisation of the root meristem. Probably, not all rules act simultaneously since there will be occasions when their respective preconditions are absent. For example, when polar auxin transport is absent, regulation of the division system is then devolved to a 'lower' level of regulation.

Table 7A.2 Summary of various rules for plant cell division which have been proposed to account for the diversity of planes of cell division in organised meristematic tissues.

Division rule	Operating conditions	Reference
1. Default	New dw attaches to the last formed dw.	Thomson & Hull (1934)
2. Errera's rule	New dw is of minimal area.	Wardlaw (1968)
3. Hofmeister's rule	New dw forms perpendicular to direction of strain, i.e. to the principal direction of growth prior to mitosis and cytokinesis.	Wardlaw (1968)
4. Lintilhac's rule	New dw forms in shear-free plane with respect to either stress or strain.	Lintilhac (1984)
5. Baluška/ Barlow's rule (modified Hofmeister's rule)	New dw forms perpendicular to direction of polarised auxin transport, as determined by the plant cell body and actin-based plant synapses.	This chapter

dw = division wall.

References

Lintilhac, P.M. (1984) Positional controls in meristem development: a caveat and an alternative, in *Positional Controls in Plant Development* (eds P.W Barlow & D.J. Carr), Cambridge University Press, Cambridge, pp. 83–105.

López-Garrido, E., Vidal-Bernabé, M.R., García de la Puerta, P. & Abadía-Fenol, F. (1993) Cell cycle in the centralmost section of *Allium cepa* L. lateral root primordia: influence on fluctuations in growth rate. ΦYTON (Buenos Aires), **54**, 81–86.

Thomson, R.B. & Hull, K.L. (1934) Physical laws and the cellular organization of plants. *Bot. Gaz.*, **95**, 511–514.

Wardlaw, CW. (1968) *Morphogenesis in Plants. A Contemporary Study*, Methuen, London.

8 Development of the shoot apical meristem

Rüdiger Simon

8.1 Architecture of the shoot meristem: layers and zones

Longitudinal sections through the meristem centre along the apical-basal axis reveal that the shoot meristem is organized into layers (Fig. 8.1). The outermost cell layer (L1) consists of progenitor cells for the epidermis that will later cover all organs that are above ground. Further inside, the next cell layer (L2) will give rise to sub-epidermal tissue, but also to the gametes. The group of cells underneath L2 is termed layer 3 (L3), although it does not consist of only a single cell layer. These cells will give rise to the majority of the plant's ground tissue. L1, L2 and L3 represent cell clones that are maintained during development; mixing of cells between the individual layers is avoided by the regularity of cell division patterns. In L1 and L2 (forming the tunica), cell divisions are anticlinal, i.e. the new cell walls are formed perpendicular to the meristem surface. In the L3 (or corpus), cells can divide in all directions. First evidence for the existence of these clonal meristem layers came from studies on plant chimeras; these are composed of tissues that differ in their genetic make-up (gr. *chimaira*, a mythological animal composed of a lion, a goat and a snake). Early markers used were chloroplast and anthocyanin pigmentation, or chromosome number; chimeras were found as spontaneous somatic mutants, or generated by meristem regeneration following grafting between genetically marked individuals (Szymkowiak & Sussex, 1996). In *Datura*, colchicine-induced changes in the ploidy level allowed the identification of (periclinal) sectors in the mature plant that could be traced back to an affected meristem layer (Satina *et al.*, 1940), confirming that the observed meristem layers represent cell clones.

Perpendicular to the apical-basal axis is the radial axis that cuts through the roughly cylindrical shoot meristem and defines central and peripheral positions or zones. Primordia that develop into lateral organs or flowers are formed exclusively from the peripheral zone of the shoot meristem. The central zone harbours non-differentiating and pluripotent stem cells. Cells in a deeper region of the meristem form the rib zone that gives rise to the majority of the ground tissue in the plant body. Chimeras have again been used to identify the fate of descendants from the three zones. In sectorial chimeras, only a part of the plant is marked with a cell clone. The size of this clone or sector along the apical-basal axis indicates how long the progenitor cell has

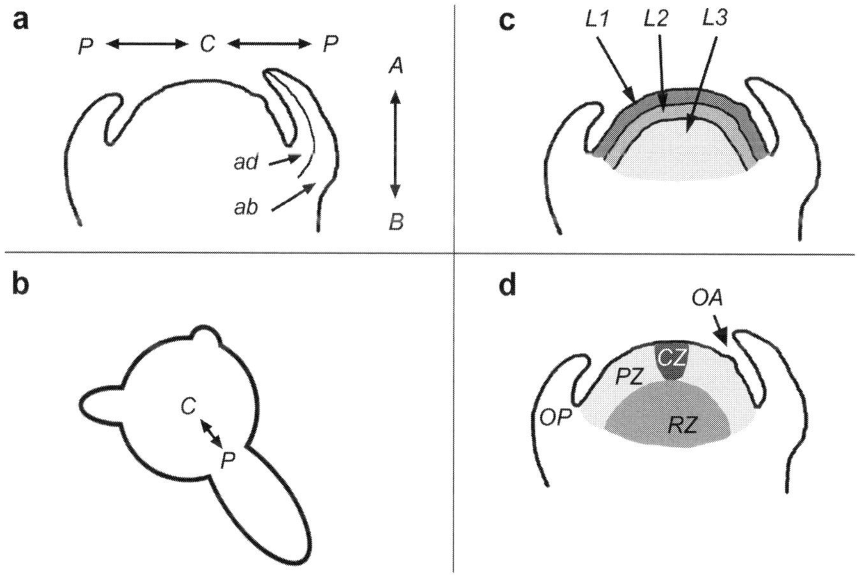

Figure 8.1 Architecture of the shoot meristem. Longitudinal (a,c,d) or cross section (b) through a shoot or floral meristem, showing the apical-basal (A to B) and central-peripheral (C to P) axes. The meristem is divided into 3 clonal layers, the L1, L2 and L3 (c). Cells from all layers contribute to the zones of the meristem (d). The central zone (CZ), containing pluripotent stem cells, is surrounded by the peripheral zone (PZ) where organ anlagen (OA) are initiated that will develop into organ primordia (OP). The rib zone (RZ) produces vasculature and stem tissue.

resided in the shoot meristem; sectors that extend from the plant's base to its tip indicate that a stem cell was marked. Such clonal analysis shows the presence of at least 3 to 4 stem cells in each meristem layer that together form the central zone of the meristem (Stewart & Dermen, 1970). Maintenance of this pool of non-differentiating cells in the central zone is essential to allow indeterminate growth of the shoot apical meristem, and for its ability to continuously provide new cells for organ initiation at the periphery.

8.2 The making of a meristem

Maintaining the indeterminate nature of meristems requires the activity of *KNOX* class homeobox genes. These are named after their founding member, the *KNOTTED1(KN1)*-gene from maize. *KN1* was first discovered as a gain-of-function mutant that produced knots, or outgrowth of undifferentiated tissue on the maize leaf surfaces (Vollbrecht *et al*., 1991). *KN1* was shown to be expressed in the SAM, but transcripts were missing at the sites of organ formation. In *Arabidopsis*, the *KNOX* gene *SHOOTMERISTEMLESS* (*STM*) is required for the formation of a shoot meristem during embryogenesis, and for its maintenance

Table 8.1 Genes controlling meristem activity in *Arabidopsis*

Name	Protein family	Mutant phenotype	Expression pattern	Activity or target genes
AG	TF (MADS-domain)	floral organ transformations, indeterminate flowers	centre of floral meristems, stamen and carpel	inhibits WUS; stamen and carpel development
AS1	TF (MYB-domain)	serrate leaves	organ primordia	inhibits KNAT1 and 2, promotes LOB expression
CLV1	MP (LRR-receptor kinase)	larger meristems, more stem cells	meristem centre	
CLV2	MP (LRR-receptor-like protein)	larger meristems, more stem protein)	all tissues	Ligand (CLV3) and receptor complex (CLV1/2), represses WUS transcription
CLV3	SP (CLE-family)	larger meristems, more stem cells	meristem stem cells	
CUC1/CUC2	TF (NAC-domain)	double mutants: no shoot meristem	embryo, organ boundaries	establish organ boundaries, required for STM expression and shoot meristem initiation
DRN/ESR1	TF (AP2/ERF-type)	aphenotypic?	central zone and apical tips of organ anlagen	promotes differentiation
KNAT1/BP	TF (Homeodomain, KNOX)	short pedicels	veg. SAM, except organ anlagen, inflorescence stems, pedicels	controls pedicel and inflorescence development, partially redundant with STM

LFY	TF	floral defects	floral meristems	promotes flower development, activates AG (and other genes)
LOB	TF (LOB-domain)	aphenotypic?	organ boundaries	interaction with KNAT genes?
PHB	TF (HD-Zip, START)	gain-of-function: adaxialized organs, ectopic meristems	meristem centre, adaxial side of organs	promotes adaxial and meristem fate
STM	TF (Homeodomain, KNOX)	no shoot meristem	SAM and FM, except organ anlagen	inhibits meristem cell differentiation, represses AtGA20ox1 expression
WUS	TF (Homeodomain)	shoot and floral meristems not maintained	meristem centre, ovules, stamen tissue	promotes stem cell activity and CLV3 expression, regulates ovule development, activates AG
PNH	AGO/PNH/PIWI	no embryonic SAM, fewer axillary meristems	vasculature, adaxial side of organs	post-transcriptional regulation (of STM?), promotes indeterminate growth

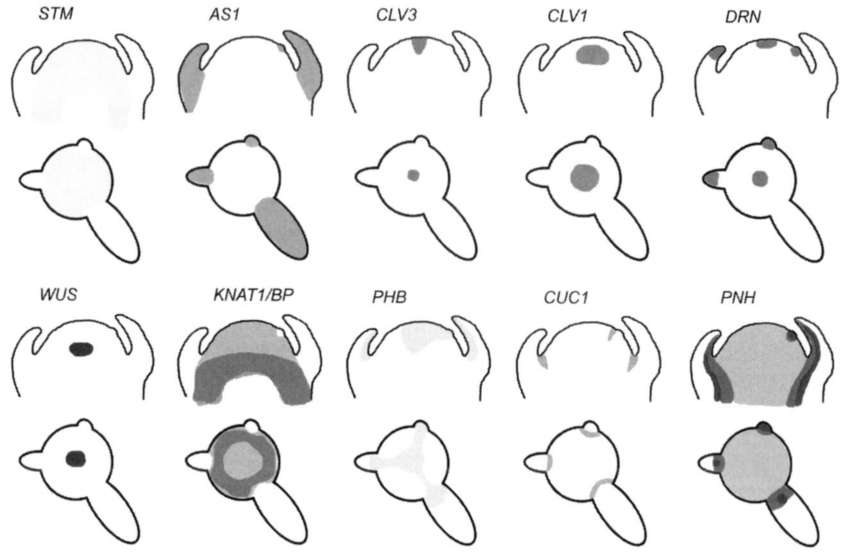

Figure 8.2 Gene expression patterns in *Arabidopsis* meristems. *STM* is expressed in the non-differentiated cells of the meristem. *AS1* RNA is found in a complementary pattern in organ anlagen and primordia. *CLV3* expression is confined to the stem cells in the central zone, with *CLV1* in an immediately adjacent domain in the L2 and L3. The *DRN* gene is expressed in stem cells, but also in organ anlagen and at the tips of organ primordia. *WUS* is expressed in the centre of the meristem, underneath the *CLV3* domain. *KNAT1/BP* is expressed in the vegetative SAM in a pattern similar to *STM,* and at higher levels in the provasculature and inflorescence stem. *PHB* expression extends from the meristem centre to the adaxial side of organ primordia. *CUC1* RNA marks the boundaries between organs and the meristem. *PNH* is expressed in the meristem, at higher levels at the adaxial side of organ primordia, and in the vasculature. Only some of the gene expression patterns discussed in the text are illustrated here.

during further development (Long *et al*., 1996; Table 8.1). Similar to maize *KN1* (and other *KNOX* genes), *STM* expression is downregulated in leaf founder cells, indicating that *STM* is required to maintain the indeterminate state of meristem cells (or to suppress differentiation) (Fig. 8.2). Indeed, misexpression of *KNOX* genes in leaf primordia can confer a partial 'meristematic identity' to them, causing the formation of serrate or compound leaves, and sometimes outgrowth of ectopic meristems (Tsiantis & Hay, 2003).

But how do *KNOX* genes act? There is no collection of target genes available yet, but several studies indicate that phytohormones may mediate some aspects of *KNOX* gene activity. *NTH15*, a *KNOX* protein from tobacco, was shown to downregulate expression of the gibberellic acid (GA) biosynthetic gene *GA20-oxidase* in tobacco meristems by binding to short target sequences in the *GA20-oxidase* promoter, thereby reducing the levels of active GA in the plant tissue (Sakamoto *et al*., 1999). The *GA20-oxidase* RNA

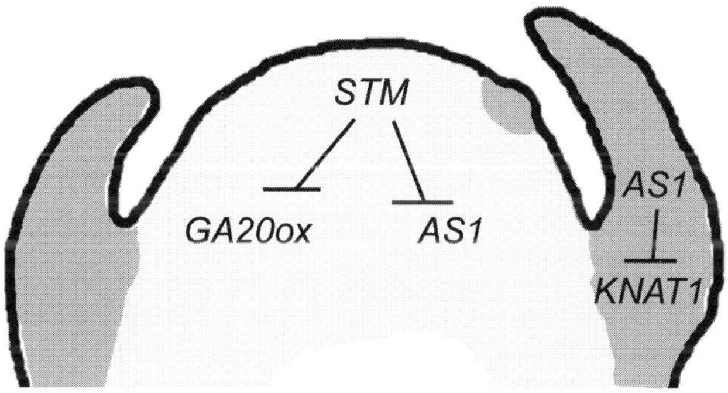

Figure 8.3 Control of the indeterminate state. *STM* expression in the indeterminate meristem inhibits expression of the GA biosynthetic gene *GA20ox* and of *AS1*. In determinate structures such as organs, *AS1* is expressed at high levels and represses expression of *KNOX* genes such as *KNAT1*.

was only found at the meristem flanks, where organs are initiated and *KNOX* gene expression is generally inhibited. Repression of *GA20-oxidase* by *KNOX* genes was also observed in several other plant species (Kusaba *et al.*, 1998a; Hay *et al.*, 2002), although the repression effect maybe more indirect in *Arabidopsis* (Hay *et al.*, 2003). Furthermore, the phenotypes induced by ectopic *KNOX* gene expression were suppressed by external application of active GA, indicating that *KNOX* genes and GA have opposite effects on the determinacy state of meristem cells (Hay *et al.*, 2002; Fig. 8.3).

Overexpression of the *KNOX* gene *OSH1* from rice was reported to increase cytokinin levels in transgenic tobacco (Kusaba *et al.*, 1998b), suggesting cytokinin production as another target of *KNOX* gene activity. Importantly, cytokinins also promote the expression of the *KNOX* genes *STM* and *KNAT1/BP* in *Arabidopsis* (Rupp *et al.*, 1999). Like strong *KNOX* gene expression, increased cytokinin production in transgenic plants causes ectopic tissue outgrowth at leaf margins, indicating that *KNOX* genes and cytokinins can act on the same pathways. Indeed, both *STM* and cytokinins can increase expression levels of the *cyclinD3* gene involved in cell cycle regulation (Soni *et al.*, 1995; Gallois *et al.*, 2002). Finally, *KNOX* genes appear to promote each other's expression: several *KNOX* genes are expressed in largely overlapping domains in *Arabidopsis* meristems, and *STM* misexpression can activate expression of the *KNOX* genes *KNAT1/BP* and *KNAT2* in the same tissues (Gallois *et al.*, 2002).

8.3 Regulation of *KNOX* gene expression

If *KNOX* gene expression distinguishes meristem from leaf, how is expression excluded from organ primordia? Genetic screens for leaf shape mutants identified several genes that are involved in restricting *KNOX* gene expression to the meristem. For example, mutations in *ASYMMETRIC LEAVES1* (*AS1*) of *Arabidopsis* allow misexpression of *KNAT1/BP* and *KNAT2* in organ primordia (Byrne *et al.*, 2000). *AS1* encodes a MYB-type transcription factor that is expressed only in organs. However, in *stm* mutants, *AS1* becomes expressed in the meristem itself, resulting in a switch to a determinate developmental programme and loss of an active shoot meristem. Significantly, *stm/as1* double mutant plants are able to maintain growth and development for an extended period, indicating that mutations in *AS1* can suppress meristem defects due to loss of *STM*. Thus, one function of *STM* is to repress *AS1* activity in the shoot meristem. In the double mutants, loss of *STM* may then be partially compensated by increased activity of *KNAT1/BP* and *KNAT2*, resulting in a restoration of meristematic activity (Fig. 8.3). Such complementary expression, mutual repression and antagonistic activity appears to be a common theme for meristem specific *KNOX* genes and organ specific *MYB* transcription factors in several plant species (Waites *et al.*, 1998; Tsiantis *et al.*, 1999).

Between meristem and organs, boundary domains become established that are marked by the expression of *LATERAL ORGAN BOUNDARY* (*LOB*) (Byrne *et al.*, 2002; Shuai *et al.*, 2002) and the *CUP-SHAPED COTYLEDONS* (*CUC1* and *2*) genes (Aida *et al.*, 1997; Takada *et al.*, 2001). It is not known if *LOB* has a function in making the boundary, since *lob* loss-of-function mutations are aphenotypic. Expression of *LOB* in the boundaries is promoted by both *AS1* and *KNAT1/BP*, and may thus be established where the expression domains of *AS1* and *KNOX* genes collide. The redundant activities of *CUC1* and *2*, encoding NAC transcription factors, first activate *STM* expression during embryogenesis (Aida *et al.*, 1999). At later stages, *STM* and *CUC* are expressed in non-overlapping domains, and *CUC* activity in a peripheral region is now required for the separation of cotyledons.

8.4 Communications arising – between periphery and centre

KNOX gene regulation illustrates the importance of interregional communication between the meristem and the organ primordia that are produced by it. Genes of the *YABBY* (*YAB*) and *KANADI* (*KAN*) families, encoding putative transcription factors, are expressed only in organ primordia, where they soon become restricted to the abaxial side during development (Siegfried *et al.*, 1999; Kerstetter *et al.*, 2001; discussed also in Chapter 10). Loss of function mutations in *YAB* genes can result in ectopic *KNOX* gene expression (Kumaran *et al.*, 2002), while misexpression of *KAN* in the central region arrests meristem activity (Kerstetter *et al.*, 2001). Thus, factors expressed at the abaxial side of primordia appear to inhibit *KNOX* gene expression and meristem fate. Conversely, gene activity at the adaxial side (closer to the meristem centre) promotes meristem fate. Further evidence for this came from studies on the dominant *phabulosa* (*phb-d*) and *phavoluta* (*phv-d*) mutants of *Arabidopsis*. Leaves of *phb-d* or *phv-d* mutants display adaxial features on all surfaces, including production of axillary meristems (McConnell & Barton, 1998). *PHB* is preferentially expressed in the meristem centre and the adaxial domain of organ primordia. Intriguingly, the *PHB* and *PHV* proteins, which are homeodomain transcription factors, were suggested to be activated by binding of an (unidentified) sterol (McConnell *et al.*, 2001); the phb-D/phv-D mutant proteins might not bind sterols, resulting in constitutive activity. Availability of the activating sterol only in central and adaxial positions could then explain why misexpression of the (non-mutant) *PHB* gene did not cause adaxialization. Thus, proteins of the PHB family would act as receptors for an adaxialising signal that ultimately promotes meristem fate. Indeed, recessive mutations in the *PHB/PHV*-related *REVOLUTA* (*REV*) gene cause a loss of adaxial domains and axillary meristem formation (Talbert *et al.*, 1995; Otsuga *et al.*, 2001). While the quest for the activator molecule has just begun, a microRNA complementary to the mRNA sequence encoding the putative sterol binding site of the PHV protein was recently identified (Rhoades *et al.*, 2002; Tang *et al.*, 2003). Spatial regulation of *PHB/PHV/REV* activity may then be achieved by controlling mRNA stability. Although the mechanism of regulation is not yet clear, it is evident that a small molecule (be it a microRNA or a sterol) could mediate the communication between the adaxial leaf domain and more central regions of the meristem.

Another gene function revealing an interdependence between central-peripheral polarity and meristem activity is *PINHEAD* (*PNH*; Moussian *et al.*, 1998; Lynn *et al.*, 1999). The SAM formed during embryogenesis of *Arabidopsis pnh* mutants is non-functional, and often terminates with the formation of a single organ. However, growth can proceed from buds that form in the axils of cotyledons. Lateral organs of *pnh* mutants are occasionally abaxialized and axillary meristems are missing, indicating that *PNH* plays a role in the polar development of the leaf and meristem initiation. The *PNH* gene belongs to a small gene family in *Arabidopsis*, and *PNH* acts partially redundantly with another family member, *ARGONAUTE* (*AGO*; Bohmert *et al.*, 1998; Lynn *et al.*, 1999). *AGO* is required for meristem indeterminacy similar to *PNH*, but also for post-transcriptional gene silencing (Fagard *et al.*, 2000; Morel *et al.*, 2002). Genes of the *AGO/PNH* family from several organisms have been implicated in post-transcriptional regulation of target genes, and a direct association between AGO/PNH-like proteins and microRNAs has been reported (Mourelatos *et al.*, 2002). It is tempting to speculate that *PNH* regulates meristem formation and activity via microRNAs that are specific for *KNOX* genes such as *STM*, or the *PHB/PHV/REV* gene family.

PNH is expressed in a roughly cylindrical domain in the young seedling, comprising the SAM and the adaxial domain of organ primordia (Lynn *et al.*,

1999). The precise expression patterns of *PNH* (and other genes) along the radial axis are first set up during embryogenesis, when the central domain giving rise to the SAM and the peripheral domain that forms the cotyledons are established. Interestingly, when the *PNH* expression domain is shifted from the adaxial to the abaxial side of organs, duplicated cotyledons are formed that produce new meristems, indicating a switch from a determinate to an indeterminate growth pattern (Newman *et al.*, 2002). Collectively, loss- and gain-of-function phenotypes show that there is a differential requirement for *PNH* activity along the radial axis to promote indeterminate growth along the apical-basal axis.

8.5 Communications along the apical-basal axis

The activity of transcription factors in the shoot meristem can also be regulated at the level of protein distribution. For example, *KN1* mRNA is found only in the L2 and L3 layers, but the KN1 protein is present in all layers of the maize shoot meristem. The KN1 protein may be transported between layers through plasmodesmata, which establish a cytoplasmic continuum between meristem cells (Lucas *et al.*, 1995). In other studies, transcription factors that were expressed in only one meristem layer were shown to act non-cellautonomously and affect gene expression in adjacent layers (Hantke *et al.*, 1995; Perbal *et al.*, 1996; Sessions *et al.*, 2000; Kim *et al.*, 2002). This indicates that the clonal origin of cells and their separation into distinct layers may be far less important than initially thought. Indeed, a recent study suggests that movement of transcription factors between layers represents the default state, and that factors have to be actively retained to restrict their activity to individual layers (Wu *et al.*, 2003). In this scenario, protein movement would be central to allow for coordination of growth and development between otherwise separate layers.

In the following paragraphs, another important example for cell communication along the apical-basal axis will be discussed: the regulation of stem cell number.

8.6 Establishment and maintenance of stem cells

Longevity and continuous activity of the shoot meristem is only possible if a pool of stem cells is maintained that will replenish and restore the meristem cell pool, since cells are constantly lost from the meristem when organs are formed at the periphery. Shoot meristem stem cells were first identified as cells that divide infrequently, and it was occasionally questioned that they divide at all. In the last 10 years, we have started to understand stem cell behaviour in shoot meristems. Most of the information was collected from studies with *Arabidopsis*, but related genes and mechanisms governing stem cell biology have been identified in other plant species such as maize.

Stem cell number in the shoot apical meristem of *Arabidopsis* is controlled by (at least) two antagonistic pathways that promote or repress stem cell activity, respectively. These two pathways involve the exchange of signals along the apical-basal axis of the plant; between stem cells at the meristem tip and a group of cells in the deeper L3 layer. A key element of the stem cell promoting pathway is the *WUSCHEL* (*WUS*) gene, encoding a homeodomain transcription factor (Laux *et al.*, 1996; Mayer *et al.*, 1998). In the absence of *WUS* function, meristems are initiated, but they fail to keep a sufficient number of stem cells to maintain prolonged growth and development. Although *WUS* activity is required for stem cell maintenance, *WUS* RNA is not found within the stem cells, but in a small group of cells underneath. Thus, *WUS* has to act non-cellautonomously for induction of stem cell fate at the meristem tip. So far, there is no indication that the WUS protein itself is moving out to adjacent cells, which has been observed for other transcription factors in plants. *WUS* activity may therefore involve the generation of a (unknown) signal that controls stem cell formation at the meristem apex.

A phenotype just opposite to that of *wus* mutants, namely accumulation of stem cells in meristems, is found when at least one out of three *CLAVATA* (*CLV1*, *2* and *3*) genes is mutated (Clark *et al.*, 1993, 1995; Kayes & Clark, 1998). *clv* mutant meristems maintain more stem cells in the central zone than required to support normal meristem growth, resulting in enlarged meristems that occasionally lose any growth restriction and start to fasciate. This size increase of the central zone also has direct consequences for organ number: the peripheral zone that surrounds the meristem centre enlarges correspondingly, allowing the formation of more organs in *clv* mutants. The *CLV* genes normally act to restrict proliferation of meristem stem cells, and the gene products of *CLV1*, *2* and *3* are likely working together in a signal transduction pathway. *CLV1* encodes a putative receptor kinase, consisting of extracellular leucine-rich-repeats, a transmembrane domain and a cytoplasmic serine/threonine kinase domain (Clark *et al.*, 1997). *CLV1* is expressed in the meristem centre. The CLV2 protein is constructed like CLV1, but lacks the kinase moiety (Jeong *et al.*, 1999). CLV1 and CLV2 may form a heterodimeric receptor and act at the core of a larger receptor complex (Trotochaud *et al.*, 1999). The receptor complex may then be activated by binding of CLV3, a small protein that is secreted from the stem cells at the meristem tip (Fletcher *et al.*, 1999). Detailed studies are still missing that show the co-localisation of these proteins in plant tisssues, but molecular genetic evidence suggests that CLV3 indeed requires both CLV1 and CLV2 for its activity, and that CLV3 dose is rate-limiting for receptor activation (Brand *et al.*, 2000). Further components of these receptor complexes have been identified. One of them is KAPP, a kinase associated protein phosphatase that may modulate signalling via the CLV1 receptor kinase (Williams *et al.*, 1997; Stone *et al.*, 1998). How the signal is transduced further is not known, but Rho-like GTPases (ROP) have been implicated in this process that were copurified with CLV complexes (Trotochaud *et al.*, 1999).

Several observations indicate that the main consequence of *CLV* signalling is transcriptional repression of *WUS*: first, *WUS* RNA is found in an expanded domain in *clv* mutant meristems (Brand *et al.*, 2000; Schoof *et al.*, 2000); and second, increased expression of *CLV3* results in decreased *WUS* expression (Brand *et al.*, 2000). However, *CLV3* expression levels are in turn dependent

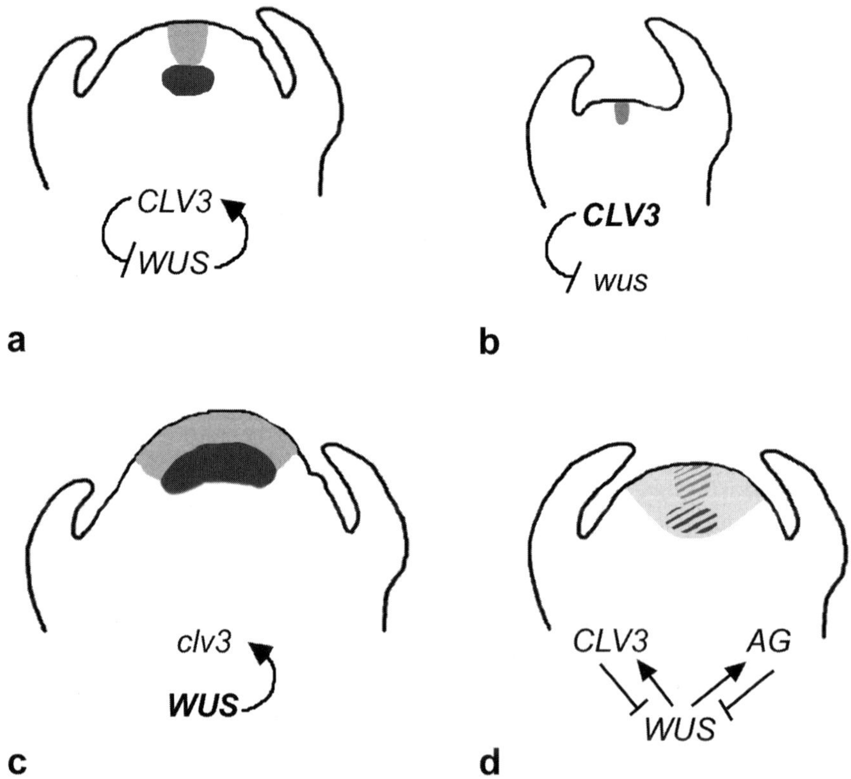

Figure 8.4 Negative feedback regulation of stem cell number in shoot and floral meristems. In the wildtype, *CLV3* is expressed and secreted from the stem cells (light grey) and restricts WUS expression in cells underneath (dark grey). *WUS* in turn promotes stem cell fate at the meristem tip and *CLV3* expression. (b) Increased *CLV3* expression (or mutations in the *WUS* gene) results in a loss of *WUS* activity and the size of the stem cell domain decreases correspondingly. (c) In *clv* mutants, *WUS* is expressed in more cells, and stem cells accumulate in the meristem. (d) In floral meristems, *WUS* activity is controlled by *CLV* signalling and *AG*. *WUS* (together with *LFY*) promotes *AG* expression (pale area around cross-hatching), which in turn switches off *WUS* expression at a later stage.

on the stem cell promoting activity of *WUS* (Brand *et al.*, 2000; Schoof *et al.*, 2000). The antagonistic activities of *CLV3* and *WUS* can thus establish a continuously acting, feedback regulated control circuitry of stem cell number: any decrease in stem cell number during growth will result in less CLV3, reduced *CLV* signalling and increased expression of the stem cell promoting factor WUS; this will in turn induce the formation of more stem cells, causing a corresponding increase in CLV3 levels, strong *CLV* signalling and again repression of *WUS* expression. Thus, the combined activities of *CLV*3 and *WUS* are integrated into a negative feedback loop, which allows to maintain a permanent number of stem cells at the meristem tip (Fig. 8.4).

8.7 Robustness and flexibility of stem cell regulation

So far, a coarse picture of the gene interaction system that controls the maintenance of a stem cell population has been sketched. Since stem cell maintenance is essential for meristem activity and to support continuous development of a plant, it is worthwhile to look at some of the ramifications of the proposed model, and investigate the molecular activities of the genes involved in more detail.

The model as described would act to keep the number of stem cells proportional to the number of *WUS* expressing cells underneath; however, the simple scheme for stem cell regulation, as depicted in Fig. 8.4, is not sufficient to account for homeostasis of absolute meristem size. Obviously, meristem size may change during development (see below), but only within a narrow range. For example, a seasonal increase in organ production rates may require that more stem cells are maintained in the meristem. The activity of the *CLV*/*WUS* stem cell regulator has to be flexible to allow for such moderate changes, but also robust enough to guarantee maintenance of an appropriate stem cell number during changing environmental conditions.

Several factors have been identified that modulate *CLV* signalling: one negative regulator of *CLV* signalling is the previously mentioned KAPP, a protein phosphatase that interacts with a variety of receptor kinases. Another candidate regulator is the POLTERGEIST (POL) protein phosphatase 2C; *pol* mutants have been identified as phenotypic suppressors of hypomorphic *clv1* mutations, indicating that POL activity interferes with *CLV* signalling (Yu *et al.*, 2000, 2003). Similar to KAPP, *POL* is not only expressed in meristems but in many plant tissues, indicating that POL may function in several signal transduction pathways. Interestingly, a recent analysis of the *pol* mutant phenotypes suggests that *CLV* signalling may act upon additional, but as yet unidentified targets besides *WUS* (Yu *et al.*, 2003).

SHEPHERD (*SHD*) affects *CLV* activity at a very different level (Ishiguro *et al.*, 2002). The *SHD* gene encodes an ER-resident HSP90–like protein that is required for the proper folding of CLV proteins. Homozygous *shp* mutants form larger shoot and floral meristems, thus resembling *clv* mutants, but they also show defects in root meristem organization and pollen tube growth. It is likely that SHD acts in the assembly of several receptor signalling complexes, and does not have an exclusive role in the regulation of the *CLV* pathway.

8.8 Specificity of the *CLV/WUS* regulator

We are now taking a closer look at the components of the *CLV/WUS* regulator, and how they are made ready for their specific tasks. There are at least 610 receptor-like kinases encoded in the *Arabidopsis* genome, many of them showing high amino acid sequence similarity with CLV1, and some of the around 220 transmembrane LRR receptors could have partially overlapping functions with CLV1 in meristem regulation. This is supported by the recent observation that all previously described *clv1* alleles that exhibit strong or intermediate phenotypes are dominant-negative alleles (Dievart *et al.*, 2003); *clv1* null alleles result in only weak phenotypes, and are clearly distinguishable from null mutants in *CLV3*. Thus, CLV3 may signal via CLV1 and additional related receptors that have not been identified so far. The mutant clv1 protein encoded by strong *clv1* alleles is then predicted to bind to and inhibit the activation of these redundant receptors.

The CLV2 protein is not necessarily involved in a specific interaction with the CLV3 ligand, but may rather promote the stability of the CLV1 receptor kinase within the complex. Stem cells accumulate in *clv2* mutants because the amount of CLV1 receptor available for signal transmission is reduced. However, the *clv2* phenotype is suppressed when plants are grown under SD conditions, which could suggest that other RLPs can substitute for CLV2 under certain conditions (Kayes & Clark, 1998; Jeong *et al.*, 1999). All *clv2* mutants display only weak meristem phenotypes compared to *clv1* or *clv3* mutants, but they show additional defects in organ development such as pedicel elongation. Consistent with functions outside the meristem, *CLV2* is expressed throughout the plant.

CLV3 was first described as a unique sequence; meanwhile, more than 40 genes have been identified in the *Arabidopsis* and other plants' genomes that encode CLV3-like (CLE) proteins (Cock & McCormick, 2001). The sequence similarities between them are confined to an N-terminal secretion signal and the CLE-motif, a short stretch of 14 amino acids near the C-terminus. Both *clv3-1* and *clv3-5* mutant alleles carry a point mutation in this region, indicating that the CLE motif is required for protein function (Fletcher *et al.*, 1999). Of the 25 Arabidopsis *CLE* genes (including *CLV3*), 17 are expressed at detectable levels in the shoot apex (Sharma *et al.*, 2003). Although detailed studies have still to be perfomed, we can assume that several CLE proteins are also expressed in the meristem, available for binding to the CLV receptor complex. Therefore, the recognition of CLV3 by CLV1 must be very specific to avoid cross-signalling by other CLE proteins; alternatively, a low binding specificity can be tolerated if only CLV3 is expressed at sufficient levels in the meristem to interact with CLV1. Indeed, misexpression of *CLE40* from the *CLV3* promoter rescued the meristem defects of *clv3* mutant plants, indicating that CLV3 is singular in its expression pattern, but not in its capacity to interact with CLV1 (Hobe *et al.*, 2003).

The loss of stem cells in *wus* mutants indicates that *WUS* activity is essential for stem cell maintenance. Is *WUS* expression also sufficient to induce stem cell fate? Misexpression of *WUS*, using a variety of different promoters, can induce the expression of the stem cell marker *CLV3* even in differentiating organs (Schoof *et al.*, 2000). However, not all cells respond to increased *WUS* activity, indicating that a specific cellular competence is required. For example, *WUS* expression in leaves was reported to activate *CLV3* expression only in leaf axils (Brand *et al.*, 2002) or in the vasculature (Lenhard *et al.*, 2002), but not in all leaf cells. Interestingly, high level *WUS* expression in roots caused the *de novo* formation of embryos from differentiated root cells, suggesting that *WUS* may have previously unsuspected roles in embryogenesis (Zuo *et al.*, 2002). During ovule development, *WUS* is expressed in the nucellus, and promotes the initiation of the integuments from the chalazal region (Gross-Hardt *et al.*, 2002); during flower development, *WUS* activates expression of homeotic genes (see below). In summary, *WUS* does not appear to provide a signal that is specific for the generation of stem cells. Instead, *WUS* and its immediate target genes seem to form a signalling module that is used repeatedly during plant development to mediate communication and coordinated growth between adjacent cell groups. The specific response of a cell to the *WUS* derived signal may then depend entirely on a cells micro-environment or developmental history.

The components of the *CLV*/*WUS* stem cell regulator appear to be used in many plant species. In petunia, *TERMINATOR* (*TER*) represents the orthologue to *WUS* of *Arabidopsis*, and *ter* mutants terminate shoot meristem activity at early stages (Stuurman *et al.*, 2002). Mutations in the *FASCIATED EAR2* (*FEA2*) gene of maize cause a massive overproliferation of the inflorescence meristem; the FEA2 protein localises to the plasma membrane and shows high sequence similarity to CLV2 of *Arabidopsis* (Taguchi-Shiobara *et al.*, 2001). Finally, chimeras that were generated by grafting between wildtype and *fasciated* (*fas*) mutant tomatoes had shown that meristem size in tomato (like in *Arabidopsis*) is controlled from the L3 layer (Szymkowiak & Sussex, 1992). The gene product of the tomato *FAS* gene is not yet known, but with analogy to *Arabidopsis*, it may encode a component of a *CLV* signalling pathway that controls expression of a tomato *WUS*-orthologue in the L3.

8.9 Determinate or indeterminate: shutting off stem cells in flower development

In floral meristems, stem cells are maintained only for a limited period, i.e. as long as new organs are formed. They are consumed during the production of the innermost floral organs, the carpels. As in the shoot meristem, the *CLV*/*WUS* circuitry also controls stem cell fate in flowers: *wus* mutant flowers form fewer stamens and lack the central carpels, indicating that stem cells were consumed prematurely; *clv* mutants accumulate stem cells in floral meristems that are then used to form additional carpels or even extra floral whorls, resulting in a misshapen, club-like gynoecium (hence the name *CLAVATA*, derived from *clava* (lat.) = club). However, there are also other regulators acting exclusively in floral meristems. One of them is *AGAMOUS* (*AG*), encoding a MADS box transcription factor that controls the production of stamens and carpels in the inner two flower whorls (Yanofsky *et al.*, 1990). The floral meristems of *ag* mutants become indeterminate, implicating *AG* in the restriction of stem cell number. In wildtype floral meristems, *WUS* expression is lost around floral stage 6, but *WUS* remains expressed in *ag* mutant flowers beyond that stage (Lenhard *et al.*, 2001; Lohmann *et al.*, 2001). Furthermore, *WUS* activity is required for the indeterminacy of *ag* mutant flowers, indicated by the lack of floral organ proliferation in *ag/wus* double mutants (Laux *et al.*, 1996). Thus, *AG* appears to control meristem stem cell proliferation by negatively regulating *WUS* expression. How *AG* performs this function is still unknown; however, it appears that *AG* acts independently of the *CLV* genes, since floral meristem indeterminacy is further enhanced in *ag/clv* double mutants, and the *WUS* expression domain expands further in *ag/clv* mutant meristems compared to those of *ag* or *clv* single mutants (Fig. 8.4; Lohmann *et al.*, 2001).

This role of *AG* in the control of meristem determinacy is restricted to flowers. In the shoot meristem, *AG* expression is not activated. One important factor that distinguishes floral meristems from shoot meristems is the expression of LFY protein. LFY is a transcription factor that can convert shoot to flower meristems by activating a set of flower specific regulator genes (Weigel *et al.*, 1992; Parcy *et al.*, 1998). Activation of *AG* in the centre of floral meristems is controlled by the combined activities of *LFY*, which is expressed throughout floral meristems, and *WUS*. First evidence for this role of *WUS* in the activation of a floral homeotic regulator gene came from misexpression experiments, where *WUS* transcription was controlled from promoters that direct organ specific gene expression in flowers (Lenhard *et al.*, 2001; Lohmann *et al.*, 2001). For example, expression of *WUS* from the *AP3* promoter (in the presumptive second and third floral whorl) resulted in the production of extra floral organs that developed into carpelloid stamens – both organ types require *AG* activity. Indeed, increased *WUS* expression caused proliferation of meristem cells, and ectopic activation of *AG* expression. A more detailed analysis at the molecular level then showed that the WUS protein can bind specific DNA target sequences of the *AG* regulatory regions that are located immediately adjacent to the binding sites for the LFY protein (Lohmann *et al.*, 2001). Both proteins can act independently to activate *AG* expression in a yeast assay system. However, since *AG* is normally only expressed in the centre of floral meristems, both *LFY* and *WUS* appear to be required for normal levels of *AG* transcription.

From stage 3 until stage 6, both *AG* and *WUS* are expressed in an overlapping set of cells, indicating that *AG* expression *per se* is not sufficient to downregulate *WUS* transcription. Furthermore, ectopic expression of *AG* throughout the plant causes early flowering and homeotic organ transformations, but does not terminate shoot meristem development (Mizukami & Ma, 1992). This indicates that either *AG* expression levels have to reach a certain threshold level, or accumulation of additional and so far unidentified factors is required to cooperate with *AG* in the regulation of *WUS*, and consequently meristem determinacy. The precise timepoint during development that this happens may be of great importance, since a premature loss of stem cells could result in sterile flowers if carpels are not formed. We may therefore expect that *AG* activity is tightly controlled, and possibly at several levels. Indeed, in addition to positive and negative regulation of *AG* transcription, there is also post-transcriptional control of *AG* activity. Several of the factors involved were first identified through sensitized genetic screens, i.e. by searching for extragenic mutations that enhance the meristem indeterminacy phenotype of a weak *ag* mutant (Chen & Meyerowitz, 1999). At least 4 genes (*HUA1* and *2, HEN2* and *4*) have so far been shown to affect the generation of functional *AG* mRNA (Li *et al.*, 2001; Chen *et al.*, 2002; Western *et al.*, 2002; Cheng *et al.*, 2003). At the molecular level, *HUA1*,*2* and *HEN4* seem to promote the processing of (or inhibit aberrant polyadenylation within) the unusually large second intron of *AG*. Interestingly, it is this second intron which also contains the key sequences for both positive and negative regulation of *AG* transcription (Sieburth & Meyerowitz, 1997; Busch *et al.*, 1999).

From what we have discussed so far, a picture of stem cell control in the meristem emerges. When meristems are first formed, *WUS* is activated in a small set of cells and signals to specify stem cells at the meristem tip. These stem cells in turn express CLV3, which acts via the CLV1/2 receptor complex to restrict *WUS* expression. After floral induction, floral meristems are formed at the flanks of the shoot apical meristem that express the LFY transcription factor. The combined activity of LFY and WUS induces transcription of *AG* in the central zone of the floral meristems. *AG* now acts at two levels: it specifies the production of stamens and carpels, in combination with several other genes, and switches off *WUS* expression. Thus, *WUS* is controlled by two, and probably independently acting feedback regulated systems. The determinate state of floral meristems, with a complete loss of stem cells, requires the combined activities of both systems, since mutations in either the *CLV* genes or *AG* result in indeterminacy.

However, to be a floral meristem is not necessarily the end of it all: occasionally, floral meristems may switch back to an indeterminate state and restart as a shoot meristem. In *Arabidopsis*, this process, called floral reversion, can be induced by manipulating the flower-inducing photoperiodic stimulus, or by reducing *LFY* or *AG* activity. When plants heterozygous for a *lfy* mutation are cultivated under short-day conditions, flowers can revert even after carpel formation and grow a new shoot from gynoecium cells that express *AG* (Okamuro *et al*., 1996). This indicates that even high levels of *AG* expression in the meristem centre are not sufficient for a permanent repression of stem cell fate. Maintenance of the determinate floral state may require not only down-regulation of *WUS* via *AG* and the *CLV* genes, but also repression of other, shoot meristem specific genes. Indeed, it has been proposed that *LFY* has a dual function in activating flower specific, but repressing shoot specific gene expression (Okamuro *et al*., 1996; Parcy *et al*., 2002).

8.10 Integration of meristem activity: new genes and mutants

We have so far just begun to understand how the control of stem cell number is integrated with the overall growth and development of the shoot system, and with the formation of organs. Genetic screens for mutants that are affected in meristem maintenance continue to reveal new gene functions that have to be incorporated into the existing regulatory networks. For example, *hairy meristem* (*ham*) mutants of petunia fail to maintain a shoot meristem, similar to *ter* mutants (the homologue of *wus* mutants in *Arabidopsis*; Stuurman *et al*., 2002). *ham/ter* double mutants display additive effects, indicating that these genes act in parallel pathways. The *HAM* gene encodes a transcription factor of the GRAS family that is expressed in the provasculature of the stem and developing organs. Other *GRAS* genes have been shown to be involved in root cell type specification and stem cell maintenance, but also in mediating gibberellic acid signalling. The exact mode of action of *HAM* is unknown, but its mutant phenotype together with its expression pattern indicates that *HAM* acts non-cellautonomously in a novel pathway that links stem cell perpetuation in meristems with the generation of differentiated cell types at the periphery.

Similar to *ham* mutants of petunia, the shoot meristems of *Arabidopsis Dornröschen* (*drn-D*) mutants terminate prematurely. The affected gene, *DRN/ESR1*, encodes an AP2/ERF type transcription factor that is mis-expressed in *drn-D* mutants, resulting in altered regulation of *CLV3*, *WUS*, *STM* and possibly other target genes (Kirch *et al*., 2003). Interestingly, *DRN/*

ESR1 is normally transcribed in the central zone of shoot and floral meristems, and in the anlagen of lateral organs, where expression is maintained for a short period at the tip of the primordium. For example, *DRN/ESR1* is expressed in single epidermal cells of the ovule anlagen and remains expressed in the most apical cell of the growing ovule. The common theme underlying the spatial distribution of *DRN/ESR1* mRNA may be 'apical position' for all meristems and organs. Cellular differentiation of lateral organs starts from the organ tip and may be promoted by *DRN* expression at this position. *DRN* expression in the central zone of meristems would then counteract or balance stem cell promoting signals by fostering the exit of cells from the central fate to differentiation.

Many other genes and functions that control meristem activity still remain to be discovered; however, these two examples discussed above already indicate that meristems and organs use various short and long range communication systems to mutually adjust their developmental pace.

8.11 Polarity in the shoot meristem: stability in changing

A shoot meristem that was formed during embryogenesis can stay active throughout a plant's lifetime. However, its cellular composition changes: there is a continuous flux of cells that leave the meristem to be incorporated into organ primordia, which are replaced by new cells after divisions. Nevertheless, the basic structure of the meristem, its division into zones, as well as gene expression patterns are stable in this changing environment. Some of the factors that contribute to this stability have been discussed here: regulation of cell division patterns, interdependence of organ formation and meristem activity, mutual control of gene expression between adjacent cell groups, and extensive intercellular communication. The picture of a meristem that can be drawn now is that of a highly self-regulating and dynamic system, consisting of interacting cell groups that act as transient task-forces, rather than individual cells that acquire a permanent fate.

Acknowledgement

Work in my laboratory is supported by the DFG.

References

Aida, M., Ishida, T. & Tasaka, M. (1999) Shoot apical meristem and cotyledon formation during Arabidopsis embryogenesis: interaction among the *CUP-SHAPED COTYLEDON* and

SHOOT MERISTEMLESS genes. *Development*, **126**, 1563–1570.
Aida, M., Ishida, T., Fukaki, H., Fujisawa, H. & Tasaka, M. (1997) Genes involved in organ separation in Arabidopsis: an analysis of the cup-shaped cotyledon mutant. *Plant Cell*, **9**, 841–857.
Bohmert, K., Camus, I., Bellini, C., Bouchez, D., Caboche, M. & Benning, C. (1998) *AGO1* defines a novel locus of Arabidopsis controlling leaf development. *EMBO J.*, **17**, 170–180.
Brand, U., Fletcher, J.C., Hobe, M., Meyerowitz, E.M. & Simon, R. (2000) Dependence of stem cell fate in Arabidopsis on a feedback loop regulated by CLV3 activity. *Science*, **289**, 617–619.
Brand, U., Grünewald, M., Hobe, M. & Simon, R. (2002) Regulation of *CLV3* expression by two homeobox genes in Arabidopsis. *Plant Physiol.*, **129**, 565–575.
Busch, M.A., Bomblies, K. & Weigel, D. (1999) Activation of a floral homeotic gene in Arabidopsis. *Science*, **285**, 585–587.
Byrne, M.E., Barley, R., Curtis, M., Arroyo, J.M., Dunham, M., Hudson, A. & Martienssen, R.A. (2000) Asymmetric leaves1 mediates leaf patterning and stem cell function in Arabidopsis. *Nature*, **408**, 967–971.
Byrne, M.E., Simorowski, J. & Martienssen, R.A. (2002) *ASYMMETRIC LEAVES1* reveals knox gene redundancy in Arabidopsis. *Development*, **129**, 1957–1965.
Chen, X. & Meyerowitz, E.M. (1999) *HUA1* and *HUA2* are two members of the floral homeotic *AGAMOUS* pathway. *Mol. Cell*, **3**, 349–360.
Chen, X., Liu, J., Cheng, Y. & Jia, D. (2002) *HEN1* functions pleiotropically in Arabidopsis development and acts in C function in the flower. *Development*, **129**, 1085–1094.
Cheng, Y., Kato, N., Wang, W., Li, J. & Chen, X. (2003) Two RNA binding proteins, HEN4 and HUA1, act in the processing of AGAMOUS pre-mRNA in Arabidopsis thaliana. *Dev. Cell*, **4**, 53–66.
Clark, S.E., Running, M.P. & Meyerowitz, E.M. (1993) CLAVATA1, a regulator of meristem and flower development in Arabidopsis. *Development*, **119**, 397–418.
Clark, S.E., Running, M.P. & Meyerowitz, E.M. (1995) CLAVATA3 is a specific regulator of shoot and floral meristem development affecting the same processes as CLAVATA1. *Development*, **121**, 2057–2067.
Clark, S.E., Williams, R.W. & Meyerowitz, E.M. (1997) The *CLAVATA1* gene encodes a putative receptor kinase that controls shoot and floral meristem size in Arabidopsis. *Cell*, **89**, 575–585.
Cock, J.M. & McCormick, S. (2001) A large family of genes that share homology with *CLAVATA3*., *Plant Physiol.*, **126**, 939–942.
Dievart, A., Dalal, M., Tax, F.E., Lacey, A.D., Huttly, A., Li, J. & Clark, S.E. (2003) CLAVATA1 Dominant-negative alleles reveal functional overlap between multiple receptor kinases that regulate meristem and organ development. *Plant Cell*, **15**, 1198–1211.
Fagard, M., Boutet, S., Morel, J.B., Bellini, C. & Vaucheret, H. (2000) AGO1, QDE-2, and RDE-1 are related proteins required for post-transcriptional gene silencing in plants, quelling in fungi, and RNA interference in animals. *Proc. Natl. Acad. Sci. USA*, **97**, 11650–11654.
Fletcher, J.C., Brand, U., Running, M.P., Simon, R. & Meyerowitz, E.M. (1999) Signaling of cell fate decisions by CLAVATA3 in Arabidopsis shoot meristems. *Science*, **283**, 1911–1914.
Gallois, J.L., Woodward, C., Reddy, G.V. & Sablowski, R. (2002) Combined SHOOT MERISTEMLESS and WUSCHEL trigger ectopic organogenesis in Arabidopsis. *Development*, **129**, 3207–3217.
Gross-Hardt, R., Lenhard, M. & Laux, T. (2002) WUSCHEL signalling functions in interregional communication during Arabidopsis ovule development. *Genes Dev.*, **16**, 1129–1138.
Hantke, S.S., Carpenter, R. & Coen, E.S. (1995) Expression of *floricaula* in single cell layers of periclinal chimeras activates downstream homeotic genes in all layers of floral meristems. *Development*, **121**, 27–35.
Hay, A., Jackson, D., Ori, N. & Hake, S. (2003) Analysis of the competence to respond to KNOTTED1 activity in Arabidopsis leaves using a steroid induction system. *Plant Physiol.*, **131**, 1671–1680.
Hay, A., Kaur, H., Phillips, A., Hedden, P., Hake, S. & Tsiantis, M. (2002) The gibberellin

pathway mediates KNOTTED1-type homeobox function in plants with different body plans. *Curr. Biol.*, **12**, 1557–1565.

Hobe, M., Muller, R., Grunewald, M., Brand, U. & Simon, R. (2003) Loss of CLE40, a protein functionally equivalent to the stem cell restricting signal CLV3, enhances root waving in Arabidopsis. *Dev. Genes Evol.*, **213**, 271–281.

Ishiguro, S., Watanabe, Y., Ito, N., Nonaka, H., Takeda, N., Sakai, T., Kanaya, H. & Okada, K. (2002) SHEPHERD is the Arabidopsis GRP94 responsible for the formation of functional CLAVATA proteins. *EMBO J.*, **21**, 898–908.

Jeong, S., Trotochaud, A.E. & Clark, S.E. (1999) The Arabidopsis *CLAVATA2* gene encodes a receptor-like protein required for the stability of the CLAVATA1 receptor-like kinase. *Plant Cell*, **11**, 1925–1934.

Kayes, J.M. & Clark, S.E. (1998) *CLAVATA2*, a regulator of meristem and organ development in Arabidopsis. *Development*, **125**, 3843–3851.

Kerstetter, R.A., Bollman, K., Taylor, R.A., Bomblies, K. & Poethig, R.S. (2001) *KANADI* regulates organ polarity in Arabidopsis. *Nature*, **411**, 706–709.

Kim, J.Y., Yuan, Z., Cilia, M., Khalfan-Jagani, Z. & Jackson, D. (2002) Intercellular trafficking of a KNOTTED1 green fluorescent protein fusion in the leaf and shoot meristem of Arabidopsis. *Proc. Natl. Acad. Sci. USA*, **99**, 4103–4108.

Kirch, T., Simon, R., Grunewald, M. & Werr, W. (2003) The *DORNROSCHEN/ENHANCER OF SHOOT REGENERATION1* Gene of Arabidopsis Acts in the Control of Meristem Cell Fate and Lateral Organ Development. *Plant Cell*, **15**, 694–705.

Kumaran, M.K., Bowman, J.L. & Sundaresan, V. (2002) *YABBY* polarity genes mediate the repression of KNOX homeobox genes in Arabidopsis. *Plant Cell*, **14**, 2761–2770.

Kusaba, S., Fukumoto, M., Honda, C., Yamaguchi, I., Sakamoto, T. & Kano-Murakami, Y. (1998a) Decreased GA1 content caused by the overexpression of *OSH1* is accompanied by suppression of GA 20-oxidase gene expression. *Plant Physiol.*, **117**, 1179–1184.

Kusaba, S., Kano-Murakami, Y., Matsuoka, M., Tamaoki, M., Sakamoto, T., Yamaguchi, I. & Fukumoto, M. (1998b) Alteration of hormone levels in transgenic tobacco plants overexpressing the rice homeobox gene *OSH1*., *Plant Physiol.*, **116**, 471–476.

Laux, T., Mayer, K.F., Berger, J. & Jurgens, G. (1996) The *WUSCHEL* gene is required for shoot and floral meristem integrity in Arabidopsis. *Development*, **122**, 87–96.

Lenhard, M., Bohnert, A., Jurgens, G. & Laux, T. (2001) Termination of stem cell maintenance in Arabidopsis floral meristems by interactions between *WUSCHEL* and *AGAMOUS*., *Cell*, **105**, 805–814.

Lenhard, M., Jurgens, G. & Laux, T. (2002) The *WUSCHEL* and *SHOOTMERISTEMLESS* genes fulfil complementary roles in Arabidopsis shoot meristem regulation. *Development*, **129**, 3195–3206.

Li, J., Jia, D. & Chen, X. (2001) *HUA1*, a regulator of stamen and carpel identities in Arabidopsis, codes for a nuclear RNA binding protein. *Plant Cell*, **13**, 2269–2281.

Lohmann, J.U., Hong, R.L., Hobe, M., Busch, M.A., Parcy, F., Simon, R. & Weigel, D. (2001) A molecular link between stem cell regulation and floral patterning in Arabidopsis. *Cell*, **105**, 793–803.

Long, J.A., Moan, E.I., Medford, J.I. & Barton, M.K. (1996) A member of the KNOTTED class of homeodomain proteins encoded by the *STM* gene of Arabidopsis. *Nature*, **379**, 66–69.

Lucas, W.J., Bouche-Pillon, S., Jackson, D.P., Nguyen, L., Baker, L., Ding, B. & Hake, S. (1995) Selective trafficking of KNOTTED1 homeodomain protein and its mRNA through plasmodesmata. *Science*, **270**, 1980–1983.

Lynn, K., Fernandez, A., Aida, M., Sedbrook, J., Tasaka, M., Masson, P. & Barton, M.K. (1999) The *PINHEAD/ZWILLE* gene acts pleiotropically in Arabidopsis development and has overlapping functions with the *ARGONAUTE1* gene. *Development*, **126**, 469–481.

McConnell, J.R. & Barton, M.K. (1998) Leaf polarity and meristem formation in Arabidopsis. *Development*, **125**, 2935–2942.

McConnell, J.R., Emery, J., Eshed, Y., Bao, N., Bowman, J. & Barton, M.K. (2001) Role of *PHABULOSA* and *PHAVOLUTA* in determining radial patterning in shoots. *Nature*, **411**, 709–713.

Mayer, K.F., Schoof, H., Haecker, A., Lenhard, M., Jurgens, G. & Laux, T. (1998) Role of *WUSCHEL* in regulating stem cell fate in the Arabidopsis shoot meristem. *Cell*, **95**, 805–815.

Mizukami, Y. & Ma, H. (1992) Ectopic expression of the floral homeotic gene *AGAMOUS* in transgenic Arabidopsis plants alters floral organ identity. *Cell*, **71**, 119–131.
Morel, J.B., Godon, C., Mourrain, P., Beclin, C., Boutet, S., Feuerbach, F., Proux, F. & Vaucheret, H. (2002) Fertile hypomorphic *ARGONAUTE* (*ago1*) mutants impaired in post-transcriptional gene silencing and virus resistance. *Plant Cell*, **14**, 629–639.
Mourelatos, Z., Dostie, J., Paushkin, S., Sharma, A., Charroux, B., Abel, L., Rappsilber, J., Mann, M. & Dreyfuss, G. (2002) miRNPs: a novel class of ribonucleoproteins containing numerous microRNAs. *Genes Dev.*, **16**, 720–728.
Moussian, B., Schoof, H., Haecker, A., Jürgens, G. & Laux, T. (1998) Role of the *ZWILLE* gene in the regulation of central shoot meristem cell fate during Arabidopsis embryogenesis. *EMBO J.*, **17**, 1799–1809.
Newman, K.L., Fernandez, A.G. & Barton, M.K. (2002) Regulation of Axis Determinacy by the Arabidopsis *PINHEAD* Gene. *Plant Cell*, **14**, 3029–3042.
Okamuro, J.K., den Boer, B.G., Lotys-Prass, C., Szeto, W. & Jofuku, K.D. (1996) Flowers into shoots: photo and hormonal control of a meristem identity switch in Arabidopsis. *Proc. Natl. Acad. Sci. USA*, **93**, 13831–13836.
Otsuga, D., DeGuzman, B., Prigge, M.J., Drews, G.N. & Clark, S.E. (2001) *REVOLUTA* regulates meristem initiation at lateral positions. *Plant J.*, **25**, 223–236.
Parcy, F., Bomblies, K. & Weigel, D. (2002) Interaction of *LEAFY*, *AGAMOUS* and *TERMINAL FLOWER1* in maintaining floral meristem identity in Arabidopsis. *Development*, **129**, 2519–2527.
Parcy, F., Nilsson, O., Busch, M.A., Lee, I. & Weigel, D. (1998) A genetic framework for floral patterning. *Nature*, **395**, 561–566.
Perbal, M.C., Haughn, G., Saedler, H. & Schwarz-Sommer, Z. (1996) Non-cell-autonomous function of the Antirrhinum floral homeotic proteins DEFICIENS and GLOBOSA is exerted by their polar cell-to-cell trafficking. *Development*, **122**, 3433–3441.
Rhoades, M.W., Reinhart, B.J., Lim, L.P., Burge, C.B., Bartel, B. & Bartel, D.P. (2002) Prediction of plant microRNA targets. *Cell*, **110**, 513–520.
Rupp, H.M., Frank, M., Werner, T., Strnad, M. & Schmulling, T. (1999) Increased steady state mRNA levels of the *STM* and *KNAT1* homeobox genes in cytokinin overproducing Arabidopsis thaliana indicate a role for cytokinins in the shoot apical meristem. *Plant J.*, **18**, 557–563.
Sakamoto, T., Nishimura, A., Tamaoki, M., Kuba, M., Tanaka, H., Iwahori, S. & Matsuoka, M. (1999) The conserved KNOX domain mediates specificity of tobacco KNOTTED1-type homeodomain proteins. *Plant Cell*, **11**, 1419–1432.
Satina, S., Blakeslee, A.F. & Avery, A.G. (1940) Demonstration of the three germ layers in the shoot apex of Datura by means of induced polyploidy in periclinal chimeras. *Am. J. Bot.*, **27**, 895–905.
Schoof, H., Lenhard, M., Haecker, A., Mayer, K.F., Jurgens, G. & Laux, T. (2000) The stem cell population of Arabidopsis shoot meristems in maintained by a regulatory loop between the *CLAVATA* and *WUSCHEL* genes. *Cell*, **100**, 635–644.
Sessions, A., Yanofsky, M.F. & Weigel, D. (2000) Cell-cell signalling and movement by the floral transcription factors LEAFY and APETALA1. *Science*, **289**, 779–782.
Sharma, V.K., Ramirez, J. & Fletcher, J.C. (2003) The Arabidopsis *CLV3-like* (*CLE*) genes are expressed in diverse tissues and encode secreted proteins. *Plant Molec. Biol.* **51**, 415–425.
Shuai, B., Reynaga-Pena, C.G. & Springer, P.S. (2002) The lateral organ boundaries gene defines a novel, plant-specific gene family. *Plant Physiol.*, **129**, 747–761.
Sieburth, L.E. & Meyerowitz, E.M. (1997) Molecular dissection of the AGAMOUS control region shows that cis elements for spatial regulation are located intragenically. *Plant Cell*, **9**, 355–365.
Siegfried, K.R., Eshed, Y., Baum, S.F., Otsuga, D., Drews, G.N. & Bowman, J.L. (1999) Members of the *YABBY* gene family specify abaxial cell fate in Arabidopsis. *Development*, **126**, 4117–4128.
Soni, R., Carmichael, J.P., Shah, Z.H. & Murray, J.A. (1995) A family of cyclin D homologs from plants differentially controlled by growth regulators and containing the conserved retinoblastoma protein interaction motif. *Plant Cell*, **7**, 85–103.
Stewart, R.N. & Dermen, H. (1970) Determination of number and mitotic activity of shoot apical

initial cells by analysis of mericlinal chimeras. *Am. J. Bot.*, **57**, 816–826.

Stone, J.M., Trotochaud, A.E., Walker, J.C. & Clark, S.E. (1998) Control of meristem development by CLAVATA1 receptor kinase and kinase-associated protein phosphatase interactions. *Plant Physiol.*, **117**, 1217–1225.

Stuurman, J., Jaggi, F. & Kuhlemeier, C. (2002) Shoot meristem maintenance is controlled by a GRAS-gene mediated signal from differentiating cells. *Genes Dev.*, **16**, 2213–2218.

Szymkowiak, E.J. & Sussex, I.M. (1992) The internal meristem layer (L3) determines floral meristem size and carpel number in tomato periclinal chimeras. *Plant Cell*, **4**, 1089–1100.

Szymkowiak, E.J. & Sussex, I.M. (1996) What chimeras can tell us about plant development. *Ann. Rev. Plant Physiol. Plant Molec. Biol.*, **47**, 351–376.

Taguchi-Shiobara, F., Yuan, Z., Hake, S. & Jackson, D. (2001) The fasciated ear2 gene encodes a leucine-rich repeat receptor-like protein that regulates shoot meristem proliferation in maize. *Genes Dev.*, **15**, 2755–2766.

Takada, S., Hibara, K., Ishida, T. & Tasaka, M. (2001) The *CUP-SHAPED COTYLEDON1* gene of Arabidopsis regulates shoot apical meristem formation. *Development*, **128**, 1127–1135.

Talbert, P.B., Adler, H.T., Parks, D.W. & Comai, L. (1995) The *REVOLUTA* gene is necessary for apical meristem development and for limiting cell divisions in the leaves and stems of Arabidopsis thaliana. *Development*, **121**, 2723–2735.

Tang, G., Reinhart, B.J., Bartel, D.P. & Zamore, P.D. (2003) A biochemical framework for RNA silencing in plants. *Genes Dev.*, **17**, 49–63.

Trotochaud, A.E., Hao, T., Wu, G., Yang, Z. & Clark, S.E. (1999) The CLAVATA1 receptor-like kinase requires CLAVATA3 for its assembly into a signalling complex that includes KAPP and a Rho-related protein. *Plant Cell*, **11**, 393–406.

Tsiantis, M. & Hay, A. (2003) Comparative plant development: the time of the leaf? *Nature Rev. Genet.*, **4**, 169–180.

Tsiantis, M., Schneeberger, R., Golz, J.F., Freeling, M. & Langdale, J.A. (1999) The maize *rough sheath2* gene and leaf development programs in monocot and dicot plants. *Science*, **284**, 154–156.

Vollbrecht, E., Veit, B., Sinha, N. & Hake, S. (1991) The developmental gene Knotted-1 is a member of a maize homeobox gene family. *Nature*, **350**, 241–243.

Waites, R., Selvadurai, H.R., Oliver, I.R. & Hudson, A. (1998) The *PHANTASTICA* gene encodes a MYB transcription factor involved in growth and dorsoventrality of lateral organs in Antirrhinum. *Cell*, **93**, 779–789.

Weigel, D., Alvarez, J., Smyth, D.R., Yanofsky, M.F. & Meyerowitz, E.M. (1992) *LEAFY* controls floral meristem identity in Arabidopsis. *Cell*, **69**, 843–859.

Western, T.L., Cheng, Y., Liu, J. & Chen, X. (2002) HUA ENHANCER2, a putative DExH-box RNA helicase, maintains homeotic B and C gene expression in Arabidopsis. *Development*, **129**, 1569–1581.

Williams, R.W., Wilson, J.M. & Meyerowitz, E.M. (1997) A possible role for kinase-associated protein phosphatase in the Arabidopsis CLAVATA1 signalling pathway. *Proc. Natl. Acad. Sci. USA*, **94**, 10467–10472.

Wu, X., Dinneny, J.R., Crawford, K.M., Rhee, Y., Citovsky, V., Zambryski, P.C. & Weigel, D. (2003) Modes of intercellular transcription factor movement in the Arabidopsis apex. *Development*, **130**, 3735–3745.

Yanofsky, M.F., Ma, H., Bowman, J.L., Drews, G.N., Feldmann, K.A. & Meyerowitz, E.M. (1990) The protein encoded by the Arabidopsis homeotic gene agamous resembles transcription factors. *Nature*, **346**, 35–39.

Yu, L.P., Miller, A.K. & Clark, S.E. (2003) *POLTERGEIST* Encodes a protein phosphatase 2C that regulates CLAVATA pathways controlling stem cell identity at Arabidopsis shoot and flower meristems. *Curr. Biol.*, **13**, 179–188.

Yu, L.P., Simon, E.J., Trotochaud, A.E. & Clark, S.E. (2000) POLTERGEIST functions to regulate meristem development downstream of the *CLAVATA* loci. *Development*, **127**, 1661–1670.

Zuo, J., Niu, Q.W., Frugis, G. & Chua, N.H. (2002) The *WUSCHEL* gene promotes vegetative-to-embryonic transition in Arabidopsis. *Plant J.*, **30**, 349–359.

9 Polar signals in vascular development

Thomas Berleth and Enrico Scarpella

9.1 Introduction

The plant vascular system is organized in networks of interconnected cells through which water and dissolved materials are distributed throughout the plant (Esau, 1965; Steeves & Sussex, 1989). Two types of conducting tissues, phloem and xylem, are specialized in the transport of dissolved photo-assimilates and minerals, respectively. Vascular tissues typically originate from common precursor cells and remain associated in organ-specific arrangements within continuous structures, which are referred to as veins (in leaves), bundles (in stems) and the vascular cylinder (in roots).

The molecular basis of vascular tissue patterning has intrigued many, in particular because any proposed mechanism has to account for the potentially conflicting features of continuity and plasticity: vascular tissues differentiate as continuous strands of perfectly aligned cells, although this precision cannot be conferred by rigid positional specification, since their routes are influenced by variable external factors.

Vascular tissue patterning seems to be integrated into the larger tissue context in at least two ways (Fig. 9.1).

First, signals reflecting organismal polarities control the route and internal organization of vascular strands. More precisely, signals defining the apical-basal polarity of the plant as a whole direct the course of vascular strand formation, and the radial arrangement of xylem and phloem within each vascular strand reflects signals specifying adaxial-abaxial polarity in the surrounding organ. In summary, vascular bundles and surrounding organ patterns are integrated through common polar cues.

Second, once established under the control of organismal cues, a vascular strand seems to serve as a source of positional information to direct patterned differentiation in adjacent cell layers. It has long been known that numerous local cell patterning events are correlated to the positions of vascular strands, suggesting that these strands serve as reference systems in local cell patterning (Nelson & Dengler, 1997). More recently, details have become apparent of the molecular mechanisms through which vascular tissues polarly supply positional information in ground tissue patterning and root meristem initiation.

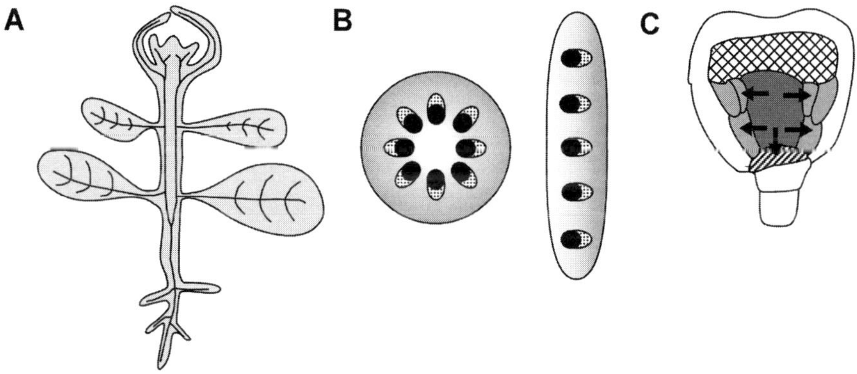

Figure 9.1 Integration of vascular tissues into the organismal context. A: Apical-basal organization: While the precise route of vascular strands cannot be predicted, vascular differentiation must be continuous and the positions and orientations of the cells within each strand must be precisely integrated to ensure conductivity. Auxin produced in shoot organs and transported basally could serve as an integrating signal along the length of a strand. B: Adaxial-abaxial polarity: The adaxial-abaxial or central-peripheral polarity of stems (schematic cross section, left) and leaves (right) is reflected in the polar organization of xylem (filled) and phloem (open). C: Vascular strands as sources of positional information: The procambium (dark grey) of an early heart-stage *Arabidopsis* embryo directs (arrow) division and differentiation events in the overlying ground tissue (light grey) and in the basal center of an incipient primary root meristem (stripes), (apical embryo region, cross-hatched; epidermis and suspensor, white).

9.2 Apical-basal polarity in vascular regeneration experiments

Evidence for an involvement of polar signals in vascular strand formation has been derived from local auxin application experiments (Fig. 9.2). Auxin has long been implicated in vascular differentiation, since indole-3-acetic acid, or IAA, the predominant form of auxin in plants, can induce (together with other factors) xylem tracheary element differentiation in suspension culture cells (reviewed in Fukuda, 1997) and auxin-overproducing transgenic plants have increased amounts of vascular tissues (Klee *et al.*, 1987). Most importantly, IAA can define the positions of new vascular strands. This is observed when auxin locally induces the transdifferentiation of parenchymatic cells to vascular cells, which differentiate in continuity basal to the site of auxin application (Fig. 9.2; reviewed in Sachs, 1981). This observation goes far beyond the mere identification of a vascular differentiation-promoting substance, as it establishes a spatial relationship between an IAA source and a patterned cellular response, which is intriguing in several ways. First, the response is local, as the newly-formed vascular strand extends from the particular site of IAA application, indicating that IAA provides positional information in this experiment. Second, the response is polar, as IAA will

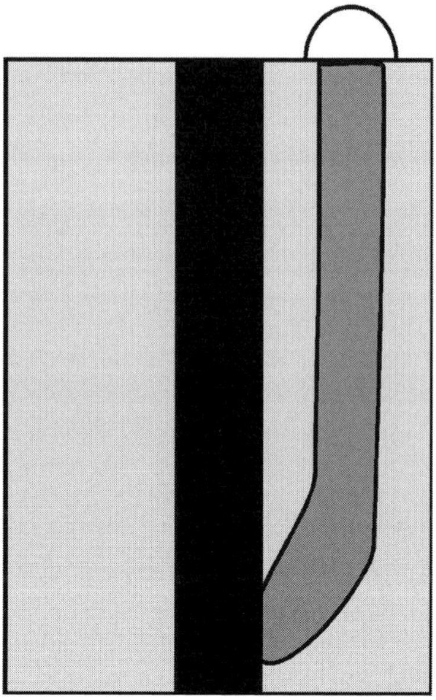

Figure 9.2 Vascular strand differentiation in response to local auxin application. Polarity of vascular strand differentiation in a hypocotyl segment after IAA application. A local source of IAA (white blob) triggers the formation of a vascular strand (dark grey) that connects the auxin source to an existing vascular strand (black). While the precise route of vascular differentiation is not necessarily reproducible or straight, the differentiation process is always restricted to a narrow strip of cells, integrates the differentiation along the entire strand and leads to continuity with existing vasculature, suggesting integrating mechanisms acting along the length of the strand. Polar auxin transport has been implicated in defining the routes of vascular strand development. The drawing summarizes findings reviewed in Sachs (1981).

induce a vascular strand only towards the basal pole of the organ. Third, the response is continuous, as the responding cells differentiate in a continuous area and eventually connect the IAA source with pre-existing vasculature. Fourth, the differentiation zone is restricted in the radial dimension, as differentiation occurs only within a narrow strip of cells. Finally, a patterned vascular differentiation response requires the use of polarly transported auxin, such as IAA, and is obstructed in the presence of auxin transport inhibitors.

Experiments of this type can be performed only in fully polarized organs, such as fragments of hypocotyls. Together, the above features suggest that the polar differentiation process co-opts internal polarity cues, which are probably identical with polar flow of auxin. In normal plant growth, IAA is predominantly produced in apical regions, such as young leaves or flowers, and is transported basally. Studies in the last few years have not only added molecular details and a cellular backbone to long-standing models for the apical-basal cell-to-cell transport of auxin (reviewed in Lomax *et al.*, 1995;

Muday & Murphy, 2002), but have also accumulated evidence for an instrumental role of polar auxin transport in defining apical-basal polarity in embryos and plant organs (reviewed in Paquette & Benfey, 2001; Jürgens, 2001; Berleth & Chatfield, 2002 and below). As a polarly transported, non-cell autonomous signal substance, auxin could mediate coordinated axis formation and polarization across many cells or entire organisms, and the local addition of the same signal substance to an already polarized system would conceivably result in a local, polar and continuous response (Fig. 9.4B on page 272).

9.3 Apical-basal polarity in embryo vascular patterning

The integration of longitudinal patterning of vascular tissues into organismal polar signaling can be best illustrated in the embryo. The embryo pattern in *Arabidopsis* can be formalized as sequences of apically-basally and centrally-peripherally arrayed elements. The main elements along the apical-basal axis are the shoot apical meristem, the cotyledons, the hypocotyl, the radicle and the root apical meristem (Fig. 9.3). Most of the embryo body takes the form of a cylinder, along which the central-peripheral pattern of concentrically-arrayed tissue layers is formed, comprising epidermis, cortex, endodermis and stele tissue (Fig. 9.3A). In the seedling, hypocotyl and radicle (embryonic root) are sections along this cylinder, with distinguishable anatomical and physiological properties, and the radial pattern of the root is perpetuated by the activity of the root apical meristem.

Does the apical-basal polarity of the embryo and the position of the procambial strand in its centre depend on auxin flow? Available experimental and more abundant genetic evidence is consistent with this view.

IAA efflux inhibitor application to *Brassica juncea* embryos dramatically interferes with embryo axis formation and occasionally results in entirely apolar, ball-shaped embryos and seedlings (Hadfi *et al.*, 1998). Equivalent genetic tests of the role of polar auxin transport in embryos are not easily accomplished because all proposed elements of the auxin transport machinery are encoded by potentially redundant members of large gene families, and the loss-of-function mutants display correspondingly subtle abnormalities (reviewed in Bennett *et al.*, 1998; Palme & Gälweiler, 1999; Friml & Palme, 2002; see also Chapter 2). However, mutations in a single *Arabidopsis* gene, *EMBRYO DEFECTIVE 30/GNOM* (*emb30/gn*), result in embryonic defects comparable to those reported for wild types treated with high doses of auxin transport inhibitors (Mayer *et al.*, 1993). Embryos and seedlings mutant for *emb30/gn* can be entirely ball-shaped with no detectable apical-basal polarity and, intriguingly, the vascular system of these mutants consists of disconnected, randomly oriented cells (Fig. 9.3B). The latter observation is the

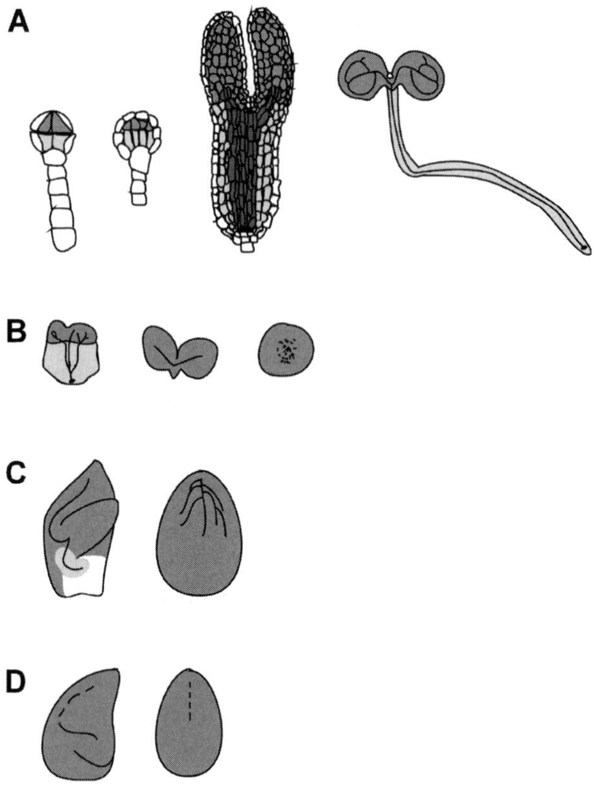

Figure 9.3 Vascular development in embryos. (A) *Arabidopsis* embryos and seedling. In the transition from the dermatogen (left) to the early globular (centre left) stage, the inner cells in the basal part of the embryo (light grey) acquire a common apical-basal axis, which then becomes radially refined in the torpedo (centre right) embryo (vascular cylinder, dark grey; ground tissue, intermediate grey; shoot apical meristem, epidermis and suspensor, white). Redrawn from Berleth and Chatfield, 2002. (B) *Arabidopsis* seedling mutants. Early in embryogenesis (not shown) all three types of mutants shown are characterized by disorganized cell patterns. However, later development reveals different primary defects. Various types of cell shape defective mutants (left, the *fass* mutant, Torres-Ruiz & Jürgens, 1994) will establish proper apical-basal polarity and continuous vascular strands. By contrast, auxin-insensitive mutants (centre, *mp*, see text for details) and auxin transport defective mutants (right, *emb30/gn*, see text for details) will show lasting apical-basal pattern and vascular continuity defects, implicating auxin as a polar signal in embryo axis and vascular strand formation (compare to wild type seedling in A, right). Redrawn from Berleth (2001). (C) Mature wild-type rice embryo. Left, median longitudinal section. Right, dorsal view. A procambial strand is positioned at the center of the C-shaped embryonic axis (light grey). The radicle and root meristem are surrounded by the coleorhiza (white), which is not present in the Arabidopsis embryo. The vascular system of the scutellum, which represents the single cotyledon, is composed of a primary procambial strand that branches at the apex into a series of secondary strands. Occasionally, tertiary strands branch from secondary strands. Redrawn from Scarpella *et al.* (2003). (D) Mature *ral1* rice embryo. Left, median longitudinal section. Right, dorsal view. The embryonic axis with its central procambial strand are absent and the vascular system of the scutellum is limited to a short and interrupted primary procambial strand where cells are misaligned. Higher-order strands are completely missing. Redrawn from Scarpella *et al.* (2003).

most remarkable, as the mere randomization of cell division planes in early embryos does not lead to apolar embryos (Fig. 9.3B; reviewed in Berleth, 2001). Instead, numerous cell biological defects can lead to apparently chaotic cell division patterns and abnormal cell shapes in early embryos without having any effect on the polarity of the embryo. Therefore, mutations in *emb30/gn* seem to identify a specific cellular mechanism underlying both organismal polarity and continuous vascular differentiation aligned with the apical-basal axis of the embryo. Recent molecular and cell biological findings directly link *EMB30/GN* function to the polar transport of auxin.

The EMB30/GN protein is highly similar to yeast guanosine exchange factor (GEF) GEA1, a member of a yeast GEF family implicated in the control of vesicle transport and sensitive to the fungal metabolite brefeldin A (Busch *et al.*, 1996; Steinmann *et al.*, 1999). *Arabidopsis EMB30/GN* can complement yeast *gea1* mutations and displays a BFA sensitive GEF activity. Moreover, mutations in *EMB30/GN* as well as BFA treatment can perturb the basal position of the presumed auxin efflux component PIN FORMED 1 (PIN1) (Steinmann *et al.*, 1999). Permanent *EMB30/GN*-dependent vesicle transport may be necessary to maintain auxin transport, because auxin efflux carrier components may not be statically anchored to the plasma membrane but seem to be subject to constant reshuffling from intracellular sites (Geldner *et al.*, 2001). *EMB30/GN*-dependent cycling between intracellular and membrane locations has been demonstrated for the PIN1 product and, if true for other redundantly acting proteins, could explain the central importance of *EMB30/GN* for auxin transport and embryo polarity. Moreover, the process seems to be dependent on microfilament guidance (Geldner *et al.*, 2001; Muday & Murphy, 2002). Thus, as in animal embryogenesis, polar orientation of cytoskeletal elements are key to organismal axis formation, and the role of auxin could be to integrate coordinated axis formation across large groups of cells.

In summary, vascular strand formation is strictly dependent on apical-basal polarity and polar auxin transport. Together with the results of local auxin-application experiments, this correlation suggests that apical-basal organismal axis establishment and vascular strand formation have a common origin in the orientation and distribution of auxin flow. While obviously necessary, auxin flow is probably not sufficient to ensure the precision of aligned vascular differentiation, which is likely to be further promoted by supplementary, unidentified mechanisms.

9.4 Auxin-signal transduction in embryo vascular patterning

In addition to the polar localization of membrane proteins ensuring the directionality of auxin flow, proper auxin perception seems to be required to promote vascular differentiation. Consistent with this expectation, vascular abnormalities have been reported for auxin perception mutants (Lincoln *et al.*, 1990). However, severe auxin insensitivity seems to be associated with embryo and/or seedling lethality and it is therefore possible that more mutants in this class will be identified in the future. Mutations in three *Arabidopsis* genes, *MONOPTEROS* (*MP*), *BODENLOS* (*BDL*) and *AUXIN RESISTANT6* (*AXR6*) interfere to varying degrees with the formation of the central vascular strand in the heart-stage embryo (as well as with vascular differentiation in general), with cell differentiation in the entire basal domain of the embryo and with the formation of a primary root meristem (Berleth & Jürgens, 1993; Hamann *et al.*, 1999; Hobbie *et al.*, 2000; Przemeck *et al.*, 1996; Fig. 9.3B). Mutations in all three genes have also been associated with other types of defective auxin responses (Hamann *et al.*, 1999; Hobbie *et al.*, 2000; Mattsson *et al.*, 2003) and the products of all three genes are directly linked to auxin signal transduction. *MP* encodes a transcriptional regulator of the 'auxin response factor' (ARF) family (Hardtke & Berleth, 1998). This family of transcription factors can confer auxin responsiveness to downstream genes by binding to conserved 'auxin response' promoter elements (Ulmasov *et al.*, 1997), apparently modulated in activity by interaction with nuclear proteins of the Aux/IAA family (recently reviewed in: Hagen & Guilfoyle, 2002; Liscum & Reed, 2002; Leyser, 2001). The dominant *bdl* mutation is presumed to confer enhanced stability and elevated abundance to the negative ARF regulator IAA12 (Hamann *et al.*, 2002). Since MP and BDL are expressed in overlapping domains and interact in yeast two-hybrid assays, it is possible that the embryonic phenotype of *bdl* mutants result from inhibition of MP activity. Finally, the two dominant *axr6* mutations affect a crucial residue in the CUL1 protein, which affects its ability to assemble into stable SCF complexes (Hellmann *et al.*, 2003). These complexes, in turn, are required for the efficient degradation of Aux/IAA negative regulators of auxin response.

The vast divergence in the appearance of angiosperm embryos justifies the question whether any general conclusion can be drawn from genetic analyses in *Arabidopsis*. With regard to dicot embryos, the answer is probably yes, because the various embryogenesis programs converge remarkably towards the end of embryo development and generate highly related seedling patterns. Monocot embryos, by contrast, are the result of an additional morphogenetic process, which generates a series of embryonic organs not present in dicot embryos. Furthermore, shoot development proceeds to a more advanced stage, with the result that a number of leaf primordia are initiated by the shoot apical meristem before embryogenesis is completed. Finally, the role of auxin in monocot patterning is still not clear. Large-scale mutant screens to unravel the mechanistic basis of monocot embryogenesis have been performed in maize and rice (Clark & Sheridan, 1991; Hong *et al.*, 1995), but the affected gene products remain to be identified. Despite the divergent appearance of monocot embryos, there may still be landmarks to be assessed for similarities with *Arabidopsis* embryos. For example, one can still recognize the presence of a procambial strand positioned in the center of the embryo axis that joins the shoot apical meristem with the root apical meristem (Fig. 9.3C). Interestingly, a recent study reports that a mutation in the *RADICLELESS1* (*RAL1*) gene of rice is characterized by defects related to those of *mp, bdl* and *axr6* mutants of *Arabidopsis* (Scarpella *et al.*, 2003). Mutant embryos fail to establish an embryonic axis, recognizable by the absence of a central vascular strand, and neither a primary root nor a root apical meristem is formed (Fig. 9.3D). Most strikingly, like the *mp, bdl* and *axr6 Arabidopsis* mutants, *ral1* mutants are auxin-insensitive. It is therefore possible that related apical-basal patterning mechanisms control vascular strand formation in dicots and monocots and that they can be identified through diagnostic vascular abnormalities.

9.5 Vascular patterning in the leaf

Historically, the formation of the intricate network of vascular strands in leaves has been a focus of interest in vascular research. From the perspective of network formation, this focus may be justified. For the understanding of general principles underlying vascular strand formation *per se*, however, it is more problematic. Long-distance connection through vascular strands has been a stringent requirement ever since multicellular plants colonized the land. One may therefore argue that leaf venation patterns merely reflect specialized adaptations of general principles, designed to ensure appropriate vein distribution within an unusually shaped organ type. Previously established mechanisms integrating aligned cell differentiation during vascular strand formation were probably not reinvented in the evolution of leaves, but more likely amended by leaf-specific controls.

Unlike vascular patterns along the main axis of the plant, leaf venation patterns are highly reticulate and form networks of hierarchically ranked veins (Hickey, 1973, 1988; Klucking, 1995). These patterns are clearly inheritable and are often used as taxonomic traits, especially since they are reproducible in many species. As this presumably reflects some selective pressure on precise vein positioning, venation patterns are probably not just readouts of apical-basal auxin flow, but subject to additional or even entirely unknown controls. Three general types of control mechanisms may be envisaged (Fig. 9.4).

First, veins may be defined invariantly as integral parts of leaf lamina formation. For example, leaf veins in species with invariant venation patterns, could be rigidly specified along with all other features of the leaf, for instance as overlaps between gene expression domains defining leaf areas (Fig. 9.4A). Therefore plasticity of leaf vein patterns, if they exist at all, could not be uncoupled from other aspects of leaf anatomy. Any change in leaf venation

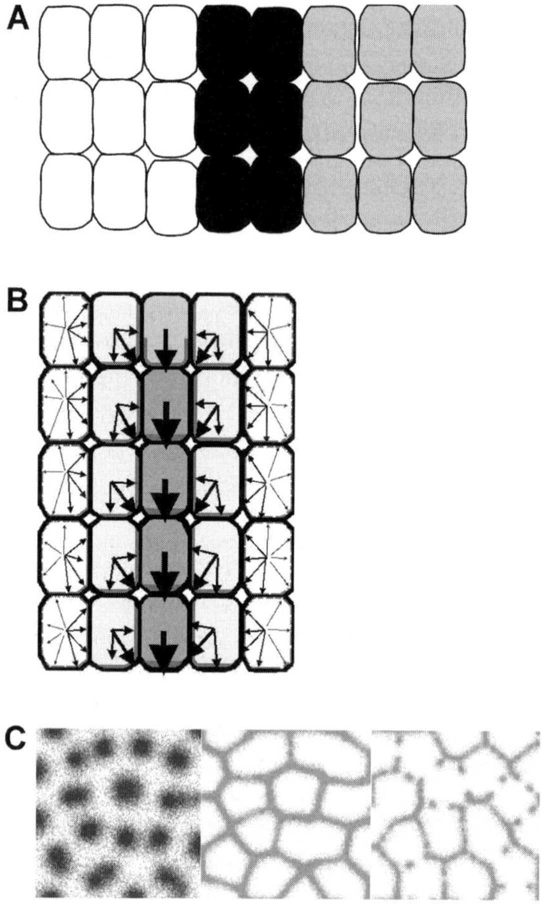

Figure 9.4 Vascular patterning models. (A) Rigid specification. Veins (black) differentiate in the position where the expression of gene A (white) overlaps with that of gene B (light grey). (B) Auxin canalization hypothesis. In this simplified view, it is assumed that intensity and direction of auxin flow is solely controlled through the quantity and distribution of auxin efflux carriers (dark grey) in the plasma membranes of cells (rectangles). Arrows of different strength indicate the intensity of auxin flow. The hypothesis proposes that auxin flow and cell polarization are connected in a positive feedback loop, illustrated here by restricting auxin efflux to the basal side of each cell as an expression of cell polarization. Through the same mechanism, cells mutually stabilize (integrate) apical-basal polarity and restrict auxin flow to preferred routes (light grey), which subsequently differentiate to vascular tissues. Reproduced with permission from Berleth & Chatfield (2002). (C) Network formation by reaction diffusion, or more precisely, autocatalysis and long-range inhibition. Local concentration maxima are generated by the interaction of a short-range autocatalytic molecule with an antagonistic reaction at long-range (left). If these signaling centers determine where vascular strands are not to be formed, a similar autocatalytic system can generate closed strands at maximum distance from the signaling centers (center). Interrupted strands could, for example, result from a mutation in the self-enhancing strand-forming reaction (right). Reproduced with permission from Berleth (2000). (Image courtesy of Hans Meinhardt.)

could occur only as a consequence of changes in gene expression domains with a corresponding impact on overall leaf architecture.

Second, veins could be flexibly defined through self-stabilizing feedback mechanisms involving a long-distance apical-basal signal flow. In its most elaborate form, this model is known as the auxin-canalization hypothesis (Sachs, 1981, 1991), wherein a positive feedback between auxin transport through a cell and the cell's auxin conductivity progressively restricts an initially dispersed auxin flow to narrow preferred canals, which finally differentiate into vascular tissues (Fig. 9.4B).

Third, self-organizing patterning mechanisms, for example those driven by autocatalysis and long-range inhibition (Meinhardt, 1982; Meinhardt & Gierer, http://www.eb.tuebingen.mpg.de/dept4/meinhardt/theory.html) could generate network patterns consistent with those observed in a variety of leaves (Fig. 9.4C).

Obviously, the importance of rigid specification of vein positions in the course of leaf development is directly reciprocal to the amount of observed plasticity of venation patterns within a given leaf. In species with variable venation patterns, such as *Arabidopsis* and other analyzed dicots, even low concentrations of auxin transport inhibitors, which do not significantly affect the shape of the leaf, result in entirely new types of venation patterns (Mattsson *et al.*, 1999; Sieburth, 1999). Therefore, in these species, venation patterns do not seem to be constrained by leaf-specific controls but may be largely self-organized, possibly by a combination of the latter two mechanisms. In monocot species with extremely organized and highly invariant venation patterns, the situation is less clear. Whereas most dicot leaves show a ramified pattern of progressively branched veins, most monocot leaves have striate venation patterns in which major veins lie parallel along the proximodistal axis of the leaf and are connected transversely by minor transverse veins. Typically, the distribution and arrangement of veins in monocot leaves is highly reproducible and may therefore be rigidly specified. Alternatively, however, it is possible that treatments that uncouple venation patterns from other aspects of leaf development in dicots, such as auxin transport inhibition, could have similar effects on monocot leaves. In a recent experiment, application of polar auxin transport inhibitors at concentrations that did not cause aberrations in leaf shape resulted in alterations of vein spacing without affecting the overall arrangements of different veins (Scarpella *et al.*, 2002). This result indicates that monocot venation patterns have a certain amount of plasticity, but simultaneously shows that this plasticity is far more constrained than in dicot leaves, presumably by genetic cues that regulate the development of the leaf. Such readily tractable genetic cues should make monocot species suitable candidates for the genetic dissection of leaf venation pattern formation.

The two remaining models (Fig. 9.4B, C) imply some degree of vascular self-organization and both are by no means mutually exclusive. Not only could they exist side-by-side, with the signal-flow ensuring long-distance continuity and local diffusion ensuring an even spread of veins across the lamina, both could also interact in a variety of ways. These models should therefore not be taken as distinct alternatives, but as mere conceptual extremes. As such they are of great value, as each makes predictions that can be tested experimentally. For example, any signal flow model involving auxin suggests elevated levels of auxin to be associated with vascular differentiation. This has indeed been shown in a variety of studies (Morris & Thomas, 1978; Wangermann, 1974; Avsian-Kretchmer *et al.*, 2002; Aloni *et al.*, 2003), and a recent report further demonstrates that elevated levels of local auxin response (reflected in reporter gene expression from a synthetic auxin-response promoter) precedes procambial differentiation in both normal and auxin transport inhibited conditions (Mattsson *et al.*, 2003). The fact that the pattern of elevated auxin response and procambial differentiation is congruent under various experimental conditions lends support to an instrumental role of auxin in specifying sites of procambial differentiation.

Conversely, signal flow should be excluded as an organizing principle in all discontinuous venation patterns. Discontinuous veins are observed in many species as well as in characterized mutants (Table 9.1). In the *cotyledon vascular pattern1* and *2* (Carland *et al.*, 1999), the *scarface* (Deyholos *et al.*, 2000) and the *vascular network* (Koizumi *et al.*, 2000) mutants of *Arabidopsis* as well as in the *ral1* mutant of rice (Scarpella *et al.*, 2003) for example, the interrupted vascular strand phenotype has been traced back to discontinuous procambium. The experimental observation, that in these mutants the procambium can differentiate in separate 'islands', may suggest that continuity is not a prerequisite for vascular strand formation. However, differentiating procambium is the result of a preceding patterning mechanism, regardless of whether this mechanism involves a continuous signal flow or not. The observation that even in discontinuous vein mutants the interrupted procambium is still formed along the same paths where it would normally differentiate continuously, suggests that the isolated stretches of procambium are visible segments within a larger invisible pattern. It therefore seems that the bottleneck in the genetic analysis of venation patterning lies in our inability to identify, visualize and manipulate implicated molecular cues. To some extent, this is still true for the flow of auxin and even more for the entirely hypothetical molecules of the reaction diffusion model.

In summary, there is a growing body of mutually supporting evidence for a decisive role of auxin as an intercellular signal mediating integrated polarization of plant cells and the differentiation of vascular strands along routes of elevated auxin concentrations. This concept does not imply that auxin is sufficient for these processes, nor does it explain how the positions of auxin sources and sinks are themselves regulated and how highly reproducible vein patterns are ensured. However, it reconciles plasticity and continuity of

Table 9.1 Genes in vascular strand formation

Gene	Species	Vascular anatomy	Organismal polarity defect	Molecular identity	References
AXR6/AtCUL1	Arabidopsis	Reduced, discontinuous and centralized vascularization in cotyledons of homozygous mutants	Apical-basal	Cullin	Hobbie *et al.* (2000); Hellmann *et al.* (2003)
BDL/IAA12	Arabidopsis	Reduced vascularization in cotyledons and leaves of homozygous mutants	Apical-basal	Member of the Aux/IAA family of short-lived nuclear proteins	Hamann *et al.* (1999, 2002)
EMB30/GN/VAN7	Arabidopsis	Disconnected, randomly oriented vascular cells in homozygous mutant seedlings	Apical-basal	Guanine exchange factor	Mayer *et al.* (1993); Shevell *et al.* (1994); Busch *et al.* (1996); Steinmann *et al.* (1999); Koizumi *et al.* (2000)
MP/ARF5	Arabidopsis	In homozygous mutants, vascularization in the stem is reduced, misaligned and discontinuous. Vascularization in cotyledons and leaves is reduced, misaligned, discontinuous and centralized. Vascularization in adventitious roots is discontinuous.	Apical-basal	Transcription factor of the ARF family	Berleth & Jürgens (1993); Przemeck *et al.* (1996); Hardtke & Berleth (1998); Mattsson *et al.* (2003)
PIN1	Arabidopsis	In homozygous mutants, vascularization in the stem is increased and discontinuous. Vascularization in leaves is increased and marginalized	Apical-basal	Putative auxin efflux carrier	Okada *et al.* (1991); Galweiler *et al.* (1998); Mattsson *et al.* (1999)

Table 9.1 (*continued*)

Gene	Species	Vascular anatomy	Organismal polarity defect	Molecular identity	References
RAL1	Rice	In homozygous mutants, vascularization in scutella and leaves is reduced, discontinuous and misaligned. Vascularization in leaves displays altered timing of development. Vascularization in the stem and adventitious roots is reduced	Apical-basal	Unknown	Scarpella *et al.* (2003)
AGO1 and ZLL/PNH	Arabidopsis	Amphicribral vascular bundles in leaves of the *ago1* homozygous mutant in the *zll/pnh* heterozygous background	Adaxial-abaxial	Piwi/PAZ-domain protein	Lynn *et al.* (1999); Cerutti *et al.* (2000)
AVB1	Arabidopsis	Amphivasal vascular bundles in the stem and leaves of homozygous and heterozygous mutants	Adaxial-abaxial	Transcription factor of the HD-Zip family	Zhong *et al.* (1999); Ye (2002)
KAN1 and 2	Arabidopsis	Absence of vasculature in cylindrical leaves of 35S-KAN1, AS1-KAN1 and AS1-KAN2 plants	Adaxial-abaxial	Transcription factors of the GARP class	Eshed *et al.* (2001); Kerstetter *et al.* (2001)
PHAB	Arabidopsis	Funnel-shaped leaves of heterozygous mutants display the presence of amphivasal vascular bundles. Phloem or both phloem and xylem are absent in cylindrical leaves	Adaxial-abaxial	Transcription factor of the HD-Zip III family (Athb14)	McConnell & Barton (1998); McConnell *et al.* (2001)

PHAN	Anthirrinum	In homozygous mutants, cylindrical leaves, which develop at the upper part of the stem, contain amphicribral vascular bundles. In the *phan hb* double mutants, all leaves are cylindrical and contain amphicribral vascular bundles	Adaxial-abaxial	Transcription factor of the MYB class	Waites & Hudson (1995); Waites *et al*. (1998); Waites & Hudson (2001)
RS1	Maize	Disordered vascular anastomosis at the blade-sheath boundary. Increased and ectopic xylem production. Occasional amphivasal vascular bundles	Adaxial-abaxial	Transcription factor of the KNOX class	Becraft & Freeling (1994); Schneeberger *et al*. (1995)
AS2	Arabidopsis	In leaves of homozygous mutants, the midvein is reduced in size or absent. Major veins run parallel and separately from the midvein. The number of minor veins is reduced. Discontinuous vascularization		Novel protein characterised by presence of cysteine repeats and leucine zipper domain	Semiarti *et al*. (2001); Iwakawa *et al*. (2002)
CVP1	Arabidopsis	Vascularization in cotyledons of homozygous mutants is reduced, misaligned and discontinuous. Vascularization in the stem is increased		Sterol Methyltransferase2	Carland *et al*. (1999, 2002)
CVP2	Arabidopsis	Vascularization in cotyledons and leaves of homozygous mutants is reduced and discontinuous		Unknown	Carland *et al*. (1999)

Table 9.1 (*continued*)

Gene	Species	Vascular anatomy	Organismal polarity defect	Molecular identity	References
FILl/YAB1	Arabidopsis	Reduced number of minor veins and discontinuous vascularization in leaves of the *fil/yab1 yab3* double mutants. Absence of vasculature in cylindrical leaves of 35S-FIL/YAB1 plants		Transcription factor of the YABBY class	Sawa *et al.* (1999)
SFC	Arabidopsis	Discontinuous and reduced vascularization in cotyledons and leaves of homozygous mutants		Unknown	Deyholos *et al.* (2000)
VAN1, 2, 4 and 6	Arabidopsis	In homozygous mutants, vascularization in cotyledons and leaves is discontinuous and reduced. In the hypocotyl, phloem is reduced		Unknown	Koizumi *et al.* (2000)
VAN3	Arabidopsis	Discontinuous and reduced vascularization in cotyledons and leaves of homozygous mutants		Unknown	Koizumi *et al.* (2000)
VEP1/AWI31	Arabidopsis	In antisense lines, vascularization in leaves is discontinuous and reduced. Vascularisation in cotyledons, stem and root is reduced		Polypeptide similar to animal proteins involved in apoptosis (death-domain proteins)	Jun *et al.* (2002); Yang *et al.* (1997)

vascular patterns in species with variable venation patterns and may direct the route of vascular strands outside of leaves even in other species. Further, the role of auxin may be restricted to the delineation of procambial tissues, while other aspects of vascular patterning may be controlled by unrelated cues (see below).

9.6 Adaxial-abaxial polarity affects vascular bundle organization

The formation of flat lateral organs involves the specification of adaxial and abaxial cell identities. Leaf anatomy and gene expression studies suggest that early leaf primordia are initially uniform and become polarized in the adaxial-abaxial dimension (reviewed in Hudson & Waites, 1998; Bowman, 2000; see also Chapter 10). A growing number of regulatory genes implicated in adaxial-abaxial patterning have recently been reviewed (Sessions & Yanofsky, 1999; Bowman *et al.*, 2002).

The adaxial-abaxial polarity of a leaf is reflected in the polar organization of leaf veins, with xylem directed towards the adaxial side of the leaf and phloem facing the abaxial side, referred to as collateral bundle organization (Figs. 9.1B, 9.5). An integrated regulation of leaf and vascular strand polarity is suggested by the phenotypes resulting from mutations in various genes. Both adaxialized and abaxialized leaves do not form laminae and therefore develop as cylindrical organs with a single central vascular strand, in which the polarity shift of the surrounding organ is reflected in a corresponding shift towards amphivasal or amphicribral organization (Fig. 9.5). Examples of leaves that develop as cylindrical organs with correspondingly altered internal vein organization have been observed in the *Arabidopsis argonaute1* mutant in heterozygous plants for *zwille/pinhead* (Lynn *et al.*, 1999), the *phantastica* (*phan*) single mutant and in the *phan handlebar* (*hb*) double mutant of *Anthirrinum* (Waites & Hudson, 1995; Waites *et al.*, 1998; Waites & Hudson, 2001), the *phabulosa/Athb14* (*phb/Athb14*) (McConnell & Barton, 1998; McConnell *et al.*, 2001) and the *amphivasal bundle1* (*avb1*) mutants of *Arabidopsis* (Zhong *et al.*, 1999) and in the *Rough sheath1* (*Rs1*) mutant of maize (Becraft & Freeling, 1994). In *phb/Athb14*, *avb1* and *Rs1*, the vascular strands become amphivasal, i.e. xylem surrounds as a ring a central phloem cylinder, whereas in *phan*, in *phan hb* and in the *ago1* mutant in the *zll/pnh* heterozygous background, the single vein is amphicribral, i.e. the phloem is organized as a sheath around a xylem cylinder. Furthermore, ectopic overexpression of genes involved in leaf abaxial polarity acquisition, such as the *KANADI* genes (Eshed *et al.*, 2001; Kerstetter *et al.*, 2001) and the *FILAMENTOUS FLOWER/YABBY1* gene (Sawa *et al.*, 1999) results in abaxially radialized leaves that do not develop any vasculature. Vascular bundles are also missing from most leaves of severely adaxialised *phb/Athb14*

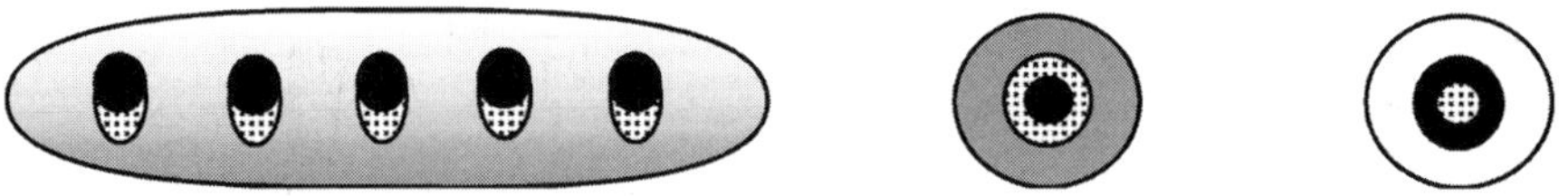

Figure 9.5 Adaxial-abaxial patterning of leaves and veins. Wild type Arabidopsis leaves (left) are polarized in the adaxial-abaxial dimension (graded grey) and comprise veins of collateral organization (phloem, dotted; xylem, black). Since the formation of the leaf lamina depends on adaxial-abaxial polarity, mutant leaves may be cylindrical (cross-sections, center and right). Abaxialized leaves contain amphicribral veins (phloem encloses xylem, center), adaxialized leaves correspondingly contain amphivasal veins (xylem encloses phloem, right).

plants (McConnell & Barton, 1998). As both extreme phenotypes are incompatible with vascular tissue formation, it is possible that either intermediate positional values or adaxial-abaxial polarity *per se* may be essential for vascular tissue formation.

Adaxial-abaxial polarity may be derived from ancestral central-peripheral polarity in cylindrical organs (Tasaka, 2001), and this seems to be reflected in the organization of vascular bundles in stems and in the central-peripheral organization of hypocotyls, which in cases of extreme adaxial-abaxial shifts may be devoid of any vascular tissues (Kerstetter *et al.*, 2001). Therefore, there is evidence for an integrated regulation of adaxial-abaxial patterns within vascular strands and surrounding organs in the majority of plant organs, and this control seems to be sufficient to account for the internal organization of vascular bundles and for their functional integration into the organ context.

9.7 Vascular tissues as sources of positional information

At least two important patterning events occur in stable polar relationship to vascular tissues: the radial subdivision of the ground tissue and the specification of quiescent center (QC) cells at the root pole.

Already during embryogenesis and continuously in the root meristem, the ground tissue cell layer is split by an asymmetric, periclinal division, followed by the acquisition of differential tissue identities as endodermis and cortical parenchyma by the newly generated layers (Fig. 9.6). Two GRAS domain transcription factors, *SHORT ROOT (SHR)* and *SCARECROW (SCR)*, have been implicated in both the periclinal division as well as in the acquisition of separate cell identities (Scheres *et al.*, 1995; Di Laurenzio *et al.*, 1996; Helariutta *et al.*, 2000). *SHR* acts upstream through regulation of *SCR* expression (Helariutta *et al.*, 2000). Turned on by *SHR*, *SCR* promotes the periclinal division of the ground tissue and suppresses the expression of cortical markers in the inner daughter layer. By genetic data, *SHR* seems to act at the top of the hierarchy that leads to the generation and differential specification of ground tissue layers. How can *SHR* gene function trigger the

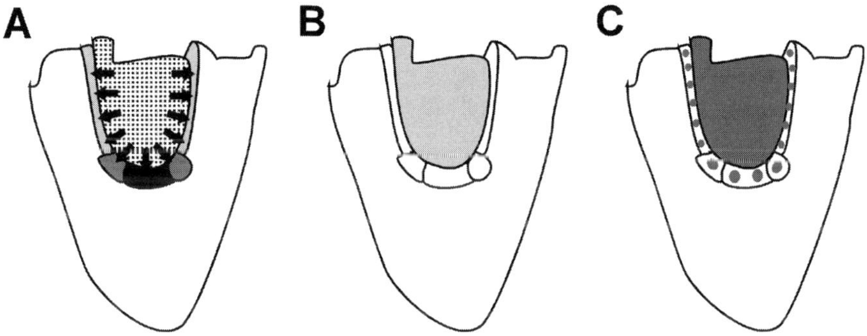

Figure 9.6 Inside-out signaling from the vascular cylinder. (A) Inside-out signaling (arrows) from the vascular cylinder (dotted) to all cells immediately flanking the vascular cylinder (endodermis, light grey; endodermal/cortical initial, dark grey; QC, black). (B) Expression domain of the *SHR* mRNA (grey). (C) Distribution of the SHR protein (black). The protein is present, but not restricted to the nuclei, in the vascular cylinder, where it does not seem to have a function. The protein is further present in cells flanking the vascular cylinder, where it is concentrated in the nuclei and required to promote formative divisions and cell specification.

periclinal division, but then remain restricted to only the innermost ground tissue layer, conferring endodermis identity? The intriguing answer is that SHR acts polarly from the centre. SHR is expressed exclusively in the vascular cylinder and the protein product acquires limited mobility to move to the adjacent neighboring layer (Fig. 9.6, Nakajima *et al.*, 2001). Through this very restricted mobility, SHR confers positional information from the vascular cylinder, using the vascular cylinder as a positional reference to ensure endodermis identity in the immediately adjacent cell layer. That the polar delivery out of the vascular cylinder is instrumental, is demonstrated by the fact that ubiquitous misexpression of *SHR* leads to disorganized division and differentiation (Helariutta *et al.*, 2000). By contrast, misexpression of *SHR* in the *SCR* domain results in extra layers of endodermis, probably by reiterating the emission of a positional signal from the center.

Interestingly not all cells adjacent to the vascular cylinder adopt endodermal fate. Those at its basal end are instead specified as QC cells of the emerging root meristem, and auxin may act as a positional signal in proximal-distal patterning in the root meristem, specifying the positions of the QC and columella initials relative to the vascular cylinder (Sabatini *et al.*, 1999). When the intensity of an auxin response was visualized through reporter gene expression under the control of an auxin-response element, an auxin-response maximum was found associated with the position of the columella initials, adjacent to the QC (Fig. 9.7). Interestingly, any change in the position of this maximum relative to the vascular cylinder, for example by inhibition of auxin transport, was associated with a correlated change in the pattern of distal cell fates, suggesting an instrumental role of auxin signals in

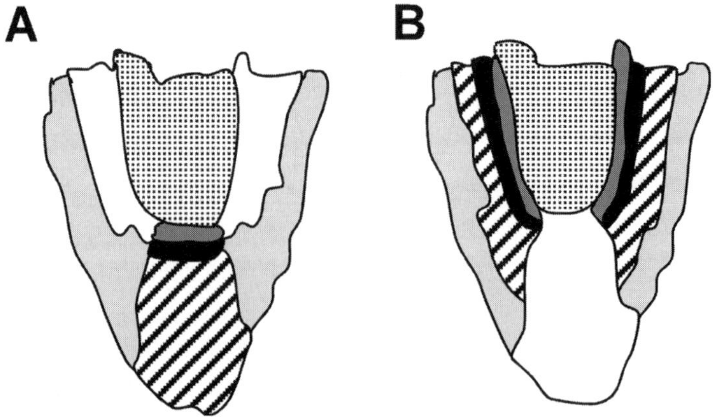

Figure 9.7 Distal cell specification at the vascular cylinder. (A) The root tip, comprising the vascular cylinder (dotted) and the distal tissues (QC, dark grey; columella, stripes; lateral root cap, light grey). An auxin-response reporter gene indicates a maximum (black) at the position of the columella initial cells. (B) Even after experimental displacement of the auxin response maximum, for example, through auxin transport inhibition, the positions of all three cell types in relation to the vascular cylinder and to the auxin-response maximum are maintained, suggesting an instrumental role of auxin distribution in root meristem patterning. Redrawn from Scheres (2000).

root meristem patterning. Auxins are known to promote root formation and, as polarly supplied signals from the shoot, could coordinate root initiation with shoot requirements. Therefore in the positioning of the QC, the vascular cylinder may not only serve as a positional reference but may also supply shoot derived auxin. Together, both influences would ensure the formation of the QC at the basal end of the vascular cylinder and, through the QC's stem cell promoting influence on neighboring cells (van den Berg *et al.*, 1997), the proper positioning of a root meristem.

The failure of mutants of the *mp, bdl, axr6* class to form primary roots could be a direct consequence of insufficient auxin sensitivity to promote QC and root formation. Alternatively, it could be an indirect consequence of the failure to produce a central vascular strand in the embryo or of both. Embryo development of *mp* and *bdl* mutants has been documented in sufficient detail to show that no central vascular strand is initiated and, as this would also serve as a prerequisite for ground tissue patterning, the entire basal domain of these embryos appears poorly differentiated.

In summary, the wealth of available embryo and root mutants supports a concept in which the central vascular cylinder serves as the origin of polarly supplied positional information in the patterning of ground tissues and in the positioning of a root meristem. Vascular tissues may have further, less pronounced influences on cell patterning events in other organs that ensure their functional integration with other tissues.

9.8 Future research

As internal structures with inherent variability, vascular tissues have not been particularly amenable to mutant surveys. Nevertheless, the lessons learned from mutants affecting embryo, leaf and root development have generated an emerging picture in which vascular tissues are patterned by co-opting apical-basal and central-peripheral polarities of the organism to then serve as origins of positional information to structure their neighborhood. Further research will include determining the molecular and cellular details of both organismal polarities, their pathways of regulating vascular tissue organization and, from this, possibilities to selectively manipulate vascular tissues. Most importantly, however, high-throughput visualization techniques and the use of suitable species may expand the options for new genetic screens. These may reveal those levels of vascular tissue organization, such as reproducible network patterns, that cannot easily be integrated in concepts of organismal polarities.

Acknowledgments

The authors would like to thank Naden Krogan for critically reading the manuscript. Further we would like to thank Elsevier Science Ltd. and the American Society of Plant Biologists for the permission to reproduce figures. We apologize to colleagues whose results could not be included in the available space. The authors' vascular tissue research is supported by grants from the Natural Science and Engineering Research Council of Canada.

References

Aloni, R., Schwalm, K., Langhans, M. & Ullrich, C.I. (2003) Gradual shifts in sites of free-auxin production during leaf-primordium development and their role in vascular differentiation and leaf morphogenesis in Arabidopsis. *Planta*, **216**, 841–853.

Avsian-Kretchmer, O., Cheng, J.C., Chen, L., Moctezuma, E. & Sung, Z.R. (2002) Indole acetic acid distribution coincides with vascular differentiation pattern during Arabidopsis leaf ontogeny. *Plant Physiol.*, **130**, 199–209.

Becraft, P.W. & Freeling, M. (1994) Genetic analysis of *Rough sheath1* developmental mutants of maize. *Genetics*, **36**, 295–311.

Bennett, J.M., Marchant, A., May, S.T. & Swarup, R. (1998) Going the distance with auxin: unravelling the molecular basis of auxin transport. *Phil. Trans. R. Soc. Lond. B*, **353**, 1511–1515.

Berleth, T. (2001) Top-down and inside-out: directionality of signaling in vascular and embryo development. *J. Plant Growth Regul.*, **20**, 14–21.

Berleth, T. & Chatfield, S. (2002) Pattern formation during embryogenesis, in *The Arabidopsis Book* (eds. C.R. Somerville & E.M. Meyerowitz), doi/10.1199/tab.0051. Rockville: American Society of Plant Biologists. <http://www.aspb.org/publications/arabidopsis/toc.cfm>.

Berleth, T. & Jürgens, G. (1993) The role of the *MONOPTEROS* gene in organising the basal body region of the *Arabidopsis* embryo. *Development*, **118**, 575–587.

Bowman, J.L. (2000) Axial patterning in leaves and other lateral organs. *Curr. Opin. Genet. Dev.*, **10**, 399–404.

Bowman, J.L., Eshed, Y. & Baum, S.F. (2002) Establishment of polarity in angiosperm lateral organs. *Trends Genet.*, **18**, 134–141.

Busch, M., Mayer, U. & Jürgens, G. (1996) Molecular analysis of the *Arabidopsis* pattern formation of gene *GNOM*: gene structure and intragenic complementation. *Mol. Gen. Genet.*, **250**, 681–691.

Carland, F.M., Berg, B.L., FitzGerald, J.N., Jianamornphongs, S., Nelson, T. & Keith, B. (1999) Genetic regulation of vascular tissue patterning in *Arabidopsis*. *Plant Cell*, **11**, 2123–2137.

Carland, F.M., Fujioka, S., Takatsuto, S., Yoshida, S. & Nelson, T. (2002) The identification of *CVP1* reveals a role for sterols in vascular patterning. *Plant Cell*, **14**, 2045–2058.

Cerutti, L., Mian, N. & Bateman, A. (2000) Domains in gene silencing and cell differentiation proteins: the novel PAZ domain and redefinition of the Piwi domain. *Trends Biochem. Sci.*, **25**, 481–482.

Clark, J.K. & Sheridan, W.F. (1991) Isolation and Characterization of 51 embryo-specific Mutations of Maize. *Plant Cell*, **3**, 935–951.

Deyholos, M.K., Cordner, G., Beebe, D. & Sieburth, L.E. (2000) The *SCARFACE* gene is required for cotyledon and leaf vein patterning. *Development*, **127**, 3205–3213.

Di Laurenzio, L., Wysocka-Diller, J., Malamy, J.E., Pysh, L., Helariutta, Y., Freshour, G., Hahn, M.G., Feldmann, K.A. & Benfey, P.N. (1996) The *SCARECROW* gene regulates an asymmetric cell division that is essential for generating the radial organization of the Arabidopsis root. *Cell*, **86**, 423–433.

Esau, K. (1965) *Vascular Differentiation in Plants*. Holt, Rinehart and Winston Inc., New York.

Eshed, Y., Baum, S.F., Perea, J.V. & Bowman, J.L. (2001) Establishment of polarity in lateral organs of plants. *Curr. Biol.*, **11**, 1251–1260.

Friml, J. & Palme, K. (2002) Polar auxin transport old questions and new concepts? *Plant Mol. Biol.*, **49**, 273–284.

Fukuda, H. (1997) Tracheary element differentation. *Plant Cell*, **9**, 1147–1156.

Gälweiler, L., Guan, C., Müller, A., Wisman, E., Mendgen, K., Yephremov, A. & Palme, K. (1998) Regulation of polar auxin transport by *AtPIN1* in *Arabidopsis* vascular tissue. *Science*, **18**, 2226–2230.

Geldner, N., Friml, J., Stierhof, Y.-D., Jürgens, G. & Palme, K. (2001) Auxin transport inhibitors block PIN1 cycling and vesicle trafficking. *Nature*, **413**, 425–428.

Hadfi, K., Speth, V. & Neuhaus, G. (1998) Auxin-induced developmental patterns in Brassica juncea embryos. *Development*, **125**, 879–887.

Hagen, G. & Guilfoyle, T. (2002) Auxin-responsive gene expression: genes, promoters and regulatory factors. *Plant Mol. Biol.*, **49**, 373–385.

Hamann, T., Benkova, E., Bäurle, I., Kientz, M. & Jürgens, G. (2002) The *Arabidopsis BODENLOS* gene encodes an auxin response protein inhibiting MONOPTEROS-mediated embryo patterning. *Genes Dev.*, **16**, 1610–1615.

Hamann, T., Mayer, U. & Jürgens, G. (1999) The auxin-insensitive *bodenlos* mutation affects primary root formation and apical-basal patterning in the *Arabidopsis* embryo. *Development*, **126**, 1387–1395.

Hardtke, C.S. & Berleth, T. (1998) The *Arabidopsis* gene *MONOPTEROS* encodes a transcription factor mediating embryo axis formation and vascular development. *EMBO J.*, **17**, 1405–1411.

Helariutta, Y., Fukaki, H., Wysocka-Diller, J., Nakajima, K., Jung, J., Sena, G., Hauser, M.T. & Benfey, P.N. (2000) The SHORT-ROOT gene controls radial patterning of the Arabidopsis root through radial signaling. *Cell*, **101**, 555–567.

Hellmann, H., Hobbie, L., Chapman, A., Dharmasiri, S., Dharmasiri, N., del Pozo, C., Reinhardt, D. And Estelle, M. (2003) Arabidopsis *AXR6* encodes CUL1 implicating SCF ubiquitin ligases in auxin regulation of embryogenesis. *EMBO J.*, **22**, 3314–3325.

Hickey, L.J.C. (1973) Classification of the architecture of dicotyledonous leaves. *Am. J. Bot.*, **60**, 17–33.

Hickey, L.J. (1988) A revised classification of the architecture of dicotyledonous leaves, in *Anatomy of the Dicoyledons* (eds. C.R. Metcalfe & L. Chalk), Oxford University Press, New York, NY, pp. 25–39.

Hobbie, L., McGovern, M., Hurwitz, L.R., Pierro, A., Liu, N. Y., Bandyopadhyay, A. & Estelle, M. (2000) The *axr6* mutants of *Arabidopsis thaliana* define a gene involved in auxin response and early development. *Development*, **127**, 23–32.
Hong, S.-W., Aoki, T., Kitano, H., Satoh, H. & Nagato, Y. (1995) Phenotypic diversity of 188 rice embryo mutants. *Dev. Gen.*, **16**, 298–310.
Hudson A. & Waites, R. (1998) Early events in leaf development. *Semin. Cell Dev. Biol.*, **9**, 207–211.
Iwakawa, H., Ueno, Y., Semiarti, E., Onouchi, H., Kojima, S., Tsukaya, H., Hasebe, M., Soma, T., Ikezaki, M., Machida, C. & Machida, Y. (2002) The *ASYMMETRIC LEAVES2* gene of *Arabidopsis thaliana*, required for formation of a symmetric flat lamina, encodes a member of a novel family of proteins characterized by a cysteine repeats and a leucine zipper. *Plant Cell Physiol.*, **43**, 467–478.
Jun, J.H., Ha, C.M. & Nam, H.G. (2002) Involvement of the *VEP1* gene in vascular strand development in *Arabidopsis thaliana. Plant Cell Physiol.*, **43**, 323–330.
Jürgens, G. (2001) Apical-basal pattern formation in Arabidopsis embryogenesis. *EMBO J.*, **20**, 3609–3616.
Kerstetter, R.A., Bollman, K., Taylor, R.A., Bomblies, K. & Poethig, R.S. (2001) *KANADI* regulates organ polarity in *Arabidopsis. Nature*, **411**, 706–709.
Klee, H.J., Horsch, R.B., Hinchee, M.A., Hein, M.B. & Hoffmann, N.L. (1987) The effects of overproduction of two *Agrobacterium tumefaciens* T-DNA auxin biosynthetic gene products in transgenic petunia plants. *Genes Dev.*, **1**, 86–96.
Klucking, E.P. (1995) *Leaf Venation Patterns*, 7 vols. J. Cramer: Berlin.
Koizumi, K., Sugiyama, M. & Fukuda, H. (2000) A series of novel mutants of *Arabidopsis thaliana* that are defective in the formation of continuous vascular network: calling the auxin signal flow canalization hypothesis into question. *Development*, **127**, 3197–3204.
Leyser, O. (2001) Auxin signalling: the beginning, the middle and the end. *Curr. Opin. Plant Biol.*, **4**, 382–386.
Lincoln, C., Britton, J.H. & Estelle, M. (1990) Growth and development of the *axr1* mutants of *Arabidopsis. Plant Cell*, **2**, 1071–1080.
Liscum, E. & Reed, J.W. (2002) Genetics of Aux/IAA and ARF action in plant growth and development. *Plant Mol. Biol.*, **49**, 387–400.
Lomax, T.L., Muday G.K. and Rubery, P.H. (1995) Auxin transport, in *Plant Hormones: Physiology, Biochemistry and Molecular Biology* (ed. P.J. Davies), Kluwer Academic Publishers: Dordrecht, pp. 509–530.
Lynn, K., Fernandez, A., Aida, M., Sedbrook, J., Tasaka, M., Masson, P., & Barton, M.K. (1999) The *PINHEAD/ZWILLE* gene acts pleiotropically in *Arabidopsis* development and has overlapping functions with the *ARGONAUTE1* gene. *Development*, **126**, 469–481.
McConnell, J.R. & Barton, M.K. (1998) Leaf polarity and meristem formation in *Arabidopsis. Development*, **125**, 2935–2942.
McConnell, J.R., Emery, J., Eshed, Y., Bao, N., Bowman, J. & Barton, M.K. (2001) Role of *PHABULOSA* and *PHAVOLUTA* in determining radial patterning in shoots. *Nature*, **411**, 709–713.
Mattsson, J., Ckurshumova, W. & Berleth, T. (2003) Auxin signaling in Arabidopsis leaf vascular development. *Plant Physiol.*, **131**, 1327–1339.
Mattsson, J., Sung, Z.R. & Berleth, T. (1999) Responses of plant vascular systems to auxin transport inhibition. *Development*, **126**, 2979–2991.
Mayer, U., Büttner, G. & Jürgens, G. (1993) Apical-basal pattern formation in the *Arabidopsis* embryo: studies on the role of the *gnom* gene. *Development*, **117**, 149–162.
Meinhardt, H. (1982) *Models of Biological Pattern Formation*. Academic Press: London.
Meinhardt, H. & Gierer, A. <http://www.eb.tuebingen.mpg.de/dept4/meinhardt/theory/html>.
Morris, D.A. & Thomas, A.G. (1978) A microautoradiographic study of auxin transport in the stem of intact pea seedlings (*Pisum sativum* L.). *J. Exp. Bot.*, **29**, 147–157.
Muday, G.K. & Murphy, A.S. (2002) An emerging model of auxin transport regulation. *Plant Cell*, **14**, 293–299.
Nakajima, K., Sena, G., Nawy, T. & Benfey, P.N. (2001) Intercellular movement of the putative transcription factor SHR in root patterning. *Nature*, **413**, 307–311.
Nelson, T. & Dengler, N. (1997) Leaf vascular pattern formation. *Plant Cell*, **9**, 1121–1135.

Okada, K., Ueda, J., Komaki, M.K., Bell, C.J. & Shimura, Y. (1991) Requirement of the auxin polar transport system in early stages of *Arabidopsis* floral bud formation. *Plant Cell*, **3**, 677–684.

Palme, K. & Gälweiler, G. (1999) PIN-pointing the molecular basis of auxin transport. *Curr. Opin. Plant Biol.*, **2**, 375–381.

Paquette, A.J. & Benfey, P.N. (2001) Axis formation and polarity in plants. *Curr. Opin. Genet Dev.*, **11**, 405–409.

Przemeck, G.K.H., Mattsson, J., Hardtke, C.S., Sung, Z.R. & Berleth, T. (1996) Studies on the role of the *Arabidopsis* gene *MONOPTEROS* in vascular development and plant cell axialisation. *Planta*, **200**, 229–237.

Sabatini, S., Beis, D., Wolkenfelt, H., Murfett, J., Guilfoyle, T., Malamy, J., Benfey, P., Leyser, O., Bechtold, N., Weisbeek, P. & Scheres, B. (1999) An auxin-dependent distal organizer of pattern and polarity in the *Arabidopsis* root. *Cell*, **99**, 463–472.

Sachs, T. (1981) The control of the patterned differentiation of vascular tissues. *Adv. Bot. Res.*, **9**, 151–262.

Sachs, T. (1991) Cell polarity and tissue patterning in plants. *Development*, Supplement **1**, 83–93.

Sawa, S., Watanabe, K., Goto, K., Kanaya, E., Morita, E.H. & Okada, K. (1999) *FILAMENTOUS FLOWER*, a meristem and organ identity gene of *Arabidopsis*, encodes a protein with a zinc finger and HMG-related domains. *Genes Dev.*, **13**, 1079–1088.

Scarpella, E., Boot, K.J.M., Rueb, S. & Meijer, A.H. (2002) The procambium specification gene *Oshox1* promotes polar auxin transport capacity and reduces its sensitivity towards inhibition. *Plant Physiol.*, **130**, 1349–1360.

Scarpella, E., Rueb, S. & Meijer, A.H. (2003) The *RADICLELESS1* gene is required for vascular pattern formation in rice. *Development*, **130**, 645–658.

Scheres, B. (2000) Non-linear signaling for pattern formation? *Curr. Opin. Plant Biol.*, **3**, 412–417.

Scheres, B., Di Laurenzio, L., Willemsen, V., Hauser, M.T., Janmaat, K., Weisbeek, P. & Benfey, P.N. (1995) Mutations affecting the radial organisation of the *Arabidopsis* root display specific defects throughout the embryonic axis. *Development*, **121**, 53–62.

Schneeberger, R.G., Becraft, P.W., Hake, S. & Freeling, M. (1995) Ectopic expression of the knox homeobox gene *ROUGH SHEATH1* alters cell fate in the maize leaf. *Genes Dev.*, **9**, 2292–2304.

Semiarti, E., Ueno, Y., Tsukaya, H., Iwakawa, H., Machida, C. & Machida, Y. (2001) The *ASYMMETRIC LEAVES2* gene of *Arabidopsis thaliana* regulates formation of a symmetric lamina, establishment of venation and repression of meristem-related homeobox genes in leaves. *Development*, **128**, 1771–1783.

Sessions, A. & Yanofsky, M.F. (1999) Dorsoventral patterning in plants. *Genes Dev.*, **13**, 1051–1054.

Shevell, D.E., Leu, W.M., Gillmor, C.S., Xia, G., Feldmann, K.A. & Chua, N.H. (1994) *EMB30* is essential for normal cell division, cell expansion, and cell adhesion in *Arabidopsis* and encodes a protein that has similarity to Sec7. *Cell*, **77**, 1051–1062.

Sieburth, L.E. (1999) Auxin is required for leaf vein pattern in *Arabidopsis*. *Plant Physiol.*, **121**, 1179–1190.

Steeves, T.A. & Sussex, I.M. (1989) *Patterns in plant development*. Cambridge University Press, Cambridge.

Steinmann, T., Geldner, N., Grebe, M., Mangold, S., Jackson, C.L., Paris, S., Gälweiler, L., Palme, K. & Jürgens, G. (1999) Coordinated polar localisation of auxin efflux carrier PIN1 by GNOM ARF GEF. *Science*, **286**, 316–318.

Tasaka, M. (2001) From central-peripheral to adaxial-abaxial. *Trends Plant Sci.*, **6**, 548–550.

Torres-Ruiz, R.A. & Jürgens, G. (1994) Mutations in the *FASS* gene uncouple pattern formation and morphogenesis in Arabidopsis development. *Development*, **120**, 2967–2978.

Ulmasov, T., Hagen, G. & Guilfoyle, T.J. (1997) ARF1, a transcription factor that binds to auxin response elements. *Science*, **276**, 1865–1868.

van den Berg, C., Willemsen, V., Hendriks, G., Weisbeek, P. & Scheres, B. (1997) Short-range control of cell differentiation in the Arabidopsis root meristem. *Nature*, **390**, 287–289.

Waites, R. & Hudson, A. (1995) *phantastica*: a gene required for dorsoventrality of leaves in *Antirrhinum majus*. *Development*, **121**, 2143–2154.

Waites, R. & Hudson, A. (2001) The *Handlebars* gene is required with *Phantastica* for dorsoventral asymmetry of organs and for stem cell activity in *Antirrhinum*. *Development*, **128**, 1923–1931.

Waites, R., Selvadurai, H.R.N., Oliver, I.R. & Hudson, A. (1998) The *PHANTASTICA* gene encodes a MYB transcription factor involved in growth and dorsoventrality of lateral organs in *Antirrhinum*. *Cell*, **93**, 779–789.

Wangermann, E. (1974) The pathway of transport of applied indolylacetic acid through internode segments. *New Phytol.*, **73**, 623–636.

Yang, K.-Y., Moon, Y.H., Choi, K.H., Kim, Y.-H., Eun, M.-Y., Guh, J.-O., Kim, K.-C. & Cho, B.-H. (1997) Structure and expression of the *AWI31* gene specifically induced by wounding in *Arabidopsis thaliana*. *Mol. Cells*, **7**, 131–135.

Ye, Z.-H. (2002) Vascular tissue differentiation and pattern formation in plants. *Annu. Rev. Plant Physiol. Plant Mol. Biol.*, **53**, 183–202.

Zhong, R., Taylor, J.J. & Ye, Z.-H. (1999) Transformation of the collateral vascular bundles into amphivasal vascular bundles in an *Arabidopsis* mutant. *Plant Physiol.*, **120**, 53–64.

10 Establishment of polarity in lateral organs of seed plants

John L. Bowman

10.1 Introduction

For the majority of the year, a walk through the bush is characterized by a view of leafy green shoots, with each plant species displaying a uniquely shaped and sized leaf. But whereas most extant land plants are characterized by conspicuous leaves that serve to harvest light for photosynthesis, this was not always the case, with early land plants consisting primarily of dichotomously branching stems. Based on the fossil record, the evolution of determinate planar leaves probably occurred many independent times – for example, in each of the lycopsid, fern and seed plants lineages. Thus, with the evolution of leaves, the body plans of early land plants changed dramatically from primarily indeterminate radial symmetric structures to a combination of radial indeterminate organs and determinate asymmetric organs. For the purposes of this review, the term 'lateral organ' will be used to refer to these determinate organs (leaves and leaf-like organs) since, whatever their evolutionary origin, they are produced from the flanks of the apical meristems in the respective plant lineages. Subsequent to their origins, leaves have evolved into an amazingly diverse array of shapes and sizes in conjunction with their physiological roles in photosynthesis and respiration.

The large majority of lateral organs are polar in nature, with distinct asymmetries in both the proximal-distal and the adaxial-abaxial axes (Fig. 10.1a,b). The proximal-distal axis is asymmetric by definition, the proximal end being the petiole that attaches to the stem and the distal end often consisting of a flattened blade. The adaxial-abaxial (ad–ab) axis is defined by the relationship of the lateral organs to the meristem from which they are derived: the adaxial side (ad – next to) being that derived from cells adjacent to the meristem, and the abaxial side (ab – away from) being derived from cells in the leaf primordium at a distance from the meristem. Asymmetry in the ad-ab axis often derives from the specialized nature of the different surfaces of the leaf lamina. Finally, some lateral organs, such as leaves of *Eucalyptus* species also exhibit asymmetry in their 'left-right' (or medial-lateral) axis. Differential and asymmetric growth along these three axes result in the variety of morphological forms observed in the lateral organs, from the elaborate petals of orchids to the reduced leaves of acacias.

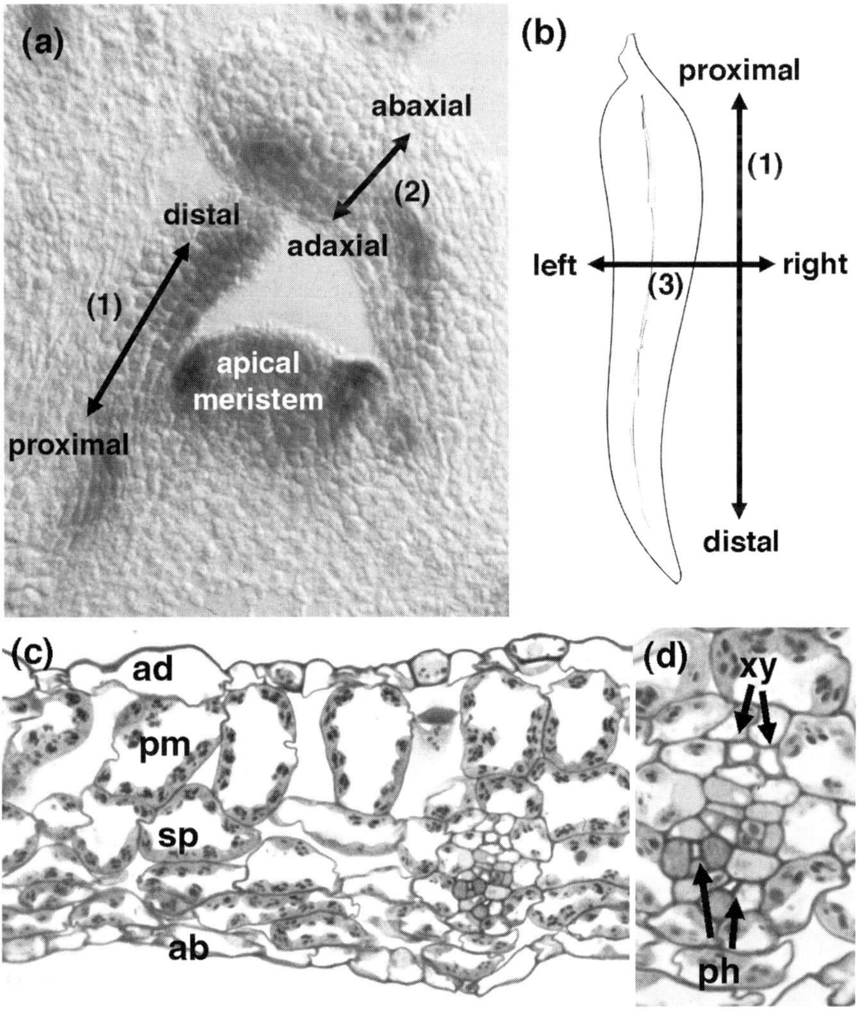

Figure 10.1 The relationship between lateral organs and the apical meristem. The three axes of leaves and other lateral organs in seed plants, at the meristem (a) and in a leaf (b). On the proximal–distal axis (1), the distal end is the leaf tip and the proximal end is attached to the stem. On the adaxial–abaxial axis (2), the adaxial side is adjacent to the meristem, and the abaxial side at a distance from the meristem. In most angiosperm leaves, polarity in the adaxial–abaxial axis is evident from differences in morphology and anatomy between upper and lower surfaces. Most leaves are symmetrical along the left–right axis (3), but some, such as the *Eucalyptus* leaf, display asymmetries along this axis (c, d). While lateral organs, such as the *Arabidopsis* leaf, are often curved over the apical meristem early in their development, later expansion results in the adaxial surface being the top of the fully differentiated leaf and the abaxial surface at the bottom. Polarity is evident in the distribution of cell types within the epidermis as well as in the mesophyll where the palisade mesophyll (pm) and xylem (xy) are located adaxially and the spongy mesophyll (sm) and phloem (ph) are located abaxially. Modified and reprinted from *Trends in Genetics* (Bowman *et al.*, 2002) with permission from Elsevier.

In many species the adaxial side eventually forms the upper surface of the leaf, and is specialized for light harvesting, with the abaxial side forming the lower surface of the leaf, often being specialized for respiration. In these species, mesophyll is often asymmetrically differentiated, with one or more layers of light-harvesting palisade mesophyll located adaxially and multiple layers of spongy mesophyll, allowing gas exchange, located abaxially (Fig. 10.1c). In addition, the vascular bundles exhibit an inherent asymmetry, with xylem developing adaxially and phloem abaxially (Fig. 10.1d). While this is a common pattern of differentiation, it is by no means universal. For example, in species in which the leaves are held vertically (e.g. *Eucalyptus*, *Jojoba*), palisade mesophyll may develop on both sides, while the vasculature remains polarized. Conversely, in some Solanaceous species an extra strand of phloem develops adaxial to the xylem, while the mesophyll develops asymmetrically. The majority of this review will focus on the ad-ab axis since this has received much attention recently, and discussion will primarily be limited to seed plants where all leaves are thought to be homologous structures.

10.2 Early dissection experiments

Lateral organs of angiosperms are derived from cells recruited from the peripheral zone of the apical meristem (Steeves & Sussex, 1989). Initially, lateral organ anlage appear to be composed of a small number of cells exhibiting relatively uniform histology, but possibly heterogeneous gene expression patterns. As lateral organ primordia emerge from the flanks of the meristem, polarity is evident both in terms of gene expression patterns and morphological differentiation, suggesting that adaxial-abaxial polarity is established in lateral organs at least as early as their transition from anlagen to morphologically distinct primordia. While fully differentiated lateral organs have many more than just two cell types, the establishment of polarity only requires the generation of two distinct populations of cells. Subsequent interactions between the two populations could then lead to the generation of a larger diversity of cellular identities.

Elegant experiments by Sussex (1954, 1955), and others (Snow & Snow, 1959; Hanawa, 1961), nearly a half-century ago, in which incipient leaf primordia were physically isolated from the shoot apical meristem by incisions, have provided a conceptual framework for the study of potential interactions between lateral organ primordia and apical meristems. If separation occurred prior to primordium formation (e.g. at the stage of an anlage), the isolated primordia developed into radially symmetric, apparently abaxialized organs. This developmental pattern suggests that the apical meristem could be the source for a signal required for proper abaxial/adaxial development of the leaf (Sussex, 1954, 1955). One interpretation of these data is that a signal(s) emanating from the apical meristem promotes adaxial identity in the cells of the lateral organ primordium closest to the meristem, and that in the absence of such a signal (e.g. in cells of the lateral organ primordium located at a distance from the meristem), abaxial cell fate would be the default pattern of differentiation. That the isolated organs developed as radially symmetric structures also implicates establishment of adaxial-abaxial polarity as a requirement for proper laminar growth (Sussex, 1954, 1955). Sussex (1954, 1955) further demonstrated that isolated leaf primordia that had already begun their morphological differentiation developed autonomously into phenotypically normal leaves. This suggests that the establishment of abaxial and adaxial domains occurs during the transition from leaf anlagen to leaf primordium. These classical experiments demonstrate that factors both intrinsic and extrinsic to lateral organ primordia contribute to the specification of cells as adaxial or abaxial.

In a landmark publication, Waites & Hudson (1995) provided the first glimpse of the genetic machinery that controls adaxial-abaxial polarity. The radially symmetric leaves of *phantastica* (*phan*) mutants of *Antirrhinum* were interpreted to be abaxialized based on both epidermal cell morphologies and the arrangement of vascular tissues. When plants carrying temperature sensitive alleles of *phan* are grown in intermediate conditions, outgrowths of tissue are produced along ectopic abaxial/adaxial boundaries leading Waites & Hudson (1995) to propose that juxtaposition of abaxial and adaxial cell fates is required for normal lamina outgrowth. This implies that signaling between the two distinct cell types is required to induce growth, consistent with the proposal of Sussex (1954, 1955) that proper establishment of polarity is required for lamina development. Thus, in the absence of any adaxial tissue, the abaxialized organs of *phan* mutants develop as filamentous radially symmetric organs as did isolated primordia in the incision experiments (Sussex 1954, 1955).

In the past few years, work in *Antirrhinum* and *Arabidopsis* has led to a better understanding of how polarity might be established at the molecular level, with the identification of additional factors promoting both adaxial and abaxial fates. A developmental paradigm has emerged, with aspects in common with pattern formation mechanisms of laminar organs found in metazoan development. In these developmental paradigms in metazoans, two distinct populations of cells are established by an initial asymmetry in positional information, perhaps mediated by interpretation of a graded concentration of a mobile signal. Subsequently, signaling between the two distinct cell populations leads to the formation of a boundary, from which signals emanate to pattern the surrounding tissues. At present, only isolated pieces of the molecular machinery involved have been identified in plants, and while an integrated view of how these factors interact is lacking, it is clear that communication between cells of differing fate is critical for both their establishment and subsequent differentiation. What follows is a description of the data that has led to one current model, with an emphasis on outstanding questions.

10.3 Next to the meristem

While the surgical experiments provide evidence for a mobile signal emanating from the meristem that induces adaxial fates in leaf primordia, the biochemical nature of the signal remains elusive. However, clues may be deduced from the nature of the genes that promote adaxial fates in lateral organs. Two major players in promoting adaxial fates have been identified in *Antirrhinum* and *Arabidopsis*, respectively: *PHAN* and several members of the Class III HD-ZIP genes. Additional genes that could promote adaxial fates have been described in other species, for example, *LEAFBLADELESS* of maize (Timmermans *et al.*, 1998), but they are not yet molecularly characterized. Unfortunately, at present, the relationship between the pathways directed by these different genes is not clear.

10.3.1 PHANTASTICA

Based on loss-of-function alleles in which leaves are abaxialized (described above), it has been proposed that *PHAN* promotes adaxial fates (Waites & Hudson, 1995). *Phan* encodes a myb transcription factor and is expressed throughout leaf anlagen and primordia, implying post-transcriptional regulation in its promotion of adaxial fates (Waites *et al.*, 1998). The activity of *PHAN* is at least partially redundant with that of the *HANDLEBARS* (*HB*) gene, with the double mutant exhibiting abaxialized radialized cotyledons as well as leaves (Waites & Hudson, 2001). While the data on *Phan* in *Antirrhinum* appears to support a role in promoting adaxial fates, loss-of-function alleles in *PHAN* orthologs in other species, such as in *Arabidopsis* (*ASYMMETRIC LEAVES1*, *AS1*) and *Zea* (*ROUGH SHEATH2*, *RS2*), do not exhibit conspicuous abaxialized phenotypes (Tsukaya & Uchimiya, 1997; Schneeberger *et al.*, 1998). For example, in contrast to the radialized leaves of *phan* mutants, loss-of-function *as1* mutants exhibit asymmetric, downward-curling epinastic leaves resembling those produced in plants ectopically expressing KNOX genes in the leaves (Lincoln *et al.*, 1994). This has led to much debate in the literature and several roles have been proposed for these genes, including establishing ab-ad leaf polarity, establishing the proximo-distal leaf axis, and suppression of KNOX gene activity in leaf primordia, each of which are discussed below (Waites & Hudson, 1995; Schneeberger *et al.*, 1998; Timmermans *et al.*, 1999; Tsiantis *et al.*, 1999; Ori *et al.*, 2000; Byrne *et al.*, 2000; Byrne *et al.* 2002; Sun *et al.*, 2002).

10.3.2 Ectopic KNOX activity

A common feature of *PHAN/AS1/RS2* loss-of-function alleles is the ectopic expression KNOX genes in the leaves (Schneeberger *et al.*, 1998; Ori *et al.*, 2000) whereas in wild-type plants KNOX activity is confined to the meristem (Smith *et al.*, 1992; Long *et al.*, 1996). The ectopic KNOX activity has not been detected in leaf anlagen, but only in differentiating leaves suggesting that *PHAN/AS1/RS2* regulation of KNOX activity may be indirect, perhaps epigenetic (Schneeberger *et al.*, 1998; Ori *et al.*, 2000; Byrne *et al.*, 2000). Ectopic KNOX gene expression in maize leaves results in characteristic knots, whereas in *Arabidopsis*, the leaves exhibit lobes and alterations in vascular patterning. While ectopic KNOX gene expression leads to a phenotype similar to *as1* loss-of-function alleles in *Arabidopsis*, it has not been conclusively shown that the *as1* phenotype is due to ectopic KNOX activity. The *Arabidopsis* genome contains four Class I KNOX genes: *SHOOT MERISTEMLESS* (*STM*), *BREVIPEDICELLUS* (*BP*; *KNAT1*), *KNAT2* and *KNAT6*. In *as1* mutants, *STM* is not ectopically expressed, but *BP* and *KNAT2* are expressed in leaves (Ori *et al.*, 2000; Byrne *et al.*, 2000). However, *as1 bp* and *as1 knat2* double mutants still exhibit the same degree of phenotypic defects as *as1* single mutants implying that if KNOX activity is responsible for the aberrant leaf phenotype there is little dose response in ectopic KNOX activity (Byrne *et al.*, 2002). Clarifying the phenotypic effects of ectopic KNOX activity in these backgrounds will require examination of the effects of loss-of-function alleles in all of the KNOX genes, including *KNAT6*.

While ectopic KNOX gene expression in lateral organs does not often alter cellular identities, it does often lead to a delay in the differentiation of the tissue and cell types of the organ (Ori *et al.*, 2000). In some cases, such as in maize leaves, there is also a shift in the proximal-distal distribution of tissues such that sheath tissue extends into the region normally occupied by the blade (Freeling, 1992; Schneeberger *et al.*, 1998). These phenotypes have led to the idea of timing-dependent cellular differentiation in developing leaves, and ectopic KNOX gene activity disrupts differentiation along the proximal-distal axis by delaying differentiation of cell types (Freeling, 1992). Since their expression is excluded from developing leaves, in most species with simple leaves KNOX genes are unlikely to play a direct role in leaf development. However, the boundary formed between KNOX gene expression in the apical meristem and *PHAN/AS1/RS2* expression in lateral organ anlage may be instructive in establishing some aspects of the proximal-distal axis during leaf primordium formation. In species with dissected (compound) leaves, the antagonism between KNOX and *PHAN/AS1/RS2* activities is not observed, with KNOX expression in leaf primordia contributing to their elaborated development (Bharathan *et al.*, 2002).

Loss of activity of YABBY genes also leads to ectopic KNOX activity in leaves, but results in a significantly different leaf phenotype compared to the phenotype observed in *as1* leaves (Kumaran *et al.*, 2002). However, these genotypes are not directly comparable as the two genes are thought to have converse roles in promoting ab–ad polarity in the leaves (*AS1* – possibly adaxial; YABBY abaxial, see below), and a different spectrum of KNOX – activity is observed in the two genotypes. For example, *STM* is ectopically expressed in *yabby* loss-of-function mutants (but not in *as1* mutants), resulting in the development of ectopic meristems from the adaxial surfaces of the leaves (Siegfried *et al.*, 1999; Kumaran *et al.*, 2002). However, a common feature between the YABBY and *AS1* negative regulation of KNOX activity is that ectopic KNOX gene expression appears only after leaf primordium formation, suggesting indirect or redundant mechanisms. Efforts to demonstrate a direct role in negative regulation of KNOX genes by either *AS1* or YABBY have been hampered by lack of appropriate promoters for localized ectopic gene expression and the requirement of associated proteins for activity (e.g. for *AS1*, the need for *AS2* to be co-expressed see – below).

Recently, an antagonistic relationship between gibberellins (GAs) and KNOX gene activity has been demonstrated (Tanaka-Ueguchi *et al.*, 1998; Sakamoto *et al.*, 2001; Hay *et al.*, 2002). Ectopic KNOX activity in the leaves leads to a down regulation of expression of a GA biosynthetic gene (*AtGA20ox1*), and conversely, constitutive signaling through the GA pathway leads to a reduction in KNOX activity in the shoot meristem (Tanaka-Ueguchi *et al.*, 1998; Sakamoto *et al.*, 2001; Hay *et al.*, 2002). Since GA biosynthesis is restricted to the lateral organs, Hay *et al.* (2002) proposed that one of the functions of KNOX gene activity is to exclude GA biosynthesis from the meristem. Mutations in *PICKLE* (*PKL*) also result in a reduction in the expression of *AtGA20ox1* (Hay *et al.*, 2002), consistent with a reduction in GA mediated responses, or perhaps GA biosynthesis, in *pkl* mutants (Ogas *et al.*, 1997). Since *PKL* encodes a chromatin remodeling enzyme (Eshed *et al.*, 1999; Ogas *et al.*, 1999), the enhancement of the *as1* mutant phenotype by *pkl* mutations (Ori *et al.*, 2000), and the increased meristematic activity in *pkl* mutants (Eshed *et al.*, 1999) could be explained by the failure to remodel chromatin structure at some targets of KNOX activity (Hay *et al.*, 2002). Alternatively, since proteins similar to PKL associate with histone deacetylases in other species, PKL probably contributes to the formation of chromatin structures repressing gene expression (Wade *et al.*, 1998; Zhang *et al.*, 1998), and could act in a parallel pathway mediating GA biosynthesis.

10.3.3 Ad-ab polarity role of Phan/AS1/RS2

While the role of *PHAN/AS1/RS2* in promoting adaxial fates has been equivocal outside of *Antirrhinum*, recent studies on *AS1* and another gene whose loss-of-function phenotype is similar, *AS2* (Semiarti *et al.*, 2001) have provided some evidence for such a role in *Arabidopsis*. As in *Antirrhinum*, *AS1* is expressed throughout the leaf anlagen (Byrne *et al.*, 2000), however, *AS2* expression is confined the adaxial regions of leaf primordia (Iwakawa *et al.*, 2002). While ectopic expression of *AS1* results in little phenotypic defects in the leaves, ectopic expression of *AS2* causes defects that could be interpreted as an adaxialization of the leaves (Iwakawa *et al.*, 2002). Since *as1 as2* double mutants resemble the *as1* and *as2* single mutants (Ori *et al.*, 2000), and the respective proteins interact *in vitro* (Iwakawa *et al.*, 2002), one scenario has AS1 and AS2 acting in a complex to promote adaxial fates. The different phenotypes associated with their ectopic expression would be a result of their differing endogenous expression patterns. Furthermore, the downward curling leaves characteristic of *as1* and *as2* mutants could be interpreted as an abaxialization since this phenotype is observed in many slightly abaxialized *Arabidopsis* genotypes (Siegfried *et al.*, 1999). The converse, a cup-shaped leaf phenotype is observed in adaxialized mutants (Eshed *et al.*, 2001; Kerstetter *et al.*, 2001). If this hypothesis is confirmed, it would provide a unification of some of the observations in *Antirrhinum* and *Arabidopsis*.

10.3.4 PHABULOSA

In a seminal publication, McConnell & Barton (1998) described the isolation of semi-dominant *phabulosa* (*phb-1d*) mutations in *Arabidopsis* that act in a dose dependent manner to cause an adaxialization of lateral organs (Fig. 10.2b). Adaxialized lateral organs of *phb1–d* homozygotes are filamentous and radially symmetric, consistent with the idea that a juxtaposition of adaxial and abaxial domains is required for lamina outgrowth (McConnell & Barton, 1998). A subsequent genetic screen resulted in the identification of similar mutations in a related gene, *PHV* (McConnell *et al.*, 2001). *PHB* and *PHAVOLUTA* (*PHV*) encode members of the plant specific Class III homeodomain/leucine zipper (HD-ZIP) family of proteins, of which there are five members in the *Arabidopsis* genome (McConnell *et al.*, 2001). In addition to the homeodomain, the proteins encoded by these genes share a domain carboxyl to the HD-ZIP domain, the START domain, that exhibits sequence similarity to mammalian sterol/lipid binding proteins (Pontig & Aravind, 1999). *PHB* is expressed in lateral organ anlagen, with its expression restricted to adaxial regions as the lateral organ primordia become morphologically apparent (McConnell *et al.*, 2001). While loss-of-function *phb* alleles are indistinguishable from wild type, this is likely due to redundancy with *PHV* and other Class III HD-ZIP genes, such as *REVOLUTA* (*REV*). *REV* is also expressed adaxially in lateral organs (Fig. 10.3d,e; Otsuga *et al.*, 2001; Eshed *et al.*, 2001) and loss-of-function *rev* alleles result in a loss of axillary meristems (Talbert *et al.*, 1995), a feature that could be interpreted

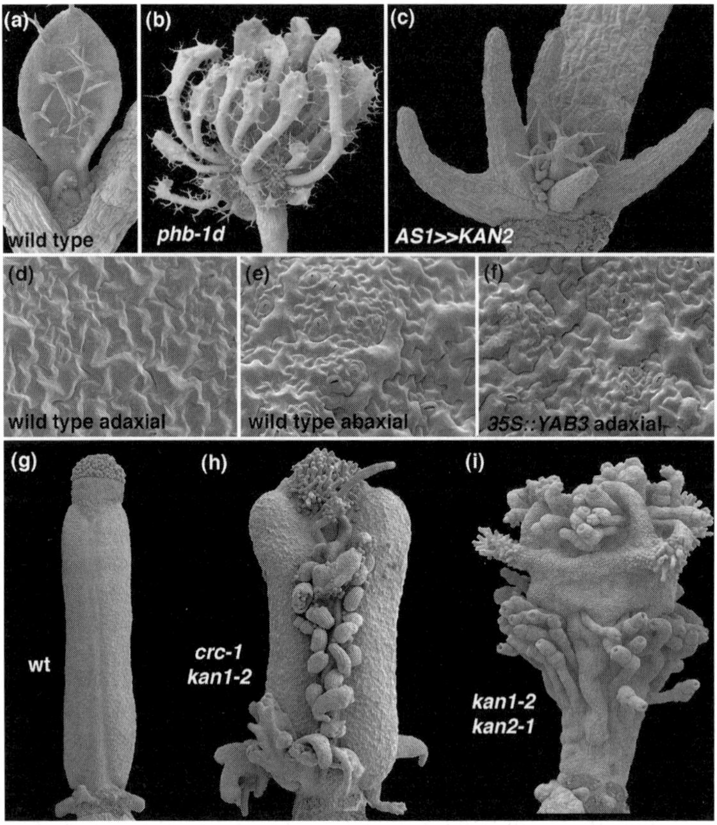

Figure 10.2 Phenotypes of *Arabidopsis* lateral organ polarity mutants. Wild-type leaves exhibit lamina expansion along the left-right axis. On the first formed leaves of *Arabidopsis*, trichomes are confined the adaxial surface (a), however, on the later formed leaves (e.g. after the fifth leaf) the density of trichomes on the abaxial surface gradually increases. The radial leaves of *phb-1d* homozygotes are adaxialized-based on epidermal morphology, color and abundance of trichomes (b). Ectopic expression of either *KAN1* or *KAN2* throughout leaf primordia, such as in *AS1>>KAN2* plants (c), causes the converse phenotype and the leaves are abaxialized. *AS1>>KAN2* leaves are radial and fail to develop trichomes on their first formed leaves. Later formed leaves do differentiate trichomes indicating the heteroblastic leaf morphology changes are maintained in these plants. Ectopic expression of members of the *YABBY* gene family in the adaxial regions of lateral organs is sufficient for adaxial epidermal cells to differentiate with abaxial fates. The epidermal cells of adaxial (d) and abaxial (e) surfaces of wild-type cotyledons display distinct morphologies. Constitutive expression of *YAB3*, as in *35S::YAB3* plants, is sufficient to cause the adaxial epidermis of cotyledons to assume an abaxial fate (f). While carpels of *crc* (the founder *YABBY* gene) single mutants do not display obvious polarity defects, carpels of *crc kanadi1* double mutants (h) exhibit a duplication of adaxial (inside) tissues in abaxial (outside) positions [compare with wild type (g)]. In *crc kanadi1* carpels adaxial tissues such as septum tissue (sp) and placentae bearing ovules (ov) are found in both adaxial and abaxial positions. The carpels of *kan1 kan2* double mutants (i) consist entirely of adaxial tissues, such as septum and placentae bearing ovules. Modified and reprinted from *Trends in Genetics* (Bowman *et al.*, 2002) with permission from Elsevier.

as a loss of adaxial characters. We refer to the collective activity of *PHB*, *PHV* and *REV* as PHABULOSA activity.

That the dominant gain-of-function mutations lie in the START domain has led to speculation that these proteins could act as receptors with the ligand being a lipid or sterol (McConnell *et al.*, 2001). However, the discovery of microRNAs complementary to the region of the semi-dominant *phb* and *phv* lesions has cast doubt on this interpretation of the gain-of-function alleles (Rhoades *et al.*, 2002; Tang *et al.*, 2002; see below). Thus, there are two, not necessarily mutually exclusive, hypotheses concerning the mechanism of action of the dominant gain-of-function alleles. First, the alterations in the putative sterol/lipid binding domain may render the protein ligand independent and constitutively active resulting in a gain of function phenotype (McConnell *et al.*, 2001). Second, the missense mutations may be acting at the level of mRNA stability, with the altered transcripts being resistant to negative regulation by the complementary microRNAs (miRNAs) (Rhoades *et al.*, 2002; Tang *et al.*, 2002). *PHB* alleles in which the miRNA binding site is changed without alteration of the amino acid sequence and miRNA alleles complementary to the *phb-1d* allele will help clarify these hypotheses. Ectopic expression of the wild-type version of *PHB* is insufficient to cause alterations in cell fate, consistent with its activity being limited by the availability of a ligand or a miRNA (McConnell *et al.*, 2001).

Regardless of the mechanism by which the semi-dominant gain-of-function alleles operates, PHB/PHV/REV could potentially encode receptors for, or their activities dependent upon, the proposed signal emanating from the apical meristem to establish adaxial fates in lateral organ primordia with PHB/PHV become 'activated' by binding a ligand whose concentration is related to proximity to the meristem (McConnell *et al.*, 2001). Several observations are consistent with this scenario. First, *PHB* mRNA is detected in a gradient in lateral organ primordia with the highest concentration in the adaxial most regions, suggesting that it may act in a dose dependent manner (McConnell *et al.*, 2001). Second, *PHB* activity may be maintained by a positive feedback loop as *PHB* mRNA is expressed at a much higher level in *phb-1d* mutants than in wild type (McConnell *et al.*, 2001). Such a feedback loop, if accompanied by the ability to positively regulate ligand synthesis and/or stability, would enable the maintenance of a new reference point for adaxial position that could function as the developing leaf becomes independent of its original source of positional information (McConnell *et al.*, 2001). This idea is consistent with primordia developing autonomously if they are isolated from the apical meristem after having become morphologically distinct from it (Sussex, 1954, 1955). Alternatively, the high level of *PHB* mRNA observed in *phb-1d* plants could be the result of greater mRNA stability due to recalcitrance of the mutant mRNA to cleavage by the miRNAs (Rhoades *et al.*, 2002; Tang *et al.*, 2002). Third, *PHB* is expressed at high levels in the central domain of the apical meristem and in groups of cells that lie between the apical meristem central domain and positions where lateral organ primordia are developing (McConnell *et al.*, 2001). It is tempting to speculate that these expression patterns reflect communication pathways between cells of the apical meristem and the lateral organs and mark the functional concentration of the putative ligand or lack of miRNA expression.

In summary, based on both loss- and gain-of-function alleles, PHABULOSA activity appears to promote adaxial identity in lateral organs and may also have a role in the apical meristem itself. Evidence that PHABULOSA activity is directly linked to the putative meristem borne signal must await further characterization of the nature of the signal and whether PHABULOSA acts as a receptor.

10.4 In the realm of microRNAs

Recently it has become apparent that small RNA molecules exert a profound influence on gene regulation in eukaryotes (for reviews see: Matzke *et al.*, 2001; Vaucheret *et al.*, 2001; Waterhouse *et al.*, 2001; Hannon, 2002; Plasterk, 2002; Zamore, 2002; Cerutti, 2003). These modes of gene regulation rely on small microRNA (miRNA) molecules that are complementary to the regulatory target and are associated with a large protein complex that mediates either mRNA degradation or inhibition of translation of the target mRNA. Of the large number of miRNA molecules identified in *Arabidopsis* (Reinhart *et al.*, 2002; Llave *et al.*, 2002), two (MIR165 and MIR166) are complementary to *PHB* and *PHV* mRNA, the region of complementarity spanning the location of the semi-dominant gain-of-function *phb/phv* alleles (Rhoades *et al.*, 2002). Since the miRNAs can direct the cleavage of *PHV* mRNA, *in vitro*, the possibility is raised that *PHB/PHV* are regulated by miRNA-directed mRNA degradation (Tang *et al.*, 2002). At present, it is not clear at what hierarchical level this might occur. If the miRNAs are leading to degradation of *PHB/PHV* mRNA, negatively regulating *PHB/PHV*, they may be acting in the abaxial domains of the developing lateral organs, possibly under control of genes such as *KANADI* that promote abaxial fate (see below). A role for the miRNAs as a component of the signal from the meristem is more difficult to envisage.

Further evidence for miRNA regulation of genes directing establishment of polarity in lateral organs comes from the analysis of *argonaute1* (*ago1*) and *pinhead* (*pnh*; also known as *zwille*) mutants (McConnell & Barton, 1995; Bohmert *et al.*, 1998; Moussian *et al.*, 1998; Lynn *et al.*, 1999). *AGO1* is required for virus induced gene silencing (Morel *et al.*, 2002), and AGO1-like proteins are part of the complex that mediates miRNA regulation of gene expression (Tabara *et al.*, 1999; Fagard *et al.*, 2000). *PNH* encodes a protein similar to that of *AGO1* and is genetically redundant with *AGO1* (Lynn *et al.*, 1999). Leaves of *ago1* plants have been interpreted as being abaxialized, whereas constitutive expression of *AGO1* leads to adaxialized leaves (Bohmert *et al.*, 1998; Lynn *et al.*, 1999). Compromised *PNH* activity enhances loss of *AGO1* activity, with the double mutant being embryo/seedling lethal (Lynn *et al.*, 1999). At present, the relationship between *AGO1*/*PNH* and MIR165/166 is not known, with the abaxialized phenotype of *ago1*/*pnh* loss-of-function alleles difficult to reconcile with the proposed negative regulation of *PHB*/*PHV* by MIR165/166.

Viral suppressors of gene silencing also implicate the involvement of miRNA regulation in lateral organ polarity establishment. In some cases, virally encoded suppressors of gene silencing appear to target the production or mobility of the silencing signal and small miRNAs (Voinnet *et al.*, 2000; Llave *et al.*, 2000; Mallory *et al.*, 2001, 2002; Hamilton *et al.*, 2002). Constitutive expression of the white clover mosaic virus protein TGB1, which is related to viral suppressers of gene silencing, in *Nicotiana benthamiana* results in a phenotype similar to that of *ago1* mutants of *Arabidopsis*, with radially symmetric leaves (Foster *et al.*, 2002). The radial leaves express a marker of abaxial cells, *NtYABBY*, in a sporadic and inconsistent manner, suggesting that polarity establishment is disrupted at an early stage. Additionally, in these transgenic plants, viruses are able to move into the shoot apex, a situation not observed in wild-type plants, indicating that communication pathways, probably via plasmodesmata, are altered in these plants (Foster *et al.*, 2002). While it is difficult to provide a coherent synthesis of these disparate data, at present, it is clear that there is ample circumstantial evidence for the involvement of miRNAs in the regulation genes directing organ polarity.

10.5 Away from the meristem

The meristem-based signaling model predicts that abaxial promoting factors are suppressed by interpreters of the signal in a concentration-dependent manner. In the absence of the signal, e.g. when the leaf anlagen is separated from the meristem, leaves develop with uniform abaxial characteristics implying that abaxial-promoting factors remain active throughout the developing leaf. Based on loss- and gain-of-function alleles, members of two transcription factor gene families, the YABBY and GARP gene families, have been proposed to promote abaxial fates (Sawa *et al.*, 1999; Siegfried *et al.*, 1999; Eshed *et al.*, 1999; Villanueva *et al.*, 1999; Eshed *et al.*, 2001; Kerstetter *et al.*, 2001).

10.5.1 KANADI

KANADI1 (*KAN1*) was first identified in a genetic screen for enhancers of *crabs claw* (*crc*) mutations (Eshed *et al.*, 1999), which disrupt carpel development (Alvarez & Smyth, 1999). The carpels of *crc kan1* plants exhibited a mirror image duplication (KANADI = mirror, Coorg) of medial adaxial tissue in abaxial positions, suggesting that both *CRC* and *KAN1* contribute to the specification of abaxial tissues in the carpel (Fig. 10.2h; Eshed *et al.*, 1999). *KAN1* is a member of the GARP transcription factor gene family in *Arabidopsis*, and *kan1* single mutants only have a subtle aberrant phenotype. Mutations in its sister gene, *KAN2*, enhance the *kan1* phenotype such that lateral organs of *kan1 kan2* plants are adaxialized (Fig. 10.2i; Eshed *et al.*, 2001). While the floral organs of *kan1 kan2* double mutants are severely adaxialized, *kan1 kan2* leaves are less so and develop ectopic abaxialized outgrowths from their abaxial sides (Eshed *et al.*, 2001). The residual abaxial character in *kan1 kan2* plants may be provided by other close members of the GARP gene family, *KAN3* and *KAN4* (Eshed *et al.*, 2001). Consistent with the loss-of-function phenotype, *KAN1* is expressed abaxially in a pattern that is the converse of that observed with *PHB/PHV/REV* (Kerstetter *et al.*, 2001). Also, gain-of-function *KAN* alleles lead to a dramatic abaxialization of lateral organs (Fig. 10.2c; Eshed *et al.*, 2001; Kerstetter *et al.*, 2001). Constitutive expression of *KAN1*, *KAN2*, and *KAN3* additionally results in the loss of apical meristem development in the embryo (Eshed *et al.*, 2001; Kerstetter *et al.*, 2001). We refer to the combined action of *KAN1*, *KAN2*, and *KAN3* as KANADI activity.

One interpretation of these results is that KANADI provides peripheral positional information and PHABULOSA provides central positional information (Kerstetter *et al.*, 2001). The loss of apical meristems in plants constitutively expressing KANADI would then be evidence of KANADI repressing PHABULOSA and the resulting plant consisting only of peripheral tissues. Another interpretation, not necessarily mutually exclusive with this, is that the boundary between KANADI and PHABULOSA activities provides positional information to the plant, and that tissues, such as leaf mesophyll and vasculature, are patterned from this boundary. Consistent with this latter hypothesis is:

1. The loss of vasculature that is observed in leaves of plants constitutively expressing KANADI (Eshed *et al.*, 2001; Kerstetter *et al.*, 2001).
2. The observation that leaves of *phb-1d* homozygotes are completely radialized and lacking vasculature.

In these organs, homogenization of gene expression (either KANADI or PHABULOSA) throughout the lateral organs may lead to a loss, not only of the expected tissues, but also those not normally associated with the respective activities.

10.5.2 YABBY

The YABBY transcription factor gene family in *Arabidopsis* consists of six members (Bowman & Smyth, 1999). Each member of the family is expressed abaxially in one or more lateral organs that are above ground, with every above ground lateral organ that exhibits asymmetry expressing at least one family member. Three members (*FILAMENTOUS FLOWER* [*FIL*], *YABBY2* [*YAB2*], *YAB3*) are expressed in a polar manner in all lateral organs produced by the apical and flower meristems (Fig. 10.3a–c; Siegfried *et al.*, 1999; Sawa *et al.*, 1999b). In contrast, *CRABS CLAW* (*CRC*) and *INNER NO OUTER* (*INO*) are more specialized in that their expression is restricted either to the carpels and nectaries or to the outer integuments, respectively (Bowman & Smyth, 1999; Eshed *et al.*, 1999; Villanueva *et al.*, 1999). Evidence for a role for these genes in polarity establishment has come from both loss- and gain-of-function alleles. For example, as mentioned above, *crc kan1* carpels have adaxial tissues in abaxial positions (Eshed *et al.*, 1999). Conversely, expression of any of *FIL*, *YAB3* or *CRC* in adaxial cells has been shown to be sufficient for those cells to differentiate as if they were in an abaxial position (Fig. 10.2d–f; Sawa *et al.*, 1999b; Siegfried *et al.*, 1999; Eshed *et al.*, 1999).

Due to extensive redundancy of the vegetatively expressed YABBY gene family members, loss-of-function alleles have so far been less informative (Sawa *et al.*, 1999b; Chen *et al.*, 1999; Kumaran *et al.*, 1999). However, as *fil* mutants have radialized floral organs, it is consistent with the hypothesis that juxtaposition of abaxial and adaxial cells is required for lamina outgrowth. Furthermore, *fil yab3* plants have narrow leaves, while the respective single mutants have wild-type leaves (Siegfried *et al.*, 1999; Kumaran *et al.*, 2002). While the abaxial cells of *fil yab3* leaves are not wild type in appearance, several aspects of leaf polarity, such as trichome density, are maintained in *fil yab3* plants (Siegfried *et al.*, 1999; Kumaran *et al.*, 2002). Activity of the KANADI genes or the other YABBY genes, *YAB2* and *YAB5*, are candidates for promoting abaxial cell identity and the residual lamina expansion in *fil yab3* leaves – and it will of interest to examine the phenotypes of plants in which all YABBY gene activity has been compromised. That *NtYABBY* is also expressed abaxially suggests that the function of these genes is conserved at least throughout the eudicots (Foster *et al.*, 2002).

The relationship between the KANADI and YABBY functions remains obscure. *FIL* expression in *kan1 kan2* lateral organs is reduced as compared to wild type indicating that wild-type KANADI activity is required for YABBY activity (Eshed *et al.*, 2001). This regulation could be direct, or indirect through repression of *PHB* since in *phb-1d* mutants, *FIL* expression is also greatly reduced (Siegfried *et al.*, 1999). In a similar manner, the proposed repression of *PHB* by KANADI activity could also be direct, or indirect, for example, by regulating MIR165/166. Ectopic YABBY activity

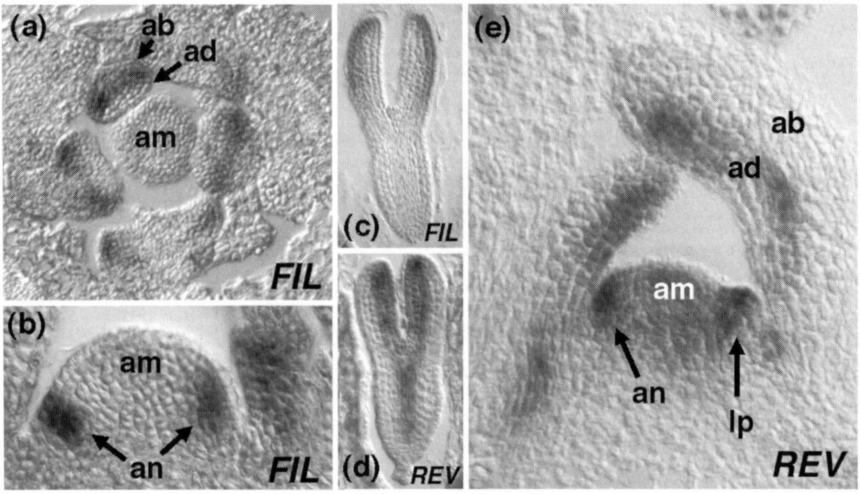

Figure 10.3 Gene expression patterns in developing lateral organs. The leaf abaxial domain is marked by *FIL* (a) which is initially expressed in anlagen (b) but as primordia emerge, it becomes restricted to the abaxial domains of developing leaves. In contrast, *REV* is expressed in the central zone of the apical meristem (am), in leaf anlagen (an), and becomes restricted to the adaxial regions of developing leaves (e). Once leaf primordia (lp) have become distinct from the apical meristem, the domains of *FIL* and *REV* expression are complementary. This is readily seen in the embryos (c, d) where *FIL* expression is restricted to the abaxial regions of the cotyledons, while *REV* expression is localized to the adaxial regions of the cotyledons, the apical meristem, and the developing vasculature. Modified and reprinted from *Trends in Genetics* (Bowman *et al.*, 2002) with permission from Elsevier.

in a *kan1 kan2* background is similar to that observed in a wild-type background, and conversely, ectopic KANADI activity in a *fil yab3* background results in the same phenotypic alterations as when in a wild-type background (Eshed *et al.*, 2001). This mutual epistasis could be the effects of the other members of the *YABBY* and *KANADI* genes remaining in these backgrounds, or alternatively a division of labor between the two pathways, such as, for example, with one activity primarily repressing adaxial activity, and the other contributing to lamina expansion. Loss-of-function alleles in all members of these gene families are required to resolve these issues.

KANADI and *YABBY* expression commences prior to the emergence of the lateral organ primordia (Siegfried *et al.*, 1999; Sawa *et al.*, 1999b; Kerstetter *et al.*, 2001). However, it is not clear whether all cells of the presumptive anlage express these genes in wild-type plants. A comparison of the number of cells in the anlage expressing *FIL* (Siegfried *et al.*, 1999; Sawa *et al.*, 1999b) with the number of cells in which *STM* is down regulated (Long *et al.*, 1996) reveals that there probably exist some cells which express neither. This suggests that the communications pathways within the meristem establishing the adaxial and abaxial domains of lateral organs may be acting prior to the emergence of primordia. However, the domains are unlikely to be irreversibly fixed until after primordium emergence, based on the phenotypes of isolated leaf primordia.

10.5.3 Lamina expansion

Sussex (1955) proposed that polarity *per se* is required for the proper development of the lamina. Waites & Hudson (1995) refined this hypothesis, stating that a juxtaposition of adaxial and abaxial cells is required for lamina expansion. That abaxialized or adaxialized leaves of various genotypes are also radialized is consistent with this hypothesis. Thus, all of the above-mentioned genes are, at least indirectly, involved in promoting lamina expansion. Since lamina expansion is associated with interactions between cells, factors intimately associated with this process are likely to exhibit non-cell autonomous action.

Mutations in the *LAM1* gene of *Nicotinana sylvestris* result in leaves that lack a lamina and results from observations of periclinal chimeras indicate that *LAM1* activity is non-cell autonomous (McHale, 1993; McHale & Marcotrigiano, 1998). While dorso-ventrality is established normally in *lam1* leaves, adaxial cell types are lacking. In periclinal chimeras, adaxial cells can be restored in a mutant L1 by a wild-type L2 and L3, and in a mutant L2 by wild-type cells in the L3. However, the signal has relatively short range as adaxial cell types in mutant L1 cells are not restored by wild-type cells in the L3. Since *lam1* leaves lack lamina growth, the mobile adaxializing factor is a candidate for the proposed signal(s) communicating between the abaxial and adaxial domains.

Another candidate for a signaling molecule is the product of the *Rolled leaf* (*Rld1*) of maize (Nelson *et al.*, 2002). Dominant antimorphic *Rld1* mutations result in loss of abaxial cell types and their replacement with adaxial ones, and in some cases a complete reversal of polarity in leaf sectors. Mosaic analysis indicates non-cell-autonomy with mutant *Rld1* product in the abaxial epidermis sufficient to induce the *Rld1* mutant phenotype in adjacent overlying wild-type cells, suggesting the abaxial epidermis sends or receives a signal that maintains polarity in the differentiating leaf (Nelson *et al.*, 2002). Strikingly, no ectopic lamina develops along the ectopic boundaries suggesting that the action of *Rld1* is subsequent to polarity establishment leading to lamina expansion.

While little data exist on the cell autonomy of the major players identified in *Arabidopsis* and *Antirrhinum*, the expression pattern of members of the YABBY gene family is consistent with a role for these genes in establishing communication pathways driving lamina expansion. For example, *FIL* and *YAB3* are expressed initially throughout the abaxial regions of developing leaves, but their expression soon becomes confined to the lamina of the developing leaves (Siegfried *et al.*, 1999; Sawa *et al.*, 1999b; Kumaran *et al.*, 2002). This pattern of expression precisely mirrors where cells remain cytoplasmically dense, presumably with the capability of further divisions, and thus, boundaries of YABBY gene expression correspond to regions of ongoing lamina expansion. Later in leaf development, the acropetal differentiation of leaves is reflected in a basipetal pattern of cell cycle associated gene expression (Donnelly *et al.*, 1999), a pattern that is also observed in YABBY gene expression (Siegfried *et al.*, 1999; Sawa *et al.*, 1999; Kumaran *et al.*, 2002). A candidate for a downstream target of the ab/ad boundary mediated lamina expansion is *AINTEGUMENTA*. Gain-of-function *ANT* alleles result in prolonged cell division in lateral organs (Krizek, 1999; Mizukami & Fischer, 2000) while loss-of-function alleles result in lateral organs of reduced size (Elliott *et al.*, 1996; Klucher *et al.*, 1996), suggesting that *ANT* expression may reflect a competence for cell division. Expression of *ANT* along the ab/ad boundary of developing leaves in a pattern that parallels that of YABBY expression (Long & Barton, 2000), stronger towards the leaf margins, is consistent with idea of boundary mediated lamina expansion. Eventually these early pathways must be linked with those coordinating cell division and leaf growth in both the abaxial-adaxial and medio-lateral axes and thus, promoting ‘flatness’ in growth (Nath *et al.*, 2003).

10.5.4 Lamina differentiation

In many species, the lamina is differentiated into four distinct cell types: the adaxial epidermis, the palisade mesophyll, the spongy mesophyll and the abaxial epidermis. While these cell types are often in successive layers from the top of the leaf to the bottom, as in *Arabidopsis*, this is not always the case. For example, in *Eucalyptus* leaves, adaxial mesophyll is found juxtaposed with the epidermis on both the adaxial and abaxial sides. Thus, it is likely that the classes of genes that are initially responsible for the establishment of polarity in lateral organs do not necessarily specify the distinct cell types, but that these later differentiation processes may be uncoupled from the early expression of these genes. A broader phylogenetic analysis of gene expression patterns is required to investigate this possibility.

While genes specifying the identity of individual cell types along the ad-ab axis have not yet been identified, the differentiation of the palisade mesophyll has been linked to environmental signals. When chloroplast differentiation is disrupted by mutation, the palisade mesophyll cells fail to assume their normal columnar shape (Reiter *et al.*, 1994; Keddie *et al.*, 1996; Chatterjee *et al.*, 1996) and wild-type plants grown under low light levels exhibit a similar alteration in palisade mesophyll cell shape (Weston *et al.*, 2000). Thus, both endogenous and environmental signals influence the ultimate cellular leaf anatomy along the ad-ab axis.

10.5.5 The other axes

Much less is known about the establishment of the proximo-distal and left-right axes of lateral organs. While it has been argued that ectopic KNOX activity reveals a timing dependent differentiation of tissues along the proximo-distal axis (Freeling, 1992), endogenous KNOX gene activity is unlikely to be involved in wild-type leaves as its expression is excluded from them at inception. However, as mentioned above, the boundary between KNOX and PHAN/AS1/RS2 activities could provide position clues to leaf anlagen at their establishment. Application of enhancer trapping techniques (Sundaresan *et al.*, 1995) to identify genes expressed distally or proximally in developing leaves may provide insight into the establishment and differentiation of this axis.

Even less is known about how left-right symmetry, or asymmetry (for example the complex margins of oak or maple leaves) is established or maintained during leaf development. In genetic sectorial chimeras of the *okra* mutation in *Gossypium* leaves, boundaries of leaf morphology coincided with genotype (Dolan & Poethig, 1991, 1997). This suggests that leaf shape is genetically determined locally, without long range communication between leaf margins through lateral signaling (Dolan & Poethig, 1991, 1997). Thus, genes expressed locally at the margins of leaves, such as *PRESSED FLOWER* which acts to delineate the marginal epidermis of lateral organs, may be formative in controlling leaf marginal growth (Matsumoto & Okada, 2001). It is likely that localized regulation of polarized leaf cell expansion or cell division plays a role in asymmetric growth (Tsuge *et al.*, 1996; Tsukaya, 2002). As most model systems exhibit rather simple leaves, studies in species in which this axis exhibits asymmetries or complex patterns would be informative.

10.5.6 Relationship between the adaxial leaf domain and axillary meristems

A striking feature of *phb-1d* gain-of-function mutants is the development of axillary meristems, normally found only adaxially in the leaf axil, around the entire circumference at the base of the adaxialized leaves, suggesting that their formation is associated with adaxial cell fate (McConnell & Barton, 1998). Conversely, the absence of axillary meristems in loss-of-function *rev* mutants could be interpreted either as a partial loss of adaxial identity or a partial loss in the competence to form meristems (Talbert *et al.*, 1995; Otsuga *et al.*, 2001). Since *REV* encodes a closely related HD-ZIP family member (Ratcliffe *et al.*, 2000) and is expressed adaxially in lateral organs (Otsuga *et al.*, 2001), it is likely that *REV* acts in conjunction with *PHB* and *PHV* to promote adaxial identity in lateral organs. This is consistent with the idea that the development of axillary meristems is linked with the differentiation of adaxial tissues of leaves (Lynn *et al.*, 1999; Long & Barton, 2000).

PHB, *PHV*, and *REV* are also expressed in the apical and axillary meristems themselves (McConnell *et al.*, 2001; Otsuga *et al.*, 2001; Eshed *et al.*, 2001), and constitutive expression of *KANADI* results in a loss of the apical meristem (Eshed *et al.*, 2001; Kerstetter *et al.*, 2001). Based on the proposed antagonism between KANADI and PHABULOSA activities, one explanation of these observed phenotypes is that *PHB*/*PHV*/*REV* activity is required for apical meristem formation.

Another factor contributing to the restriction of axillary meristems to the adaxial side of leaves, may lie in the antagonism between GA and KNOX activity, with the expression of some GA biosynthetic genes limited to the abaxial regions of developing leaves (Hay *et al.*, 2002). Thus, even when ectopic KNOX gene expression appears restricted to the abaxial side as in *fil yab3* double mutants (Kumaran *et al.*, 2002), ectopic meristems only develop adaxially, perhaps due to movement of KNOX proteins from the abaxial to the adaxial cells (Lucas *et al.*, 1995; Kim *et al.*, 2002). Thus, a requirement for PHABULOSA activity and the antagonism of GA with respect to meristematic activity provide an explanation for the restriction of ectopic meristem development to the adaxial regions of leaves when genes controlling meristem development are ectopically expressed in leaves (Sinha *et al.*, 1993; Lincoln *et al.*, 1994; Chuck *et al.*, 1996; Sentoku *et al.*, 2000; Takada *et al.*, 2001; Kumaran *et al.*, 2002).

In *lateral suppressor* (*ls*) mutants of tomato, lateral meristems fail to form while the apical meristem is phenotypically normal (Malayer & Guard, 1964; Schumacher *et al.*, 1999). As the morphology of *ls* leaves appears normal, the action of *ls* may be downstream of the action of PHABULOSA to establish adaxial identity. *LS* encodes a member of the GRAS gene family (Schumacher *et al.*, 1999) and analysis of its expression pattern and whether such mutants influence the distribution of ectopic meristems in plants ectopically-expressing *KNOX* genes should give clues to role in the relationship with adaxial identity and establishment of the axillary meristem.

10.6 From the leaf back to the meristem

Most models of phyllotaxy propose that communication from the newly initiated leaf primordia back to the meristem influences the position of the next leaf primordium to be initiated (Steeves & Sussex, 1989). In a simplified scenario, leaf primordia recently initiated could produce an inhibitory signal, such that the next leaf primordium to form would be furthest from those already present, with stability and mobility influencing the resulting spiral phyllotaxy. While little is known about the biochemical nature of such proposed signals, the phytohormone auxin has long been associated with this developmental process (Snow & Snow, 1937). Recent work has also implicated a role for auxin in the promotion of the initiation of lateral organ primordia (Reinhardt *et al.*, 2000; Stieger *et al.*, 2002). Application of auxin to apices in which polar auxin transport was abolished, resulted in the development of lateral organ primordia from the peripheral zone at the presumed site of the highest auxin concentration, suggesting that auxin maxima may mark the sites of primordia initiation providing a chemical basis for phyllotaxy.

While it is not known whether any of the genes involved in establishing polarity in lateral organs can be implicated in regulating phyllotaxy, at least some of them appear to involved in communication between the lateral organ primordia and the apical meristem. The most striking example of this phenomenon is the failure to maintain an apical meristem in *phan* plants grown at low temperatures (Waites *et al.*, 1998). Since *PHAN* expression is limited to the lateral organs, this activity is non-cell autonomous, indicating that communication between the lateral organs and the meristem is bi-directional. One attractive hypothesis is that interactions between *PHAN*- and KNOX-expressing cells, defining the lateral organ anlagen and the apical meristem, respectively, result in the reinforcement of the identity of each domain.

10.7 Phylogenetic perspective

10.7.1 Developmental paradigms: of wings and leaves

Organizing centers, small groups of cells that pattern the differentiation of surrounding cells, that are formed at boundaries between two adjacent cell types have been shown to provide positional information in the development of appendages during embryogenesis in metazoans (reviewed in Brook *et al.*, 1996; Lawrence & Struhl, 1996; Basler, 2000). This developmental paradigm involves three steps:

1. The establishment of asymmetry to produce two distinct cell populations adjacent to each other.
2. Asymmetric short-range signaling between the two cell populations creating a third type of cell at or near the boundary.
3. A symmetric long-range organizing signal, produced by the third cell type, that patterns the differentiation of surrounding tissues.

Such systems provide a powerful mechanism to generate complex patterns from an initial simple asymmetry.

The wings of *Drosophila* are laminar structures not unlike the leaves of seed plants. Two independent organizing centers pattern the *Drosophila* wing in the anterior-posterior (A/P) and dorsal-ventral (D/V) axes, respectively, and provide instructive examples (simplified in the description below) of the developmental paradigm (reviewed in Brook *et al.*, 1996; Lawrence & Struhl, 1996; Basler, 2000). In the case of the A/P axis, the two distinct populations are established by the asymmetric distribution of a homeodomain transcription factor, *engrailed*: the posterior cells are specified by the expression of *engrailed*, with the anterior cells characterized by the lack of *engrailed* expression. Engrailed activates the expression of a short-range signaling molecule, hedgehog. However, the receptor for hedgehog is only active in anterior cells, such that only anterior cells near the boundary of the posterior cells is receptive to the hedgehog signal. This situation could be created by engrailed repressing the hedgehog receptor. This creates a line of cells near the boundary with a fate distinct from that of either the anterior or posterior cells. In this third cell type, the expression of a long-range signal, *decapentaplegic*, is activated. The long range signal, decapentaplegic, appears to activate gene expression (indirectly) in a concentration-dependent manner such that from the original two cell types, several are generated along the A/P axis. The establishment of the D/V axis is similar, with another homeodomain transcription factor, apteros, establishing the dorsal domain, a NOTCH-based short-range signaling molecule providing asymmetric communication, and wingless, acting as the long-range signal. While the systems described are more complicated *in vivo*, the paradigm provides an example of how simple asymmetries can be elaborated to pattern complex organs.

While it is unknown whether there is a paradigm utilized in the development of lateral organs in seed plants, there are some striking similarities in that a juxtaposition of two cell types is required for the proper patterning of leaves. In this case, the initial asymmetry could be provided by the antagonism between the KANADI and PHABULOSA activities, creating two distinct populations of cells defined by the expression of the respective transcription factors (Fig. 10.4). Unfortunately, most of the genes identified to date, encode transcription factors, rather than encoding molecules active in signaling pathways from the meristem to the lateral organs and between the adaxial and abaxial cells of the lateral organs. In the case of the signal from the apical meristem to initiating lateral organs, PHABULOSA could act as a receptor, either through a sterol or miRNA based signal, and this initial signaling event is required to establish the two distinct cell types. Even less is known about the biochemical nature of the signal(s) between the juxtaposed adaxial and abaxial domains. It is also not known if these signals are involved only in lamina expansion directly, or in patterning the differentiating of cell types in both domains. Candidates for short range signals directing lamina expansion include *LAM1*, while a candidate for a long range patterning signal directing differentiation of cell types is *Rld1*. Unfortunately, the molecular identity of these genes is not yet known. Intercellular movement of a transcription factor, *SHORT ROOT*, patterns radial differentiation in

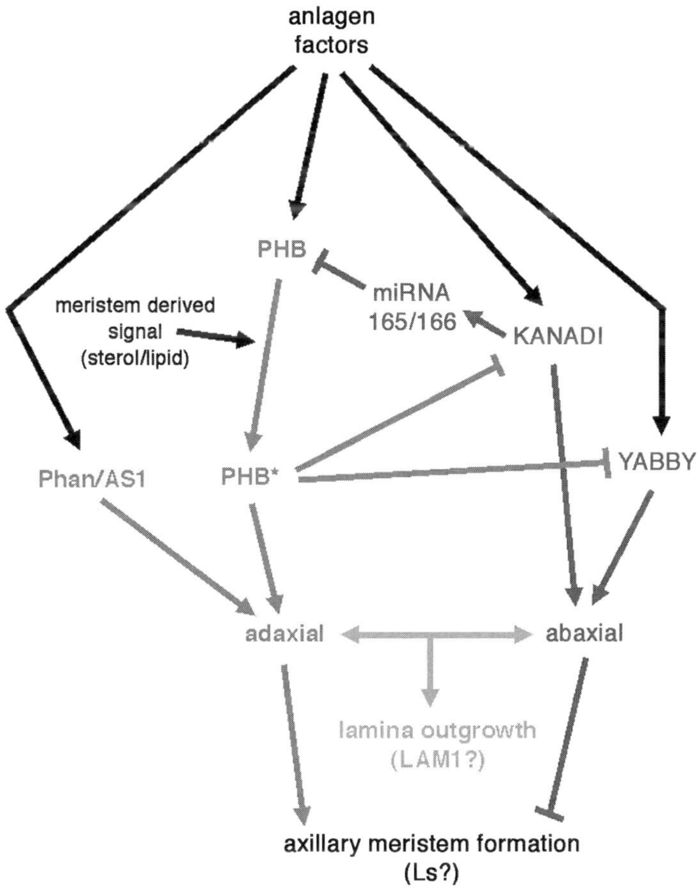

Figure 10.4 Signaling establishes cell fates in lateral organs. The emerging picture from classical and molecular genetic analyses is that as incipient lateral organ primordia develop from the flanks of the shoot apical meristem, factors both intrinsic and extrinsic to the organ primordia contribute to the specification of cells as adaxial (medium grey) or abaxial (dark grey). The apical meristem itself probably provides a signal(s) that promotes adaxial cell fate, whose perception may be mediated through PHB (activated PHB protein is shown as PHB*). Abaxial cell fate may be a 'default' in the absence of signal, for instance, if the lateral organ primordia are separated from the apical meristem. This 'default' state could be the result of the failure to repress genes promoting abaxial identity (e.g. *YABBY* and *KANADI* genes) which are initially activated in the anlagen. Based on epistasis experiments, KANADI activity may mediate between PHB and YABBY activities, perhaps through the regulation of miRNA165/166. Expression of *PHAN/AS1/Rs2* is also activated in lateral organ anlagen, and in at least some species, contributes to the promotion of adaxial fates. The relationship between the pathways mediated by *PHAB/AS1/Rs2* and *PHB* is not known. Subsequent interactions between the juxtaposed adaxial and abaxial domains (light gray), mediated by relative levels of KANADI and YABBY activities, are required for lamina outgrowth. The nature of the communication pathways between the abaxial and adaxial domains required for lamina expansion is also a mystery. Modified and reprinted from *Trends in Genetics* (Bowman *et al.*, 2002) with permission from Elsevier.

Arabidopsis roots (Nakajima *et al.*, 2001), so that a precedent exists for a role for transcription factors in intercellular signaling. In this respect, non-cell autonomous effects of ectopic YABBY expression suggest that YABBY proteins could act as a short range signal (Eshed & Bowman, unpublished).

It is tempting to speculate that this developmental paradigm may be in operation in other processes in plant development, with a conspicuous candidate being the establishment of lateral organ anlagen. In this case, the two domains are defined by expression, or lack of *STM*, with the apical meristem expressing the transcription factor *STM*, and the lateral organ anlage expressing a combination of transcription factors, including *AS1*. The populations of cells defining the lateral organ anlage may initially be defined by an asymmetric distribution of auxin, as described above (Reinhardt *et al.*, 2000). The boundary between these two populations of cells is defined by the expression of other transcription factors, such as *CUP-SHAPED COTYLEDON* (*CUC1* and *CUC2*), and *LATERAL ORGAN BOUNDARIES* (*LOB*) genes (Aida *et al.*, 1997, 1999; Shuai *et al.*, 2002). The boundary could act, both as a barrier to isolate developmental competence, and as a conduit for communication between the two domains.

10.7.2 Evolution of leaves

Despite common positional, anatomical, and functional aspects of leaves in vascular plants, interpretation of the fossil record indicates that leaves evolved independently in the major extant groups of plants (lycophytes, ferns and seed plants) from leafless ancestors in each clade (see Stewart & Rothwell, 1993; Doyle, 1998; Kenrick & Crane, 1997; Foster & Gifford, 1974). Thus, a phylogenetic interpretation of the fossil evidence implies that developmental mechanisms for producing leaves evolved several times in vascular plants.

Results of experimental manipulations of developing leaf primordia reveal both similarities and differences among different vascular plant groups. For example, when young leaf primordia of *Dryopteris austriaca* and *Osmunda* (ferns) were separated from the apical meristem, they often developed as shoots (Wardlaw, 1965). However, when older primordia were separated from the shoot apical meristem they sometimes developed as radial leaf-like organs (Wardlaw, 1965), reminiscent of young *Solanum tuberosum* (seed plant) leaves separated from the SAM. In contrast, isolated *Solanum tuberosum* leaves never developed as shoots (Sussex, 1955). These differences in determinacy may reflect differences in leaf development, where the apices of angiosperms leaves differentiate soon after inception, while the apical cell of fern leaves remains meristematic for a considerable time. While these differences are consistent with the independent evolution of leaves in these taxa, the existence of meristem to leaf, and leaf to meristem, signaling systems may be a common feature of vascular plants although the players in the signaling pathways may be different.

The identification in other vascular plants of genes orthologous to those directing polarity establishment in seed plants should provide insight into the evolution of leaves in all vascular plants. It is of particular interest that all of the Class III HD-ZIP genes are expressed in developing vascular tissues, in some cases limited to the xylem (Baima *et al.*, 1995; Zhong & Ye, 1999; McConnell *et al.*, 2001; Otsuga *et al.*, 2001; Baima *et al.*, 2001; Kang & Dengler, 2002) and that mutations in these genes can alter vascular patterning (Zhong & Ye, 1999; Zhong *et al.*, 1999; Ye, 2002). Since Class III HD-ZIP genes are present in ferns and moss (Aso *et al.*, 1999; Sakakibara *et al.*, 2001), they are likely to be found universally throughout green plants. One attractive hypothesis is that patterning of vasculature and leaf polarity are intricately linked, and that the genetic programs involved are ancient, dating to the origin of green plants.

Acknowledgements

I thank Yuval Eshed, David Smyth and members of my laboratory for thoughtful discussions and critical reading of the manuscript. The photographs in Figs 10.1–3 were taken by former members of the lab, Yuval Eshed and Stuart Baum. Work in my laboratory is supported by the US National Science Foundation.

References

Aida, M., Ishida, T., Fukaki, H., Fujisawa, H.. & Tasaka, M. (1997) Genes involved in organ separation in *Arabidopsis*: an analysis of the *cup-shaped cotyledon* mutant. *Plant Cell*, **9**, 841–857.

Aida, M., Ishida, T. & Tasaka, M. (1999) Shoot apical meristem and cotyledon formation during Arabidopsis embryogenesis: interaction among the *CUP-SHAPED COTYLEDON* and *SHOOT MERISTEMLESS* genes. *Development*, **126**, 1563–1570.

Alvarez, J. & Smyth, D.R. (1999) *CRABS CLAW* and *SPATULA* two *Arabidopsis* genes that control carpel development in parallel with *AGAMOUS*. *Development*, **126**, 2377–2386.

Aso, K., Kato, M., Banks, J.A. & Hasebe, M. (1999) Characterization of homeodomain-leucine zipper genes in the fern *Ceratopteris richardii* and the evolution of the homeodomain-leucine zipper gene family in vascular plants. *Mol. Biol. Evol.*, **16**, 544–552.

Baima, S., Nobili, F., Sessa, G., Lucchetti, S., Ruberti, I. & Morelli, G. (1995) The expression of the *Athb-8* homeobox gene is restricted to provascular cells in *Arabidopsis thaliana*. *Development*, **121**, 4171–4182.

Baima, S., Possenti, M., Matteucci, A., Wisman, E., Altamura, M.M., Ruberti, I. & Morelli, G. (2001) The *Arabidopsis ATHB-8* HD-zip protein acts as a differentiation-promoting transcription factor of the vascular meristems. *Plant Physiol.*, **126**, 643–655.

Basler, K. (2000) Waiting periods, instructive signals and positional information. *EMBO J.*, **19**, 1169–1175.

Bharathan, G., Goliber, T.E., Moore, C., Kessler, S., Pham, T. & Sinha, N.R. (2002) Homologies in leaf form inferred from KNOXI gene expression during development. *Science*, **296**, 1858–1860.

Bohmert, K., Camus, I., Bellini, C., Bouchez, D., Caboche, M. & Benning, C. (1998) *AGO1* defines a novel locus of *Arabidopsis* controlling leaf development. *EMBO J.*, **17**, 170–180.

Bowman, J.L., Eshed, Y. & Baum, S.F. (2002) Establishment of abaxial-adaxial polarity in lateral organs. *Trends Genet.*, **17**, 134–141.

Bowman, J.L. & Smyth, D.R. (1999) *CRABS CLAW* a gene that regulates carpel and nectary development in *Arabidopsis* encodes a novel protein with zinc finger and helix-loop-helix domains. *Development*, **126**, 2387–2396.

Brook, W.J., Diaz-Benjumea, F.J. & Cohen, S.M. (1996) Organizing spatial pattern in limb development. *Annu. Rev. Cell Dev. Biol.*, **12**, 161–180.

Byrne, M.E., Barley, R., Curtis, M., Arroyo, J.M., Dunham, M., Hudson, A. & Martienssen, R.A. (2000) *Asymmetric leaves1* mediates leaf patterning and stem cell function in *Arabidopsis*. *Nature*, **408**, 967–971.

Byrne, M.E., Simorowski, J. & Martienssen, R.A. (2002) *ASYMMETRIC LEAVES1* reveals knox gene redundancy in *Arabidopsis*. *Development*, **129**, 1957–1965.

Cerutti, H. (2003) RNA interference: traveling in the cell and gaining functions? *Trends Genet.*, **19**, 39–46.

Chatterjee, M., Sparvoli, S., Edmunds, C., Garosi, P., Findlay, K. & Martin, C. (1996) *DAG* a gene required for chloroplast differentiation and palisade development in *Antirrhinum majus*. *EMBO J.*, **15**, 4194–4207.

Chen, Q., Atkinson, A., Otsuga, D., Christensen, T., Reynolds, L. & Drews, G.N. (1999) The *Arabidopsis FILAMENTOUS FLOWER* gene is required for flower formation. *Development*, **126**, 2715–2726.

Chuck, G., Lincoln, C. & Hake, S. (1996) *KNAT1* induces lobed leaves with ectopic meristems when overexpressed in *Arabidopsis*. *Plant Cell*, **8**, 1277–1289.

Dolan, L. & Poethig, R.S. (1991) Genetic analysis of leaf development in cotton (*Gossypium barbadense* L.) *Development* (Supplement), **1**, 39–46.

Dolan, L. & Poethig, R.S. (1997) The *OKRA* leaf shape mutation in cotton is active in all cell layers of the leaf. *Am. J. Bot.*, **85**, 322–327.

Donnelly, P.M., Bonetta, D., Tsukaya, H., Dengler, R.E. & Dengler, N.G. (1999) Cell cycling and cell enlargement in developing leaves of *Arabidopsis*. *Dev. Biol.*, **215**, 407–419.

Doyle, J.A. (1998) Phylogeny of vascular plants. *Ann. Rev. Ecol. System.*, **29**, 567–599.

Elliott, R.C., Betzner, A.S., Huttner, E., Oakes, M.P., Tucker, W.Q., Gerentes, D., Perez, P. & Smyth, D.R. (1996) *AINTEGUMENTA*, **an** *APETALA2*-like gene of *Arabidopsis* with pleiotropic roles in ovule development and floral organ growth. *Plant Cell*, **8**, 155–168.

Eshed, Y., Baum, S.F. & Bowman, J.L. (1999) Distinct mechanisms promote polarity establishment in carpels of *Arabidopsis*. *Cell*, **99**, 199–209.

Eshed, Y., Baum, S.F., Perea, J.V. & Bowman, J.L. (2001) Establishment of polarity in lateral organs of plants. *Curr. Biol.*, **11**, 1251–1260.

Fagard, M., Boutet, S., Morel, J.B., Bellini, C. & Vaucheret, H. (2000) *AGO1*, *QDE-2*, and *RDE-1* are related proteins required for post-transcriptional gene silencing in plants, quelling in fungi, and RNA interference in animals. *Proc. Nat. Acad. Sci. USA*, **97**, 11650–11654.

Foster, A.S. & Gifford, E.M. (1974) *Comparative Morphology of Vascular Plants*. W.H. Freeman and Company: San Francisco.

Foster, T.M., Lough, T.J., Emerson, S.J., Lee, R.H., Bowman, J.L., Forster, R.L.S. & Lucas, W.J. (2002) Reversible establishment of organ polarity as a probe for signaling in plant development. *Plant Cell*, **14**, 1497–1508.

Freeling, M. (1992) A conceptual framework for maize leaf development. *Dev. Biol.*, **153**, 44–58.

Hamilton, A., Voinnet, O., Chappell, L. & Baulcombe, D. (2002) Two classes of short interfering RNA in RNA silencing. *EMBO J.*, **21**, 4671–4679.

Hanawa, J. (1961) Experimental studies on leaf dorsiventrality in *Sesamum indicum* L. *Bot. Mag. Tokyo*, **74**, 303–309.

Hannon, G.J. (2002) RNA interference. *Nature*, **418**, 244–251.

Hay, A., Kaur, H., Phillips, A., Hedden, P., Hake, S. & Tsiantis, M. (2002) The gibberellin pathway mediates *KNOTTED1*-type homeobox function in plants with different body plans. *Curr. Biol.*, **12**, 1557–1565.

Iwakawa, H., Ueno, Y., Semiarti, E., Onouchi, H., Kojima, S., Tsukaya, H., Hasebe, M., Soma, T., Ikezaki, M., Machida, C. & Machida, Y. (2002) The *ASYMMETRIC LEAVES2* gene of

Arabidopsis thaliana required for formation of a symmetric flat leaf lamina, encodes a member of a novel family of proteins characterized by cysteine repeats and a leucine zipper. *Plant Cell Physiol.*, **43**, 467–478.

Kang, J. & Dengler, N. (2002) Cell cycling frequency and expression of the homeobox gene *ATHB-8* during leaf vein development in *Arabidopsis*. *Planta*, **216**, 212–219.

Keddie, J.S., Carroll, B., Jones, J.D. & Gruissem, W. (1996) The *DCL* gene of tomato is required for chloroplast development and palisade cell morphogenesis in leaves. *EMBO J.*, **15**, 4208–4217.

Kenrick, P. & Crane, P.R. (1997) *The origin and early diversification of land plants: a cladistic study*. Smithsonian Institution Press: Washington DC.

Kerstetter, R.A., Bollman, K., Taylor, R.A., Bomblies, K. & Poethig, R.S. (2001) *KANADI* controls organ polarity in *Arabidopsis*. *Nature*, **411**, 706–709.

Kim, J.Y., Yuan, Z., Cilia, M., Khalfan-Jagani, Z. & Jackson, D. (2002) Intercellular trafficking of a KNOTTED1 green fluorescent protein fusion in the leaf and shoot meristem of *Arabidopsis*. *Proc Natl Acad Sci USA*, **99**, 4103–4108.

Klucher, K.M., Chow, H., Reiser, L. & Fischer, R.L. (1996) The *AINTEGUMENTA* gene of *Arabidopsis* required for ovule and female gametophyte development is related to the floral homeotic gene *APETALA2*. *Plant Cell*, **8**, 137–153.

Krizek, B.A. (1999) Ectopic expression of *AINTEGUMENTA* in *Arabidopsis* plants results in increased growth of floral organs. *Dev. Genet.*, **25**, 224–236.

Kumaran, M.K., Bowman, J.L. & Sundaresan, V. (2002) *YABBY* Genes Mediate Down-regulation of *KNOTTED*-like Genes During Leaf Development in *Arabidopsis thaliana*. *Plant Cell*, **14**, 2761–2770.

Kumaran, M.K., Ye, D., Yang, W.C., Griffith, M.E., Chaudhury, A.M. & Sundaresan, V. (1999) Molecular cloning of *ABNORMAL FLORAL ORGANS*: a gene required for flower development in *Arabidopsis*. *Sex. Plant Reprod.*, **12**, 118–122.

Lawrence, P.A. & Struhl, G. (1996) Morphogens, compartments and pattern: lessons from *Drosophila*. *Cell*, **85**, 951–961.

Lincoln, C., Long, J., Yamaguchi, J., Serikawa, K. & Hake, S. (1994) A knotted1-like homeobox gene in *Arabidopsis* is expressed in the vegetative meristem and dramatically alters leaf morphology when overexpressed in transgenic plants. *Plant Cell*, **6**, 1859–1876.

Llave, C., Kasschau, K.D. & Carrington, J.C. (2000) Virus-encoded suppressor of post-transcriptional gene silencing targets a maintenance step in the silencing pathway. *Proc. Natl. Acad. Sci. U S A.*, **97**, 13401–13406.

Llave, C., Kasschau, K.D., Rector, M.A. & Carrington, J.C. (2002) Endogenous and silencing-associated small RNAs in plants. *Plant Cell*, **14**, 1605–1619.

Long, J. & Barton, M.K. (2000) Initiation of axillary and floral meristems in *Arabidopsis*. *Dev. Biol.*, **218**, 341–353.

Long, J.A., Moan, E.I., Medford, J.I. & Barton, M.K. (1996) A member of the KNOTTED class of homeodomain proteins encoded by the *STM* gene of *Arabidopsis*. *Nature*, **379**, 66–69.

Lucas, W.J., Bouche-Pillon, S., Jackson, D.P., Nguyen, L., Baker, L., Ding, B. & Hake, S. (1995) Selective trafficking of KNOTTED1 homeodomain protein and its mRNA through plasmodesmata. *Science*, **270**, 1980–1983.

Lynn, K., Fernandez, A., Aida, M., Sedbrook, J., Tasaka, M., Masson, P. & Barton, M.K. (1999) The *PINHEAD/ZWILLE* gene acts pleiotropically in *Arabidopsis* development and has overlapping functions with the *ARGONAUTE1* gene. *Development*, **126**, 1–13.

McConnell, J.R. & Barton, M.K. (1995) Effect of mutations in the *PINHEAD* gene of *Arabidopsis* on the formation of shoot apical meristems. *Dev. Genet.*, **16**, 358–366.

McConnell, J.R. & Barton, M.K. (1998) Leaf polarity and meristem formation in *Arabidopsis*. *Development*, **125**, 2935–2942.

McConnell, J.R., Emery, J.F., Eshed, Y., Bao, N., Bowman, J. & Barton, M.K. (2001) Role of *PHABULOSA* and *PHAVOLUTA* in determining radial patterning in shoots. *Nature*, **411**, 709–713.

McHale, N.A. (1993) *LAM1* and *FAT* genes control development of the leaf blade in *Nicotiana slyvestris*. *Plant Cell*, **5**, 1029–1038.

McHale, N.A. & Marcotrigiano, M. (1998) *LAM1* is required for dorsoventrality and lateral growth of the leaf blade in *Nicotiana*. *Development*, **125**, 4235–4243.

Malayer, J.C. & Guard, A.T. (1964) A comparative developmental study of the mutant sideshootless and normal tomato plants. *Am. J. Bot.*, **51**, 140–143.

Mallory, A.C., Ely, L., Smith, T.H., Marathe, R., Anandalakshmi, R., Fagard, M., Vaucheret, H., Pruss, G., Bowman, L. & Vance, V.B. (2001) HC-Pro suppression of transgene silencing eliminates the small RNAs but not transgene methylation or the mobile signal. *Plant Cell*, **13**, 571–583.

Mallory, A.C., Reinhart, B.J., Bartel, D., Vance, V.B. & Bowman, L.H. (2002) A viral suppressor of RNA silencing differentially regulates the accumulation of short interfering RNAs and micro-RNAs in tobacco. *Proc. Natl. Acad. Sci. U S A*, **99**, 15228–15233.

Matsumoto, N. & Okada, K. (2001) A homeobox gene, *PRESSED FLOWER* regulates lateral axis-dependent development of *Arabidopsis* flowers. *Genes Dev.*, **15**, 3355–3364.

Matzke, M., Matzke, A.J. & Kooter, J.M. (2001) RNA: guiding gene silencing. *Science*, **293**, 1080–1083.

Mizukami, Y. & Fischer, R.L. (2000) Plant organ size control: *AINTEGUMENTA* regulates growth and cell numbers during organogenesis. *Proc. Natl. Acad. Sci. USA*, **97**, 942–947.

Morel, J.B., Godon, C., Mourrain, P., Beclin, C., Boutet, S., Feuerbach, F., Proux, F. & Vaucheret, H. (2002) Fertile hypomorphic *ARGONAUTE* (*ago1*) mutants impaired in post-transcriptional gene silencing and virus resistance. *Plant Cell*, **14**, 629–639.

Moussian, B., Schoof, H., Haecker, A., Jurgens, G. & Laux, T. (1998) Role of the *ZWILLE* gene in the regulation of central shoot meristem cell fate during *Arabidopsis* embryogenesis. *EMBO J.*, **17**, 1799–1809.

Nakajima, K., Sena, G., Nawy, T. & Benfey, P.N. (2001) Intercellular movement of the putative transcription factor *SHR* in root patterning. *Nature*, **413**, 307–311.

Nath, U., Crawford, B.C., Carpenter, R. & Coen, E. (2003) Genetic control of surface curvature. *Science*, **299**, 1404–1407.

Nelson, J.M, Lane, B. & Freeling, M. (2002) Expression of a mutant maize gene in the ventral leaf epidermis is sufficient to signal a switch of the leaf's dorsoventral axis. *Development*, **129**, 4581–4589.

Ogas J., Cheng J.C., Sung Z.R. & Somerville C. (1997) Cellular differentiation regulated by gibberellin in the *Arabidopsis thaliana pickle* mutant. *Science*, **277**, 91–94.

Ogas, J., Kaufmann, S., Henderson, J. & Somerville, C. (1999) *PICKLE* is a CHD3 chromatin-remodeling factor that regulates the transition from embryonic to vegetative development in *Arabidopsis*. *Proc. Natl. Acad. Sci. USA*, **96**, 13839–13844.

Ori, N., Eshed, Y., Chuck, G., Bowman, J.L. & Hake, S. (2000) Mechanisms that control knox gene expression in the *Arabidopsis* shoot. *Development*, **127**, 5523–5532.

Otsuga, D., DeGuzman, B., Prigge, M.J., Drews, G.N. & Clark, S.E. (2001) *REVOLUTA* regulates meristem initiation at lateral positions. *Plant J.*, **25**, 223–236.

Plasterk, R.H. (2002) RNA silencing: the genome's immune system. *Science*, **296**, 1263–1265.

Pontig, C.P. & Aravind, L. (1999) START: a lipid-binding domain in StAR, HD-ZIP and signalling proteins. *Trends Biochem. Sci.*, **24**, 130–132.

Ratcliffe, O.J., Riechmann, J.L. & Zhang, J.Z. (2000) *INTERFASCICULAR FIBERLESS1* is the same gene as *REVOLUTA*. *Plant Cell*, **12**, 315–317.

Reinhardt, D., Mandel, T. & Kuhlemeier, C. (2000) Auxin regulates the initiation and radial position of plant lateral organs. *Plant Cell*, **12**, 507–518.

Reinhart, B.J., Weinstein, E.G., Rhoades, M.W., Bartel, B. & Bartel, D.P. (2002) MicroRNAs in plants. *Genes Dev.*, **16**, 1616–1626.

Reiter, R.S., Coomber, S.A., Bourett, T.M., Bartley, G.E. & Scolnik, P.A. (1994) Control of leaf and chloroplast development by the *Arabidopsis* gene *pale cress*. *Plant Cell*, **6**, 1253–1264.

Rhoades, M.W., Reinhart, B.J., Lim, L.P., Burge, C.B., Bartel, B. & Bartel, D.P. (2002) Prediction of plant microRNA targets. *Cell*, **110**, 513–520.

Sakakibara, K., Nishiyama, T., Kato, M. & Hasebe, M. (2001) Isolation of homeodomainleucine zipper genes from the moss *Physcomitrella patens* and the evolution of homeodomainleucine zipper genes in land plants. *Mol. Biol. Evol.*, **18**, 491–502.

Sakamoto, T., Kamiya, N., Ueguchi-Tanaka, M., Iwahori, S. & Matsuoka, M. (2001) KNOX homeodomain protein directly suppresses the expression of a gibberellin biosynthetic gene in the tobacco shoot apical meristem. *Genes Dev.*, **15**, 581–590.

Sawa, S., Ito, T., Shimura, Y. & Okada, K. (1999) *FILAMENTOUS FLOWER* controls the formation and development of *Arabidopsis* inflorascences and floral meristems. *Plant Cell*,

11, 69–86.
Sawa, S., Watanabe, K., Goto, K., Liu, Y.G., Shibata, D., Kanaya, E., Morita, E.H. & Okada, K. (1999b) *FILAMENTOUS FLOWER* a meristem and organ identity gene of *Arabidopsis* encodes a protein with a zinc finger and HMG-related domains. *Genes Dev.*, **13**, 1079–1088.
Schneeberger, R., Tsiantis, M., Freeling, M. & Langdale, J.A. (1998) The *rough sheath2* gene negatively regulates homeobox gene expression during maize leaf development. *Development*, **125**, 2857–2865.
Schumacher, K., Schmitt, T., Rossberg, M., Schmitz, G. & Theres, K. (1999) The Lateral suppressor (Ls) gene of tomato encodes a new member of the VHIID protein family. *Proc. Natl. Acad. Sci. U S A*, **96**, 290–295.
Semiarti, E., Ueno, Y., Tsukaya, H., Iwakawa, H., Machida, C. & Machida, Y. (2001) The *ASYMMETRIC LEAVES2* gene of *Arabidopsis thaliana* regulates formation of a symmetric lamina, establishment of venation and repression of meristem-related homeobox genes in leaves. *Development*, **128**, 1771–1783.
Sentoku, N., Sato, Y. & Matsuoka, M. (2000) Overexpression of rice OSH genes induces ectopic shoots on leaf sheaths of transgenic rice plants. *Dev. Biol.*, **220**, 358–364.
Shuai, B., Reynaga-Pena, C.G. & Springer, P.S. (2002) The *lateral organ boundaries* gene defines a novel, plant-specific gene family. *Plant Physiol.*, **129**, 747–761.
Siegfried, K.R., Eshed, Y., Baum, S.F., Otsuga, D., Drews, G.N. & Bowman, J.L. (1999) Members of the YABBY gene family specify abaxial cell fate in *Arabidopsis*. *Development*, **126**, 4117–4128.
Sinha, N.R., Williams, R.E. & Hake, S. (1993) Overexpression of the maize homeo box gene, *KNOTTED-1* causes a switch from determinate to indeterminate cell fates. *Genes Dev.*, **7**, 787–795.
Smith, L.G., Greene, B., Veit, B. & Hake, S. (1992) A dominant mutation in the maize homeobox gene, Knotted-1, causes its ectopic expression in leaf cells with altered fates. *Development*, **116**, 21–30.
Snow, M. & Snow, R. (1937) Auxin and leaf formation. *New Phytol.*, **36**, 1–18.
Snow, M. & Snow, R. (1959) The dorsiventrality of leaf primordia. *New Phytol.*, **58**, 188–207.
Steeves, T.A. & Sussex, I.M. (1989) *Patterns in Plant Development*, 2nd edn. Cambridge University Press: Cambridge.
Stewart, W.N. & Rothwell, G.W. (1993) *Paleobotany and the Evolution of Plants.* Cambridge University Press: New York.
Stieger, P.A., Reinhardt, D. & Kuhlemeier, C. (2002) The auxin influx carrier is essential for correct leaf positioning. *Plant J.*, **32**, 509–517.
Sun, Y., Zhou, Q., Zhang, W., Fu, Y. & Huang, H. (2002) *ASYMMETRIC LEAVES1*, an *Arabidopsis* gene that is involved in the control of cell differentiation in leaves. *Planta*, **214**, 694–702.
Sundaresan, V., Springer, P., Volpe, T., Haward, S., Jones, J.D., Dean, C., Ma, H. & Martienssen, R. (1995) Patterns of gene action in plant development revealed by enhancer trap and gene trap transposable elements. *Genes Dev.*, **9**, 1797–1810.
Sussex, I.M. (1954) Experiments on the cause of dorsiventrality in leaves. *Nature*, **174**, 351–352.
Sussex, I.M. (1955) Morphogenesis in *Solanum tuberosum* L.: Experimental investigation of leaf dorsoventrality and orientation in the juvenile shoot. *Phytomorphol.*, **5**, 286–300.
Tabara, H., Sarkissian, M., Kelly, W.G., Fleenor, J., Grishok, A., Timmons, L., Fire, A. & Mello, C.C. (1999) The *rde-1* gene, RNA interference, and transposon silencing in *C. elegans*. *Cell*, **99**, 123–132.
Takada, S., Hibara, K., Ishida, T. & Tasaka, M. (2001) The *CUP-SHAPED COTYLEDON1* gene of *Arabidopsis* regulates shoot apical meristem formation. *Development*, **128**, 1127–1135.
Talbert, P.B., Adler, H.T., Parks, D.W. & Comai, L. (1995) The *REVOLUTA* gene is necessary for apical meristem development and for limiting cell divisions in the leaves and stems of *Arabidopsis thaliana*. *Development*, **121**, 2723–2735.
Tanaka-Ueguchi, M., Itoh, H., Oyama, N., Koshioka, M. & Matsuoka, M. (1998) Overexpression of a tobacco homeobox gene, NTH15, decreases the expression of a gibberellin biosynthetic gene encoding GA 20–oxidase. *Plant J.*, **15**, 391–400.
Tang, G., Reinhart, B.J., Bartel, D.P. & Zamore, P.D. (2002) A biochemical framework for RNA silencing in plants. *Genes Dev.*, **17**, 49–63.

Timmermans, M.C., Schultes, N.P., Jankovsky, J.P. & Nelson, T. (1998) *Leafbladeless1* is required for dorsoventrality of lateral organs in maize. *Development*, **125**, 2813–2823.

Timmermans, M.C., Hudson, A., Becraft, P.W. & Nelson, T. (1999) *ROUGH SHEATH2*: a myb protein that represses knox homeobox genes in maize lateral organ primordia. *Science*, **284**, 151–153.

Tsiantis, M., Schneeberger, R., Golz, J.F., Freeling, M. & Langdale, J.A. (1999) The maize *rough sheath2* gene and leaf development programs in monocot and dicot plants. *Science*, **284**, 154–156.

Tsuge, T., Tsukaya, H. & Uchimiya, H. (1996) Two independent and polarized processes of cell elongation regulate leaf blade expansion in *Arabidopsis thaliana* (L. Heynh.) *Development*, **122**, 1589–1600.

Tsukaya H. (2002) Leaf anatomy of a rheophyte, *Dendranthema yoshinaganthum* (Asteraceae), and of hybrids between *D. yoshinaganthum* and a closely related non-rheophyte, *D. indicum*. *J. Plant Res.*, **115**, 329–333.

Tsukaya, H. & Uchimiya, H. (1997) Genetic analyses of the formation of the serrated margin of leaf blades in Arabidopsis: combination of a mutational analysis of leaf morphogenesis with the characterization of a specific marker gene expressed in hydathodes and stipules. *Mol. Gen. Genet.*, **256**, 231–238.

Vaucheret, H., Beclin, C. & Fagard, M. (2001) Post-transcriptional gene silencing in plants. *J. Cell Sci.*, **114**, 3083–3091.

Villanueva, J.M., Broadhvest, J., Hauser, B.A., Meister, R.J., Schneitz, K. & Gasser, C.S. (1999) *INNER NO OUTER* regulates abaxial-adaxial patterning in *Arabidopsis* ovules. *Genes Devel.*, **13**, 3160–3169.

Voinnet, O., Lederer, C. & Baulcombe, D.C. (2000) A viral movement protein prevents spread of the gene silencing signal in *Nicotiana benthamiana*. *Cell*, **103**, 157–167.

Wade, P.A., Jones, P.L., Vermaak, D. & Wolffe, A.P. (1998) A multiple subunit Mi-2 histone deacetylase from *Xenopus laevis* cofractionates with an associated Snf2 superfamily ATPase. *Curr. Biol.*, **8**, 843–846.

Waites, R. & Hudson, A. (1995) *phantastica*: a gene required for dorsiventrality of leaves in *Antirrhinum majus*. *Development*, **121**, 2143–2154.

Waites, R. & Hudson, A. (2001) The *Handlebars* genes is required with *Phantastica* for dorsoventral asymmetry of organs and for stem cell activity in *Antirrhinum*. *Development*, **128**, 1923–1931.

Waites, R., Selvadurai, H.R., Oliver, I.R. & Hudson, A. (1998) The *PHANTASTICA* gene encodes a MYB transcription factor involved in growth and dorsiventrality of lateral organs in *Antirrhinum*. *Cell*, **93**, 779–789.

Wardlaw, C.W. (1965) *Organization and Evolution in Plants*. Longmans, Green and Co. Ltd., London.

Waterhouse, P.M., Wang, M.B. & Lough, T. (2001) Gene silencing as an adaptive defence against viruses. *Nature*, **411**, 834–42.

Weston, E., Thorogood, K., Vinti, G. & Lopez-Juez, E. (2000) Light quantity controls leaf-cell and chloroplast development in *Arabidopsis thaliana* wild type and blue-light-perception mutants. *Planta*, **211**, 807–815.

Ye, Z.H. (2002) Vascular tissue differentiation and pattern formation in plants. *Ann. Rev. Plant Biol.*, **53**, 183–202.

Zamore, P.D. (2002) Ancient pathways programmed by small RNAs. *Science*, **296**, 1265–1269.

Zhang, Y., Leroy, G., Seelig, H., Lane, W.S. & Reinberg, D. (1998) The dermatomyositis-specific autoantigen Mi2 is a component of a complex containing histone deacetylase and nucleosome remodeling activities. *Cell*, **95**, 279–289.

Zhong, R. & Ye, Z.H. (1999) *IFL1* a gene regulating interfascicular fiber differentiation in *Arabidopsis* encodes a homeodomainleucine zipper protein. *Plant Cell*, **11**, 2139–2152.

Zhong, R., Taylor, J.J. & Ye, Z.H. (1999) Transformation of the collateral vascular bundles into amphivasal vascular bundles in an *Arabidopsis* mutant. *Plant Physiol.*, **120**, 53–64.

11 Polarity in floral development

Martin Kieffer and Brendan Davies

11.1 Introduction

The origins of flowering plants (angiosperms) are still largely shrouded in a mystery described by Darwin as 'abominable' (Crepet, 2000). How could they suddenly arise in the late Jurassic and, by the early Cretaceous (130–90 million years ago), have become so successful that they challenged the domination of the gymnosperms such as conifers and cicads? Today, with at least 250,000 extant species, flowering plants have evolved a breathtaking diversity of flower shapes and structures, which has undoubtedly contributed to their success (Endress, 1999). It is likely that this floral extravaganza was largely driven by co-evolution with pollinators, culminating in the beautiful complexity of the orchid flower (Ren, 1998; Neal *et al.*, 1998). The way in which genes control the three dimensional architecture of an organism is the new 'abominable mystery' of biology. In order to specify the three dimensional shape of a flower, various axes of polarity are required. We can view the polarity of flowers at several different levels and we have therefore structured the chapter to reflect this (Fig. 11.1). At the level of the whole plant, flowers are produced in defined spatial patterns, reflecting the organisation of the inflorescence. Individual flowers also have a polarity, both internally, along their radial axis and in respect of their orientation relative to the apical-basal axis of the shoot. Finally the morphology of the floral organs reflects additional polarity.

Recent molecular evidence has revealed that the establishment of polarity in flowers and floral organ primordia is a prerequisite for their normal development. Mechanisms underpinning polarity of primordia appear largely conserved between leaves and floral organs (see Chapter 10). Extensive analysis has also demonstrated that flower and floral organ polarity is established by the integration of signals initiated at different levels, from the whole plant to the flower and organ primordia themselves. In response, flowers and floral organs develop as highly specialised structures whose reproductive efficiency has ensured the evolutionary success of angiosperms. In this review we focus primarily on the control of polarity, at each level of flower development, in the two model species *Arabidopsis thaliana* and *Antirrhinum majus*.

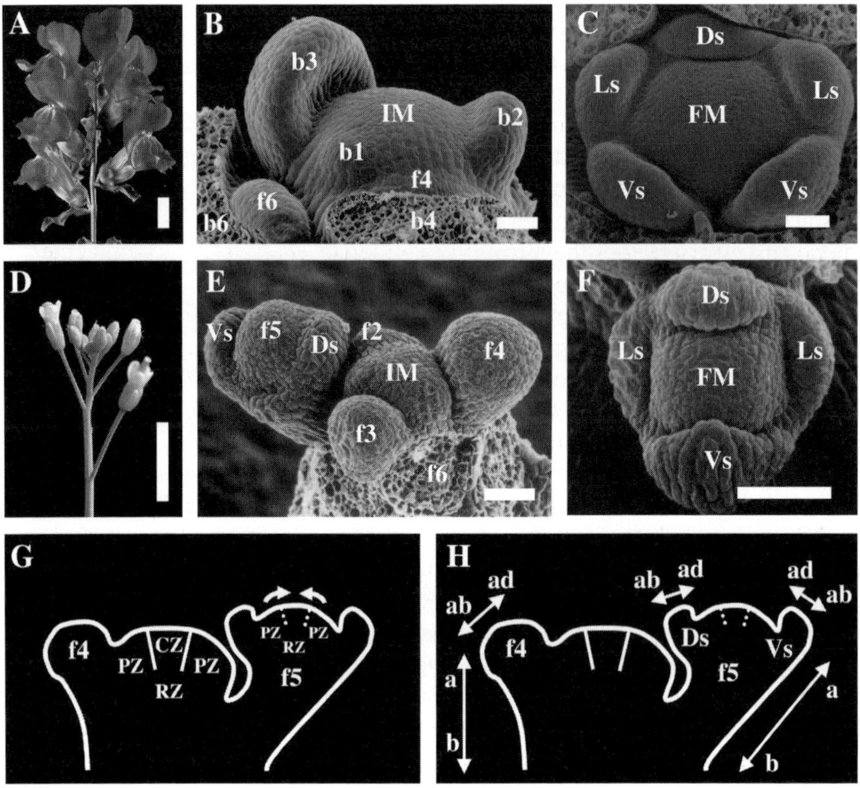

Figure 11.1 Different levels of polarity in the flower. Inflorescences of *Antirrhinum* (A) and *Arabidopsis* (D) are shown next to SEM pictures of their inflorescence meristems (B and E). The inflorescence meristems (IM) form primordia on their flanks and the floral primordia are numbered from youngest to oldest (e.g. f4). Each *Antirrhinum* flower is initiated sequentially in close association with a bract (B) which are also numbered by age (e.g. b3). In *Arabidopsis*, floral primordia are initiated directly by the IM (E). Developing flowers are compared in C and F, just after initiation of the sepals. In keeping with the notation used in this chapter, the sepals are designated as dorsal (nearest to the IM) (Ds), lateral (Ls) and ventral (furthest from IM) (Vs). Schematic longitudinal sections through an *Arabidopsis* inflorescence meristem show the different zones of the apical and floral meristems (G) and the different axes of polarity discussed in this chapter (H). In G, the determinacy of the floral meristem is indicated by two arrows which signify differentiation of the Central Zone (CZ). In H, the apical basal polarity of the inflorescence shoot and floral primordium is shown (a↔b). The adaxial-abaxial polarity of the floral primordium corresponding to f4 in (G) is also indicated (ad↔ab). The flower corresponding to F5 in (G) has a dorso-ventral axis identified by the dorsal and ventral sepal primordia (Ds and Vs). The developing floral organs have an adaxial-abaxial polarity with respect to the floral meristem (ad↔ab). Peripheral Zone (PZ), Rib Zone (RZ). Bar, A: 1cm, D: 0.5cm, B, C, E, F: 30μm.

11.2 Polarity in inflorescence architecture

The initiation of primordia on the flanks of inflorescence meristems follows various strict geometrical distributions, or phyllotaxis, defining a large diversity of inflorescence architectures (Weberling, 1989). These patterns often involve elaborate production of secondary inflorescences and changes in phyllotaxy during the transition between vegetative and floral development. For example, the decussate organisation of *Antirrhinum* leaves contrasts with the spiral arrangement of flowers on the inflorescence (Bradley *et al.*, 1996b). The polar organisation of the inflorescence meristem reflects the arrangement established in the shoot apical meristem of the embryo (Fig. 11.1) (see Chapter 8). The apical central zone (CZ) harbours undifferentiated cells, which divide slowly and gradually displace the most peripheral cells to adjacent domains. The rib zone (RZ) lies immediately below the central zone and produces the stem pith. Laterally, the peripheral zone (PZ) surrounds the central zone and is the site of floral primordium initiation as well as generating most of the stem tissue. Cells, therefore, become incorporated into differentiating tissues in response to positional signals radiating from, and revolving around, the central zone of the meristem. In effect, the central zone acts as the organising point of the inflorescence.

11.2.1 Flower initiation

Flower initiation in *Arabidopsis* has been studied by analysis of sectoral chimeras. The initiation of each flower starts with the recruitment of blocks of four founding cells on the flanks of the inflorescence meristem. The mechanism of recruitment is still unclear but growing evidence implicates the synthesis and transport of the phytohormone, auxin. The floral primordia are polarised from the time of inception, in response to their position on the flanks of the inflorescence meristem (Bossinger & Smyth, 1996). This field is the subject of intense research and exciting new discoveries are unravelling the mechanisms controlling the demarcation of the adaxial (facing the inflorescence meristem) and abaxial (facing away from the inflorescence meristem) regions of the floral primordium (Fig. 11.1). In addition, the early establishment of primordial polarity appears to condition further floral apical-basal development.

The involvement of auxin in floral initiation is illustrated by a series of *Arabidopsis* mutants which affect auxin distribution and display a reduced ability to flower; a phenotype which can be reproduced by treatment of wild type plants with auxin polar transport inhibitors (Okada *et al.*, 1991). Inflorescences of strong *pin-formed 1* (*pin1*) mutants often completely lack flower primordia. The organisation of the inflorescence meristem appears normal but its peripheral zone is unable to co-ordinate the initiation of primordia (Vernoux *et al.*, 2000). Early markers of floral primordium initiation (*LEAFY* and *AINTEGUMENTA*) are expressed in a ring like domain covering the entire peripheral zone instead of the expected discrete expression sites, which precede primordium initiation. In addition, the expression of the NAC domain gene *CUP-SHAPED COTYLEDON2* (*CUP2*), which normally defines the boundaries of flower primordia, is also extended to the whole peripheral zone. *PIN1* encodes a putative auxin transporter (efflux carrier; Gälweiler *et al.*, 1998). Local application of auxin on the flanks of mutant inflorescence meristems induces local outgrowth, mimicking primordium initiation (Reinhard *et al.*, 2000) and suggesting that a shortage of auxin may be responsible for the lack of primordia initiation in the mutant. Localised variations in auxin levels in the peripheral zone of the inflorescence meristem therefore have the ability to promote outgrowth, influence phyllotaxy and establish the boundaries of flower primordia.

Similar inflorescence defects are seen in the *pinoid* (*pid*) and *monopteros* (*mp*) mutants (Alvarez *et al.*, 1994; Bennett *et al.*, 1995; Przemeck *et al.*, 1996). PID is a serine-threonine protein kinase, which either regulates polar auxin transport or its function (Christensen *et al.*, 2000; Benjamins *et al.*, 2001; DeLong *et al.*, 2002). Consistent with the role of auxin in floral initiation, *MP* encodes a member of the auxin response factor (ARF) family of transcription factors. As expected *PIN1*, *PID*, and *MP* are all expressed early in floral primordia, before tissue outgrowth.

The precise role of auxin in flower initiation remains to be elucidated. It would be particularly interesting to determine whether auxin is a central part of the elusive mechanism controlling phyllotaxis or if it is merely responding to it. The *FLOOZY* (*FZY*) gene of Petunia might provide a clue to this (Tobena-Santamaria *et al.*, 2002). FZY is a flavin mono-oxygenase-like protein which is likely to be involved in the synthesis of auxin. Flowers produced on *fzy* mutant inflorescences consist of a pistil and a single sepal-like organ. *FZY* is the petunia orthologue of *YUCCA* genes in *Arabidopsis* (Zhao *et al.*, 2001b) and it will be interesting to determine the exact inter-relationship between the *YUCCA* genes, *PIN1*, *PID*, *MP* and localised auxin levels in the developing flower.

11.2.2 The adaxial-abaxial axis

In inflorescences of *filamentous flower* (*fil*) mutants, a proportion of flowers arrest at a more advanced developmental stage than in *pin1* and consist of elongated filaments (Chen *et al.*, 1999; Sawa *et al.*, 1999). FIL is a member of the YABBY family of transcription factors and is expressed in all shoot lateral organ primordia, including emerging flowers. Two other members of this family, *YABBY2* (*YAB2*) and *YABBY3* (*YAB3*), are expressed in a similar manner (Siegfried *et al.*, 1999). Disruption of both *FIL* and *YAB3* induces a more severe phenotype; with most flowers reduced to filamentous structures, indicating that the two genes redundantly control flower development. In addition, cotyledons and leaves also develop as partly radialised structures (Siegfried *et al.*, 1999; Kumaran *et al.*, 2002) revealing a more general function in primordium development. *FIL* is initially expressed throughout the young floral primordium, but later becomes restricted to the abaxial side. The *YABBY* gene family specifies abaxial identity in both leaf and floral primordia (Bowman, 2000; Chapter 10). Disruption of the adaxial-abaxial polarity in leaves of both *Arabidopsis* and *Antirrhinum* induces production of radially symmetrical organs (Waites & Hudson, 1995; McConnell & Barton, 1998; Siegfried *et al.*, 1999). It was therefore hypothesised that establishment of adaxial-abaxial polarity is required for lateral expansion of the leaf blade (Waites & Hudson, 1995). The abnormal filamentous flowers formed in the *fil* mutant indicates that the establishment of adaxial-abaxial polarity in the floral primordium is also necessary for appropriate development of floral organs.

Floral primordia share strong similarities with lateral shoot primordia, since both require the activation of genes controlling meristem initiation and maintenance, such as *SHOOTMERISTEMLESS* (*STM*), *WUSCHEL* (*WUS*) and the *CLAVATA* (*CLV*) genes. Mutations affecting *REVOLUTA* (*REV*) induce the loss of most lateral shoot meristems and the conversion of some flowers to filamentous structures (Talbert *et al.*, 1995; Otsuga *et al.*, 2001). In inflorescences of plants homozygous for both *fil* and *rev*, all the flowers are reduced to filamentous structures (Chen *et al.*, 1999). *REV* transcripts are detectable at a very early stage and their expression predicts the initiation of floral primordia. Since *REV* is expressed before the meristem initiation and maintenance genes, and its expression pattern matches the initial expression patterns of *WUS* and *STM*, it is possible that *REV* activates *WUS* and *STM* at the positions destined to become floral meristems (Otsuga *et al.*, 2001). REV is a member of the homeodomain-leucine zipper (HD-ZIP) family of transcription factors and shares similarities with *PHABULOSA* (*PHB*) and *PHAVOLUTA* (*PHV*), two key regulators of leaf adaxial identity (McConnell *et al.*, 2001).

11.3 Polarity of flowers

Initiation and adaxial-abaxial patterning of floral primordia are tightly controlled by the inflorescence meristem and echo the apical-basal and radial polarities of the shoot. However, as flowers form, a new radial patterning becomes apparent with the initiation of floral organs. Angiosperm flowers share a conserved distribution of organs (Endress, 2001). The centre of the flower is occupied by carpels, sheltering concealed ovules. Following fertilisation, ovules develop as seeds enclosed in carpel-derived fruit. Carpels are surrounded by stamens (the male component of the flower) carrying four pollen sacs in two symmetrical thecae. In dicot species, petals and sepals usually complete the structure of the flower, but the outer organs of monocot flowers can contain a variety of different types of perianth organs (Ambrose *et al.*, 2000). Flowers are therefore highly polarised structures developing complex patterns of organs, tissues and cell types. We will discuss floral polarity by reference to two axes; the radial axis, which is apparent in most plants and the dorso-ventral (or adaxial-abaxial) axis which is only obvious in certain species (Fig. 11.1).

11.3.1 The dorso-ventral axis of flowers

In flowers such as *Antirrhinum*, the dorso-ventral axis is clear, because the dorsal petals are so obviously different in shape to the ventral petals, producing a flower which is bilaterally symmetrical (zygomorphic; Fig. 11.2). However, the dorso-ventral axis is not so apparent in radially symmetrical (actinomorphic) flowers, such as those of *Arabidopsis*. As a result of this, the dorso-ventral axis has been extensively studied in *Antirrhinum* although, as we will see later, it is likely that a distinction between the dorsal and ventral sides of the flower also exists in apparently radially symmetrical flowers.

As noted above, floral primordia acquire polarity from their spatial relationship to the inflorescence meristem. An obvious indication of this is provided by the *centroradialis* (*cen*) mutant of *Antirrhinum* (Bradley *et al.*, 1996a). Wild type *Antirrhinum* flowers consist of five sepals, five petals, four stamens and two fused carpels. Each of the five petals falls into one of three morphologically distinct types; two dorsal, two lateral and one ventral. In addition, the single dorsal stamen aborts, leaving a staminode and reducing the number of fully-developed stamens to four (Carpenter *et al.*, 1995; Endress, 1999). *Antirrhinum* inflorescences are normally indeterminate, with the meristem continuously producing floral primordia on its flanks until senescence. However, in the *cen* mutant the apical inflorescence meristem differentiates into a floral primordium and produces a single terminal flower. In contrast to the bilaterally symmetrical flowers produced on the flanks of the inflorescence, the terminal flower produced in *cen* mutants is radially symmetrical (Bradley *et al.*, 1996a). The symmetry of the flower reflects its apical position and the radial organisation of the stem. Therefore, the inflorescence meristem acts as a reference point in the establishment of dorso-ventral asymmetry of the flower.

The establishment or perception of the dorso-ventral axis of the flower is compromised in the case of certain mutants which produce radially symmetrical flowers, even when they arise in the normal positions on the flanks of the apical meristem. Such radially symmetrical, or peloric mutants are known in a range of species including *Antirrhinum* (Luo *et al.*, 1996;

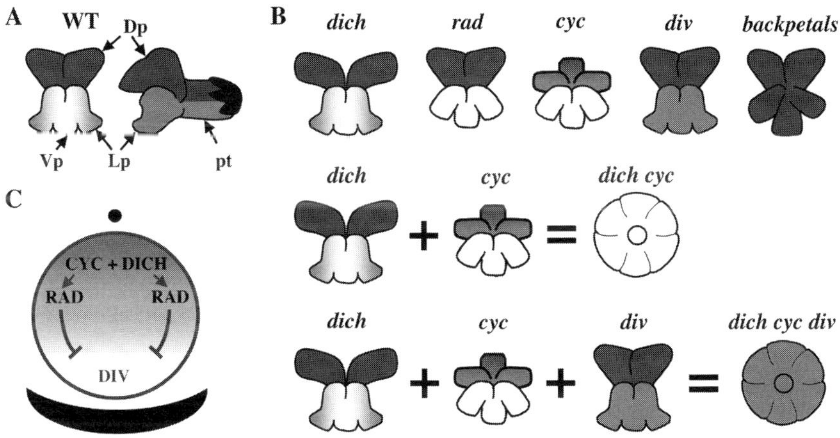

Figure 11.2 The dorso-ventral axis of *Antirrhinum* flowers. Schematic *Antirrhinum majus* flower (A): left, frontal view; right, lateral view. The five petals of the wild-type corolla are fused along their basal end in a petal tube (pt) and display a free petal lobe at their apical end. Petals adopt three different morphologies shaded in the figure. Dark grey: two dorsal petals (Dp); light grey: two lateral petals (Lp); white: the single ventral petal (Vp). Sepals are represented in black. A range of mutations disrupts the normal dorso-ventral development of the petals (B). In *dichotoma* dorsal petals fusion is less extensive. Both *radialis* and *divaricata* mutants have normal dorsal petals but the other petals adopt a single identity: ventral in *rad* and lateral in *div*. In *cycloidea* the corolla includes six petals, three adopt a ventral identity, the others have a mixture of lateral-dorsal identity. Overexpression of *CYC* in *backpetals* induces the development of all petals with a dorsal identity. Radially symmetrical flowers are produced in the *dich cyc* double mutant and in the *dich cyc div* triple mutant. In the first case, all the petals adopt a ventral identity. In the second case, a lateral identity is adopted. These interactions are the basis of the model describing the genetic control of floral symmetry in *Antirrhinum* flower primordia (C): arrows represent up-regulation, barred lines show repression. The black spot indicates the position of the inflorescence meristem. The bract subtending the flower is represented in black.

Cubas *et al.*, 1999b; Rudall & Bateman, 2003). Perfectly radially symmetrical flowers are produced when two partially redundant genes, *CYCLOIDEA* (*CYC*) and *DICHOTOMA* (*DICH*), are mutated (Luo *et al.*, 1996). In *cyc dich* double mutants, all of the petals adopt a ventral identity and the staminode develops into a normal stamen (Luo *et al.*, 1996; Almeida *et al.*, 1997). Single mutants in either *CYC* or *DICH* result in only incomplete loss of dorso-ventral asymmetry (Fig. 11.2). Both genes belong to the TCP family of transcription factors (Cubas *et al.*, 1999a). In young floral primordia both *CYC* and *DICH* are expressed in a dorsal domain, reflecting their function in establishing the identity of the dorsal petals and stamen (Luo *et al.*, 1999). However, the expression patterns of *CYC* and *DICH* differ within the dorsal domain, with *CYC* being expressed over a slightly wider region than *DICH* at later stages of organ development. This is in agreement with the phenotype of the two single mutants, since *cyc* mutants affect all petals, whereas only the dorsal petals are affected in *dich* mutants (Fig. 11.2). The dorsalising role of *CYC* is apparent in *backpetals*, the semidominant allele of *cyc* (Luo *et al.*, 1999). In this mutant, *CYC* becomes ectopically expressed in the lateral and ventral regions of the flower at later stages, resulting in the lateral and ventral petals resembling the lateral half of the wild-type dorsal petals. A third semi-peloric mutant, *radialis* (*rad*), is also known and the phenotype of strong *rad* alleles is similar to that of *cyc* (Carpenter & Coen, 1990). It is thought that *RAD* is likely to act downstream of *CYC* (Fig. 11.2).

CYC appears to exert its dorsalising function by repressing a ventralising function provided by a fourth gene, *DIVARICATA* (*DIV*; Fig. 11.2). In contrast to the ventralised flowers of *cyc dich* double mutants, *div* mutants show increased dorsality; with the ventral and lateral petals becoming more dorsalised (Almeida *et al.*, 1997; Galego & Almeida, 2002). *DIV* also encodes a MYB transcription factor, but its expression is not confined to the ventral part of the flower. Analysis of double mutants demonstrated that *CYC* and *DICH* control the suppression of *DIV* activity in the dorsal and lateral domains of the floral primordium, although this regulation appears to be post-transcriptional (Galego & Almeida, 2002). The *cyc dich div* triple mutant produces radially symmetrical flowers in which all petals have a lateral identity, indicating that lateral petals could be the default petal morphology (Galego & Almeida, 2002).

CYC expression is not confined to the floral primordia. *CYC* is also expressed in an adaxial position in some shoot axillary meristems (Clark & Coen, 2002). Intriguingly, this is similar to the observation that *TCP1,* the likely *Arabidopsis CYC* orthologue, is expressed in the adaxial domain of all flowers and shoot axillary meristems (Cubas *et al.*, 2001). Therefore, *CYC* and *TCP1* are both expressed in response to patterning shared between vegetative and reproductive primordia (Clark & Coen, 2002). Since zygomorphy is thought to have evolved independently several times, it is possible that different mechanisms are used in different species to establish dorso-ventral polarity (Donoghue *et al.*, 1998). It is already known that dorso-ventral polarity in *Linaria* is dependent on a *CYC* homologue (*LCYC*; Cubas *et al.*, 1999b). The finding that the *CYC*-like gene *TCP1* shows a polar expression in *Arabidopsis* might suggest that the same mechanism of underlying asymmetry could have been used to elaborate dorso-ventral asymmetry in many species. In fact, even *Arabidopsis*, which at maturity is radially symmetrical, exhibits an early transitory dorso-ventral asymmetry. The ventral sepal (the most distant from the inflorescence meristem) is always the first primordium organ to arise, followed by the dorsal and two lateral primordia, defining reference points from which the position of the inner organs can be derived (Smyth *et al.*, 1990; Fig. 11.1).

It is not known how the polar expression of *CYC/TCP1* is established early in primordial development, but adaxial-abaxial polarity of the floral primordia is a prime candidate. To address this hypothesis it will be critical to determine if *TCP1* expression is altered in the many *Arabidopsis* mutants affecting primordial adaxial-abaxial polarity. A current model describing the establishment of adaxial-abaxial polarity in leaf primordia involves the interactions of the adaxialising factors *PHB/PHV/REV* with the abaxialising factors *KANADI1/2 (KAN1/2)* and the YABBY family (Eshed *et al.*, 2001; McConnell *et al.*, 2001). Gain-of-functions induced by mutations affecting the sterol lipid binding (START) domain of *PHB/PHV* (Ponting & Aravind, 1999), are characterised by the adaxialisation of leaf primordia. In wild type plants, both genes are assumed to be activated in response to a gradient of an unidentified sterol-like ligand emanating from the shoot apical meristem. A recent alternative theory suggests that micro RNAs, complementary to the region predicted to encode the START domain, could be involved in the signalling from the meristem to these genes (Rhoades *et al.*, 2002; Tang *et al.*, 2003). Whatever the mechanism, the result of dominant mutations in the START domain is that the genes become activated in the whole primordium and their functions are no longer restricted to the adaxial side (McConnell *et al.*, 2001). Mutants where the two *KANADI* genes are compromised also produce adaxialised leaves indicating that *KAN1/KAN2* may downregulate *PHB/PHV/REV* functions in the abaxial side of the leaf. This is supported by experiments in which *KAN* genes are ectopically expressed, leading to abaxialisation of the leaves (Eshed *et al.*, 2001). Therefore early leaf primordial polarity relies on mutually antagonistic adaxialising and abaxialising functions in close relation with the shoot apical meristem. Establishment of adaxial-abaxial polarity in floral primordia is likely to involve a similar mechanism (Kerstetter *et al.*, 2001). This question can be addressed by detailed analysis of the expression of these genes in young flower primordium, in wild type and mutant backgrounds.

11.3.2 The radial axis of flowers

Recent experiments have shown that expression, or lack of expression, of a defined set of organ identity MADS-box genes, is sufficient to switch the developmental fate of leaves to that of floral organs and floral organs to that of leaves (Goto *et al.*, 2001). The well known ABC model provides a basis for understanding how expression of the organ identity genes in overlapping pairs of whorls, results in specification of the appropriate floral organ identity in each whorl (Coen & Meyerowitz, 1991). The addition of a fourth class of identity gene, the SEP or Im class, to the ABC model provides sufficient information to confer floral organ identity on a lateral organ (Gutierrez-Cortines & Davies, 2000). Specification of appropriate organ identity by the organ-identity genes requires precise spatial and temporal control of their expression. This means that specific combinations of organ-identity genes need to be active at the correct time in the appropriate organ primordia. There are two recent reviews dealing with specification of floral organ identity (Kieffer & Davies, 2001; Lohmann & Weigel, 2002). Here we intend to concentrate on how the radial polarity of the floral primordium is interpreted, to give rise to restricted expression of the organ-identity genes. This is only partly understood, but clearly involves mechanisms acting at several levels.

11.3.2.1 Initiation of floral organs

The distribution of floral organ primordia indicates that they are initiated in response to positional signals from the floral meristem and the pre-existing pattern established during floral initiation. The mechanism regulating the initiation of organ primordia has been investigated using mutants which produce abnormal numbers of organs (Running & Hake, 2001). Once again, auxin may play an important role in the initiation of floral organ primordia. Petunia *floozy* (*fzy*) mutants, which affect auxin biosynthesis, produce reduced flowers consisting of a single sepal and two abnormal carpels (Tobena-Santamaria *et al.*, 2002). The expression pattern of *FZY* confirms its role in the development of the second and third whorls, since it is expressed in young floral primordia and later becomes restricted to rings of cells surrounding petal and stamen primordia, as well as sepal internal tissues. FZY does not appear to act via the organ-identity genes, since the expression of the petal/stamen identity gene *FLORAL BINDING PROTEIN1* (*FBP1*) is normal in *fzy* meristems, although its expression level is reduced. Therefore, auxin may act to initiate organ primordia, but spatial restriction of the organ identity genes appears to utilise other mechanisms. Further evidence for the role of polar auxin transport in organ initiation is provided by *Arabidopsis* mutants affecting the putative auxin transporter PIN1, the protein kinase PID and the auxin response factor ETTIN (ETT), all of which cause various degrees of changes in organ number (Bennett *et al.*, 1995; Session *et al.*, 1997). Mutants in the bZIP-encoding gene *PERIANTHIA* (*PAN*) cause an increase in sepal and petal organ number (Chuang *et al.*, 1999). The increased number of organs is correlated with a reduced distance between emerging primordia as meristem size is unaltered. Double mutants affecting *ETT* and *PAN* show a synergistic increase in petal number, indicating that the two genes have redundant functions on partitioning whorl 2 (Session *et al.*, 1997). In contrast, mutant flowers of the uncharacterised gene *petal loss* (*ptl*) and the YABBY encoding gene *filamentous flower* (*fil*), usually lack petals (Griffith *et al.*, 1999; Sawa *et al.*, 1999). Finally, *tousled* (*tsl*) mutants are characterised by the random loss of organs from the three outer whorls (Roe *et al.*, 1993). TSL is a member of a family of nuclear protein kinases that are highly conserved between the plant and animal kingdoms. The human *TOUSLED-like* genes are cell-cycle regulated and may be involved in chromatin remodelling (Silljé *et al.*, 1999). Therefore, TSL may be required for the co-ordination of cell proliferation in the initial primordia.

It is intriguing that these mutants have different effects on different whorls. This may reflect the existence of functionally redundant genes (or pathways) but could also point to the existence of whorl specific mechanisms. Indeed, the recruitment of organ founder cells is not uniform between whorls. Sepals and carpels are derived from a leaf-like initiation pattern involving eight cells in a row; petals depend on smaller groups of cells, while stamens are initiated from blocks of four cells (Bossinger & Smyth, 1996).

Another class of mutants is characterised by the excessive production of floral organs in all whorls and, in some cases, loss of determinacy and the production of extra whorls. This class of mutants, which includes *clavata1/2/3*, *wiggum/era1* and *ultrapetala*, produce larger floral meristems, indicating that the genes affected are part of pathways regulating meristem size (Brand *et al.*, 2000; Ziegelhoffer *et al.*, 2000; Fletcher, 2001). The increase in organ number in these mutants is therefore simply a response to the larger number of cells available in each whorl. This suggests that the process of organ initiation is very flexible and probably involves a primordium spacing mechanism. Analysis of the organ identity mutants has suggested that, in most cases, the identity of the developing floral organ does not influence the organ number in that whorl. However, there is a connection at some level, since some mutant combinations affect organ number. For example, the fourth whorl of the *Antirrhinum* C-function mutant *plena* (*ple*) consists of two mosaic organs initiated at the position where the two carpels would form in the wild type. Thus a loss of carpel identity does not affect organ number in the fourth whorl. However, if *ple* mutants are combined with mutants in the *FARINELLI* (*FAR*) gene, another C-function gene that is required for male fertility and to repress the B-function genes, the fourth whorl develops as five petals (Davies *et al.*, 1999).

11.3.2.2 Regulation of the organ-identity genes

The ABC organ identity genes become restricted to their domains of expression by a complicated mechanism that is not yet fully understood. Initially the meristem identity transcription factor genes *LEAFY* (*LFY*) and *APETALA1* (*AP1*) are activated throughout the floral meristem, as a result of the earlier activation of the floral induction signals such as *SOC1* and *CONSTANS* (*CO*) (Simpson & Dean, 2002). *LFY* and *AP1* contribute to the activation of the B- and C-function MADS-box genes by working together with *UNUSUAL FLORAL ORGANS* (*UFO*) (Wilkinson & Haughn, 1995; Parcy *et al.*, 1998; Ng & Yanofsky, 2001). Initially *UFO* is expressed in the central dome of the floral meristem. During the course of floral development, this expression pattern becomes restricted to the second and third whorls and finally to the base of the petals. *UFO* does not encode a transcription factor, but rather encodes an F-box containing protein (Ingram *et al.*, 1995; Samach *et al.*, 1999). It is likely that UFO results in the activation of the B-function organ identity genes in the following way. UFO forms part of an SCF (Skp1–cullin-F-box protein) E3 ubiquitin ligase complex known as SCF^{UFO} (Samach *et al.*, 1999; Zhao *et al.*, 2001a; Wang *et al.*, 2003). SCF complexes mark out target proteins for degradation by the 26S proteasome and the specificity of target choice is determined by the F-box protein. In the case of SCF^{UFO} it is likely that UFO recruits an inhibitor of B function expression for degradation, thereby causing the B-function genes to become activated. SCF^{UFO} associates *in vivo* with the COP9 signalosome (CSN), a multiprotein complex which also interacts with various transcription factors and kinases (Wang *et al.*, 2003). CSN is known to interact with other SCF complexes and to regulate their modification. CSN might therefore provide a link between UFO and signal transduction pathways involved in the control of flower development. Consistent with this, CSN is required for UFO mediated activation of the B-function genes.

Following on from the activation of the organ-identity genes by the meristem-identity genes and UFO, their expression patterns become refined and enhanced by regulatory interactions between the organ-identity genes themselves and other regulatory genes. The refinement of expression patterns takes place at both the transcriptional and post-transcriptional level. For example, the B-function factors APETALA3 (AP3) and PISTILLATA (PI) interact to enhance the expression of both the *AP3* and *PI* genes, but the proteins are only stable when present together in the same cell (Krizek & Meyerowitz, 1996). In this way, initial overlapping low-level expression of the two genes can become refined to increased gene expression only in the regions of overlap. Expression of the C-function gene *AGAMOUS* (*AG*) is restricted to the inner two whorls by the activity of *APETALA2* (*AP2*). However, *AP2* is expressed in all whorls of the flower and so cannot account for the spatial restriction of *AG* alone (Jofuku *et al.*, 1994). Several other genes have been identified which control the spatial expression of the B and C genes. These include *SUPERMAN* (*SUP*) which regulates the expression domain of the B-function and *LEUNIG* (*LUG*) and *SEUSS* (*SEU*) which restrict the expression of the C-function (Sakai *et al.*, 1995; Conner & Liu, 2000; Franks *et al.*, 2002). The activation of the organ-identity genes in the correct spatial and temporal domains is thus, a product of early activation signals and complex interactions between the organ identity genes themselves and several other factors.

Expression of the organ-identity genes is required for long periods of organ development. Once their expression is established in the appropriate domains, it becomes necessary to maintain the correct expression domains. The *Arabidopsis curly leaf* (*clf*) mutant provided the first indication that similar epigenetic mechanisms are used to regulate the spatial expression of floral organ-identity genes and the expression domains of the unrelated *Drosophila* Hox segmentation genes (Goodrich *et al.*, 1997). In *clf* mutants *AG* becomes ectopically expressed. *CLF* encodes a polycomb group protein related to enhancer of zeste, which, in animals, acts in a complex to repress Hox genes. Proteins related to other members of the polycomb group complex also exist in the *Arabidopsis* genome, raising the possibility that analogous polycomb group complexes regulate the spatial expression patterns of homeotic determinants in both animals and plants (Wagner, 2003). The use of such chromatin-based mechanisms to establish states of activation and repression of homeotic genes would confer mitotic stability and result in the fixation of transiently established expression patterns.

11.3.2.3 Floral determinacy

In most plant species, the flower is a determinate structure and so a mechanism is required to limit the activity of the floral meristem to the production of a defined number of whorls. The mechanism of control of determinacy involves the regulation of the meristem maintenance gene *WUS* (Lenhard *et al.*, 2001; Lohmann *et al.*, 2001). Expression of *WUS* increases in the apical meristem throughout the development of the plant, reaching a peak in the floral meristems. WUS acts together with LFY to activate expression of the C-function gene *AG* in the centre of the flower, potentially by directly binding to the *AG* gene. Once activated, *AG* represses WUS expression and, since WUS expression is required for meristem maintenance, the meristem becomes unable to maintain itself. In this way the cells of the floral meristem differentiate and become incorporated into the fourth whorl carpels and further development is precluded. In *ag* mutants, expression of *WUS* in the centre of the flower is not reduced and the flowers are indeterminate.

It is very probable that a similar mechanism controls determinacy in other species, but comparative studies between *Antirrhinum* and *Arabidopsis* have revealed that the regulation of floral determinacy, while achieving the same ends with similar genes, is organised in a slightly different way. If the C-function gene is responsible for the termination of floral development, by repression of *WUS*, why do flowers have four whorls? Since C-function genes are expressed in both the third and fourth whorls, it might be expected that repression of *WUS* would terminate flower development in the third whorl. An easy answer to this question is provided by the B-function mutants of *Antirrhinum*, *deficiens* (*def*) and *globosa* (*glo*). These *def* and *glo* mutants do indeed terminate after just three whorls (Tröbner *et al.*, 1992), indicating that the termination of meristem development by the C-function gene *PLENA* (*PLE*) is antagonised in the wild type by the action of *DEF* and *GLO* in the third whorl. In *Arabidopsis*, however, the B-function mutants *ap3* and *pi* have four whorls, so something else must act to prevent *AG* from terminating flower development in the third whorl. In fact, *sup ap3* and *sup pi* double mutants terminate floral development after three whorls (Schultz *et al.*, 1991), like *def* or *glo* single mutants in *Antirrhinum*, indicating that, in *Arabidopsis*, *sup* also plays a role in preventing *AG* from terminating floral meristem activity prematurely. Analogous differences between the *ag* and *ple* mutant phenotypes can also be explained by differences in the regulatory networks controlling determinacy in the two species (Davies *et al.*, 1999).

11.4 Polarity of floral organs

Each of the four types of floral organs also show polarity. Ectopic co-expression of organ identity and Im MADS-box genes, validates Goethe's hypothesis that floral organs are modified leaves (Goto *et al.*, 2001) and suggests that conserved mechanisms might be involved in the establishment of the apical-basal, dorso-ventral and lateral axes in leaves and floral organs. This is supported, in general, by the observation that many of the leaf polarity mutants also affect the floral organs. However, some mutants cause more severe effects in certain tissues than in others and this indicates divergence in the regulatory pathways, presumably to enable different organs to develop distinctive morphologies. In this section we will look at two examples where polarity in floral organ development has been studied in more detail: asymmetry in the *Antirrhinum* petal and dorso-ventral polarity in the *Arabidopsis* carpel.

11.4.1 Asymmetry of Antirrhinum petals

Asymmetry is not restricted to the overall shape of the *Antirrhinum* flower. Within the flower, the floral organs also display an asymmetry, which is related to their position (Fig. 11.2). The single ventral petal is bilaterally symmetrical, whereas the two dorsal and lateral petals are asymmetric. Clearly, the asymmetry of the dorsal and lateral petals is related to the establishment of overall flower asymmetry, since the petals formed in radially symmetric mutants all become bilaterally symmetrical. The way in which the dorso-ventral floral signal is used to establish petal asymmetry is not currently understood, but recent experiments have begun to shed some light on this process. Using clonal analysis, Rolland-Lagan *et al.* (2003) were able to follow marked groups of cells through development of the asymmetric dorsal petal. Three parameters could affect the overall shape of the mature petal: the rate of growth, the degree to which directional growth is preferred and the angle of the main axis of growth. Since these parameters can vary in space and developmental time, interpretation of the patterns obtained from clonal analysis and confirmation that the parameters thus obtained can account for the observed shape, requires a dynamic growth model. This analysis revealed that asymmetry of petal shape results from a 45° rotation of the petal lobe, caused by preferential growth of the dorsal part of the petal tube. Surprisingly, despite the rotation of the lobe relative to the tube, the direction of growth remains fixed, parallel to the proximo-distal axis of the organ. This suggests that a long range signal maintains the proximo-distal axis. Since asymmetric growth in the tube can cause rotation of the lobe and hence asymmetric development of the petals, it is easy to see a link between dorso-ventral asymmetry of the flower and asymmetric organ growth.

In addition to left-right asymmetries derived from the dorso-ventral axis, the petals of some flowers are further complicated by the presence of localised asymmetric outgrowths or spurs. Such outgrowths are seen on the ventral petal of *Linaria* flowers. Two dominant mutants of *Antirrhinum*, *hirzina* (*hirz*) and *invaginata* (*ina*), show duplications of the petal tube growing out from the existing tube, in a manner reminiscent of the spurs of *Linaria* (Cubas *et al.*, 1999b). *HIRZ* and *INA* encode partially redundant homeobox genes and changes in the regulation of *HIRZ* and *INA*-like genes could explain the acquisition and loss of spurs in different species during the course of evolution (Golz *et al.*, 2002). Spurs have often been recruited into pollination systems involving insects or hummingbirds (Endress, 2001).

11.4.2 Polarity in the Arabidopsis *carpel*

The *Arabidopsis* gynoecium consists of two carpels, congenitally fused along their margins and presenting a complex three-dimensional pattern of tissue differentiation. Here, we will consider the apical-basal axis of the gynoecium and the adaxial-abaxial axes of the individual carpels. The apical-basal organisation involves, from apex to base: stigmatic papillae overlying style tissues which cap an ovary connected to the flower receptacle by a short internode (Fig. 11.3). This vertical distribution is disturbed when flower primordia are treated with inhibitors of polar-auxin transport (Nemhauser *et al.*, 2000) and in *pin-1*, *pid*, *mp* and *ett*; four mutants which affect auxin transport and signalling (Bennett *et al.*, 1995; Sessions & Zambryski, 1995; Przemeck *et al.*, 1996). Mutations affecting the auxin response factor *ETTIN* are the best documented and are characterised by an allele strength dependant phenotype (Sessions *et al.*, 1997). In weak alleles, the ovary is almost normal but stronger mutants show a concomitant reduction in ovary size. In the strong *ett-1* allele, ovary valves are often completely missing. An elegant model suggests the existence of an apical-basal auxin gradient in the gynoecium which is translated by *ETTIN* into the establishment of the apical and basal ovary boundaries (Sessions, 1997). *ETT* is expressed early in gynoecium development in an abaxial ring and later becomes restricted to differentiating vascular strands. Therefore, the auxin-induced carpel pre-patterning is probably established early in gynoecium development (Sessions *et al.*, 1997). In addition to its apical-basal phenotype, the *ett* mutant also shows abnormalities of the adaxial-abaxial axis, connecting the establishment of the two axes.

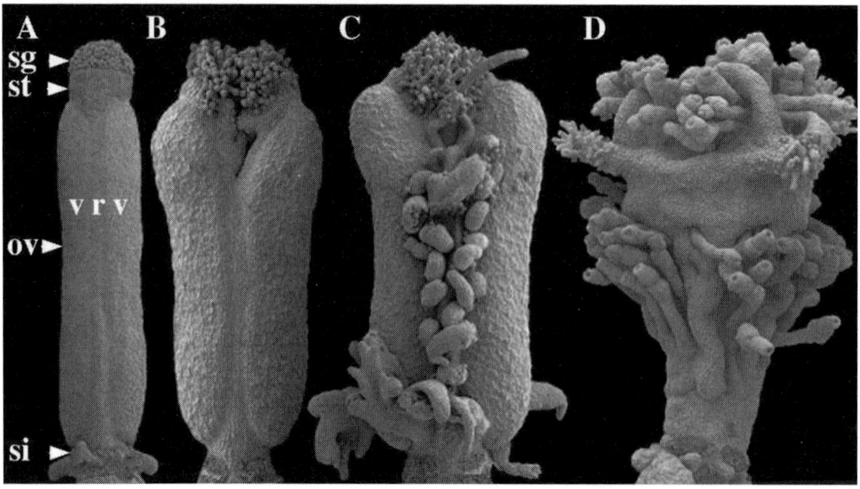

Figure 11.3 Adaxial-abaxial polarity of the *Arabidopsis* carpel. The wild-type *Arabidopsis* gynoecium (A) consists of two fused carpels and includes an apical stigma (sg) made of many papillae resting on the style (st). The ovary (ov) is delimited by the two carpel valves (v) separated by the replum (r) and is attached to the floral receptacle by a short internode (si). Gynoecium of the *crabs claw* mutant shows only partial carpel fusion but no alteration of their adaxial-abaxial development (B). However, the double mutant *crc kan1* displays the production of adaxial internal tissues on the abaxial external carpel surface, essentially in place of the replum (C). This phenotype revealed the partial functional redundancy of the two genes in the establishment of carpel adaxial-abaxial polarity. A more severe phenotype is obtained in mutants disrupting two of the *KANADI* genes (*kan1 kan2*) with abaxial production of adaxial tissue over the whole surface of the gynoecium (D).

The adaxial-abaxial axis of carpels can be distinguished by the formation of placental tissue and ovules exclusively on the adaxial side. By analogy with leaves, it might be expected that adaxial-abaxial polarity of the carpels is controlled by similar genes. At least five of the six *Arabidopsis YABBY* genes show polar expression patterns in lateral organs (Bowman *et al.*, 2002). *FIL*, *YAB2* and *YAB3* are expressed in the abaxial domains of both leaf and floral organs (Siegfried *et al.*, 1999). However, the founder member of the *YABBY* family, *CRABS CLAW* (*CRC*), is only expressed in the carpels and nectaries (Bowman *et al.*, 1999). *crc* was identified on the basis of its phenotype during carpel development (Bowman & Smyth, 1999). Although *crc* carpels are wider and shorter than wild-type carpels and are unfused towards the apex, they do not show obvious changes in polarity (Fig. 11.3). The *crc* mutant was used as the starting point in a second site modifier screen which resulted in the isolation of the *KANADI1* gene (Eshed *et al.*, 1999). *crc kan1* double mutants have a clear effect on carpel polarity, with ectopic ovules and placental tissue forming on the abaxial face of the carpel. The same modifier screen identified *PICKLE* (*PKL*) as another gene involved with *CRC* in defining the polarity of carpels. *crc pkl* double mutants also produce ectopic placental tissue and ovules, but, since PKL encodes a putative CHD3 chromatin remodelling factor, it is suggested that *PKL* and *KAN1* act in different pathways. Modifier screens have identified three further enhancers of *kan1*: *hasty* (*hst*), *splayed* (*syd*) and *kanadi2* (*kan2*), in addition to *crc* (Eshed *et al.*, 2001). In particular the *kan1 kan2* double mutant shows a striking loss of polarity in the carpel and other lateral organs (Fig. 11.3). Loss of function results are confirmed by ectopic expression experiments which show that both the YABBYs and the KANADIs promote the abaxialisation of adaxial tissues (Siegfried *et al.*, 1999; Eshed *et al.*, 1999, 2001). These results show that, just as in leaves, the *YABBY* and *KANADI* gene families control floral organ abaxial identity. In addition, the gain of function mutant *phb-1d*, also disrupts the adaxial-abaxial polarity of all floral organs (McConnell & Barton, 1998). Sepals and petals are mostly filamentous, stamens and carpels are less severely affected. All the defects are consistent with the partial adaxialisation of the organs, showing that, just as in vegetative meristems, members of the HD-ZIP family control the adaxial identity of floral organs. It is therefore reasonable to conclude that the meristem-based signalling model proposed to control leaf polarity also applies to the floral meristem and floral organ polarity.

11.5 Conclusion

The establishment of polarity in the flower has clear and unsurprising similarities to the establishment of leaf polarity, since flowers can be considered to be shortened reproductive stems and floral organs are modified leaves. However, even if their origins are similar, flowers are highly derived structures and this is likely to be reflected in increased complexity of the mechanisms controlling development. Many of the genes discussed in this review are members of gene families, in *Arabidopsis* and other plants. The presence of such gene families leads to genetic redundancy, where the effects of loss of function mutants in any one gene are masked by the presence of other members of the gene family. Multiple mutant combinations and modifier screens, such as the ones that identified the *KAN* genes, will be the key to unravelling this puzzle. We are just beginning to understand how the positional signals provided by the different meristems can be interpreted and refined to generate a complex three dimensional structure, such as the flower. A great many questions still remain. Control of phyllotaxy and the role of auxin and bio-mechanical forces in the establishment of initiation patterns remains ‘mysterious’. A plausible mechanism for the establishment of adaxial-abaxial asymmetry has been developed, but the presumed signal from the meristem has not yet been discovered. Important patterning genes have been identified, but the way in which their pattern is established is still being researched and we have a long way to go to explain the regulation of organ size and shape. These and other questions are set to ensure that polarity in flower development remain a fruitful topic for further research.

Acknowledgements

We are grateful to John Bowman for providing Fig. 11.3. We thank Barry Causier and the members of the Davies group for critical discussion of the manuscript. Martin Kieffer was supported by the European Community REGIA initiative (project, no. QLRT-1999-876).

References

Almeida, J., Rocheta, M. & Galego, L. (1997) Genetic control of flower shape in *Antirrhinum majus*. *Development*, **124**, 1387–1392.

Alvarez, J., Bossinger, G. & Smyth, D.R. (1994) The *PINOID* gene, in *ARABIDOPSIS, An Atlas of Morphology and Development* (ed. J. Bowman), Springer-Verlag, New York.

Ambrose, B.A., Lerner, D.R., Ciceri, P., Padilla, C.M., Yanofsky, M.F. & Schmidt, R.J. (2000) Molecular and genetic analyses of the *silky1* gene reveal conservation in floral organ specification between eudicots and monocots. *Mol. Cell*, **5**, 569–579.

Benjamins, R., Quint, A., Weijers, D., Hooykaas, P. & Offringa, R. (2001) The PINOID protein kinase regulates organ development in *Arabidopsis* by enhancing polar auxin transport. *Development*, **128**, 4057–4067.

Bennett, S.R.M., Alvarez, J., Bossinger, G. & Smyth, D.R. (1995) Morphogenesis in *pinoid* mutants of *Arabidopsis thaliana*. *Plant J.*, **8**, 505–520.

Bossinger, G. & Smyth, D.R. (1996) Initiation patterns of flower and floral organ development in *Arabidopsis thaliana*. *Development*, **122**, 1093–1102.

Bowman, J.L. (2000) The YABBY gene family and abaxial cell fate. *Curr. Opin. Plant Biol.*, **3**, 17–22.

Bowman, J.L. & Smyth, D.R. (1999) *CRABS CLAW*, a gene that regulates carpel and nectary development in *Arabidopsis*, encodes a novel protein with zinc finger and helix-loop-helix domains. *Development*, **126**, 2387–2396.

Bowman, J.L., Baum, S.F., Eshed, Y., Putterill, J. & Alvarez, J. (1999) Molecular genetics of gynoecium development in *Arabidopsis*. *Curr. Top. Dev. Biol.*, **45**, 155–205.

Bowman, J.L., Eshed, Y. & Baum, S.F. (2002) Establishment of polarity in angiosperm lateral organs. *Trends Genet.*, **18**, 134–141.

Bradley, D., Carpenter, R., Copsey, L., Vincent, C., Rothstein, S. & Coen, E. (1996a) Control of inflorescence architecture in *Antirrhinum*. *Nature*, **379**, 791–797.

Bradley, D., Vincent, C., Carpenter, R. & Coen, E. (1996b) Pathways for inflorescence and floral induction in *Antirrhinum*. *Development*, **122**, 1535–1544.

Brand, U., Fletcher, J.C., Hobe, M., Meyerowitz, E.M. & Simon, R. (2000) Dependence of stem cell fate in *Arabidopsis* on a feedback loop regulated by *CLV3* activity. *Science*, **289**, 617–619.

Carpenter, R. & Coen, E.S. (1990) Floral homeotic mutations produced by transposon mutagenesis in *Antirrhinum majus*. *Genes Dev.*, **4**, 1483–1493.

Carpenter, R., Copsey, L., Vincent, C., Doyle, S., Magrath, R. & Coen, E. (1995) Control of flower development and phyllotaxy by meristem identity genes in *Antirrhinum*. *Plant Cell*, **7**, 2001–2011.

Chen, Q., Atkinson, A., Otsuga, D., Christensen, T., Reynolds, L. & Drews, G.N. (1999) The *Arabidopsis FILAMENTOUS FLOWER* gene is required for flower formation. *Development*, **126**, 2715–2726.

Christensen, S.K., Dagenais, N., Chory, J. & Weigel, D. (2000) Regulation of auxin response by the protein kinase PINOID. *Cell*, **100**, 469–478.
Chuang, C.F., Running, M.P., Williams, R.W. & Meyerowitz, E.M. (1999) The *PERIANTHIA* gene encodes a bZIP protein involved in the determination of floral organ number in *Arabidopsis thaliana*. *Genes Dev.*, **13**, 334–344.
Clark, J.I. & Coen, E.S. (2002) The *cycloidea* gene can respond to a common dorsoventral prepattern in *Antirrhinum*. *Plant J.*, **30**, 639–648.
Coen, E.S. & Meyerowitz, E.M. (1991) The war of the whorls: genetic interactions controlling flower development. *Nature*, **353**, 31–37.
Conner, J. & Liu, Z. (2000) *LEUNIG*, a putative transcriptional corepressor that regulates *AGAMOUS* expression during flower development. *Proc. Natl. Acad. Sci. USA*, **97**, 12902–12907.
Crepet, W.L. (2000) Progress in understanding angiosperm history, success, and relationships: Darwin's abominably 'perplexing phenomenon'. *Proc. Natl. Acad. Sci. USA*, **97**, 12939–12941.
Cubas, P., Lauter, N., Doebley, J. & Coen, E. (1999a) The TCP domain: a motif found in proteins regulating plant growth and development. *Plant J.*, **18**, 215–222.
Cubas, P., Vincent, C. & Coen, E. (1999b) An epigenetic mutation responsible for natural variation in floral symmetry. *Nature*, **401**, 157–161.
Cubas, P., Coen, E. & Zapater, J.M. (2001) Ancient asymmetries in the evolution of flowers. *Curr Biol.*, **11**, 1050–1052.
Davies, B., Motte, P., Keck, E., Saedler, H., Sommer, H. & Schwarz-Sommer, Z. (1999) *PLENA* and *FARINELLI*: redundancy and regulatory interactions between two *Antirrhinum* MADS-box factors controlling flower development. *EMBO J.*, **18**, 4023–4034.
DeLong. A., Mockaitis, K. & Christensen, S. (2002) Protein phosphorylation in the delivery of and response to auxin signals. *Plant Molec. Biol.*, **49**, 285–303.
Donoghue, M.J., Ree, R.H. & Baum, D.A. (1998) Phylogeny and the evolution of flower symmetry in the *Asteridae*. *Trends Plant Sci.*, **3**, 311–317.
Endress, P.K. (1999) Symmetry in Flowers: Diversity and Evolution. *Int. J. Plant Sci.*, **160**, S3–S23.
Endress, P.K. (2001) Origins of flower morphology. *J. Exp. Zool.*, **291**, 105–115.
Eshed, Y., Baum. S.F. & Bowman, J.L. (1999) Distinct mechanisms promote polarity establishment in carpels of *Arabidopsis*. *Cell*, **99**, 199–209.
Eshed, Y., Baum, S.F., Perea, J.V. & Bowman, J.L. (2001) Establishment of polarity in lateral organs of plants. *Curr Biol.*, **11**, 1251–1260.
Fletcher, J.C. (2001) The *ULTRAPETALA* gene controls shoot and floral meristem size in *Arabidopsis*. *Development*, **128**, 1323–1333.
Franks, R. G., Wang, C., Levin, J.Z. & Liu, Z. (2002) *SEUSS*, a member of a novel family of plant regulatory proteins, represses floral homeotic gene expression with *LEUNIG*. *Development*, **129**, 253–263.
Galego, L. & Almeida, J. (2002) Role of *DIVARICATA* in the control of dorsoventral asymmetry in *Antirrhinum* flowers. *Genes Dev.*, **16**, 880–891.
Gälweiler,L., Guan, C., Muller, A., Wisman, E., Mendgen, K., Yephremov, A. & Palme, K. (1998) Regulation of polar auxin transport by AtPIN1 in *Arabidopsis* vascular tissue. *Science*, **282**, 2226–2230.
Golz, J.F., Keck, E.J. & Hudson, A. (2002) Spontaneous mutations in KNOX genes give rise to a novel floral structure in *Antirrhinum*. *Curr. Biol.*, **12**, 515–522.
Goodrich, J., Puangsomlee, P., Martin, M., Long, D., Meyerowitz, E.M. & Coupland, G. (1997) A polycomb-group gene regulates homeotic gene expression in *Arabidopsis*. *Nature*, **386**, 44–51.
Goto, K., Kyozuka, J. & Bowman, J.L. (2001) Turning floral organs into leaves, leaves into floral organs. *Curr Opin Genet Dev.*, **11**, 449–456.
Griffith, M.E., da Silva Conceição, A. & Smyth, D.R. (1999) *PETAL LOSS* gene regulates initiation and orientation of second whorl organs in the *Arabidopsis* flower. *Development*, **126**, 5635–5644.
Gutierrez-Cortines, M. E. & Davies, B. (2000) Beyond the ABCs: ternary complex formation in the control of floral organ identity. *Trends Plant Sci.*, **5**, 471–476.

Ingram, G.C., Goodrich, J., Wilkinson, M.D., Simon, R., Haughn, G.W. & Coen E.S. (1995) Parallels between *UNUSUAL FLORAL ORGANS* and *FIMBRIATA*, genes controlling flower development in *Arabidopsis* and *Antirrhinum*. *Plant Cell.*, **7**, 1501–1510.
Jofuku, K.D., den Boer, B.G., Van Montagu, M. & Okamuro, J.K. (1994) Control of Arabidopsis flower and seed development by the homeotic gene *APETALA2*. *Plant Cell*, **6**, 1211–1225.
Kerstetter, R.A., Bollman, K., Taylor, R.A., Bomblies, K. & Poethig, R.S. (2001) *KANADI* regulates organ polarity in *Arabidopsis*. *Nature*, **411**, 706–709.
Kieffer, M. & Davies, B. (2001) Developmental programmes in floral organ formation. *Semin Cell Dev Biol.*, **12**, 373–380.
Krizek, B. A. & Meyerowitz, E.M. (1996) The *Arabidopsis* homeotic genes *APETALA3* and *PISTILLATA* are sufficient to provide the B class organ identity function. *Development*, **122**, 11–22.
Kumaran, M.K., Bowman, J.L. & Sundaresan, V. (2002) *YABBY* polarity genes mediate the repression of *KNOX* homeobox genes in *Arabidopsis*. *Plant Cell*, **14**, 2761–2770.
Lenhard, M., Bohnert, A., Jurgens, G. & Laux, T. (2001) Termination of stem cell maintenance in *Arabidopsis* floral meristems by interactions between *WUSCHEL* and *AGAMOUS*. *Cell*, **105**, 805–814.
Lohmann, J.U., Hong, R.L., Hobe, M., Busch, M.A., Parcy, F., Simon, R. & Weigel, D. (2001) A molecular link between stem cell regulation and floral patterning in *Arabidopsis*. *Cell*, **105**, 793–803.
Lohmann, J.U. & Weigel, D. (2002) Building beauty: the genetic control of floral patterning. *Dev Cell*, **2**, 135–142.
Luo, D., Carpenter, R., Vincent, C., Copsey, L. & Coen, E. (1996) Origin of floral asymmetry in *Antirrhinum*. *Nature*, **383**, 794–799.
Luo, D., Carpenter, R., Copsey, L., Vincent, C., Clark, J. & Coen, E. (1999) Control of organ asymmetry in flowers of *Antirrhinum*. *Cell*, **99**, 367–376.
McConnell, J.R. & Barton, M.K. (1998) Leaf polarity and meristem formation in *Arabidopsis*. *Development*, **125**, 2935–2942.
McConnell, J.R., Emery, J., Eshed, Y., Bao, N., Bowman, J. & Barton, M.K. (2001) Role of *PHABULOSA* and *PHAVOLUTA* in determining radial patterning in shoots. *Nature*, **411**, 709–713.
Ng, M. & Yanofsky, M.F. (2001) Activation of the *Arabidopsis* B class homeotic genes by *APETALA1*. *Plant Cell*, **13**, 739–53.
Neal, P.R., Dafni, A. & Giurfa, M. (1998) Floral symmetry and its role in plant-pollinator systems: Terminology, Distribution, and Hypotheses. *Ann. Rev. Ecol. Syst.*, **29**, 345–373.
Nemhauser, J.L., Feldman, L.J. & Zambryski, P.C. (2000) Auxin and *ETTIN* in *Arabidopsis* gynoecium morphogenesis. *Development*, **127**, 3877–3888.
Okada, K., Ueda, J., Komaki, M.K., Bell, C.J. & Shimura, Y. (1991) Requirement of the Auxin Polar Transport System in Early Stages of *Arabidopsis* Floral Bud Formation. *Plant Cell*, **3**, 677–684.
Otsuga, D., DeGuzman, B., Prigge, M.J., Drews, G.N. & Clark, S.E. (2001) *REVOLUTA* regulates meristem initiation at lateral positions. *Plant J.*, **25**, 223–236.
Parcy, F., Nilsson, O., Busch, M.A., Lee, I. & Weigel, D. (1998) A genetic framework for floral patterning. *Nature*, **395**, 561–566.
Ponting, C.P. & Aravind L. (1999) START: a lipid-binding domain in StAR, HD-ZIP and signalling proteins. *Trends Biochem Sci.*, **24**, 130–132.
Przemeck, G.K., Mattsson, J., Hardtke, C.S., Sung, Z.R. & Berleth, T. (1996) Studies on the role of the *Arabidopsis* gene *MONOPTEROS* in vascular development and plant cell axialization. *Planta*, **200**, 229–237.
Reinhardt, D., Mandel, T. & Kuhlemeier, C. (2000) Auxin regulates the initiation and radial position of plant lateral organs. *Plant Cell*, **12**, 507–518.
Ren, D. (1998) Flower-associated brachycera flies as fossil evidence for jurassic angiosperm origins. *Science*, **280**, 85–88.
Rhoades, M.W., Reinhart, B.J., Lim, L.P., Burge, C.B., Bartel, B. & Bartel, D.P. (2002) Prediction of plant microRNA targets. *Cell*, **110**, 513–520.
Roe, J.L., Rivin, C.J., Sessions, R.A., Feldmann, K.A. & Zambryski, P.C. (1993) The *Tousled* gene in *A. thaliana* encodes a protein kinase homolog that is required for leaf and flower

development. *Cell*, **75**, 939–950.

Rolland-Lagan, A.G., Bangham, J.A. & Coen. E. (2003) Growth dynamics underlying petal shape and asymmetry. *Nature*, **422**, 161–163.

Rudall, P.J. & Bateman, R.M. (2003) Evolutionary change in flowers and inflorescences: evidence from naturally occurring terata. *Trends Plant Sci*., **8**, 76–82.

Running, M.P. & Hake, S. (2001) The role of floral meristems in patterning. *Curr. Opin. Plant Biol*., **4**, 69–74.

Sakai, H., Medrano, L.J. & Meyerowitz, E.M. (1995) Role of *SUPERMAN* in maintaining *Arabidopsis* floral whorl boundries. *Nature*, **378**, 199–203.

Samach, A., Klenz, J.E., Kohalmi, S.E., Risseeuw, E., Haughn, G.W. & Crosby, W.L. (1999) The *UNUSUAL FLORAL ORGANS* gene of *Arabidopsis thaliana* is an F-box protein required for normal patterning and growth in the floral meristem. *Plant J*., **20**, 433–445.

Sawa, S., Ito, T., Shimura, Y. & Okada, K. (1999) *FILAMENTOUS FLOWER* controls the formation and development of *Arabidopsis* inflorescences and floral meristems. *Plant Cell*, **11**, 69–86.

Schultz, E.A., Pickett, F.B. & Haughn, G.W. (1991) The *FLO10* Gene Product Regulates the Expression Domain of Homeotic Genes *AP3* and *PI* in Arabidopsis Flowers. *Plant Cell*, **3**, 771–781.

Sessions, R.A. (1997) *Arabidopsis* (Brassicaceae) flower development and gynoecium patterning in wild type and *ETTIN* mutant. *Am. J. Bot*., **84**, 1179–1191.

Sessions, A., Nemhauser, J.L., McColl, A., Roe, J.L., Feldmann, K.A. & Zambryski, P.C. (1997) *ETTIN* patterns the *Arabidopsis* floral meristem and reproductive organs. *Development*, **124**, 4481–4491.

Sessions, R.A. & Zambryski, P.C. (1995) *Arabidopsis* gynoecium structure in the wild and in *ettin* mutants. *Development*, **121**, 1519–1532.

Siegfried, K.R., Eshed, Y., Baum, S.F., Otsuga, D., Drews, G.N. & Bowman, J.L. (1999) Members of the *YABBY* gene family specify abaxial cell fate in *Arabidopsis*. *Development*, **126**, 4117–4128.

Sillјé, H.H., Takahashi, K., Tanaka, K., Van Houwe, G. & Nigg, E.A. (1999) Mammalian homologues of the plant *Tousled* gene code for cell-cycle-regulated kinases with maximal activities linked to ongoing DNA replication. *EMBO J*., **18**, 5691–5702.

Simpson, G.G. & Dean, C. (2002) *Arabidopsis*, the Rosetta stone of flowering time? *Science*, **296**, 285–289.

Smyth, D.R., Bowman, J.L. & Meyerowitz, E.M. (1990) Early flower development in *Arabidopsis*. *Plant Cell*, **2**, 755–767.

Talbert, P.B., Adler, H.T., Parks, D.W. & Comai, L. (1995) The *REVOLUTA* gene is necessary for apical meristem development and for limiting cell divisions in the leaves and stems of *Arabidopsis thaliana*. *Development*, **121**, 2723–2735.

Tang, G., Reinhart, B.J., Bartel, D.P. & Zamore, P.D. (2003) A biochemical framework for RNA silencing in plants. *Genes Dev*., **17**, 49–63.

Tobena-Santamaria, R., Bliek, M., Ljung, K., Sandberg, G., Mol, J.N., Souer, E. & Koes, R. (2002) FLOOZY of petunia is a flavin mono-oxygenase-like protein required for the specification of leaf and flower architecture. *Genes Dev*., **16**, 753–763.

Tröbner, W., Ramirez, L., Motte, P., Hue, I., Huijser, P., Lönnig, W.E., Saedler, H., Sommer, H. & Schwarz-Sommer, Z. (1992) *GLOBOSA*: a homeotic gene which interacts with *DEFICIENS* in the control of *Antirrhinum* floral organogenesis. *EMBO J*., **11**, 4693–4704.

Vernoux, T., Kronenberger, J., Grandjean, O., Laufs, P. & Traas, J. (2000) *PIN-FORMED 1* regulates cell fate at the periphery of the shoot apical meristem. *Development*, **127**, 5157–5165.

Wagner, D. (2003) Chromatin regulation of plant development. *Curr. Opin. Plant. Biol*., **6**, 20–28.

Waites, R. & Hudson, A. (1995) *Phantastica*: a gene required for dorsoventrality of leaves in *Antirrhinum*. *Development*, **121**, 2143–2154.

Wang, X., Feng, S., Nakayama, N., Crosby, W.L., Irish, V., Deng, X.W. & Wei, N. (2003) The COP9 Signalosome Interacts with SCF(UFO) and Participates in *Arabidopsis* Flower Development. *Plant Cell*, **15**, 1071–1082.

Weberling, F. (1989) *Morphology of Flowers and Inflorescences*, Cambridge University Press, Cambridge.

Wilkinson, M.D. & Haughn, G.W. (1995) *UNUSUAL FLORAL ORGANS* Controls Meristem Identity and Organ Primordia Fate in *Arabidopsis*. *Plant Cell*, **7**, 1485–1499.
Zhao, D., Yu, Q., Chen, M. & Ma, H. (2001a) The *ASK1* gene regulates B function gene expression in cooperation with *UFO* and *LEAFY* in *Arabidopsis*. *Development*, **128**, 2735–2746.
Zhao, Y., Christensen, S.K., Fankhauser, C., Cashman, J.R., Cohen, J.D., Weigel, D. & Chory, J. (2001b) A role for flavin monooxygenase-like enzymes in auxin biosynthesis. *Science*, **291**, 306–309.
Ziegelhoffer, E.C., Medrano, L.J. & Meyerowitz, E.M. (2000) Cloning of the *Arabidopsis WIGGUM* gene identifies a role for farnesylation in meristem development. *Proc. Natl. Acad. Sci. USA*, **97**, 7633–7638.

Index